Lecture Notes in Civil Engineering

Volume 226

Lecture Notes in Civil Engineering (LNCE) publishes the latest developments in Civil Engineering - quickly, informally and in top quality. Though original research reported in proceedings and post-proceedings represents the core of LNCE, edited volumes of exceptionally high quality and interest may also be considered for publication. Volumes published in LNCE embrace all aspects and subfields of, as well as new challenges in, Civil Engineering. Topics in the series include:

- Construction and Structural Mechanics
- Building Materials
- Concrete, Steel and Timber Structures
- Geotechnical Engineering
- Earthquake Engineering
- Coastal Engineering
- Ocean and Offshore Engineering; Ships and Floating Structures
- Hydraulics, Hydrology and Water Resources Engineering
- Environmental Engineering and Sustainability
- Structural Health and Monitoring
- Surveying and Geographical Information Systems
- Indoor Environments
- Transportation and Traffic
- Risk Analysis
- Safety and Security

To submit a proposal or request further information, please contact the appropriate Springer Editor:

- Pierpaolo Riva at pierpaolo.riva@springer.com (Europe and Americas);
- Swati Meherishi at swati.meherishi@springer.com (Asia - except China, and Australia, New Zealand);
- Wayne Hu at wayne.hu@springer.com (China).

All books in the series now indexed by Scopus and EI Compendex database!

More information about this series at https://link.springer.com/bookseries/15087

Paulo Mendonça · Nuno Dinis Cortiços
Editors

Proceedings of the 7th International Conference on Architecture, Materials and Construction

ICAMC 2021

Springer

Editors
Paulo Mendonça
University of Minho
Guimarães, Portugal

Nuno Dinis Cortiços
Faculdade de Arquitetura
University of Lisbon
Lisbon, Portugal

ISSN 2366-2557 ISSN 2366-2565 (electronic)
Lecture Notes in Civil Engineering
ISBN 978-3-030-94513-8 ISBN 978-3-030-94514-5 (eBook)
https://doi.org/10.1007/978-3-030-94514-5

This Springer imprint is published by the registered company Springer Nature Switzerland AG
The registered company address is: Gewerbestrasse 11, 6330 Cham, Switzerland

Preface

The International Conference on Architecture, Materials and Construction (ICAMC) is a scientific event on building cutting-edge improvements, from nanoparticles to engineering infrastructures. This year, the 7th ICAMC was hosted by Universidade de Lisboa, Faculdade de Arquitetura, Portugal. Due to the COVID-19 pandemic, the conference was held virtually on October 27–29 via Zoom's platform. The Lecture Notes in Civil Engineering (ISSN: 2366-2557) was selected to publish this proceedings book volume.

ICAMC 2021 aims to impart research skills to researchers and faculty members by improving the quality of research in architecture, materials, and construction. Once again, this conference brings positive transformation among students and faculty members through their research work sharing, allowing participants to understand the scientific advances and ongoing research in those fields.

Due to strict travel restrictions imposed by different governments, the conference committee decided to conduct the conference entirely in virtual mode. The conference program comprised seven moments, including a plenary speech session, three online sessions, two video sessions, and one poster session. Each plenary speech lasted 40 minutes, including 5 minutes for Q&A, while for other sessions, there were 15 minutes for each presentation, including 2 minutes for Q&A. The event enabled interaction between keynote speakers, participant talks, and attendees, promoting the experience and the debate between researchers, students, and delegates.

We would like to acknowledge all of those who supported ICAMC 2021. Each individual and institutional help was pivotal for the success of the conference, in particular, the organizing committee for their valuable advice and the helpful paper's peer reviews. We sincerely hope that ICAMC 2021 will be recognized as an outstanding discussion forum and a space for new ideas and the promotion of collaborative research. We are confident that the proceedings will serve as an essential reference source for scientific, engineering progress, and future products and processes.

Nuno Dinis Cortiços
Paulo Mendonça

Organization

Conference Chair

Nuno Dinis Cortiços	University of Lisbon, Portugal

Conference Co-chairs

Eric J. Strauss	Michigan State University, USA
Paulo Mendonça	University of Minho, Portugal

Program Chairs

Marco Casini	Sapienza University of Rome, Italy
João Pedro Costa	University of Lisbon, Portugal

Steering Committee

José Afon	University of Lisbon, Portugal
Marina Rynkovskaya	Peoples' Friendship University of Russia (RUDN University), Russia

Organizing Committee

Carlos C. Duarte	University of Lisbon, Portugal
Diogo Basto	University of Lisbon, Portugal
Paulo Dinis	University of Lisbon, Portugal
Gianni Montagna	University of Lisbon, Portugal

Publicity Committee

Ana Almerich-Chulia	Universitat Politecnica de Valencia, Spain
Fouad K. Kaddah	Université Saint Joseph de Beyrouth, Lebanon
Mohammad Arif Kamal	Aligarh Muslim University, India
Shady Salem	The British University in Egypt, Egypt
Silvia Mazzetto	IUAV University of Architecture in Venice, Italy
Sonia Lupica Spagnolo	Politecnico di Milano, Italy

International Technical Committee

Ayse Ozcan	Michigan State University, USA
Claudia Casapulla	University of Napoli Federico II, Italy
Ana Almerich-Chulia	Universitat Politecnica de Valencia, Spain
Djamel Boussaa	Qatar University, Qatar
Dorota Anna Krawczyk	Bialystok University of Technology, Poland
Elise Zgheib	Saint Joseph University, Lebanon
Huynh Trong-Phuoc	Can Tho University, Vietnam
João Carmo Fialho	University of Lisbon, Portugal
Katherine Wendy Toralva Garay	Continental University, Peru
Maria Kaszyńska	West Pomeranian University of Technology in Szczecin, Poland
Qi Zhou	Southeast University, China
Raffaella Maddaluno	Lisbon School of Design—Ciaud, Portugal
Sahnoune Tayeb	Université de Mostaganeme, Algeria
Salim Ferwati	Qatar University, Qatar
Samuel I. Egwunatum	Federal University of Technology, Owerri Imo State, Nigeria
Suaad Ridha	Mustansiriyah University, Iraq
Syed Ishtiaq Ahmad	Bangladesh University of Engineering and Technology, Bangladesh
Uzma Khan	School of Planning and Architecture, India
Vinod Kushvaha	Indian Institute of Technology, India

Contents

Architectural Design and Building Surveying

Building Materials and Structures

Structural Engineering and Structural Mechanics

Preliminary Design and Motion Response Analysis of VLFS

Zhao Luo[1,2,3,4](✉), Qi-He Wu[1,2,3,4], Xian-Dao Feng[1,2,3,4], and Hong-Xing Lin[1,2,3,4]

[1] CCCC Second Harbor Engineering Company, Wuhan, China
[2] Key Laboratory of Large-Span Bridge Construction Technology, Wuhan, China
[3] Research and Development Center of Transport Industry of Intelligent Manufacturing Technologies of Transport Infrastructure, Wuhan, China
[4] CCCC Highway Bridge National Engineering Research Centre Co. Ltd., Beijing, China

Abstract. VLFS is very large floating structure, whose scale is measured in kilometers. As it is constructed on factory, VLFS can be sea city in a short time. The hydrodynamics and stability is one of the most important thing for the VLFS. VLFS is used to study as a mobile military base or floating airport, which the length of ship is bigger than the ship's width for the maneuverability like ship. Consider the comfort of sea city, 3 kind of VLFS are proposed with smaller waterline area and stability. Based on potential flow theory, motion response are simulated. The results prove that the proposed 9-leg VLFS is more reliable.

1 Introduction

Very Large Floating Structure (i.e. VLFS) refers to those floating ocean structures whose scale is measured in kilometers. Due to its huge size, the material strength is difficult to meet the requirements, and it generally needs to be connected by multiple units using flexible connectors. According to the difference of unit shape, it can be divided into pontoon structure (Pontoon type) and semisubmersible type (Semisubmersible type). The shape of the pontoon structure is similar to that of a barge or a pontoon. A typical representative is the floating airport built by Japan at Yokosuka Port in Kanagawa Prefecture in 1999. The airport is welded by 6 modules with a length of 380 m, a width of 60 m, and a height of 3 m (Lamas-Pardo M 2005). The semi-submersible structure will provide the main body of buoyancy below the wave influence range, and has the advantages of weak wave load, small motion response, long rolling period, and strong seaworthiness. After the end of World War I, some Americans proposed this type of VLFS based on semi-submersible modules. They envisioned building such a large floating platform in the Atlantic Ocean as a base for aircraft refueling or staying. In 1992, the U.S. Department of Defense initiated the research on the Mobile Offshore Base (MOB), developed a VLFS composed of six semi-submersible modules, and verified its reliability.

After the completion of the floating airport in Japan, more and more scholars have conducted research on VLFS. In 2000, Cui (2000) pointed out six key technical issues

P. Mendonça and N. D. Cortiços (Eds.): ICAMC 2021, LNCE 226, pp. 3–8, 2022.
https://doi.org/10.1007/978-3-030-94514-5_1

in VLFS research, the first two of which are the conceptual design and hydrodynamic prediction of super large floating structures. In 2003, Li (2003) introduced the mooring layout, mooring pile selection and its motion equations in the VLFS design. In 2010, Kou (2010) proposed a conceptual design of VLFS adapted to the deep sea. In 2016, Li (2016) analyzed the strength of the VLFS single module and concluded that the joints of the upper desk, the column and the submerged body have greater stress.

2 Basis Principles

2.1 Potential Flow Theory

When the ratio of the wave height to the size of the structure is less than 10, the effect of viscous force can be ignored, and the potential flow theory is used to solve the load of the structure. (Faltinsen 1993) The theory of potential flow assumes that the fluid is an ideal fluid with no turbulence, inviscidity and incompressibility, and the motion and excitation force of the wave current are expressed as potential functions. When obtaining the additional mass, damping coefficient, hull motion response transfer function and other hydrodynamic parameters of the platform at different wave frequencies and directions, the hull motion equation in the frequency domain is:

$$[-\omega^2(M + A(\omega)) + i\omega(B(\omega)_p + B_v) + C + C_e]X(\omega, \beta) = F(\omega, \beta) \tag{1}$$

In the formula, ω and β refer to the angular frequency and incident angle of the wave. M *is* the mass matrix. $A(\omega)$ is the additional mass matrix. $B(\omega)_p$ is the potential flow damping matrix, B_v is the linear viscous damping matrix. C and C_e are the static water restoring force matrix and the external restoring force matrix respectively. $X(\omega, \beta)$ is the displacement vector of the hull with 6 degrees of freedom, and $F(\omega, \beta)$ is the wave excitation force vector.

2.2 Short-Term Forecast Theory

The potential flow theory solves the motion response of a structure under the action of a regular wave. In practice, the waves encountered are composed of an infinite number of small amplitude harmonics with different frequencies, different directions, varying amplitudes and disordered phases. The method of probability and random theory is used to superimpose the influence of different waves. The statistical law of these wave distributions can be reflected by the Pierson Moskowitz wave spectrum. The motion response and wave load generated by the VLFS under the action of random waves can be represented by the response spectrum under the corresponding sea state. The response spectrum is equal to the wave spectrum multiplied by the square of the response amplitude operator (RAO) of the system, namely:

$$S_R(\omega) = S(\omega)|H(\omega)|^2 \tag{2}$$

In the formula, $S_R|\omega|$ is the response spectral density function; $S(\omega)$ the wave spectral density function; $H(\omega)$ the transfer function. Under the assumption of narrowband spectrum, short-term sea state wave amplitude, VLFS motion amplitude, load amplitude, etc. all obey the Rayleigh distribution. If a probability of exceeding is given, the response value under this probability can be calculated.

3 Basic Design

The traditional semi-submersible VLFS is mainly designed as a movable base. It is generally composed of 4–6 semi-submersible modules connected by flexible connectors in structure. Each module is about 300 m long and 75–130 m wide. Below the deck is two submerged body and a number of columns, as shown in Fig. 1. The length can meet the taxiing requirements of the aircraft, and the small ship width can ensure the strong seaworthiness of the base.

Fig. 1. Schematic diagram of traditional semi submersible VLFS

This kind of VLFS, also known as mobile floating base, has strong military applications and has been studied by many scholars. Based on the good hydrological conditions, this paper puts forward two kinds of platform schemes suitable for living. The new scheme has a continuous submerged pontoon, which can provide greater carrying capacity and meet the needs of more people. According to the difference of structure size and column number, 4-column platform scheme and 9-column platform scheme are defined, and the advantages and disadvantages between them and traditional semi-submersible scheme are analyzed. For convenience of expression, the traditional ship-type scheme, 4-column platform scheme and 9-column platform scheme are scheme 1, scheme 2 and scheme 3. The design parameters are shown in Table 1. In order to avoid collision when multiple VLFS are connected, the length and width of the submerged pontoon are smaller than that of the upper body. The length and width of different columns and column spacing refer to the stability of the floating body.

Table 1. VLFS module design parameters

Parameter		Scheme 1	Scheme 2	Scheme 3
Superstructure	Length (m)	300	100	200
	Wide (m)	130	100	200
	Height (m)	9	9	9
Column	Length (m)	21	16	20
	Wide (m)	21	16	20
	Height (m)	20	20	20
	Lateral spacing (m)	65	58	60
	Longitudinal spacing (m)	64	58	60
	Number	8	4	9
Floating body	Length (m)	280	90	180
	Wide (m)	35	90	180
	Height (m)	10	7	6
	Number	2	1	1
Other parameters	Designed draft (m)	20	17	16
	Designed displacement (t)	231280	68613.5	240000
	Initial stability (m)	4.5	1.25	4.7

4 Extreme Working Sea Conditions

VLFS is a large-scale structure, which mainly bears the inertial force and diffraction force of waves. The potential flow theory can be used to calculate the motion response and wave load of VLFS. In this paper, the marine engineering software SESAM is used, Genie module is used to complete the modeling, HydroD module is used to simulate based on panel method, and Postresp module is used to analyze the response of wave spectrum and short-term forecast. The calculation grid of the three schemes is shown in the Fig. 2, and the grid spacing is 1 m.

(a) Scheme 1 (b) Scheme 2 (c) Scheme 3

Fig. 2. Calculation grid of design scheme

Generally, the extreme working sea state is defined as the corresponding wave height 3m, and the period 7s. The short-term prediction of the motion response of the floating structure under different waves lasts for 3 h. During this period, the maximum values of the deflection and displacement along the x, y and z directions are shown in Table 2 and Table 3, and the maximum wave load and wave load are shown in Table 4.

Table 2. Rotation motion under extreme working sea condition (unit: degree)

Wave direction	Roll			Pitch			Raw		
	Scheme1	Scheme2	Scheme3	Scheme1	Scheme2	Scheme3	Scheme1	Scheme2	Scheme3
0°	0.46	1.26	0.22	0.11	0.72	0.13	0.15	0.49	0.21
30°	0.36	1.19	0.21	0.12	0.92	0.16	0.18	0.65	0.27
60°	0.22	0.92	0.16	0.18	1.19	0.21	0.18	0.65	0.27
90°	0.20	0.72	0.13	0.22	1.26	0.22	0.13	0.49	0.21

Table 3. Displacement motion in m under extreme working sea condition (unit: m)

Wave direction	Roll			Pitch			Raw		
	Scheme1	Scheme2	Scheme3	Scheme1	Scheme2	Scheme3	Scheme1	Scheme2	Scheme3
0°	0.23	0.3	0.12	0.17	0.55	0.26	0.19	0.33	0.11
30°	0.25	0.37	0.14	0.14	0.5	0.22	0.17	0.33	0.1
60°	0.44	0.5	0.22	0.09	0.37	0.13	0.24	0.33	0.1
90°	0.57	0.55	0.26	0.08	0.3	0.12	0.3	0.33	0.11

From Table 2 and Table 3, firstly, the motion response of the 9-column platform is smaller, and the roll and pitch values are within 0.22°. The motion response of the 4-column platform is larger due to the smaller displacement, and the maximum deflection is 1.26°. Secondly, the motion response of ship structure is greatly affected by the wave direction angle, and the response of the design is smaller near 0°, which is the bow and stern direction of the ship is consistent with the wave direction. Thirdly, the deflection and translation motion of the 9-column platform is smaller than that of the ship type. The short-term wave load prediction of VLFS under the working sea condition is shown in Table 4.

Table 4. Wave loads under extreme working sea condition (unit: MN)

Wave direction	F_x			F_y			F_z		
	Scheme1	Scheme2	Scheme3	Scheme1	Scheme2	Scheme3	Scheme1	Scheme2	Scheme3
0°	49	11	22	26	24	48	102	45	155
30°	53	12	25	22	20	39	91	45	137
60°	91	20	39	15	12	25	132	45	137
90°	118	24	48	14	11	22	167	45	155

From Table 4, firstly, the maximum wave loads of VLFS module under the extreme working sea conditions are the z-direction loads when the wave direction angle is 90°, and the maximum value is 167 MN of the ship-type scheme 1. Seconly, the x-direction load of the ship type scheme is also larger than that of other schemes, which is 118 MN. Secondly, the 4-column scheme 2 has smaller structure size and larger motion response, but the wave load is the smallest, and the maximum wave load is 45MN. Thirdly, the 9-column scheme 3 has a smaller force in x-direction than ship-type scheme 1 with almost same displacement.

To sum up, The results show that the motion response and force of the proposed 9-column platform VLFS are smaller than those of the ship type, and the performance is better.

5 Summary

In this report, 2 kind of platform structures VLFS scheme are proposed, and compared with traditional ship-type VLFS. Considering the initial stability of each VLFS scheme, the motion response of each scheme is simulated, and the short-term prediction of the motion response and wave load under the extreme working sea condition (wave height 3 m, period 7 s) is carried out for 3 h. The simulation results show that the motion and force of the 9-column platform structure VLFS are the least and the performance is the best under the working condition.

References

1. Lamas-Pardo, M., Iglesias, G., Carral, L.: A review of very large floating structures (VLFS) for coastal and offshore uses. Ocean Eng. **109**, 677–690 (2015)
2. Weicheng, C., Yousheng, W., Runpei, L.: Technical problems in the development of very large floating structures. Ocean Eng. **18**(3), 1–8 (2000)
3. Wenlong, L., Jiahua T.: A study on the design of mooring system for very large floating structures. China Offshore Platform (02), 15–20 (2003)
4. Kou, Y.F., Xiao, L.F., Yu, H.: Research on the concept and module type of very large floating offshore base. China Offshore Platform **04** (2010)
5. Liangbi, L., Jiahuan, D., Guangen, L.: Global strength analysis of very large floating structure single module in typical wave loading cases. J. Jiangsu Univ. Sci. Technol. (Nat. Sci. Edn.) **30** (02), 109–114 (2016)
6. Faltinsen, O.: Sea Loads on Ships and Offshore Structures, vol. 1. Cambridge University Press, Cambridge (1993)

Study on Optimization Design of Portal Frame Structure

Zheyu Li(✉)

College of Civil and Transportation Engineering, Shenzhen University, Shenzhen 518060, China

Abstract. The rapid growth of the national economy has promoted the continuous development of China's construction engineering to a large extent. Light steel building is one of the common ways of modern construction. Light steel structure is widely used in industrial and civil buildings because of its relatively light quality, reliability and good impact resistance in actual use. At present, most factory buildings in China are constructed with light steel structure, which improves the standardization and professionalism of architectural design. In this paper, the main points of steel structure design of light-weight building with portal steel frame are analyzed comprehensively, which provides valuable reference for the development of construction industry. This paper investigates the design method of variable cross-section portal frame and make comparison between the design respectively following China's *Steel structure specification* and *Technical specification for steel structure of portal frame light house*.

Keywords: Steel structural design · Portal steel frame · Variable cross-section calculation

1 Research Background

The steel structure system of lightweight building with portal frame is composed of steel frame, roof, envelope system and corresponding supporting system. The main frame in the lightweight structure of the portal steel frame mainly includes the supporting system and the transverse steel frame, etc. The secondary structure consists of wall purlins and roof purlins, etc., while the envelope consists of wall panels and roof panels. According to the cross section, the portal steel frame can be divided into equal section and variable section. The portal frame can be divided into steel pipe, ordinary steel, steel plate, thin-walled steel, cold-formed steel, etc. According to the construction system, the portal frame is divided into lattice type and solid abdomen type. The main advantages of portal steel frame are: (1) reasonable structure and good performance. The portal frame has the characteristics of reasonable structure, clear force and light weight, and the portal frame has large stiffness in the bent direction and vertical bent direction, which can ensure that the structure has good performance under the action of seismic waves. (2) High standardization, fast construction speed and short construction period. Compared with other types of buildings, the prefabricated components of portal frame are uniformly manufactured in the processing plant and transported to the site for installation. The connection method of prefabricated components is generally high-

P. Mendonça and N. D. Cortiços (Eds.): ICAMC 2021, LNCE 226, pp. 9–18, 2022.
https://doi.org/10.1007/978-3-030-94514-5_2

strength bolt connection, and the column foot is connected with the foundation by the foot bolt. Therefore, this kind of structure has a series of advantages such as ensuring the quality of components and reliable installation nodes. At the same time, the components do not need to be made on site, and only the installation work needs to be completed according to the drawings. Therefore, the construction period can be shortened to the greatest extent. (3) Strong adaptability, relatively convenient disassembly and easy resettlement. Usually the connection of portal frame components, whether beam splicing, support connection, or the connection of column foot and foundation are generally connected by high strength bolts or anchor bolts. The biggest advantage of such splicing is easy to disassemble and assemble, which is very beneficial to the reuse of the structure. (4) Economic, beautiful and practical. Due to the short design cycle frame structure is short, the types of raw materials are single, the components are manufactured by advanced automation equipment, and the transportation is convenient, the engineering cycle of portal frame structure is short, the return on funds is fast, and the investment benefit is relatively high. With its outstanding advantages, portal steel frame is widely used in single-story industrial buildings, large-span exhibition centers, three-dimensional parking lots, modern agricultural greenhouses and other architectural forms [1–3].

At present, the research on portal steel frame is comprehensive and multi-angle. Research on conventional portal frame design includes: fire protection design of ordinary portal frame: it is found that for steel portal frame supported on unprotected steel roof, most portal frames fail within 15 to 20 min, and the type of collapse mode depends on the fixed mode of foundation and the constraint mode of purlin [4]. Through the analysis of the stability of light portal frame, it is found that the main design principles to ensure the stability of portal frame are: (1) to avoid bending or torsional deformation of components by necessary measures, (2) to strengthen the setting of support system and corner brace to ensure the integrity of the structure, (3) to take necessary protective measures to prevent corrosion, (4) plastic design is not suitable for portal frame with variable cross section design [5], according to the research on portal steel frames under special loads or environment, the seismic research on portal steel frames with steel curve damper is carried out. It is found that the steel curve damper plays an important role in improving the seismic performance of symmetric portal frame with inclined roof and single slope portal frame, but the use of steel bending damper and fasteners may increase the steel consumption of such portal steel frames [6], the research on portal steel frame with special steel and special technology includes: portal steel frame with C-section cold-formed steel. The test results show that the failure mode of the eave connection is mainly caused by the bending of the eave bracket and the collapse caused by the fracture of the screw, and the failure mode of the vertex connection is mainly the inelastic local buckling of the compression wing-web connection of one c section [7], The finite element aided design of the eave joint of cold-formed steel portal frame with single channel section is studied. The rotational stiffness and bending capacity of the eave joint can be predicted by using the test device. The axial effect and shear effect on the rotational stiffness and bending strength are about 13% and 5%, respectively [8], numerical study on bolt connection of cold-formed steel with variable top portal frame shows that plastic design method is not suitable for portal frame with cold-formed steel [9].

In general, there are relatively few studies on the design optimization of portal frame with variable cross-section in the existing research field, and the portal frame with variable cross-section has obvious advantages compared with the ordinary portal frame. Compared with the ordinary portal frame, the variable cross-section portal frame increases the cross-section at the position of large internal force, and decreases the cross-section at the position of small internal force. Variable cross-section design can reduce the self-weight of the structure, weaken the influence of earthquake on the structure, reduce the load transmitted by the column foot to the foundation, and form a virtuous circle in the optimization design process, so as to make full use of the performance of the material and make the design more economical and reasonable.

Starting from China's Technical specification for steel structure of portal frame light house and Steel structure specification, this paper summarizes the practical methods of conventional portal frames. Combined with calculation examples, the design methods and design ideas of variable cross-section portal frames under different specifications are introduced and analyzed in detail.

2 Design Method of Portal Frame

2.1 Design Process

It can greatly simplify the design work and reduce unnecessary waste to carry out the timing of portal rigid erection and follow a certain design process. Common design process is as follows:

- Carry on the overall structure layout, determine the column distance and span;
- Selecting calculation unit, establishing calculation model, internal force analysis;
- According to the results of internal force analysis to determine the size of the component section, bearing capacity, stability, deformation check calculation;
- Design the link node of the component, and check the bearing capacity;
- Design support system, including purlins, corner struts, wall beams, etc.

2.2 Determine the Span and Column Spacing

According to the span, the portal frame can be divided into single span, double span and multi-span rigid frame. Generally speaking, the principle of length greater than width should be followed when arranging the column net, that is, the length direction is the direction of column distance, and the width direction is the direction of span, so as to reduce the amount of steel used in the overall structure [10]. The span of the frame is usually determined according to the use requirements of the building. Generally speaking, the span of the portal frame is about 9–36 m. For standard portal frame without hanging crane or other hanging load, its economic span is about 21–36 m; column spacing is about 6–9 m, when above 9 m, the amount of steel used in the roof purlin and wall beam system will increase considerably; the amount of steel used for the main frame changes relatively little with the increase of column spacing. If larger column spacing is needed, lattice purlin can also be used to reduce the amount of steel used [11].

2.3 Preliminary Estimation of Beam-Column Section

According to the empirical formula, the section height h of the rigid frame inclined beam is about L/40−L/60, and the section width b is about h/2−h/3. The section height h is estimated to be h/15−h/20, and the section width b satisfies h/b = 2–3. According to different spans, heights and loads, the beams and columns of portal frames can adopt variable cross-section or equal-section solid-web welded I-section or rolled H-section. If the rigid frame adopts variable section design, so that the section is proportional to the bending moment, the height and thickness of the web and the width of the flange can be changed according to the need, so as to maximize the economic effect. The maximum section height of the column with variable section is 2–3 times the minimum section height as the better section. The length of the variable section of the rigid frame beam is generally 1/6−1/5 of the span, which is more economical. For the portal frame, after the internal force calculation is completed, the variable cross-section design can be carried out according to the internal force distribution, so the characteristics of the reverse bending point and the negative bending moment zone at the beam end can be fully utilized.

2.4 Support System and Check Calculation

For portal frame structure, a complete support system should be set up in each temperature section, which forms a space stable system together with the frame structure. In the whole structure system, the main three functions of the supporting system are: longitudinal rigid tie rod transfers longitudinal horizontal force (wind load, longitudinal seismic force, temperature effect and crane longitudinal braking force, etc.) or reduces the out-of-plane calculation length of rigid frame components to ensure the out-of-plane stability of components; the truss stability system formed by roof lateral support and column support directly transmits longitudinal force to foundation; the coordination of roof purlin and corner braces restrains the compression flange plate of inclined beam, and the coordination of wall beam and corner braces restrains the flange plate of steel column. The corner braces play the role of out-of-plane support points of rigid frame components, and reduce the out-of-plane calculation length of flange plate, so as to ensure the overall stability of inclined beam or steel column. Generally speaking, the spacing between columns without crane should be 30–45 m, and it should be arranged in the first or second span of the end. When there is a crane, the inter-column support should be set in layers at the ham, the support spacing should not be greater than 50 m, and the lower column support should be set only in the middle of the section. At the same time, the inter-column support should be set at the same span as the roof support as far as possible. Due to the limitation of building use, the inter-column support can be set at the temporary span. The cross point of the roof support should be corresponding to the wind resistant column and should be arranged at the turning point of the roof beam (such as the roof ridge). When there is a removing column, the roof longitudinal support should be arranged along the length of the bracket; For the workshop with large lifting weight, longitudinal support should be set on the roof edge; For the roof with high and low span, longitudinal support should be provided on the low-span roof. The strength, deformation and stability of the support system are also checked after the preliminary design.

3 Practical Project Application

Based on the design method above, the design steps and design methods are further explained through the analysis of examples, and the design points of the portal frame are respectively introduced according to the Technical specification for steel structure of portal frame light house and the Steel structure specification. The project is a single-layer shopping supermarket built by a commercial group in Beijing. The total area of the building is about 7000–9000 square meters, and the column spacing is 7 m–9 m. The horizontal section of the building is shown in the attached map. The three-span and two-slope portal frame structure is required. The profiled steel plate insulation roof is used. The external wall is 370 mm hollow brick wall below 1.2 m in elevation. The two sides are plastered and the external tile is affixed. The profiled steel plate insulation wall is above 1.2 m in elevation. The elevation of the south facade eaves is 7.2 m. The outer ditch drainage is carried out. The roof slope is 5%. The sectional view of the main frame is shown in Fig. 1.

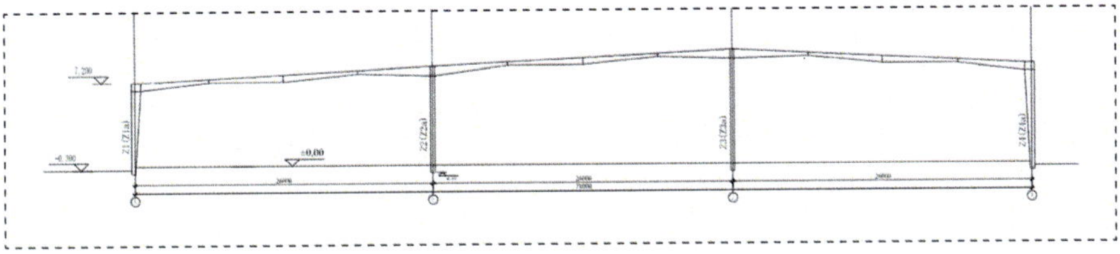

Fig. 1. Sectional view

3.1 Technical Specification for Steel Structure of Portal Frame Light House Calculation

According to the above section estimation method, the internal force and deformation under various load conditions are calculated in SAP2000 software, and the deformation envelope diagram is as follows (Fig. 2):

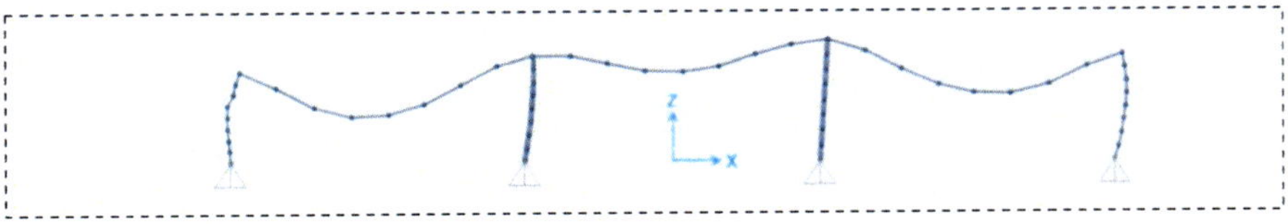

Fig. 2. Frame deformation envelope diagram

According to the requirements of the *Technical specification for steel structure of portal frame light house*, the deflection limit of the inclined beam is L/180 = 144 mm and the lateral displacement limit of the top of the column is H/60 $\leq$ 130 mm under normal use. According to the calculation results, the maximum mid-span deflection of the inclined beam is 104.98 mm, and the maximum lateral displacement of the top of the column is 20.47 mm, which meets the requirements of the code (Table 1).

Table 1. Table of maximum section displacement

Section	Beam 12	Beam 22	Beam32	Column 11	Column 21	Column 31	Column 41
x (mm)				20.47	19.92	19.42	18.94
z (mm)	104.98	53.71	93.33				

Calculation of Beam-Column Strength. Shear check based on post-buckling strength utilization of web:

When the height of the section does not change by more than 60 mm/m, the shear bearing capacity of the I-section member web can be calculated as follows:

$$V_d = \chi_{tap}\phi_{ps}h_{w1}t_wf_v \tag{1}$$

$$\chi_{tap} = 1 - 0.35\alpha^{0.2}\gamma^{2/3} \tag{2}$$

Checking the cross-section strength when bearing N, M and V:

$$M_e^N = W_ef - NW_e/A_e \tag{3}$$

$$M \leq M_e^N \tag{4}$$

The checking results of beam section strength are summarized as follows (Table 2):

Table 2. Checking table of beam section strength

Beam section	Stress ratio	ρ	M′	M	V	N
Beam 11	**0.923**	**1.000**	**495.465**	**457.4493**	**126.795**	**65.06**
Beam 14	0.542	1.000	191.625	103.8785	52.012	61.321
Beam 12	0.544	1.000	451.195	245.6728	8.438	58.299
Beam 15	0.031	1.000	192.305	5.9488	68.888	55.276
Beam 13	**0.994**	**1.000**	**654.819**	**650.9862**	**129.338**	**52.254**
Beam 21	**0.904**	**1.000**	**653.339**	**590.3437**	**120.4**	**58.068**
Beam 24	0.018	1.000	192.331	3.4721	59.95	55.046
Beam 22	0.420	1.000	452.469	189.9837	0.5	52.023
Beam 25	0.052	1.000	193.011	9.9764	60.95	49.001
Beam 23	**0.919**	**1.000**	**656.416**	**603.3523**	**121.4**	**45.978**
Beam 31	**0.924**	**1.000**	**656.395**	**606.7012**	**126.149**	**46.063**
Beam 34	0.091	1.000	193.001	17.5838	65.699	49.086
Beam 32	0.549	1.000	452.452	248.453	5.249	52.108
Beam 35	0.447	1.000	192.321	85.9063	55.201	55.131
Beam 33	**0.956**	**1.000**	**497.923**	**475.9786**	**122.476**	**58.494**

The checking results of column section strength are summarized as follows (Table 3):

Table 3. Checking table of column section strength

	Stress ratio	ρ	M′	M	V	N
Column 11	0.734	0.987	623.241	457.449	58.647	129.886
Column 12	0.000	0.954	219.831	0.000	58.647	164.986
Column 21	0.333	1.000	182.255	60.643	6.664	249.717
Column 22	0.000	1.000	182.255	0.000	6.664	249.717
Column 31	0.019	1.000	175.006	3.349	0.322	242.644
Column 32	0.000	1.000	175.006	0.000	0.322	242.644
Column 41	0.761	0.992	625.419	475.979	52.305	125.244
Column 42	0.000	0.954	219.664	0.000	52.305	166.194

Beam and Column Stability Checking. Since the inclined beam is a beam with variable cross-section, it is necessary to conduct sectional treatment in the checking calculation. The sectional nodes include the section change and the corner brace support. Since the stability is the overall property of the inclined beam, the maximum bending moment element is selected as the representative element in the checking calculation.

According to the suggestion of the code, check the stability of the beam according to the bending member. The following checking formula was used:

$$\frac{N_1}{\eta_{ty}\varphi_y A_{e1} f} + \left(\frac{M_1}{\varphi_b \gamma_x W_{e1} f}\right)^{1.3-0.3k_0} \leq 1 \tag{5}$$

For the corner brace, due to the existence of corner brace, value is calculated in accordance with Article 7 of *Technical specification for steel structure of portal frame light house* 7.1.6, that is, the calculated length outside the plane is three times the distance between the corner brace, and the corresponding size end section is calculated in accordance with the corresponding position.

For the stability calculation of columns, the in-plane stability checking formula is:

$$\frac{N_1}{\eta_t \varphi_x A_{e1}} + \frac{\beta_{mx} M_1}{1 - (N_1/N_{cr}) W_{e1}} \leq f \tag{6}$$

Out-of-plane stability checking formula:

$$\frac{N_1}{\eta_{ty}\varphi_y A_{e1} f} + \left(\frac{M_1}{\varphi_b \gamma_x W_{e1} f}\right)^{1.3-0.3k_0} \leq 1 \tag{7}$$

3.2 Calculation of Steel Structure Design Specification

Since Steel structure specification and Technical specification for steel structure of portal frame light house are approximately similar in calculation, only the differences between them are introduced here.

Preliminary Estimation of Beam-Column Section. On the basis of the portal frame section, the section is optimized for *Steel structure specification* to ensure the requirements of deflection and lateral displacement of column top. Following are the main points:

- Increasing the section area of the top of the column and reducing the lateral displacement of the top of the column;
- The maximum cross section of the middle span of the beam is increased to ensure the requirements of deflection and make the distribution of bending moment more reasonable.

Deformation Check. The internal force and deformation under various load conditions are calculated in SAP2000 software, and the deformation envelope diagram is as follows (Fig. 3):

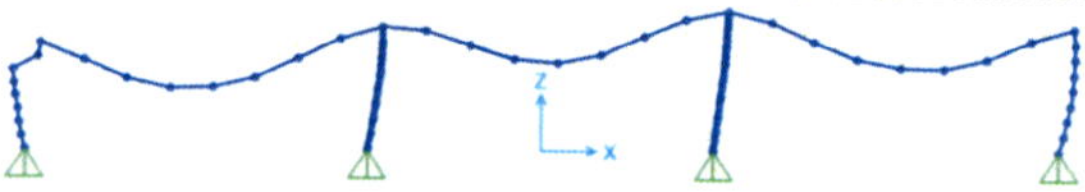

Fig. 3. Frame deformation envelope diagram

Table 4. Table of maximum section displacement

Section number	Beam 12	Beam 22	Beam 32	Beam 11	Beam 21	Beam 31	Beam 41
x (mm)				20.11	19.80	19.52	19.29
z (mm)	64.76	54.58	59.73				

Table 4 shows the maximum section displacement for each beam. According to the requirements of Appendix A of the *Steel structure specification*, the mid-span deflection of the flexural member shall not be greater than L/400. The limit of H/150 in *Steel structure specification* shall be used for the top side movement of the column.

Calculation of Beam-Column Strength. The strength checking formula of bending member is:

$$\frac{N}{A_n} + \frac{M}{\gamma_x W_{nx}} \leq f \tag{8}$$

The checking results of beam-column section strength are summarized as follows (Table 5):

Table 5. Summary of beam-column section strength checking calculation

Beam section	Beam 11	Beam 12	Beam 13	Beam 21	Beam 22	Beam 23	Beam 31	Beam 32	Beam 33
Stress ratio	**0.805**	0.479	**0.970**	**0.932**	0.505	**0.930**	**0.922**	0.459	**0.871**
Column section	Column 11	Column 12	Column 21	Column 22	Column 31	Column 32	Column 41	Column 42	
Stress ratio	0.485	0.078	0.288	0.158	0.176	0.153	0.523	0.078	

Beam and Column Stability Checking. The in-plane stability of the beam can be well solved by structural measures. The stability problem mainly exists outside the plane, and the out-of-plane stability of the beam is checked.

The first is the out-of-plane calculation length of the beam. Since the Steel structure specification does not take a special description of the knee brace limitation, the calculation takes two times of the knee brace spacing as the out-of-plane calculation length of the beam.

According to the calculation formula of flexural member, the out-of-plane stability of beam needs to meet the following equation:

$$\frac{N_0}{\varphi_y A} + \eta \frac{\beta_t M_x}{\varphi_b W_{1x}} \leq f \tag{9}$$

For the stability calculation of columns, the in-plane stability checking formula is:

$$\frac{N}{\varphi_x A} + \frac{\beta_{mx} M_x}{\gamma_x W_{1x}\left(1 - 0.8\frac{N}{N'_{EX}}\right)} \leq f \tag{10}$$

The out-plane stability checking formula is:

$$\frac{N_0}{\varphi_y A} + \eta \frac{\beta_t M_x}{\varphi_b W_{1x}} \leq f \tag{11}$$

4 Conclusions

The main objective of this paper is to verify the applicability of variable section portal frame in the design of actual plant. In this paper, the design method of variable section portal frame is studied, and the design ideas and methods are sorted out according to an actual engineering project using two different design codes respectively. Through calculation and data analysis, it is found that the variable section design in the design of portal frame is in line with the codes requirements and practical application. Through the above analysis, the following two conclusions are drawn:

- Variable cross-section design can effectively reduce the weight of the structure and reduce the amount of steel.
- The steel consumption of the main steel frame designed by two specifications is calculated and the steel consumption of the main steel frame designed by *Steel structure specification* is not much different from that designed by *Technical specification for steel structure of portal frame light house*. However, in most designs, the steel consumption of *Technical specification for steel structure of portal frame light house* is more economical than that of *Steel structure specification*, which indicating that the variable cross-section design of beam-column section can indeed greatly reduce the steel consumption and save costs.
- Steel *structure specification* has stricter requirements on deflection deformation than *Technical specification for steel structure of portal frame light house*.
- The beam-column section should be optimized according to the specification requirements. According to the requirements of *Technical specification for steel structure of portal frame light house*, the deflection limit of inclined beam is L/180 and the lateral displacement limit of column top is H/60 under normal service condition. According to Steel structure specification, the mid-span deflection of flexural members should not be greater than L/400. For the lateral displacement of column top, the limit value H/150 of *Steel structure specification* is adopted. The *Steel structure specification* requirement is clearly more stringent and should be taken into account in the calculation.

References

1. Yang, N.: Research on the advantages and application of portal frame contemporary chemical engineering research, vol. 02, pp. 11–12 (2018)
2. Chen, G.: Discussion on steel structure design of light-weight buildings with gabled steel frames. Jushe **34**, 59–60 (2020)
3. Huang, J.: Research on portal frame structure design. Hous. Ind. **12**, 77–79 (2020)
4. Moss, P.J., Dhakal, R.P., Bong, M.W., Buchanan, A.H.: Design of steel portal frame buildings for fire safety. J. Constr. Steel Res. **65**(5), 1216–1224 (2009)
5. Ma, B., Dai, G.: Stability analysis of light steel portal frame structure. Mod. Decorat. (Theory) (07), 194–196 (20120)
6. Ghabussi, A., Marnani, J.A., Rohanimanesh, M.S.: Improving seismic performance of portal frame structures with steel curved dampers. Structures **24**, 27–40 (2020)
7. Rinchen, Rasmussen, K.J.R.: Behaviour and modelling of connections in cold-formed steel single C-section portal frames. Thin-Walled Struct. **143**, 106233 (2019)
8. Pouladi, P., Ronaldson, J., Clifton, G.C., Ingham, J.M., Wrzesien, A.M., Lim, J.B.P.: Finite-element assisted design of eaves joint of cold-formed steel portal frames having single channel-sections. Structures **20**, 452–464 (2019)
9. Galatanu, T.-F., Cazacu, C., Radu, D., Muntean, R., Tamas, F.-L.: Numerical study of cold-formed steel bolted joints in pitch-roof portal frame. Proc. Manuf. **22**, 135–140 (2018)
10. Hui, G., Dong, G.: On the design of light steel buildings with gabled frames. Archit. Build. Mater. Decora. **24**, 192–193 (2017)
11. Zheng, X.: Discussion on steel structure design of single story portal frame factory building. Theor. Res. Urban constr. (Electron. Version) **5**(36), 759 (2015)

Response Prediction of Random Structure Based on Bayesian Neural Network

Tianyu Wang[1,2](✉), Huile Li[1,2], and Mohammad Noori[1,2,3]

[1] Key Laboratory of Concrete and Prestressed Concrete Structures of the Ministry of Education, School of Civil Engineering, Southeast University, Nanjing 211189, China
ty_wang@seu.edu.cn

[2] National and Local Joint Engineering Research Center for Intelligent Construction and Maintenance, Southeast University, Nanjing 211189, China

[3] Department of Mechanical Engineering, California Polytechnic State University, San Luis Obispo, CA, USA

Abstract. Randomness widely exists in civil structure. However, the randomness is difficult to be measured especially for the structure which is already built. This paper utilizes Bayesian neural network to build a surrogate model to predict the dynamic response of random structure. Statistical indicators including the mean value and standard deviation value are obtained by the proposed model. A numerical example of single floor shearing building with random parameter is presented to verify the capacity of proposed model. The results show that the proposed Bayesian model can well predict the response of random structure. This surrogate model has good potentials in the application of structural health monitoring and quick earthquake damage assessment.

Keywords: Random structure · Structural response prediction · Bayesian neural network

1 Introduction

In civil structure, randomness widely exists in both the external excitation and the model internal parameter. Therefore, in structural dynamic analysis, the randomness must be taken into consideration to obtain a better result. In addition, random dynamic responses of structural can also serve as the foundation for further probabilistic structural assessment and structural reliability evaluation [1]. For the above reasons, the problem of random structural dynamic response analysis has attracted more and more attentions since 1960s with the development of random vibration analysis methods such as probability density evolution analysis method [2] and pseudo excitation method [3]. However, most of these methods are based on the structure of which all the structural parameter distribution is known. For the civil structure which has already been constructed, the measurement of parameter will become very difficult. Data-driven surrogate models [4] are developed to overcome this problem. Artificial neural network [5] as a widely used machine learning approach is regarded as a powerful approximator and has been applied in the structural response prediction [6]. However, a

P. Mendonça and N. D. Cortiços (Eds.): ICAMC 2021, LNCE 226, pp. 19–25, 2022.
https://doi.org/10.1007/978-3-030-94514-5_3

trained neural network is with deterministic parameters, which means that the neural network surrogate model can only result in deterministic dynamic response output. Conventional neural network models need to be modified to make it suitable for random structure.

In this paper, we utilize Bayesian neural network (BNN) with random parameter in the response prediction of random structure. Bayesian inference is used to obtain the estimation of random parameter distribution. Finally, a simplified model of three floor concrete building with random parameters is built to generate the training data and verify the response prediction capacity of BNN.

2 Framework of Proposed Approach

The framework of proposed approach can be described as in Fig. 1.

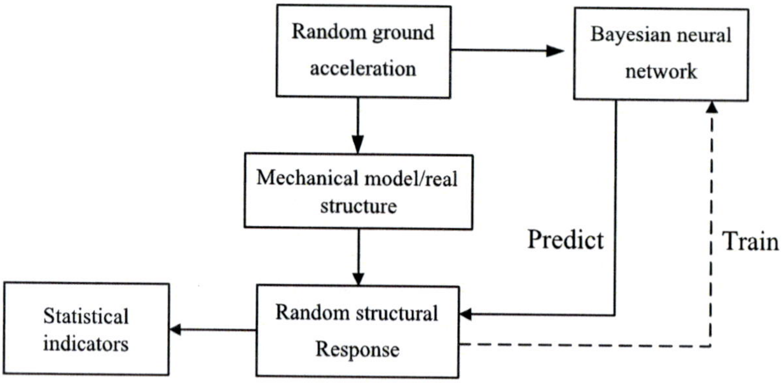

Fig. 1. The framework of proposed approach

First, the random ground acceleration is inputted to a mechanical model or work on the real structure. The response of random structural is obtained. After that the random ground acceleration and structural response are utilized as the input and output data of Bayesian neural network to train the Bayesian neural network. The trained Bayesian neural network model can be applied as the surrogate to predict the random response. Some key statistical indicator such as mean, standard deviation and power spectral density can be calculated through the Monto-Carlo simulation.

2.1 Bayesian Neural Network (BNN)

Artificial neural network is a well known machine learning model. As it is shown in Fig. 2(a), a neural network contains an input layer, an output layers and several hidden layers. In each layers of neural network, a linear transformation is carried out and activation function add the nonlinear to the output. The neural network model can be expressed as in Eq. (1).

$$\mathbf{y} = \phi^o(\cdots \mathbf{W}^i\phi^i(\cdots \phi^2(\mathbf{W}^2(\phi^1(\mathbf{W}^1\mathbf{x} + \mathbf{b}^1) + \mathbf{b}^2)\cdots) + \mathbf{b}^i\cdots) \tag{1}$$

Where $\mathbf{W}^i$ and $\mathbf{b}^i$ are the weight and bias parameter matrix of i layers and $\phi^i()$ are the corresponding activation function. In conventional neural network, back propagation algorithm [5] is utilized to train the neural network model. A trained neural network is with determined parameters. However such a neural network can not be utilized to describe a random mechanical system. Therefore, the Bayesian inference is introduced into neural network model and the Bayesian neural network model [7] is proposed.

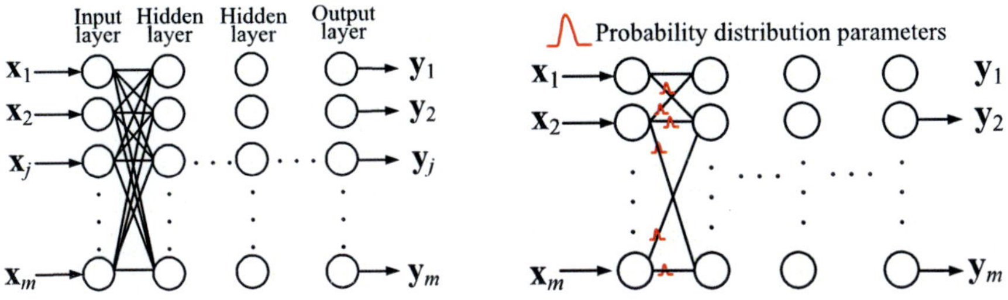

(a) The basic architecture of artificial neural network

(b) The basic architecture of Bayesian neural network

Fig. 2. The architecture of ANN and BNN

As it is shown in Fig. 2(b), the parameter of Bayesian neural network are not determined but a probability distribution. The network parameters are usually assumed to obey the Gaussian distribution and control by their mean value and standard deviation.

2.2 Training of Bayesian Neural Network

According to Bayes' theorem, the posterior probability of random parameters (weight $\mathbf{W}$ and bias $\mathbf{b}$) can be estimated by Bayesian inference as:

$$p(W, b|\mathbf{S}) = \frac{p(W, b|\boldsymbol{\theta})p(\boldsymbol{\theta})}{p(\mathbf{S})} = \frac{p(\mathbf{S}|W, b)p(W, b)}{\int p(\mathbf{S}|W, b)p(W, b)\mathbf{d}(W, b)} \tag{2}$$

where $p(\mathbf{W}, \mathbf{b})$ is the prior distribution of model parameters which can be determined according to experience or assumption; $p(\mathbf{S}|\mathbf{W}, \mathbf{b})$ is the likelihood function; and $p(\mathbf{S})$ is the distribution of training data. $\mathbf{S}$ represent the training data set. Equation (2) cannot be directly utilized to obtain the posterior distribution $p(\mathbf{W}, \mathbf{b}|\mathbf{S})$. Consequently, a new distribution $q(\mathbf{W}, \mathbf{b})$ is assumed to approximate the target distribution $p(\mathbf{W}, \mathbf{b}|\mathbf{S})$.

Herein, Kullback-Leible (KL) divergence is employed to quantify the difference between the two probability distributions as

$$\mathbf{KL}(q(\mathbf{W},\mathbf{b})||p(\mathbf{W},\mathbf{b}|\mathbf{W},\mathbf{b})) = \int q(\mathbf{W},\mathbf{b})\mathbf{log}\frac{q(\mathbf{W},\mathbf{b})}{p(\mathbf{W},\mathbf{b}|\mathbf{S})}\mathbf{d}(\mathbf{W},\mathbf{b}) \tag{3}$$

Considering $KL(q(W,b)||p(\theta W,b|S))$ as a function of $q(W,b)$, the best $q(W,b)$ will be reached at the minimum value of KL divergence. It is the so-called variational inference. In most application, we will use Bayes backprop algorithm [8] to update the random weight and bias. The key of the algorithm is that introduce KL as an additional loss function into neural network and the model parameters can be updated by:

$$\boldsymbol{\mu} = \boldsymbol{\mu} - \alpha\left(\boldsymbol{g}_{\boldsymbol{\theta}}^{mse} + \boldsymbol{g}_{\boldsymbol{\theta}}^{KL} + \boldsymbol{g}_{\boldsymbol{\mu}}^{KL}\right) \tag{4}$$

$$\boldsymbol{\sigma} = \boldsymbol{\sigma} - \alpha\left(\left(\boldsymbol{g}_{\boldsymbol{\theta}}^{mse} + \boldsymbol{g}_{\boldsymbol{\theta}}^{KL}\right)\epsilon + \boldsymbol{g}_{\boldsymbol{\sigma}}^{KL}\right) \tag{5}$$

Where $\boldsymbol{\theta}$ is the network parameter with mean $\boldsymbol{\mu}$ and standard deviation $\boldsymbol{\sigma}$. $\boldsymbol{g}_{\boldsymbol{\theta}}^{mse}$ means the gradient of $\boldsymbol{\theta}$ to MSE (mean square error) loss function while $\boldsymbol{g}_{\boldsymbol{\theta}}^{KL}$ mean gradient of $\boldsymbol{\theta}$ to KL loss function.

3 Numerical Example

3.1 Single Floor Shearing Structure

To verify the proposed Bayesian neural network model, we choose a numerical example of single floor shearing structure under random stationary ground motion as it is shown in Fig. 3. The selection of random parameters is presented in Table 1.

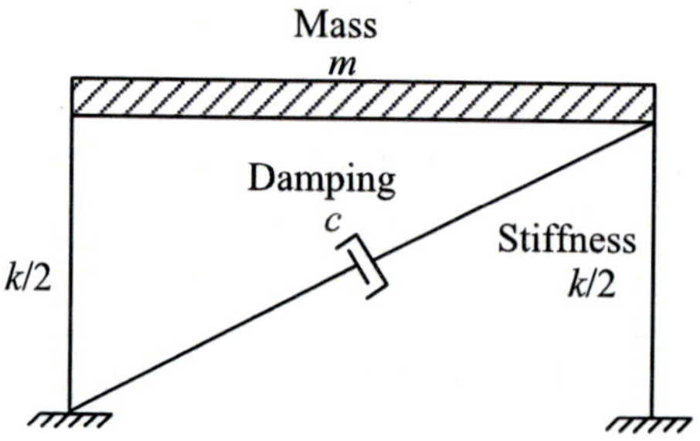

Fig. 3. Single floor shearing structure

Table 1. The random parameter of single floor shearing structure

Parameter	Mean value	Variable coefficient	Type of distribution
Mass $m\,\mathrm{kN}\cdot\mathrm{s}^2/\mathrm{cm}$	0.16	0.1	Gaussian distribution
Damping rate ζ	0.05	0.2	Gaussian distribution
Stiffness $k\,\mathrm{kN/cm}$	21.9	0.05	Gaussian distribution

3.2 Random Ground Motion Generated by Kanai-Tajimi Spectrum Model

Kanai-Tajimi spectrum [9] is wide used model to describe the random ground motion which can be described as:

$$S_{\ddot{x}_g}(\omega) = \omega_g^4 + \frac{4\zeta_g^2\omega_g^2\omega^2}{\left(\omega^2 - \omega_g^2\right)^2 + 4\zeta_g^2} S_0 \tag{6}$$

Where S_0 is the spectral intensity which reflect the strength of earthquake. ω_g and ζ_g are characteristic frequency and characteristic damping rate. In the example, these parameters are chosen as follows: $\omega_g = 15.71\,\mathrm{rad/s}, \zeta_g = 0.72$, $S_0 = 4.7478\,\mathrm{cm^2/s^3}$. Figure 4 shows a sample of random ground motion generated by Kanai-Tajimi spectrum model.

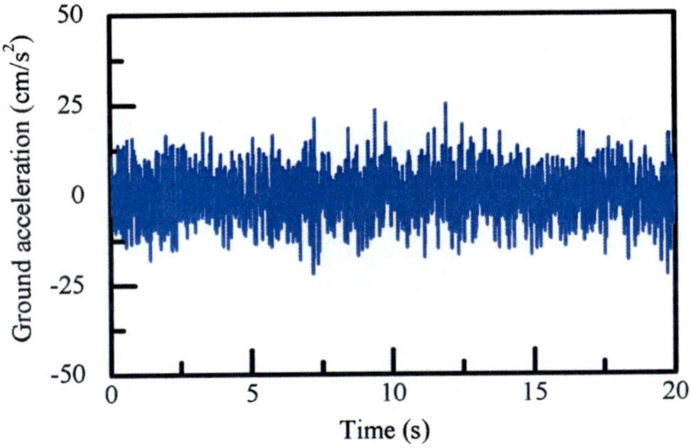

Fig. 4. A sample of ground motion generated by Kanai-Tajimi spectrum model

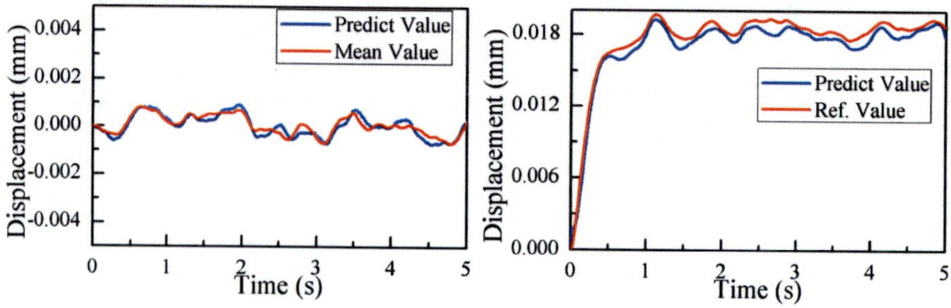

(a) Mean Value

(b) Standard deviation

Fig. 5. The mean and standard deviation of displacement response varies with time

3.3 Prediction Result

A Bayesian neural network model is trained to predict the random response of proposed single floor shearing model. The used BNN model includes three hidden layer with the 1000 node in each layer. React linear unit (ReLU) is utilized as the activation function. Bayes backprop algorithm is utilized as the training algorithm to update the model parameter. The input of neural network is the random ground acceleration while the output is the displacement and acceleration response. The trained model is utilized to predict the model response and the mean, standard deviation (stdv.) and distribution of max response are calculated from the prediction data. The mean value and stdv. result are shown in Fig. 5.

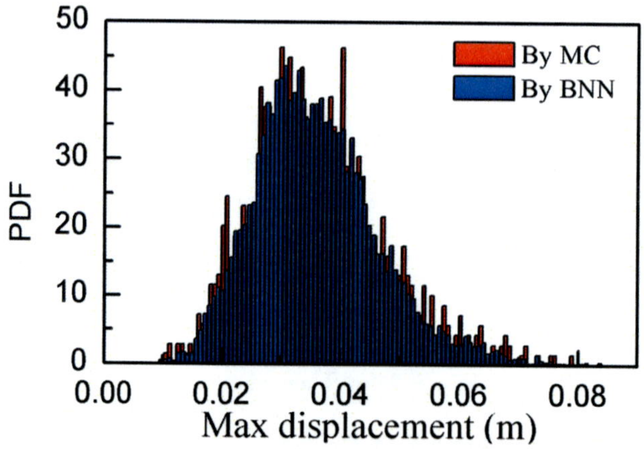

Fig. 6. The PDF of max displacement

Figure 6 gives the probability distribution function (PDF) generated by Monte Carlo simulation and BNN of max displacement response. As it is shown, the two results are closed. Therefore, the proposed method can well simulate the random response.

4 Conclusion

This paper proposed a novel method to predict the response of random structure system based on Bayesian neural network. A multilayer Bayesian neural network model is built. To verify the effectiveness of the proposed method, a single-layer shearing frame structure is established. The proposed model can be utilized to predict the structure response and get the mean value and standard deviation of random structure and the PDF of maximum response value. The proposed method can be a promising method in the dynamic response analysis of random structure.

References

1. Li, J., Chen, J.B.: Probability density evolution method for dynamic response analysis of structures with uncertain parameters. Comput. Mech. **34**(5), 400–409 (2004)
2. Melchers, R.E., Beck, A.T.: Structural Reliability Analysis and Prediction. Wiley, New Jersey (2018)
3. Lin, J.H., Zhang, Y.H., Zhao, Y.: Pseudo excitation method and some recent developments. Proc. Eng. **14**, 2453–2458 (2011)
4. Fan, C., Huang, Y., Wang, Q.: Sparsity-promoting polynomial response surface: a new surrogate model for response prediction. Adv. Eng. Softw. **77**, 48–65 (2014)
5. Goodfellow, I., Bengio, Y., Courville, A., et al.: Deep Learning. MIT Press, Cambridge (2016)
6. Kim, T., Songa, J., Kwon, O.-S.: Probabilistic evaluation of seismic responses using deep learning method. Struct. Saf. **84**, 101913 (2020)
7. Wang, H., Yeung, D.Y.: A survey on Bayesian deep learning. ACM Comput. Surv. (CSUR) **53**(5), 1–37 (2020)
8. Fortunato, M., Blundell, C., Vinyals, O.: Bayesian recurrent neural networks. arXiv preprint arXiv:1704.02798 (2017)
9. Rofooei, F.R., Mobarake, A., Ahmadi, G.: Generation of artificial earthquake records with a nonstationary Kanai-Tajimi model. Eng. Struct. **23**(7), 827–837 (2001)

Displacement Response Analysis of Multi-degree-of-Freedom System

Chenfeng Fu[1(✉)], Zirui Luo[2], Qiyuan Luo[3], and Zuhui He[3]

[1] School of Engineering, University of Liverpool, Liverpool L69 3BX, UK
sgcfu2@liverpool.ac.uk
[2] International Department of Shude High School, Chengdu 610000, China
[3] International School, Wuhan University of Science and Technology, Wuhan 430081, China

Abstract. This research project has two challenges, both of which study the solution method of displacement response in different situations of multi-degree-of-freedom system. The first challenge focuses on the displacement response of an ideal two-story structure. At the beginning, only the stiffness coefficient was introduced into the structure, and no external forces were applied to the two-story structure. This is a system with two degrees of freedom. The motion equation can be sorted out through force analysis. After the initial conditions are given, the motion equation can be solved by using mode superposition method, and the position response of the system can be determined. In the latter part, damping and external forces applied on the structure are taken into consideration, making the problem studied in this project more like the real situation. When analyzing a real problem, it is common to assume a large number of degrees of freedom to make the result more accurate, such as 100 degrees of freedom or more. In this case, the mode superposition method is still applicable. The reasons for its application will be discussed. The second challenge mainly solves the problem of a static undamped system with impact loads. The reason for choosing no damping is that under the influence of impact load, the maximum response of the structure can be reached very quickly and it would be too late for damper to absorb energy. Then the position response of the structure under changing external force will be analyzed. At 3 different phases Duhamel integral was used and the approximate expressions will be written down.

Keywords: Mechanical vibration · Multi-degree-of-freedom system · Mode superposition · Dynamic analysis

1 Introduction

In the first challenge, a two-degree-of-freedom system with no external force applied and no damping was first investigated. Firstly, the force analysis of the structure is carried out to list the equation of motion of each degree of freedom. By using the natural frequency formula in the dynamic system, the frequency of two degrees of freedom can be obtained, and then the vibration mode of each degree of freedom, namely the mode shape, can be obtained [1]. After determining the initial displacement and velocity, the position response of the system can be calculated by using the mode

P. Mendonça and N. D. Cortiços (Eds.): ICAMC 2021, LNCE 226, pp. 26–36, 2022.
https://doi.org/10.1007/978-3-030-94514-5_4

superposition method [2]. This method solves the position response of complex multi-degree of freedom system through orthogonality of mass and stiffness, and it is applicable to almost all undamped systems and some damped systems. When external forces and damping are added to the system, the position response can be calculated in the same way. In general, the mode superposition method is an excellent method, which can simplify the complex multi-degree of freedom problem into several simple single-degree of freedom problems. The position response of single degree of freedom can be obtained by solving each single degree of freedom problem, so as to integrate the overall position response. The scope and advantages of this approach will be discussed in detail in the following discussion section.

In the second challenge, the system is also a two-degree-of-freedom system. The stiffness matrix and mass matrix need to be solved at first. The stiffness coefficients and stress distribution of each story are given in the problem. According to the stress distribution the vibration equation of stiffness matrix can be obtained. The mass matrix is obtained by using orthogonality [2]. For the second question the natural frequency and mode shape need to be obtained. They can be easily figured out by looking at the answer from the first question.

The last question of challenge 2 involves an undamped structure at rest which is subjected to dynamic impulsive loads. Duhamel integral is used solve this problem. In vibration theory, Duhamel integral is a method to solve the response of linear system under arbitrary external excitation. And mode superposition method is also very useful to calculate mass m, stiffness k and external force p. Mode superposition method uses the undamped mode of the system as the space base, decouples the original dynamic equations through coordinate transformation, and solves n mutually independent equations to obtain the modal displacement, and then the response of the system is obtained by superimposing the contribution of each mode [2]. After finding equation of motion and replacing the m, k and p with its own matrix, the displacement can be calculated.

2 Procedure and Results

2.1 Challenge 1

The first challenge is to solve the positional response of the two-story structure as shown in Fig. 1. The mass of each floor m and the stiffness between each floor k are given. In part (a), the equations of motion of the system will be formulated. In part (b), the nature frequency and mode shape will be calculated. In part (c), the initial speed and displacement of each floor are given, and the undamped displacement response will be sorted out. In part (d), the force of constant magnitude applied to the first floor is given, equations for lateral floor displacement, story drift and story shear will be obtained. In part (e), the damping ratio of magnitude proportional to the stiffness is added into consideration, the first mode is given, the damping matrix and second mode will be calculated.

The values are given as:

Mass $M = 50\,\text{kips}/\text{g}$.

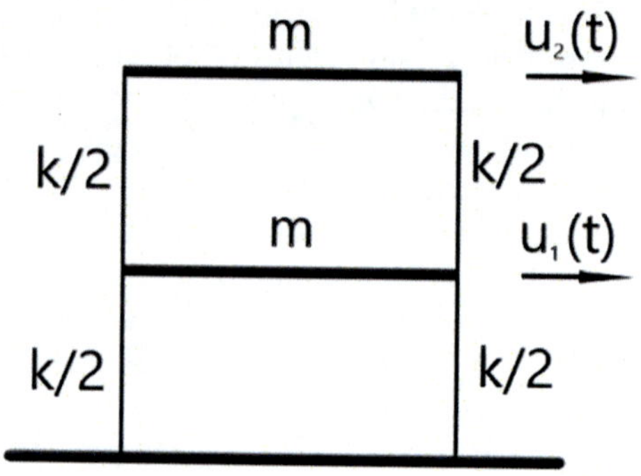

Fig. 1. Two-story structure

Stiffness coefficient $k = 15.77\,\text{kips/g}$.

Initial displacement $u(0) = \begin{bmatrix} 1 \\ 0 \end{bmatrix}$.

Initial velocity $\dot{u}(0) = \begin{bmatrix} 0 \\ 2 \end{bmatrix}$.

External force applied $p_0 = 10\,\text{kips/s}$.

Damping ratio of first story $\xi_1 = 0.05$.

(a)

First of all, it is clear that a basic equation of motion is composed of the force provided by acceleration, damping force, and elastic force. These can be written as [3]:

$$ma(t) + cv(t) + kd(t) = p(t)$$

Since it is an undamped system, only the force provided by acceleration and elastic force should be taken into consideration. For the first floor, it experiences two elastic forces, one is relative to the ground, and one is relative to the second floor. For the second floor, it experiences only the elastic force relative to the first floor. If we assume that the displacement of the two floors relative to the ground are u_1 and u_2 respectively, so we can have the equation below:

$$ku_2 - ku_1 + ma_2 = 0 \tag{1}$$

$$-ku_2 + 2ku_1 + ma_1 = 0 \tag{2}$$

Then, plug in all the values provided, it comes to a matrix below.

$$\begin{bmatrix} 15.77 & -15.77 \\ -15.77 & 31.54 \end{bmatrix}\begin{bmatrix} u_2 \\ u_1 \end{bmatrix} + \begin{bmatrix} 50 & 0 \\ 0 & 50 \end{bmatrix}\begin{bmatrix} a_2 \\ a_1 \end{bmatrix} = \begin{bmatrix} 0 \\ 0 \end{bmatrix} \tag{3}$$

(m = mass, a = acceleration, c = damping coefficient, k = elastic coefficient, v = velocity, d = displacement, p = external force).

(b)

To begin with, combined with formula k-ω^2m = 0, the matrix can be established.

$$\begin{bmatrix} k-\omega^2 m & -k \\ -k & 2k-\omega^2 m \end{bmatrix}\begin{bmatrix} u_2 = 1 \\ u_1 \end{bmatrix} \tag{4}$$

For the convenience of the calculation, assume that $B = \frac{\omega^2 m}{k}$

Therefore, the representation of the determinant of the matrix is:

$$B^2 - 3B + 1 = 0 \tag{5}$$

The two solutions are $B_1 = \frac{3-\sqrt{5}}{2}$, $B_2 = \frac{3+\sqrt{5}}{2}$ respectively.

When $B = \frac{3-\sqrt{5}}{2}$, the equation below can be established.

$$\begin{bmatrix} k-\omega^2 m & -k \\ -k & 2k-\omega^2 m \end{bmatrix}\begin{bmatrix} u_2 = 1 \\ u_1 \end{bmatrix} = 0 \tag{6}$$

Therefore, $u_1 = \frac{\sqrt{5}-1}{2}$, $\omega_1 = \sqrt{\frac{k(3-\sqrt{5})}{2m}}$

So, the mode shape can be represented as $\begin{bmatrix} 1 \\ \frac{5^{\frac{1}{2}}-1}{2} \end{bmatrix}$

When $B = \frac{3+\sqrt{5}}{2}$, the equation below can be established.

$$\begin{bmatrix} k-\omega^2 m & -k \\ -k & 2k-\omega^2 m \end{bmatrix}\begin{bmatrix} u_2 = 1 \\ u_1 \end{bmatrix} = 0 \tag{7}$$

Therefore $u_2 = \frac{-\sqrt{5}-1}{2}$, $\omega_2 = \sqrt{\frac{k(3-\sqrt{5})}{2m}}$

So the mode shape can be represented as $\begin{bmatrix} 1 \\ -\frac{\sqrt{5}-1}{2} \end{bmatrix}$.

(c)

Initial conditions are given as $u(0) = \begin{bmatrix} 1 \\ 0 \end{bmatrix}$, $\dot{u}(0) = [\begin{smallmatrix} 0 \\ 2 \end{smallmatrix}]$

There are two degrees of freedom in the system

The total displacement can be decided through mode super position method:

$$u = \varphi_1 Y_1 + \varphi_2 Y_2 \tag{8}$$

From last question, the undamped mode shape $\varphi 1$ and $\varphi 2$ have been calculated as $\varphi_1 = \begin{bmatrix} 1 \\ -\frac{1-\sqrt{5}}{2} \end{bmatrix}$, $\varphi_2 = \begin{bmatrix} 1 \\ -\frac{1+\sqrt{5}}{2} \end{bmatrix}$

Hence the mode shape matrix of the problem can be determined as $\Phi = \begin{bmatrix} 1 & 1 \\ -\frac{1-\sqrt{5}}{2} & -\frac{1+\sqrt{5}}{2} \end{bmatrix}$.

Using orthogonality to obtain generalized mass M1 and M2

$$M = \Phi^T m \Phi = \begin{bmatrix} 1 & 1 \\ -\frac{1-\sqrt{5}}{2} & -\frac{1+\sqrt{5}}{2} \end{bmatrix} \begin{bmatrix} m & 0 \\ 0 & m \end{bmatrix} \begin{bmatrix} 1 & -\frac{1-\sqrt{5}}{2} \\ 1 & -\frac{1+\sqrt{5}}{2} \end{bmatrix} = \begin{bmatrix} \frac{5-\sqrt{5}}{2} & 0 \\ 0 & \frac{5+\sqrt{5}}{2} \end{bmatrix} m \tag{9}$$

Hence $M_1 = \frac{5-\sqrt{5}}{2}m$, $M_2 = \frac{5+\sqrt{5}}{2}m$
The equations of motion of the system can be written as

$$M_n \ddot{Y}_n + K_n Y_n = 0 \tag{10}$$

The solution of this differential equation is

$$Y(t) = Y_n(0)cos\omega_n t + \frac{\dot{Y}_n(0)}{\omega_n} sin\omega_n t \tag{11}$$

Using orthogonality, the initial normal coordinate Y and its first order derivative with respect to time can be calculated as $Y_n(0) = \frac{\varphi_n^T mu(0)}{M_n}$, $\dot{Y}_n(0) = \frac{\varphi_n^T m\dot{u}(0)}{M_n}$
According to initial conditions given in the question

$$Y_1(0) = \frac{\varphi_1^T mu(0)}{M_1} = \frac{1}{\frac{5-\sqrt{5}}{2}m} \begin{bmatrix} 1 & -\frac{1-\sqrt{5}}{2} \end{bmatrix} \begin{bmatrix} m & 0 \\ 0 & m \end{bmatrix} \begin{bmatrix} 1 \\ 0 \end{bmatrix} = \frac{5+\sqrt{5}}{10} \tag{12}$$

$$\dot{Y}_1(0) = \frac{\varphi_1^T m\dot{u}(0)}{M_1} = \frac{1}{\frac{5-\sqrt{5}}{2}m} \begin{bmatrix} 1 & -\frac{1-\sqrt{5}}{2} \end{bmatrix} \begin{bmatrix} m & 0 \\ 0 & m \end{bmatrix} \begin{bmatrix} 0 \\ 2 \end{bmatrix} = \frac{2\sqrt{5}}{5} \tag{13}$$

$$Y_2(0) = \frac{\varphi_2^T mu(0)}{M_2} = \frac{1}{\frac{5+\sqrt{5}}{2}m} \begin{bmatrix} 1 & -\frac{1+\sqrt{5}}{2} \end{bmatrix} \begin{bmatrix} m & 0 \\ 0 & m \end{bmatrix} \begin{bmatrix} 1 \\ 0 \end{bmatrix} = \frac{5-\sqrt{5}}{10} \tag{14}$$

$$\dot{Y}_2(0) = \frac{\varphi_2^T m\dot{u}(0)}{M_2} = \frac{1}{\frac{5+\sqrt{5}}{2}m} \begin{bmatrix} 1 & -\frac{1+\sqrt{5}}{2} \end{bmatrix} \begin{bmatrix} m & 0 \\ 0 & m \end{bmatrix} \begin{bmatrix} 0 \\ 2 \end{bmatrix} = -\frac{2\sqrt{5}}{5} \tag{15}$$

The frequency of each floor ω_1 and ω_2 has been calculated in last question, $\omega_1 = \sqrt{\frac{3-\sqrt{5}}{2}\frac{k}{m}}$, $\omega_2 = \sqrt{\frac{3+\sqrt{5}}{2}\frac{k}{m}}$
Hence undamped displacement response can be written as

$$u(t) = \varphi_2 Y_2 + \varphi_1 Y_1 = \begin{bmatrix} 1 \\ -\frac{1+\sqrt{5}}{2} \end{bmatrix} (Y_2(0) cos\omega_2 \mathrm{t} + \frac{\dot{Y}_n(0)}{\omega_n} sin\omega_2 \mathrm{t})$$
$$+ \begin{bmatrix} 1 \\ -\frac{1-\sqrt{5}}{2} \end{bmatrix} (Y_1(0) cos\omega_1 \mathrm{t} + \frac{\dot{Y}_n(0)}{\omega_n} sin\omega_1 \mathrm{t}) = \begin{bmatrix} 1 \\ -\frac{1+\sqrt{5}}{2} \end{bmatrix}$$
$$\left(\frac{5-\sqrt{5}}{10} cos\sqrt{\frac{3+\sqrt{5}}{2}\frac{k}{m}}t + \frac{\frac{5-\sqrt{5}}{10}}{\sqrt{\frac{3+\sqrt{5}}{2}\frac{k}{m}}} sin\sqrt{\frac{3+\sqrt{5}}{2}\frac{k}{m}}t \right) \tag{16}$$
$$+ \begin{bmatrix} 1 \\ -\frac{1-\sqrt{5}}{2} \end{bmatrix} \left(\frac{5+\sqrt{5}}{10} cos\sqrt{\frac{3-\sqrt{5}}{2}\frac{k}{m}}t + \frac{\frac{5+\sqrt{5}}{10}}{\sqrt{\frac{3-\sqrt{5}}{2}\frac{k}{m}}} sin\sqrt{\frac{3-\sqrt{5}}{2}\frac{k}{m}}t \right)$$

(d)
The external force is given as $p_1(t) = p_0$
Equations of motion can be written as:

$$M_2 \ddot{Y}_2 + kY_2 = 0 \tag{17}$$

$$M_1 \ddot{Y}_1 + kY_1 = p_0 \tag{18}$$

The force is suddenly applied to the system, the initial conditions should be at rest

$$Y(0) = 0, \dot{Y}(0) = 0$$

The displacement can be obtained through solving the differential equations with zero initial conditions, the results are shown as $Y_2 = 0, Y_1 = -\frac{p_0}{k}(cos\omega_1 \mathrm{t} - 1) = -\frac{p_0}{k}\left(cos\sqrt{\frac{3-\sqrt{5}}{2}\frac{k}{m}}\mathrm{t} - 1\right)$
The response of the system is

$$u(t) = \Phi Y = \begin{bmatrix} 1 & 1 \\ -\frac{1-\sqrt{5}}{2} & -\frac{1+\sqrt{5}}{2} \end{bmatrix} \begin{bmatrix} Y_1 \\ 0 \end{bmatrix} = \begin{bmatrix} 1 & 1 \\ -\frac{1-\sqrt{5}}{2} & -\frac{1+\sqrt{5}}{2} \end{bmatrix} \begin{bmatrix} -\frac{p_0}{k}\left(cos\sqrt{\frac{3-\sqrt{5}}{2}\frac{k}{m}}\mathrm{t} - 1\right) \\ 0 \end{bmatrix} \tag{19}$$

From the equation

$$u(t) = \Phi Y = \begin{bmatrix} 1 & 1 \\ -\frac{1-\sqrt{5}}{2} & -\frac{1+\sqrt{5}}{2} \end{bmatrix} \begin{bmatrix} Y_1 \\ 0 \end{bmatrix} \tag{20}$$

The displacement in each floor can be obtained as $u_1(t) = Y_1$, $u_2(t) = -\frac{1-\sqrt{5}}{2} Y_1$
Hence the story drift can be calculated as

$$u_2(t) - u_1(t) = \frac{3-\sqrt{5}}{2} Y_1 = \frac{3-\sqrt{5}}{2} (\frac{p_0}{k} \left(cos\sqrt{\frac{3-\sqrt{5}}{2}\frac{k}{m}}t - 1 \right)) \tag{21}$$

The story shear can then be calculated as

$$k(u_2(t) - u_1(t)) = k\frac{3-\sqrt{5}}{2} Y_1 = \frac{3-\sqrt{5}}{2} (p_0 \left(cos\sqrt{\frac{3+\sqrt{5}}{2}\frac{k}{m}}t - 1 \right)) \tag{22}$$

(e)
The damping proportional to the stiffness is one kind of a Rayleigh damping.
In Rayleigh damping, $C = a_0 M + a_1 K$ [4]
In this case, $a_0 = 0$, $a_1 = \alpha$
By combining Rayleigh damping expression $(C = a_0 M + a_1 K)$ with damping expression $(C_n = 2\xi_n \omega_n M_n)$ and frequency expression $(\omega_n = \sqrt{\frac{K_n}{M_n}})$, the relation between damping ratio and frequency can be obtained:

$$\xi_n = \frac{a_0}{2\omega_n} + \frac{a_1 \omega_n}{2} \tag{23}$$

Since $a_0 = 0$, $a_1 = \alpha$, $\xi_1 = 0.05$, $\omega_1 = \sqrt{\frac{3-\sqrt{5}}{2}\frac{k}{m}}$

$$\alpha = \frac{2\xi_1}{\omega_1} = \frac{0.1}{\sqrt{\frac{3-\sqrt{5}}{2}\frac{k}{m}}} \tag{24}$$

Hence,

$$\xi_2 = \frac{\alpha\omega_2}{2} = \frac{0.05\sqrt{\frac{3+\sqrt{5}}{2}\frac{k}{m}}}{\sqrt{\frac{3-\sqrt{5}}{2}\frac{k}{m}}} \tag{25}$$

The damping matrix C can be determined as:

$$C = \alpha K = \frac{0.1}{\sqrt{\frac{3-\sqrt{5}}{2}\frac{k}{m}}} \begin{bmatrix} k & 0 \\ 0 & k \end{bmatrix} \tag{26}$$

2.2 Challenge 2

The second challenge was also to study the positional response of a two-story structure as shown in Fig. 2.

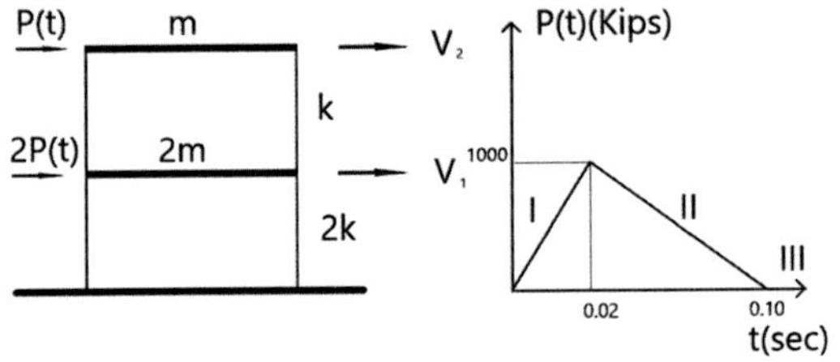

Fig. 2. Two-story structure and graph of external load

This problem is mainly about the analyzation of the position response of the structure under the condition that external forces varied over time. The stiffness between each floor and mass of each floor are given in part (a) and (b), the mass matrix m and stiffness matrix k will be calculated. In part (c), an impulsive externalload shown above will be add into consideration, and the displacement of eachstory will be calculated.

The values are given as:

Mass $m = 10\,\text{kips} \cdot \text{sec}^2/\text{in}$.

Stiffness coefficient $k = 500\,\text{kips}/\text{in}$.

(a)

$$M\ddot{X} + KX = P(t) \tag{27}$$

Since $m = 10\,\text{kips} - \text{sec}^2/\text{in}$, $k = 500\,\text{kips}/\text{in}$

Top floor/2nd floor: $k_a = 500\,\text{kips}/\text{in}$, $m_a = 10\,\text{kips} - \text{sec}^2/\text{in}$

First floor/ground: $k_b = 1200\,\text{kips}/\text{in}$, $m_b = 20\,\text{kips} - \text{sec}^2/\text{in}$

According to stress analysis:

$$-F_a + P_a = m_{top}\ddot{\mu}_a \tag{28}$$

$$-F_a - F_b + P_b = m_{st}\ddot{\mu}_b \tag{29}$$

$$500\mu_a - 500\mu_b + 10\ddot{\mu}_a = P_a \tag{30}$$

$$-500\mu_a + 1500\mu_b + 20\ddot{\mu}_b = P_b \tag{31}$$

$$\begin{bmatrix} 500 & -500 \\ -500 & 1500 \end{bmatrix} \begin{Bmatrix} \mu_a \\ \mu_b \end{Bmatrix} + \begin{bmatrix} 10 & \cdot \\ \cdot & 20 \end{bmatrix} \begin{Bmatrix} \ddot{\mu}_a \\ \ddot{\mu}_b \end{Bmatrix} = \begin{bmatrix} P_a \\ P_b \end{bmatrix} \tag{32}$$

Put in Stiffness matrix k Mass matrix m

$$k - \omega^2 m = (500\,\text{kips}/\text{in}) \begin{bmatrix} 1 - B & -1 \\ -1 & 3 - 2B \end{bmatrix} \tag{33}$$

Hence $B = \frac{\omega^2}{500}$ (b)

$$2B^2 - 5B + 2 = 0 \tag{34}$$

Hence $B_1 = 2,\ B_2 = 0.5$

$$\begin{Bmatrix} {\omega_1}^2 \\ {\omega_2}^2 \end{Bmatrix} = \begin{Bmatrix} 250 \\ 1000 \end{Bmatrix} \tag{35}$$

Since $\begin{Bmatrix} \omega_1 \\ \omega_2 \end{Bmatrix} = \begin{Bmatrix} 15.81 \\ 31.62 \end{Bmatrix}$rad/sec.
Solving the equation gives the answer $B_1 = 2,\ \mu_a = 1$

$$\begin{bmatrix} -1 & -1 \\ -1 & -1 \end{bmatrix} \begin{Bmatrix} \mu_a = 1 \\ \mu_b \end{Bmatrix} = 0 \tag{36}$$

Since $\varphi_1 = \begin{Bmatrix} 1 \\ -1 \end{Bmatrix}$.
Solving the equation gives the answer $B_2 = 0.5,\ \mu_a = 1$

$$\begin{bmatrix} 0.5 & -1 \\ -1 & 2 \end{bmatrix} \begin{Bmatrix} \mu_a = 1 \\ \mu_b \end{Bmatrix} = 0 \tag{37}$$

Hence $\varphi_2 = \begin{Bmatrix} 1 \\ 0.5 \end{Bmatrix}$ (c)
As the modal superposition method:

$$\mathrm{M} = \Phi^{\mathrm{T}} \mathrm{m} \Phi \tag{38}$$

$$\mathrm{K} = \Phi^{\mathrm{T}} \mathrm{k} \Phi \tag{39}$$

$$\mathrm{P} = \Phi^{\mathrm{T}} \mathrm{P(t)} \tag{40}$$

The results can be calculated as $\Phi_1 = \begin{Bmatrix} 1 \\ -1 \end{Bmatrix}, \Phi_2 = \begin{Bmatrix} 1 \\ 0.5 \end{Bmatrix}, \Phi = \begin{bmatrix} 1 & 1 \\ -1 & 0.5 \end{bmatrix}, \Phi^{\mathrm{T}} = \begin{bmatrix} 1 & -1 \\ 1 & 0.5 \end{bmatrix}$, $\mathrm{M} = \begin{bmatrix} 1 & 1 \\ -1 & 0.5 \end{bmatrix} \mathrm{m} \begin{bmatrix} 1 & -1 \\ 1 & 0.5 \end{bmatrix} = \begin{bmatrix} 2 & -0.5 \\ -0.5 & 1.25 \end{bmatrix} \mathrm{m}$, $\mathrm{K} = \begin{bmatrix} 2 & -0.5 \\ -0.5 & 1.25 \end{bmatrix} \mathrm{k}$,
$\mathrm{P} = \begin{bmatrix} 1 & -1 \\ 1 & 0.5 \end{bmatrix} \mathrm{p(t)}$
The undamped system means $\xi = 0$
Use Duhamel integral:
So when $\mathrm{t} \in (0, 0.02)$

$$p(t) = \frac{F_0}{t_0} t \tag{41}$$

$$v_1 = \frac{1}{m_n \omega_n} \int_0^t p_{1n}(\tau) sinw_n(t - \tau) d\tau \tag{42}$$

When $t \in (0.02, 0.1)$

$$v_1 = \frac{1}{m_n \omega_n} \int_0^{0.02} p_{1n}(\tau) sinw_n(t - \tau) d\tau + \frac{1}{m_n \omega_n} \int_{0.02}^t p_{2n}(\tau) sinw_n(t - \tau) d\tau \tag{43}$$

When $t > 0.1$

$$v_1 = \frac{1}{m_n \omega_n} \int_0^{0.02} p_{1n}(\tau) sinw_n(t - \tau) d\tau + \frac{1}{m_n \omega_n} \int_{0.02}^t p_{2n}(\tau) sinw_n(t - \tau) d\tau + 0 \tag{44}$$

3 Discussion

This part mainly discusses the advantages and applicable conditions of modal superposition method. It is very difficult to solve the problem of multiple degrees of freedom by directly inserting the mass and stiffness coefficient matrices into the equation of motion. For example, to solve the position response of a system with N degrees of freedom, we use the mass, the stiffness coefficient, and the mode shape matrix of size N times N. A typical value of N is 100, which can make the solution relatively accurate [5]. In order to solve the equation of motion, 100 times 100 formulas need to be solved, which is very inefficient. Modal superposition method makes use of the orthogonality of mass matrix and stiffness matrix to multiply its front by the transpose of mode shape matrix and the back by mode shape matrix, so that it becomes a diagonal matrix. By this method, the N degree of freedom problem can be reduced to N simple single degree of freedom problems, and the overall position response can be obtained only by solving the position response of each degree of freedom separately [2]. However, the damping of the system does not necessarily have orthogonality. When the damping does not have orthogonality, the Modal superposition method cannot be used. The problem may have to be worked out by directly solving large number of equations. Some kinds of damping do have orthogonality, such as Rayleigh damping and Caughey damping [6]. When studying and solving practical problems, if the damping of the system is close to the expressions of these two damping, they can be approximately regarded as having orthogonality, and then solved by using the Modal superposition method.

4 Conclusion

In this research project, an idealized dynamic structural system is analyzed using modal superposition method. Different external force conditions are taken into account: the magnitude is constant or varies with time. A special damping is introduced so that the position response of the system can still be solved by the mode superposition method. Finally, the advantages and application range of modal superposition method are analyzed. In general, this project has studied almost all the major elements in the dynamic structure system. Overall, this research project discussed the ways to simplify multi-degree-of-freedom system and can help solve the problems of dynamic structure system in reality by using modal superposition method.

References

1. Chen, P.C., Soroka, W.W.: Multidegree dynamic response of a system with statistical properties. J. Sound Vibr. **37**, 547–556 (1974)
2. Rao, S.S.: Mechanical Vibrations. Pearson, London (2016)
3. Ruzicka, J.E.: Fundamental concepts of vibration control. Sound Vibr. **5**(7), 16–23 (1971)
4. Saeed, M.A.: Analysis of proportional damping or Rayleigh damping on damped and undamped systems. In: 2017 Fifth International Conference on Aerospace Science & Engineering (ICASE), pp. 1–13 (2017)
5. Gladwell, I., Hanson, P.M.: Some error bounds and numerical experiments in modal methods for dynamics of systems. Earthq. Eng. Struct. Dyn. **12**, 9–36 (1984)
6. Luco, J.E., Lanzi, A.: Optimal Caughey series representation of classical damping matrices. Soil Dyn. Earthq. Eng. (0267-7261) **92**, 253–265 (2017)

The Influence of Geometry on the Frictional Sliding of $\wedge$ and $\vee$ Shaped Interlocking Joints in Masonry Assemblages

Elham Mousavian[1], Claudia Casapulla[1(✉)], and Katalin Bagi[2]

[1] Department of Structures for Engineering and Architecture, University of Napoli Federico II, Via Forno vecchio, 36, 80134 Napoli, Italy
casacla@unina.it

[2] Department of Structural Mechanics, Budapest University of Technology and Economics, Budapest, Hungary

Abstract. Applications of interlocking blocks improving the sliding resistance of masonry joints have recently been of significant interest to architects and structural engineers. In this framework, several works have studied the load capacity improvement of different joint shapes using various numerical and experimental analyses. However, applications of simplified and yet accurate analytical methods like limit analysis to interlocking blocks are rather unexplored and limited to a few interlocking joint shapes. To develop such a simplified analytical approach for interlocking assemblages with $\wedge$ and $\vee$ shaped joints, this paper explores relationships between the shear capacities of flat and interlocking joints, allowing to abstract an assemblage of interlocking blocks to its equivalent assemblage with flat joints. This work specifically studies the interaction among friction, dilation, and inclination angles of the joints. To this aim, first the relationships between a flat and an interlocking joint of two stacked blocks are fully explored and then, the associative solution for an assemblage of two wedge blocks with flat faces is compared with the non-associative solution of the same assemblage with $\wedge$ and $\vee$ shaped interlocking joints. Discrete element models using 3DEC are implemented to carry out the comparative analyses, in which the blocks are modelled as rigid units and the ultimate lateral load capacity of the assemblages is recorded.

Keywords: $\wedge$ and $\vee$ shaped interlocking joints · Associative and non-associative sliding capacities · Friction and dilation angles · Discrete element analysis · 3DEC

1 Introduction

Joints between rigid units of a discrete assembly usually act as planes of weakness due to their low shear and tensile resistances [1]. Especially for masonry structures, the interlocking among units tries to reduce such a weakness, affecting the in-plane and the out-of-plane behaviors, as generally studied through several approaches [2–5], also considering the type of wall to diaphragm connections [6]. Basically, units can be kept together using fasteners, glue, mortar, etc., or they can be interlocked to each other

P. Mendonça and N. D. Cortiços (Eds.): ICAMC 2021, LNCE 226, pp. 37–45, 2022.
https://doi.org/10.1007/978-3-030-94514-5_5

using non-planar interlocking joints. Recently, a few works have addressed the load capacity improvement of assemblages with various interlocking joint shapes including corrugated joints with rectangular, circular, and trapezoidal cross sections [7, 8], cross-shaped joints [9], and osteomorphic blocks [10, 11], through experimental tests and numerical analyses using, e.g., finite element and discrete element methods. Along the same line and for the first time in the literature, the lower-bound approach of limit analysis has been extended to the interlocking corrugated joints with rectangular cross sections within the MCSA_IF project SiDMACIB No. 791235. In that extension, an interlocking face is abstracted to an equivalent flat face with a number of contact points located over the center lines of the locks (projections on the blocks) at which the internal forces are distributed [12]. The proposed abstraction, for an interlocking block with a single lock, has been validated using experimental tests and discrete element analysis [13].

Following the same concept, this work is aimed to find simple relationships to assess the load capacity of interlocking assemblages with ∧ and ∨ shaped joints by using standard limit analysis of equivalent assemblages with flat joints. To this aim, the sliding capacity of ∧ and ∨ shaped interlocking joints is investigated and compared with the sliding capacity of the similar assemblages with flat joints.

As Drucker identified in [14], standard limit analysis involves associative frictional sliding at the joints according to which, sliding displacement has two components normal (d^n) and tangential (d^t) to the joints so that $d^n = d^t \tan \varphi^d$, where φ^d is the dilation angle considered equal to the friction angle φ^f. Since then, several works, starting from [15, 16], have proposed various methods to introduce non-associative frictional rules into the limit analysis in which φ^d is set to be zero, i.e., sliding displacement is only along the joint, that is usually more realistic compared with the associative sliding displacement.

Gilbert et al. in [17] demonstrated that the associative sliding displacement of a flat joint is similar to the non-associative sliding displacement of a 2D ∧ or ∨ shaped interlocking joint, provided that a normal displacement is considered accordingly. That observation, also experimentally validated [18], has become the main motivation of this paper to explore the relationship between the non-associative load capacity of assemblages with ∧ and ∨ shaped interlocking joints and the associative load capacity of the equivalent assemblages with flat joints.

While the associative sliding capacity of a flat joint is governed by the friction angle φ^f (and the dilation angle $\varphi^d = \varphi^f$), the non-associative sliding capacity of a ∧ (and ∨) shaped interlocking joint is governed by the friction angle φ^f and the joint inclination angle $\varphi^\wedge$ (and $\varphi^\vee$).

In the following, first, the interaction among φ^f and $\varphi^\wedge$ (and $\varphi^\vee$) is investigated with reference to a single interlocking interface between two stacked rigid blocks, when the upper one is subjected to a lateral load. The relations between the sliding capacity of the interlocking and flat joints are presented graphically through free body diagrams (for static problems) and numerically by finding the ultimate lateral load through the discrete element analysis using 3DEC.

Interactions between φ^f and $\varphi^\wedge$ (and $\varphi^\vee$) is then explored for a more complex example of two wedge shaped blocks with two interlocking side joints, when one is subjected to an uplift load. And the relation between the associative load capacity of the flat joints and non-associative load capacity of the interlocking joints is investigated using the discrete element analysis by 3DEC.

2 Sliding Capacity of a ∧ and ∨ Shaped Interlocking Joint

Figure 1a shows an assemblage of two stacked blocks with a shared horizontal flat joint, and while the lower block is fixed, the upper one is subjected to the self-weight W^u and the horizontal force L. When the inclination of the resultant R is greater than or equal to arctan (L^u/W^u) = 75.96°, the upper block will rock. Instead, for any friction angle φ^{f-f} smaller than 75.96°, the upper block starts to slide when the internal force R equilibrating the lateral load L^u and the block weight W^u coincides with the sliding yield edges of the joint. Instead, Fig. 1b illustrates the same assemblage with a ∨ shaped interlocking joint between the blocks, having friction angle φ^{f-i} = 0° and the inclination angle $\varphi^{\vee}$ equal to the friction angle of the flat joint of Fig. 1a (i.e., $\varphi^{\vee} = \varphi^{f-f}$). As shown in the free body diagram of Fig. 1b, the sliding yield domain of either of the frictionless edges of the ∨ shaped joint can be represented by a red line normal to the edge.

Since the ∨ shaped joint does not allow the block to displace downwards, the overall sliding domain of the interlocking joint can be considered the gray area between the two lines normal to the inclined edges of the joint (Fig. 1c). Based on this assumption, the upper block starts to slide when R coincides with the illustrated red yield edges.

In other words, the ultimate lateral force L^u of a flat joint with friction coefficient φ^{f-f} and that of an inclined joint with the inclination angle $\varphi^{\vee} = \varphi^{f-f}$ and the friction angle φ^{f-if} = 0° are the same, taking also into account that the associative and non-associative sliding capacity in such an assemblage are the same.

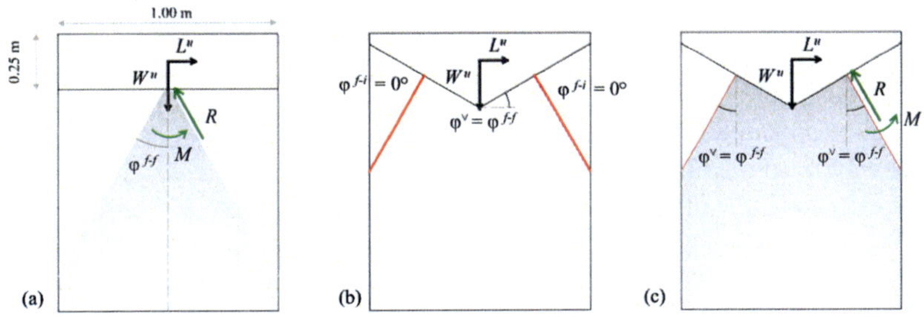

Fig. 1. Sliding yield domains of the flat joint and frictionless ∨ shaped interlocking joint.

This observation can be generalized as illustrated in Figs. 2b to 2c, showing a ∨ shaped interlocking joint having friction angle φ^{f-in} and the inclination angle $\varphi^{\vee}$ so that $\varphi^{\vee} + \varphi^{f-i} = \varphi^{f-f}$. In such an interlocking model, the inclination of the sliding yield edges with respect to the vertical axis becomes $\varphi^{\vee} + \varphi^{f-i} = \varphi^{f-f}$. This means that the sliding yield domain of this interlocking joint is identical to the sliding yield domain of a flat face with friction coefficient $\varphi^{f\text{-}f}$ (Fig. 2).

In other words, the ultimate lateral force of a flat joint with friction angle $\varphi^{f\text{-}f}$ is equivalent to the ultimate lateral force of an inclined joint with the inclination and friction angles $\varphi^{\vee}$ and $\varphi^{f\text{-}i}$ when $\varphi^{\vee} + \varphi^{f\text{-}i} = \varphi^{f\text{-}f}$.

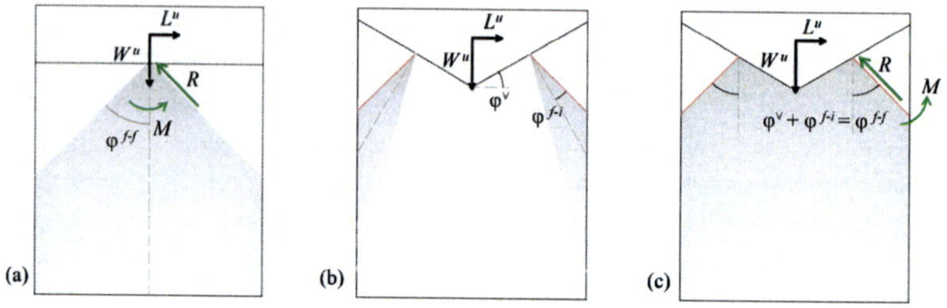

Fig. 2. Sliding yield domains of the flat joint with $\varphi^{f\text{-}f}$ and $\vee$ shaped interlocking joint with $\varphi^{\vee} + \varphi^{f\text{-}I} = \varphi^{f\text{-}f}$.

This general finding has also been verified by the discrete element analysis using 3DEC. To do so, the lower interlocking block has been modelled by two rigid convex polyhedrons that are joined together to generate a concave block in 3DEC. Then, two rigid blocks (concave lower and convex upper) are stacked over each other with shared deformable contacts in between having shear and tensile stiffness of 10^{20} N/m. Changing the inclination angle, the flat, $\wedge$ and $\vee$ shaped joints can be modelled.

To investigate the sliding capacity of the joints, the lower block is fully fixed, while the upper block is subjected to uniformly distributed forces applied laterally on the block face using the 'boundary stress' command in 3DEC. Since the ultimate lateral force is dependent on the upper block weight, to compare the sliding capacity of the interlocking blocks with different inclination angles their weight must be kept equal.

Figure 1a shows the dimensions of the assemblage of two conventional blocks with flat faces whose densities are set to be 2500 kg/m^3. Changing the inclination angle, the volume of the upper interlocking block changes and to keep the same weight, the upper block density must also be changed.

Figure 3 shows the ultimate lateral forces for a flat joint with $\varphi^{f\text{-}f} = \{0°, 10°, 20°, 30°, 40°\}$ as well as $\vee$ shaped joints with different combinations of $\varphi^{\vee}$ and $\varphi^{f\text{-}i}$ so that $\varphi^{\vee} + \varphi^{f\text{-}I} = \{0°, 10°, 20°, 30°, 40°\}$. The results prove that the sliding capacity of a flat face with friction angle $\varphi^{f\text{-}f}$ and a $\vee$ shaped joint $\varphi^{\vee} + \varphi^{f\text{-}I} = \varphi^{f\text{-}f}$ are identical for all the tested combinations of friction and inclination angles.

On the other hand, there is a small discrepancy between the ultimate lateral forces of the $\wedge$ and $\vee$ shaped joints, with $\varphi^{\wedge} = \varphi^{\vee} = 10°$ as illustrated in Fig. 4. For the other inclination angles, however, the recorded ultimate lateral forces are identical.

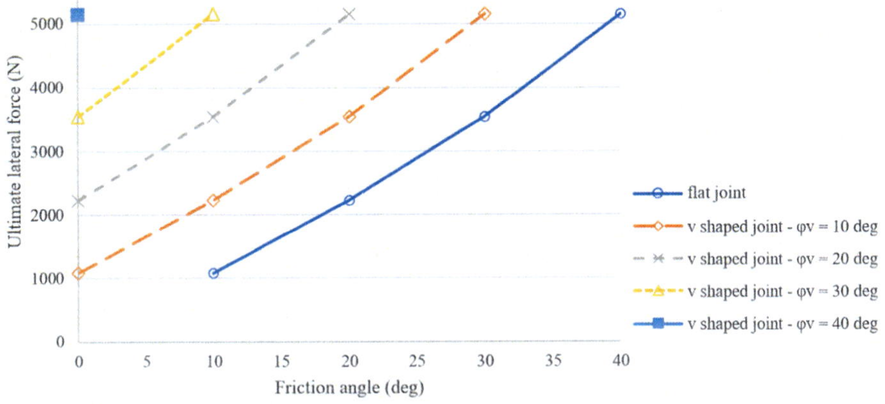

Fig. 3. Sliding capacity of flat and ∨ shaped interlocking joints with various $\varphi^f$and $\varphi^\vee$.

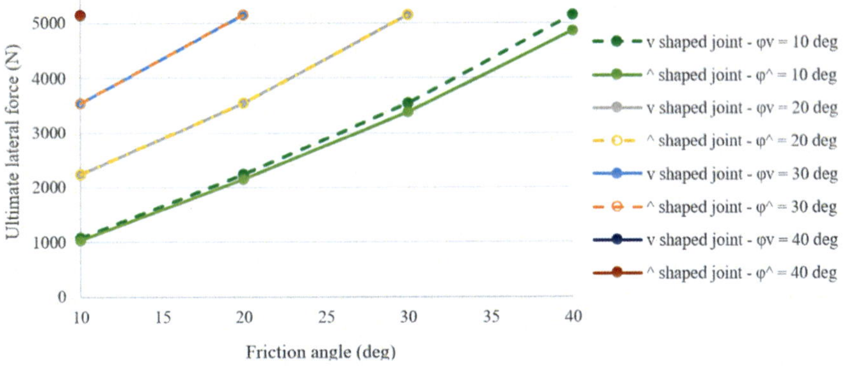

Fig. 4. Sliding capacity of ∧ and ∨ shaped interlocking joint with various φ^f, $\varphi^\wedge$, and $\varphi^\vee$.

3 Two Wedge Shaped Blocks

As explained above, the associative sliding displacement of a flat joint is similar to the non-associative displacement of ∧ and ∨ shaped interlocking joints [17], provided that proper inclination and friction angles are considered. Motivated by that observation, the chosen example of this section allows exploring the relationship between the non-associative load capacity of assemblages with ∧ and ∨ shaped interlocking joints and the associative load capacity of the equivalent assemblages with flat joints.

The example (known as Livesley's classic two block problem [16, 17]) comprises two wedge shaped blocks with weights equal to W, (blue and yellow in Fig. 5) supported laterally by two fixed support blocks (pink in Fig. 5), while the left-hand side block (blue in Fig. 5) is subjected to an uplift load U. According to [17], considering the friction angle of all joints to be 9°, the limit value of U leading to failure found by the associative and non-associative solutions are 1.8 W and 1.28 W, respectively.

Figures 5a and 5b show how the example with flat and interlocking joints are modelled in 3DEC as rigid blocks. The uplift load is applied as equivalent distributed loads on the upper face of the left-hand side block.

For the assemblage with flat joints with the friction angle of 9° (Fig. 5a), when the dilation angle is set to be 9° (associative), the ultimate uplift load becomes 1.795 W that is very close to the solution obtained in [17]; and for dilation angle 0° (non-associative), the calculated ultimate uplift load is 1.36 W with a small discrepancy compared with the solution found by [17].

As shown in Fig. 5b, the interlocking assemblage is modelled by two interlocking side joints with zero dilation, and the middle flat joint with dilation angle equal to the friction angle ($\varphi^f = \varphi^d$). Considering the inclination $\varphi^\wedge$ and friction φ^f angles of the two interlocking side joints to be 9° and 0°, respectively, and φ^f of the middle joint to be 9°, the ultimate uplift load becomes 1.857 W, that is a little bit larger than the associative solution of the assemblage with flat joints.

Same considerations have been done to compare the associative ultimate uplift load for an assemblage with all flat joints having $\varphi^f = \varphi^d = a$, where $a = \{0°, 1° \ldots 8°, 9°\}$, with the ultimate uplift load of the interlocking assemblage with two interlocking side joints having $\varphi^\wedge = a$ and $\varphi^f = \varphi^d = 0°$, and a middle flat joint with $\varphi^f = \varphi^d = a$ (Fig. 5). Considering the ultimate uplift load to be $U = C{\cdot}W$, Fig. 6 depicts the coefficient C versus the inclination angle a (also representing the friction angle of flat joints). This figure shows that the results of the two types of assemblages are very close to each other, while increasing a, the discrepancy between the results slightly increases.

On the other hand, it can be observed that, when $\varphi^f + \varphi^\wedge = 9°$ and $\varphi^d = 0°$ at the interlocking side joints, and $\varphi^f = \varphi^d = 9°$ at the flat middle joint, the ultimate loads for the interlocking assemblage are in good agreement (low percentage differences) with the associative ultimate load of the assemblage with flat joints, as reported in Table 1. This means that the non-associative solution of an assemblage of interlocking blocks can be obtained by exploiting the associative solution of an equivalent assemblage of blocks with flat joints.

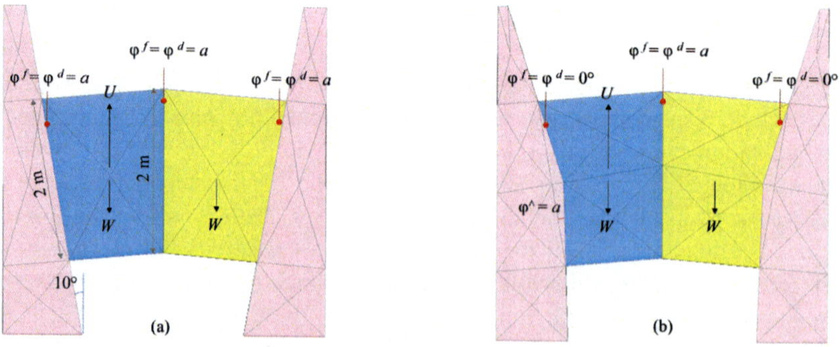

Fig. 5. Assemblages of two wedge blocks modelled in 3DEC.

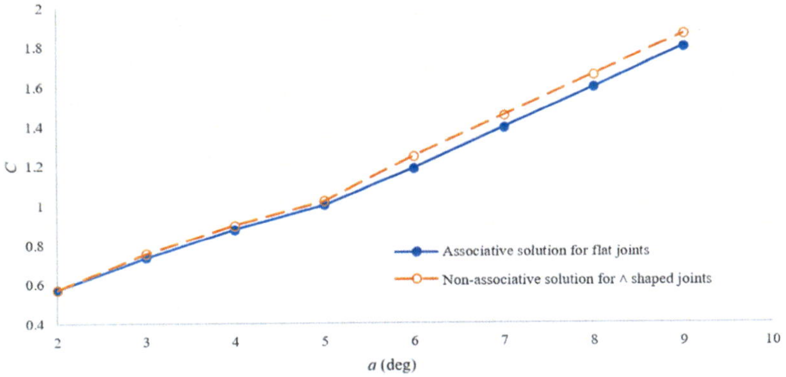

Fig. 6. Ultimate uplift load multiplier C for interlocking assemblages with various interlocking inclinations and their equivalent assemblages with flat joints.

Table 1. Percentage differences of C for various combinations of φ^f and $\varphi^\wedge$.

$\varphi^\wedge$	2	3	4	5	6	7	8	9
φ^f	7	6	5	4	3	2	1	0
C	1.795	1.836	1.857	1.857	1.877	1.877	1.897	1.857
Diff	0%	2.284%	3.454%	3.454%	4.568%	4.568%	5.682%	3.454%

4 Potential Applications

Previous studies on the contact behavior have shown that the non-associative solutions are usually more realistic compared to the associative ones, as well as conservative [14, 16, 17]. However, methods developed to find reliable solutions for non-associative problems of assemblages with arbitrary geometry can be computationally expensive, especially if involving highly non-linear mathematical programming [19]. This difficulty largely increases when the analysis is referred to assemblages with interlocking blocks of complex geometry, for which reliable shear resistances for interlocking joints should also be provided [12].

Therefore, a first attempt to overcome this difficulty is represented by the investigated relation between the non-associative solution for an assemblage with ∧ and ∨ shaped interlocking joints and the associative one for the same assemblage with flat joints (conventional assemblage). This enables to study the interlocking assemblage as a standard linear programming (LP) problem, which can be solved rapidly using well developed and widely available LP solvers.

Figure 7 illustrates common assemblages that can be modelled with the ∧ and ∨ shaped interlocking joints including barrel vaults, hemispherical domes, and logarithmic skewed arches. The investigated relation will be studied in the future for these complex models.

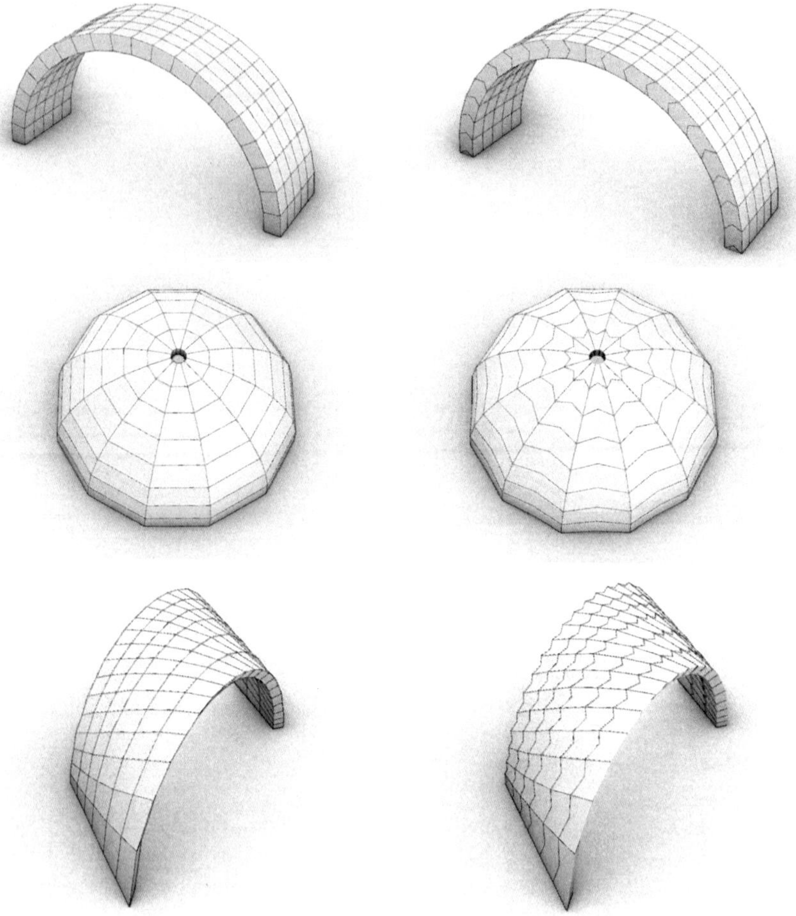

Fig. 7. Complex masonry assemblages composed of flat (left) and interlocking (right) joints.

5 Conclusions

This paper has explored a simplified method to assess the load capacity of an interlocking assemblage with $\wedge$ and $\vee$ shaped interlocking joints, through abstracting them to the equivalent assemblage with flat joints. The executed investigations have shown that the associative sliding capacity of an assemblage with flat joints is similar to the non-associative sliding capacity of the same assemblage with $\wedge$ and $\vee$ shaped interlocking joints, provided that the summation of their inclination and friction angles are equal to the friction angle of the flat joints. This observation has been obtained through two examples of two stacked blocks and two wedge shaped blocks, using discrete element method and modelling the blocks as rigid units in 3DEC. In the future, this exploration will be extended to more complex 2D assemblages with sliding failure and different associative and non-associative solutions.

References

1. Puppio, M.L., Giresini, L.: Estimation of tensile mechanical parameters of existing masonry through the analysis of the collapse of Volterra's urban walls. Fratire Integr. Strutturale **13** (49), 725–738 (2019)
2. Sarhosis, V., Lemos, J.V., Bagi, K.: Discrete element modeling. In: Ghiassi, B., Milani, G. (eds.) Numerical Modeling of Masonry and Historical Structures: From Theory to Application, pp. 469–5011. Elsevier (2019)
3. Casapulla, C., Argiento, L.U., Maione, A., Speranza, E.: Upgraded formulations for the onset of local mechanisms in multi-storey masonry buildings using limit analysis. Structures **31**, 380–394 (2021)
4. Giresini, L., Butenweg, C.: Earthquake resistant design of structures according to Eurocode 8. In: Meskouris, K., Butenweg, C., Hinzen, K.G., Höffer, R., Structural Dynamics with Applications in Earthquake and Wind Engineering, chap. 4, pp. 226–373, Springer, Germany (2019)
5. Pantò, B., Cannizzaro, F., Caliò, I., Lourenço, P.B.: Numerical and experimental validation of a 3D macro-model for the in-plane and out-of-plane behavior of unreinforced masonry walls. Int. J. Archit. Heritage **11**(7), 946–964 (2017)
6. Solarino, F., Oliveira, D.V., Giresini, L.: Wall-to-horizontal diaphragm connections in historical buildings: a state-of-the-art review. Eng. Struct. **199**, 109559 (2019)
7. Liu, H., Liu, P., Lin, K., Zhao, S.: Cyclic behavior of mortarless brick joints with different interlocking shapes. Materials **9**(3), 166 (2016)
8. Jiang, H.B., Chen, L., Ma, Z.J., Feng, W.X.: Shear behavior of dry joints with castellated keys in precast concrete segmental bridges. J. Bridg. Eng. **20**(2), 04014062 (2015)
9. Ali, M., Gultom, R.J., Chouw, N.: Capacity of innovative interlocking blocks under monotonic loading. Constr. Build. Mater. **37**, 812–821 (2012)
10. Dyskin, A.V., Estrin, Y., Pasternak, E.: Topological interlocking materials. In: Estrin, Y., Bréchet, Y., Dunlop, J., Fratzl, P. (eds.) Architectured Materials in Nature and Engineering. SSMS, vol. 282, pp. 23–49. Springer, Cham (2019). https://doi.org/10.1007/978-3-030-11942-3_2
11. Weizmann, M., Amir, O., Grobman, Y.J.: The effect of block geometry on structural behavior of topological interlocking assemblies. Autom. Constr. **128**, 103717 (2021)
12. Mousavian, E., Casapulla, C.: Structurally informed design of interlocking block assemblages using limit analysis. J. Comput. Design Eng. **7**(4), 448–468 (2020)
13. Casapulla, C., Mousavian, E., Argiento, L.U., Ceraldi, C., Bagi, K.: Torsion-shear behaviour at the interfaces of rigid interlocking blocks in masonry assemblages. Exp. Invest. Anal. Approaches Mater. Struct. **54**(3), 134 (2021)
14. Drucker, D.C.: Coulomb friction, plasticity, and limit loads. J. Appl. Mech. **21**(1), 71–74 (1954)
15. Michalowski, R., Mroz, Z.: Associated and non-associated sliding rules in contact friction problems. Arch. Mech. **30**(3), 259–276 (1978)
16. Livesley, R.K.: Limit analysis of structures formed from rigid blocks. Int. J. Numer. Meth. Eng. **12**(12), 1853–1871 (1978)
17. Gilbert, M., Casapulla, C., Ahmed, H.M.: Limit analysis of masonry block structures with non-associative frictional joints using linear programming. Comput. Struct. **84**(13–14), 873–887 (2006)
18. Bahaaddini, M., Sharrock, G., Hebblewhite, B.K.: Numerical direct shear tests to model the shear behaviour of rock joints. Comput. Geotech. **51**, 101–115 (2013)
19. Ferris, M., Tin-Loi, F.: Limit analysis of frictional block assemblies as a mathematical program with complementarity constraints. Int. J. Mech. Sci. **43**, 209–224 (2001)

The Influence of Soil-Pile-Structure Interaction on Fundamental Period of Burj Khalifa

Zihan Liu[1], Shiheng Jin[2], Yuqian Qiu[3(✉)], Jennifer Hong[4], and Fanxi Xia[5]

[1] Department of Civil and Environmental Engineering, Northeastern University, Boston 02120, USA
[2] Department of Physics, University of California, Santa Barbara 93106, USA
[3] Vanke Meisha Academy, Shenzhen 518000, China
qiuyuqian@stu.vma.edu.cn
[4] Culver Academics, Culver, Indiana 46511, USA
[5] YK PAO School, Shanghai 201620, China

Abstract. Burj Khalifa is currently standing in Dubai, United Arab Emirates (UAE) as the tallest building in the world. Even though Dubai is not in an active seismic zone, it borders the northern Emirates, which experience most of the seismic activities in UAE. Since soil-pile-structure interaction (SPSI) influences the fundamental natural period (T) of a building, it may also impact its seismic-resistant ability. This study intends to get a basic understanding of SPSI's impact on the T of a building and estimate the maximum deformation resulted from resonance. We hypothesize that the building without pile foundation will have a shorter T, while the building with pile foundation will have a longer T due to the flexibility resulted from SPSI. Modal analysis is employed to find the natural frequency of the building, and harmonic motion analysis is employed to find the maximum amplitude of the frequency response of the building. The results from both analyses show high precision, and total deformations of the 2 buildings are visualized. The results suggest that SPSI has a considerable impact on T of a building, can result in greater amplitude of response frequency, and lead to greater total deformation of the building.

Keywords: Soil-pile-structure interaction · Fundamental period · Total deformation · Modal analysis · Harmonic motion analysis

1 Introduction

Burj Khalifa located in Dubai, UAE, bordering the Northern Emirates, a seismic-active region. Despite having 163 floors and a height of 828 m, Burj Khalifa is designed to be wind-resistant and is able to withstand a seismic activity up to magnitude 7. The Y-shaped floor plan and the setting back of one of the three wings on each tier creates a spiraling shape that allows the building to be increasingly narrow at the top. From Baker et al., this spiraling shape can effectively reduce wind forces by "confusing the wind" [1]. Because the building shape on each tier is different, the wind never gets the

P. Mendonça and N. D. Cortiços (Eds.): ICAMC 2021, LNCE 226, pp. 46–57, 2022.
https://doi.org/10.1007/978-3-030-94514-5_6

chance to form an organized vortex shedding along one side of the building. In Baker et al., the reduced weight at the top also shifts the center of mass and allows for sound gravitational support [1]. Baker and Pawlikowski proposes that its 3.7 m thick pile-supported raft foundation helps reduce seismic shocks by allowing as little displacement at the base as possible [2]. The foundation is consisted of 192 piles, each of a 1.5 m diameter and capable of withstanding 3,000 metric tons of weight. Kumar et al. performed pseudostatic and dynamic analyses and showed that the combined pile–raft foundation (CPRF) is capable of sharing 62 to 76% pseudostatic load [3]. Although the design may be functional for wind and earthquake resistance, there is still a great risk of collapse during earthquake due to effects of soil-pile-structure interaction (SPSI). The magnitude of its impact depends on soil flexibility/stiffness.

2 Literature Review

2.1 Soil-Pile-Structure Interaction (SPSI) and Fundamental Natural Period (T)

Fundamental Natural Period (T) is an intrinsic property of a building that is dependent on factors such as flexibility and mass. Generally, the more flexible the building and the more mass the building possesses, the longer the T.

During earthquakes, the ground will be shaken by a mixture of different sinusoidal waves that have different frequencies ($f = \frac{1}{T}$). Knowing the fundamental natural period of a building is of particular interest since it offers an insight into the seismic response of a building. Murty demonstrates that if the building has a short period, it will respond more strongly to seismic waves that also have short periods, and vice versa [4].

Even though the traditional approach to analyzing the seismic response of a building is to fix the base of the building model and obtain a fixed-base T for simplicity's sake, soil-pile-structure interaction (SPSI) or soil-structure interaction (SSI) plays an important role in affecting the T of the structure and should not be ignored. Kramer's study shows that SPSI may lengthen the T of a structure [5]. Sharma et al. conducted analyses to obtain the T of structures under different pile foundations and soil types [6]. They concluded that the building frames supported by looser soil exhibit higher SSI effects [6]. Khalil et al.'s analysis and modeling using discrete springs yield a similar result: the higher the soil structure relative rigidity (Kss), the less evident the SSI influence [7]. Using a 3-D finite element model supported by pile foundation and nonlinear soil model, Chu and Truman demonstrated that SPSI extends T while degrading the stiffness of soil [8]. Wei & Jiang employed the Dunkerley's formula developed by Luco and reported that the increase in structure-to-soil relative stiffness is coupled by the increase of period lengthening ratio, which means that the more flexible the soil, the larger the flexible-base T than the fixed-based T [9].

3 Methology

3.1 Rendering Approach

The aim of our study is to use ANSYS (Finite Element Analysis) to show the difference between the T of the tower with pile foundation and the T of the tower without pile foundation, thus demonstrating the effect of SPSI on T of a building. We have constructed a detailed model of Burj Khalifa using the commercial application ANSYS and assigned actual material properties to each portion of the building, soil base, and pile foundation in accordance with details specified by O. Gendy & M. Gendy [10]. We then fix the base of the tower without pile foundation, while allowing flexibility in the tower with pile foundation, and employed mode analysis and harmonic motion analysis. We hypothesize that the building without pile foundation will have a shorter T, while the building with pile foundation will have a longer T due to the flexibility resulted from SPSI.

3.2 How to Simulate Burj Khalifa in ANSYS

An accurate and comprehensive model simulation of Burj Khalifa is needed for advanced analyses. The model of Burj Khalifa is built using Design Modeler, a modeling software affiliated to ANSYS Workbench.

Main Building. The construction of Burj Khalifa main building (tower) is based precisely on a detailed side view of Burj Khalifa by Richard Braddish and an AutoCAD file, so that the scale and dimension of our model are close to the real building [11]. The entire complete model of Burj Khalifa's main building is 852.53 m high and has a radius of 99.3 m.

The main building is constructed tier by tier. For each tier, the Y-shape floor plan will be set back so that each subsequent tier is narrower that the last tier. In the first 3 tiers (Fig. 1), 3 wings are set back together. Thus, the first 3 tiers are represented by similar shapes. Starting from tier 4, only 1 of the 3 wings will be set back. Overall, the entire building appears in a spiraling shape (Fig. 2).

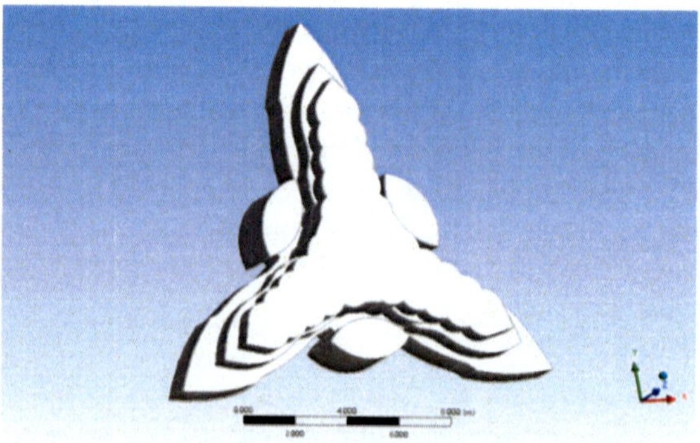

Fig. 1. First three tiers of 1:10 scaled model

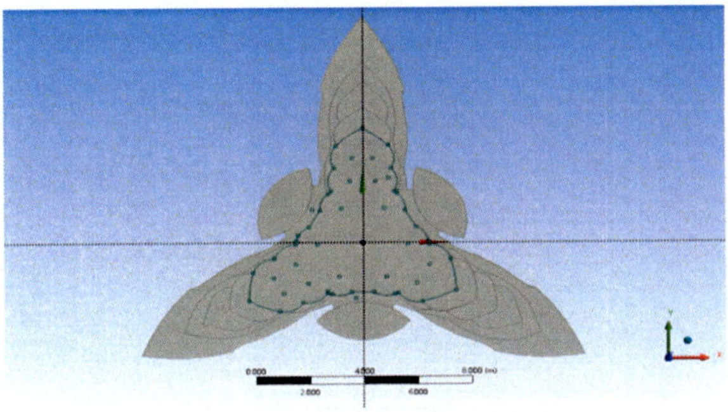

Fig. 2. Sketch of a shrink tier of 1:10 scaled model

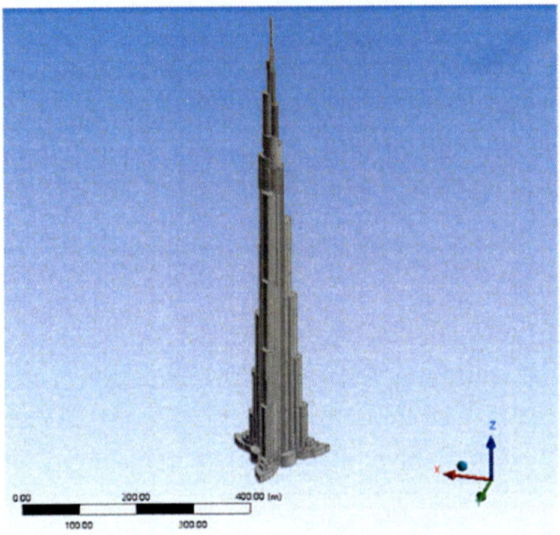

Fig. 3. Geometry model of entire main building

The structure and scale of Burj Khalifa's pile foundation model is modeled according to details specified by Poulos and Bunce (Fig. 3) [12]. The foundation is comprised of a 3.7 m thick raft, which is supported by 192 piles, each with a diameter of 1.5 [m] and a length of 47.45 [m]. Due to the mesh limitation of the academic version of ANSYS, the pile foundation prototype has to be simplified by reducing the number of piles from 192 to 15. This will not cause a huge impact on the final result. According to Sharma et al. the "number of piles have relatively lower influence on the value of MF," which is an index used to predict T [6].

Fig. 4. Initial pile foundation of 1:10 scaled model

Meanwhile, the diameter and location of each pile are adjusted (Fig. 4) so that the total cross-section area and length of piles remain unchanged, and the function of the simplified foundation stays rational (Fig. 5).

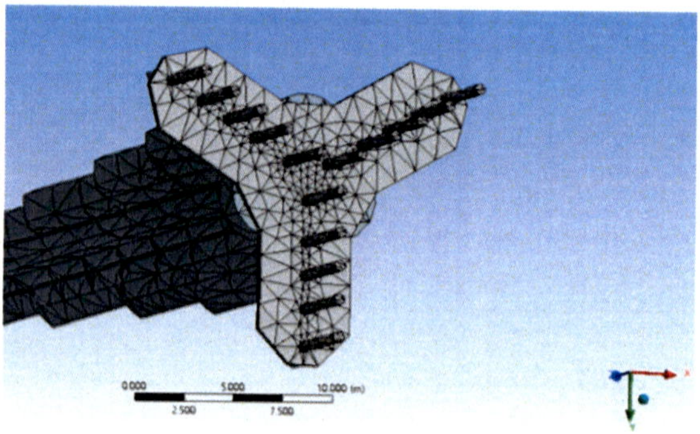

Fig. 5. Simplified pile foundation of 1:10 scaled model

Soil Stratum. The soil stratum model is 70 m thick and has a radius of 500 m. According to M. El Gendy's table of the subsurface structure under Burj Khalifa's pile foundation, the model of Soil Stratum is divided into four layers, and each layer reflects a strata group with homogeneous or similar material type [10]. The first layer, extending from the earth's surface to 3.7 m underground, mainly consists of silt or soil. The second layer consists of limestone, stretching from 3.7 to 9.8 m beneath the surface. The third layer is composed of sandstone and stretches from 9.8 m to 31.0 m underground. And limestone comprises the last 31 to 70 m underground.

3.3 How ANSYS Work on the Structure

In order to perform relatively accurate advanced analyses such modal and harmonic motion analysis, material properties need to be assigned to the building, foundation, and subsurface stratum. The connection between main building, pile foundation, and strata should be considered, and meshes should be generated.

Material Used. According to the side view of Burj Khalifa by Richard Braddish, the lower portion of the main building, from the bottom of the tower to 586 m, consists of concrete structure. And the upper part, from 586 m to the top of the spire, is composed of steel frame. To prevent the density differences between the model and the real building from creating significant errors, the density of a solid geometry tower model is adjusted. Through dividing the total mass of the real building (about 500,000 tons) by the volume of geometry model ($1.75 * 10^6\ [m^3]$), the density of the entire building can be estimated. The estimated density ($285.7\ [\frac{kg}{m^3}]$) is assigned to both lower and upper portions of the main building.

Since the pile foundation itself is a solid composed of concrete, engineering data of concrete in ANSYS is directly applied.

Various materials need to be specified for different layers of soil stratum geometry model (Table 1). Due to constraints in ANSYS material variety, the soil model is simplified into 4 strata. By employing weighted calculation of the thickness and Young's modulus of each sub-strata in a strata group, the ideal Young's modulus for a strata group can be obtained. Four weighted calculation results are applied to assign Young's modulus to the four layers of the geometry model.

Table 1. Material properties assigned to each of the 4 stratum

Strata	Subsurface material	Thickness (m)	Young's modulus (MPa)
1	Soil	3.7	21
2	Limestone	6.1	373
3	Sandstone	21.2	153
4	Limestone	39.0	373

Connection Methods. In order to render the simulation as rational as possible, different connection situations between each contact region need to be considered. For the Burj Khalifa model with pile foundation, frictional contact type with frictional coefficient 0.3 is used on contact regions between the main building, soil layer, pile foundation, and subsurface layer. For the comparative geometry model without pile foundation, the frictional method is applied on the contact region between the bottom of main building and soil layer.

For both prototypes, one with pile foundation and another without pile foundation, fix support is set on the bottom of Soil Stratum. To eliminate SPSI in the prototype without pile foundation, a fix support is also placed on the bottom of the main building of the prototype without pile foundation.

Mesh. Meshes on the mechanical model are automatically generated by ANSYS, and the size of each mesh is equivalent.

Modal Analysis and Harmonic Analysis. Modal analysis and harmonic motion analysis are conducted by ANSYS to analyses fundamental period (T) of the prototype structure.

The first step of the modal analysis is to import building model geometry into the ANSYS mechanical enterprise. The model geometry then is inserted into mechanical enterprise and mesh element size is adjusted to be X inch everywhere. After that, ANSYS software would apply arbitrary harmonic motions to the building model to determine the frequency at which the model produces maximum response. After modal analysis, 6 T can be determined according to the 6 mode statues.

Harmonic motion analysis procedure is similar to the modal analysis. Two structure model geometries including with pile foundation system and without pile foundation system are import into the harmonic response enterprise to determine the total deformation of the structure. After importing the building model geometry into the harmonic response enterprise and setting up the connection details, the 3.724 $\left[\frac{m}{s^2}\right]$ acceleration is applied, and the total deformation of the structure model is obtained. Since Dubai borders Sharjah, which is part of the northern Emirates (the most seismic-active region in UAE), according to Abdalla's seismic hazard assessment, the maximum peak ground acceleration (PGA) of this region is predicted to be 0.38 [g], which is 3.724 $\left[\frac{m}{s^2}\right]$ in the longitudinal direction [13]. After harmonic motion analysis, the maximum response amplitude of the model structure corresponding to its designated period can be determined.

4 Result and Analysis

To have a better understanding of the SPSI effects, modal analysis and harmonic motion analysis are conducted to determine the T of the prototype structures (with/without pile foundation).

4.1 Modal Analysis

Based on Table 2 and 3, the frequency of the model structure without pile and the frequency of the model structure with pile differs by 0.012 [Hz] under mode 1 and 0.016 [Hz] under mode 2. However, as the mode increases, the differences increase in

Table 2. Natural frequency and fundamental period of the model structure without pile foundation under six modes

Mode	Frequency (Hz)	Fundamental period (s)
1	0.072	14.0
2	0.079	12.7
3	0.273	3.7
4	0.295	3.4
5	0.620	1.6
6	0.656	1.5

Table 3. Natural frequency and fundamental period of the model structure with pile foundation under six modes

Mode	Frequency (Hz)	Fundamental period (s)
1	0.060	16.8
2	0.063	15.8
3	0.216	4.6
4	0.232	4.3
5	0.503	2.0
6	0.522	1.9

an increasing rate. Especially under mode 5 and 6, the differences exceed 0.1 [Hz]. This finding indicates that SPSI has a relatively minor influence on the natural frequency of the building under modes 1 and 2, but the influence is increasing evident as the mode increases. The results suggest that SPSI have a relatively great influence on the natural frequency of the building under modes 5 and 6. Our result differs from the finding of Baker et al., who reported that the T of Khalifa Tower is 11.5 [s] under mode 1. This discrepancy may be due to the differences in model structures, pile group configurations, and mesh sizes.

The model structure deformation under mode 1 to 6 is shown in Fig. 6 and Fig. 7 below. The result of the modal analysis shows that SPSI would decrease the natural frequency and increase the T of the Khalifa tower.

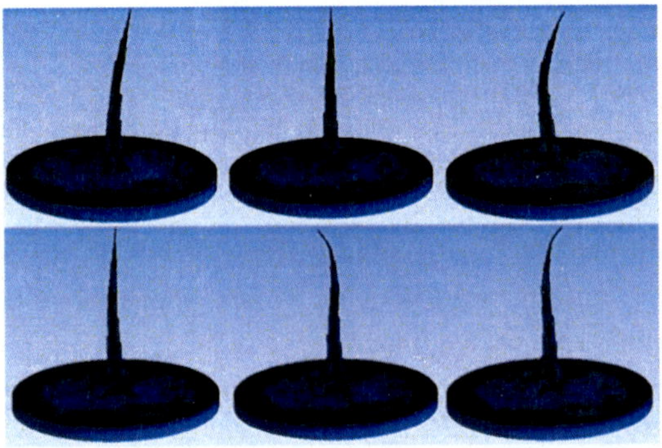

Fig. 6. Modal analysis deformation under mode 1 to 6 (without pile foundation)

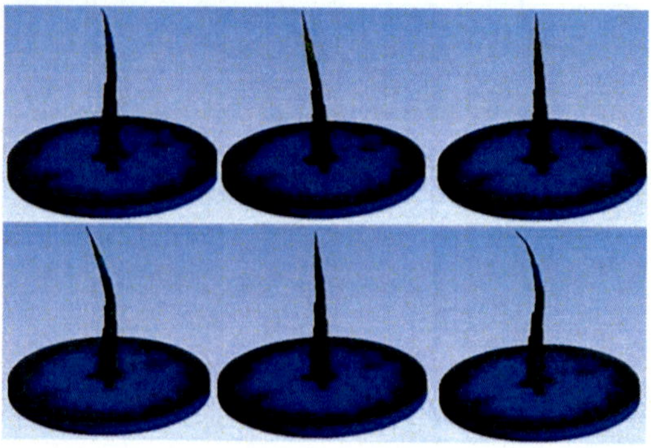

Fig. 7. Modal analysis deformation under mode 1 to 6 (with pile foundation)

4.2 Harmonic Motion Analysis

Table 4. Maximum response amplitude and corresponded period (model without pile foundation)

Period (s)	Frequency (Hz)	Amplitude (m)
23	0.043	12
13	0.076	49

Table 5. Maximum response amplitude and corresponded period (model with pile foundation)

Period (s)	Frequency (Hz)	Amplitude (m)
17.2	0.058	155
16.7	0.060	457

Table 4 and 5 shows the 2 greatest amplitudes of frequency response of each building (Greatest amplitude results when the frequency of the applied load corresponds with or approaches the frequency of the building).

In the modal analysis, under Mode 1 and Mode 2, the Ts of the building without pile foundation are 14.0 [s] and 12.7 [s] respectively, and the frequencies are 0.072 [Hz] and 0.079 [Hz] respectively. The accurate period may be between 14.0 [s] and 12.7 [s]. The result of the harmonic motion analysis validates this: the maximum amplitude occurs at a period of 13 s, which is between 14.0 [s] and 12.7 [s]; the frequency is 0.076 [Hz], which is between 0.072 [Hz] and 0.079 [Hz].

In the modal analysis, under Mode 1 and Mode 2, the Ts of the building with pile foundation are 16.8 [s] and 15.8 [s] respectively, and the frequencies are 0.060 [Hz] and 0.063 [Hz] respectively. The accurate period may be between 16.8 [s] and 15.8 [s]. The result of the harmonic motion analysis validates this: the maximum amplitude occurs at a period of 16.7 [s], which is between 16.8 [s] and 15.8 [s]; the frequency is 0.06 [Hz], which corresponds with 0.060 [Hz] obtained in modal analysis.

Based on the magnitude of the maximum amplitudes of 2 buildings (with/without pile foundation), we can see that the amplitude of the building with the pile is larger (457 [m]) than that of the building without pile (49 [m]). The total deformation corresponds with the amplitude magnitude as shown in Fig. 8 and 9 below. The building without pile foundation (has fixed support between the building and the soil and thus has no SPSI) exhibits a maximum deformation of 3.8143 [m], while the building with pile foundation (does not have fixed support between the building and the soil and thus subjected to the influence of SPSI) exhibits a maximum deformation of 177.59 [m]. Such a result suggests that SPSI may result in susceptibility of a building to seismic activity. If the soil stratum is loose or subject to conditions such as soil liquefication, the SPSI can have detrimental effect on building.

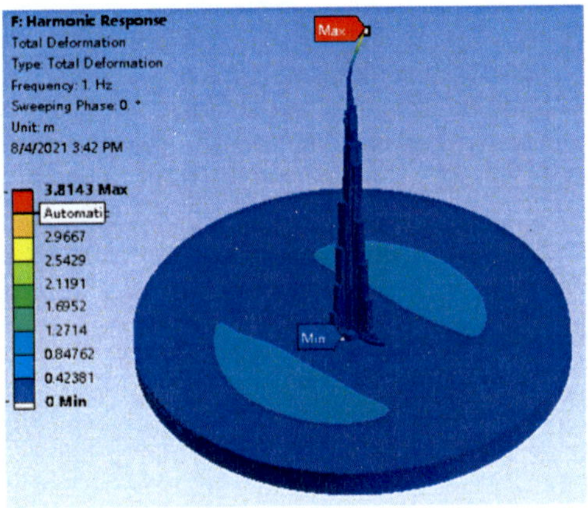

Fig. 8. Total deformation of the main building without pile foundation (No SPSI)

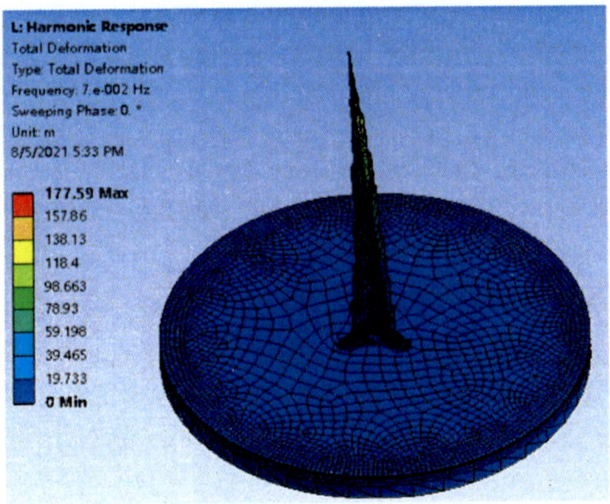

Fig. 9. Total deformation of the main building with pile foundation (Impacted by SPSI)

5 Conclusion and Discussion

5.1 Conclusion

In order to visualize the effect of SPSI on T of Burj Khalifa, two models are used, one with pile foundation and the other without pile foundation. Modal analysis serves to find the T of 2 buildings, while harmonic motion analysis serves to find the frequency response amplitude of the building. The result of harmonic motion analysis indicates that the maximum amplitude occurs when the T of the force applied corresponds with the T of the building. Thus, the two analyses are in accordance with each other. Similar to the results of previous studies, the results of this study show that SPSI will decrease the fundamental frequency and increase the T of the building. The more flexible the soil basement, the more evident and detrimental the SPSI impacts.

5.2 Limitations

The modeling focuses on the external building features, pile foundation, and subsurface profile. Thus, the model fails to capture the interior structure of the tower. This may affect the T of the building model, since T is partially dependent on mass density. Therefore, we compute an equivalent mass density by dividing the total weight of the building by the total volume of the building.

A generalized subsurface profile is created to capture the geotechnical conditions of the site, because some materials are not provided in ANSYS 2021R1 student academic version.

Each soil layer boundary is assumed to be in regular shape, and a cylinder is used to model each soil stratum. Soil strength parameter including unit weight, friction angle, and undrained shear strength are not represented in ANSYS software. To compensate for this drawback, we calculated the modulus of elasticity for each soil stratum.

Pile foundation system for this model structure included only 15 piles due to ANSYS mesh constraint, while the real tower has 192 piles. This will result in discrepancy, since the skin friction will be different given the huge reduction in surface area.

ANSYS software cannot simulate seismic events since the earthquake is not considered as a harmonic motion. Seismic loading analysis on the structure required a hazard response spectrum.

Further improvements might include obtaining ANSYS full edition. Further study direction could be determining how pile configuration influences the anti-seismic performance of the Burj Khalifa tower.

6 Reflection and Acknowledgement

2-week-period passes swiftly. The busy and fulfilling life of scientific research is coming to an end. Such academic research experience expands our knowledge of computational mechanics and also enhances our research ability.

We appreciate Professor Borja for teaching us the principles behind Finite Difference Method, Finite Element Method, and the difference in errors between the two methods. Professor also provided invaluable advices for us during office hours, suggesting us to take into consideration the SPSI and skin friction reduction caused by reduced pile numbers.

We also thank the teaching assistants, Blair Geng and Curtis Yu, for helping us to grasp all the lecture contents, teaching us the application of ANSYS, and providing timely assistance for us during the learning and modeling processes.

References

1. Baker, W.F., Pawlikowski, J.J., Young, B.S.: The Burj Khalifa triumphs: reaching toward the heavens. Civ. Eng. Mag. Arch. **80**(3), 48–55 (2010). https://doi.org/10.1061/ciegag.0000274
2. Baker, B., Pawlikowski, J.: The design and construction of the world's tallest building: the Burj Khalifa, Dubai. Struct. Eng. Int. **25**(4), 389–394 (2015). https://doi.org/10.2749/101686615x14355644770857
3. Kumar, A., Choudhury, D., Katzenbach, R.: Effect of earthquake on combined pile-raft foundation. Int. J. Geomech. **16**(5), 04016013 (2016). https://doi.org/10.1061/(asce)gm.1943-5622.0000637
4. Murty, C.: How flexibility of buildings affects their earthquake response? (2004)
5. Kramer, S.I.: Geotechnical Earthquake Engineering. Prentice-Hall Inc., Englewood Cliffs (1996). https://www.amazon.com/Geotechnical-Earthquake-Engineering-Steven-Kramer/dp/0133749436
6. Sharma, N., Dasgupta, K., Dey, A.: Natural period of reinforced concrete building frames on pile foundation considering seismic soil-structure interaction effects. Structures **27**, 1594–1612 (2020). https://doi.org/10.1016/j.istruc.2020.07.010
7. Khalil, L., Sadek, M., Shahrour, I.: Influence of the soil–structure interaction on the fundamental period of buildings. Earthq. Eng. Struct. Dyn. **36**(15), 2445–2453 (2007). https://doi.org/10.1002/eqe.738
8. Chu, D., Truman, K.Z.: Modeling of unbounded domain in seismic soil-pile-structure interaction. In: Proceedings of the 6th Technical Council on Lifeline Earthquake Engineering, Long Beach, CA (2003). https://doi.org/10.1061/40687(2003)99
9. Xiong, W., Jiang, L.Z., Li, Y.Z.: Influence of soil–structure interaction (structure-to-soil relative stiffness and mass ratio) on the fundamental period of buildings: experimental observation and analytical verification. Bull. Earthq. Eng. **14**, 139–160 (2016). https://doi.org/10.1007/s10518-015-9814-2
10. Gendy, M., Gendy, O.: Analysis of piled raft of Burj Khalifa in Dubai by the program ELPLA. GEOTEC Software Inc., PO Box 14001 Richmond Road PO, Calgary AB, Canada T3E 7Y7 (2018). https://www.researchgate.net/publication/328649931
11. Braddish, R.: Detail of Burj Dubai. Blueprints (2008). https://drawingdatabase.com/burj-khalifa/AutoCAD. (n.d.) Burj Khalifa (Khalifa Tower) 3D DWG Model for AutoCAD. DesignsCAD. https://designscad.com/downloads/burj-khalifa-khalifa-tower-3d-dwg-model-for-autocad/
12. Poulos, G., Bunce, G.: Foundation design for the Burj Dubai - the world's tallest building. In: International Conference on Case Histories in Geotechnical Engineering, p. 14 (2008). https://scholarsmine.mst.edu/icchge/6icchge/session_01/14
13. Abdalla, J.: Seismic hazard assessment of United Arab Emirates and its surroundings. J. Earthq. Eng. **8**(6), 817 (2004). https://doi.org/10.1142/s1363246904001778. Engineering **14**(1), 139–160. https://doi.org/10.1007/s10518-015-9814-2

Internal Force Solution of Plane Rigid Frame Based on Matrix Displacement Method

Haiping Liu(✉)

Chongqing University, Chongqing, China

Abstract. In this paper, the program is designed on Matlab for solving continuous beam and rigid frame. The demonstration of how to matrix displacement method of the idea, the principle of programming in MATLAB. In order to facilitate the computer to solve the internal force of the structure, some skills are included, such as "multiplying large numbers" and "visual design" which will be shown in the paper. In this paper, according to the relevant content of the matrix displacement method, the basic information of the structure is input on the basis of MATLAB programming, and the bending moment and other internal forces of the structure are obtained, and the internal forces are drawn in the form of images.

Keywords: MATLAB · Matrix displacement method · Internal force of rigid frame · Plane bar system structure

1 Introduction

Structural mechanics is an ancient and rapidly developing subject which is the basic knowledge of structural design. And through the application of mechanical concepts, structural schemes can be optimized, component forces can be improved, and structural investment can be reduced, which has extraordinary significance for engineering construction [1].

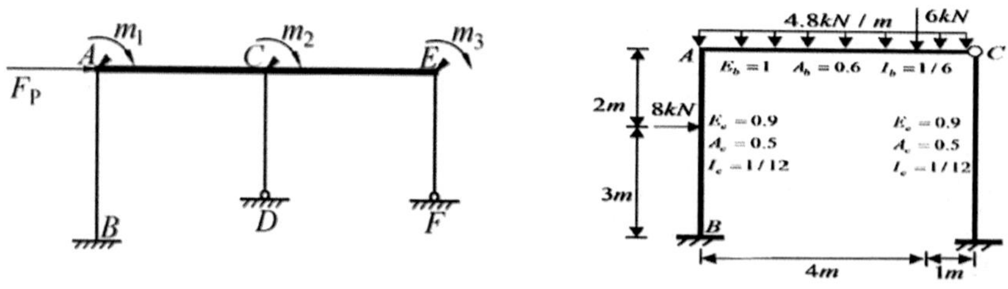

Fig. 1. Practical application of plane rigid frame

As one of the classic models of structural mechanics, the truss structure has a wide range of applications in the numerous aspects which include transportation of transmission equipment for large machine tools of hoisting machinery, automobile bracket,

P. Mendonça and N. D. Cortiços (Eds.): ICAMC 2021, LNCE 226, pp. 58–64, 2022.
https://doi.org/10.1007/978-3-030-94514-5_7

construction machinery, and curtain wall support systems [2]. Therefore, the analysis of the force and displacement of the truss structure under different loads can effectively improve the design of mechanical equipment and architectural structures in reality as shown in Fig. 1.

Finite element method (FEM) is the most widely applied numerical method in engineering analysis nowadays which can solve complex problems with high credibility. It has been utilized in many fields such as machinery, aviation, vehicles, and civil engineering [3]. In order to comprehend and make use of the FEM, it is necessary to master this method scientifically from shallow to deep (specifically, from spring element, rod element to plane rigid frame element, and lastly to continuum).

In the field of solid mechanics, material mechanics provides the basic knowledge for structural mechanics [4]. Besides, elastic mechanics and plastic mechanics are the theoretical foundations of structural mechanics. At the same time, the appearance of linear algebra greatly simplifies the calculation steps of structural mechanics and increases the understanding of structural mechanics and linear algebra. The appearance of finite element methods and computers makes the use of programming languages to realize the organically combination between structural mechanics and finite element, which expands the subject area of structural mechanics.

In structural mechanics, the force method is often used to solve the statically indeterminate problem as shown in Fig. 2. When the number of statically indeterminate is large, the traditional force method will be complicated [5]. Based on the software characteristics of MATLAB which is better at matrix operations.

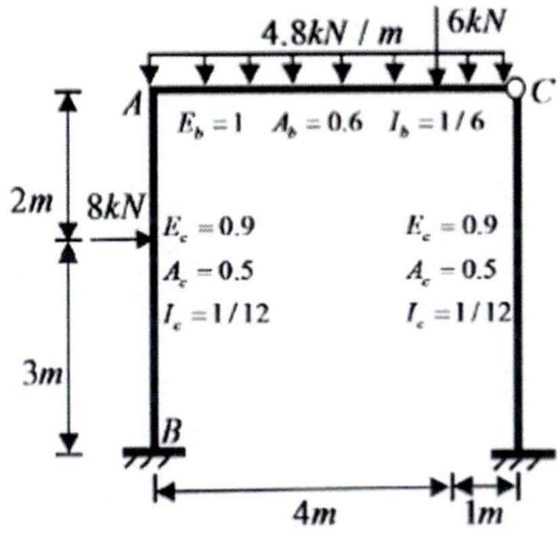

Fig. 2. Structural properties and load diagram

2 Program Flow

2.1 Information Input

To complete the calculation of the plane rigid frame, some basic variables need to be input, so the "typeinformation input" –"TypeInInfomation input" function is constructed to enter the basic information of each member. And the function includes

EI-current member bending stiffness, EA-current member tensile stiffness; X1-the abscissa of the first node of the element, Y1-the ordinate of the first node of the element, X2-he abscissa of the first node of the unit, Y2-the ordinate of the first node of the unit.

And according to the horizontal and vertical coordinates of the left and right nodes of the element to obtain length–the length of the current member, COS–the cosine value of the current member's corner, sin–the sine value of the current member's corner "using the radian system, define the horizontal right is 0, clockwise is positive", **forcetype**–the load type acting on the current member "**0**"–no load; "**1**"–uniform load; "**2**"–concentrated load (consider it act at half the length of the rod), **forcesize**–load acting value "if the load is uniformly distributed–the unit is m/N; if the load is concentrated–the unit is N".

After inputting the above basic variables, **F0**—equivalent nodal load vector can be calculated. **F0** is a six-row and one-column vector based on the type of the previously defined function "**forcetype**" and "**forcesize**" as shown in Figs. 3 and 4.

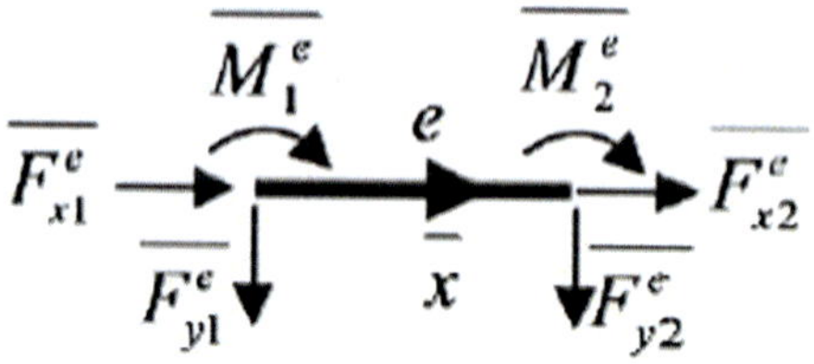

Fig. 3. Force vector diagram of plane rigid frame

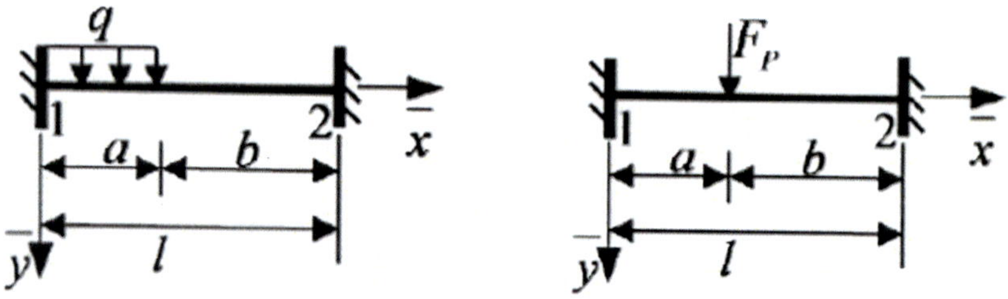

Fig. 4. Equivalent force diagram

2.2 Formation of the Stiffness Matrix

The structure is divided into finite elements, solving the internal forces and displacements of each element separately for superposition. Specifically, in the matrix displacement method, the element stiffness matrix of each independent element needs to be established firstly. The force vector F0 is calculated as shown above. With the displacement method, mathematical relation which is "$\Delta \rightarrow F$" (the corresponding relationship between displacement and force) can be deduced: $\overline{F}^e = \overline{k}^e \overline{\Delta}^e$ and $\overline{k}^e$ is the element stiffness matrix in the local coordinate system with six rows and six columns.

The element stiffness matrix and force in the local coordinate system of each element are defined by defining the cell matrix. By the same way, the definition of "tyepinformation input" function "TypeInInfomation" in all the required input variable (including length, EI, EA, and forcetype, forcevalue) for "zeros[n, 1]" (the number of column vector is one, the number of rows is n, namely the number of truss is n and each truss stores one of the above information). Angle = "zeros (2 * n,1)"(the number of rows is n, the number of column vector is one, each truss stores two information of the cosine value and sine value of the truss).

In the program, the function "TypeInInfoMation" is called n times to input the element information into the storage space with the set value. After the input is complete, the loop is repeated n times.

The element stiffness matrices generated in the previous section are the respective cases in the local coordinate system. The local coordinate system of each rigid frame is different, and a global coordinate system needs to be set uniformly to complete the calculation of the whole model as follows:

$$k^e = T^T \overline{k}^e T$$

$$F^e = -P^e = T^T \overline{k}^e T \overline{\Delta}^e$$

In the program, "Kegeneral = cell(n,1)", "P0General = cell(n,1)", "T = cell(n, 1)" are set to record the stiffness matrix with the form of cell matrix, the force vector and the transformation matrix of each element.

The n rods form n + 1 nodes, and the final result is a K (3n + 1 row, 3n + 1 column), the overall element stiffness matrix. Therefore, $\overline{k}^e$ is placed in K according to the λ positioning vector matching method, that is, the storage space is set "Ktotal = zeros(3 * n + 3,3 * n + 3)" and Ktotal = $\sum K^e$, each element stiffness matrix $\overline{k}^e$ only affects the elements in the corresponding position in Ktotal, and this is the element integration method to calculate Ktotal.

"P = Zeros (3 * n + 3,1)" is set using the same principle to assemble the P matrix which is called the total force matrix.

"λ = cell(n,1)" is defined for positioning the element. The same cycle is repeated n times to obtain the λ positioning vector. The λ positioning vector is nested with three FOR loops to obtain the overall stiffness matrix and total force matrix.

2.3 The Formation of Δ

The basic equations to solve the force are required to solve statically indeterminate structure as follows:

$$\mathrm{ktotal} * \Delta + F_P = 0$$

$$\mathrm{ktotal} * \Delta = -F_P = P$$

The displacement Δ can be solved using the equation

$$\Delta = \text{ktotal}^{-1}\text{P}$$

In the process of dealing with the boundary conditions, based on the premise that the matrix dimension is not changed, the global stiffness matrix is dealt with by the method of multiplying large numbers. Specifically, if a certain direction of node m, a large number A is multiplied by the corresponding position, and the element P (m,1) of the total physical force matrix is 0. For example,if axial direction and transverse direction are constrained at the same time, Then multiply A large number A by Ktotal (3 * m, 3 * m) and (3 * m + 1, 3*m + 1), so that the m + 1 element P (m + 1,1) of the total physical force matrix is 0, and the rest of the constraints are so on.

2.4 Solving Δ

After processing and p by multiplying large numbers method, a new overall stiffness matrix and a new total physical force matrix p are obtained. According to the formula Ktotal*Δ = p, Δ = Ktotal\p can be obtained, which is the column vector of 3*n + 3 rows and 1 column.

2.5 The Formation of F^e&$\overline{F}^e$

After the whole is obtained, it is compared with pe, and it can be found that it is defined by Δ for nodes. But both ke and ke are concerned with the definition of the member element. This requires the conversion of Δ to Δ^e. And then, the force in the global coordinate system and the force on each bar (in the local coordinate system) F^e can be obtained.

2.6 Convert Δ to Δ^e

u is defined for nodes, and a "uele = cell(n,1)" cell matrix needs to be created by comparing with Pe to correspond to the number of elements storage variables. According to the result of solution, the force vector can be solved with the displacement.

3 Visual Design

Design visualizations can also be designed on demand. For example, draw a bending moment diagram. The bending moment diagram can be designed to intuitively judge the size of the bending moment. The design of the drawing function, DrawM, is based on the previously calculated FE and other information under the local coordinates of each bar to draw the bending moment diagram. It is consisted of two steps: location + connection. Firstly, locate the locationOfUnit of the element position (which involves the length and Angle of the bar), and then locate the locationOfM to record the bending moment diagram. In addition, the number can be got according to the forcetype. 0 (no load) is directly connected to the front and back two control points, 1 (uniform load) is connected to the parabola of three control points, and 2 (concentrated load) is connected to the polyline of three control points. Finally, the bending moment diagram of the structure is obtained to analyze the internal force of the structure.

4 Numerical Example

As shown above, the matrix displacement method is used to calculate the internal force of the structure as follows:

(1) Carried out local coding and overall coding for the unit and rigid frame
(2) The element stiffness matrix in the local coordinate system is formed
(3) Forming the element stiffness matrix in the global coordinate system
(4) The integral stiffness matrix K and force vector are formed by the element integration method
(5) Solve the equation to find the node displacement vector
(6) Calculate the internal force vector of the rod end.

4.1 Specific Example (Plane Rigid Frame)

In this paper, according to a specific example, the frame is composed of three elements and four nodes. This model sets the length of the three elements, the extramural rigidity – EI and tensile rigidity – EA are both 1. The relative positions and basic information of the rods are shown in Fig. 5. As is shown in, coordinate information: coordinates of four nodes are (0,0), (0,1), (1,1), (1,0); the load condition: the first bar receives uniform load of size 1, while the second and third bar have no load; constraint condition, the first and fourth nodes are fixed end constraints, while the second and third nodes are unconstrained. At last, the final bending moment is shown in Fig. 6.

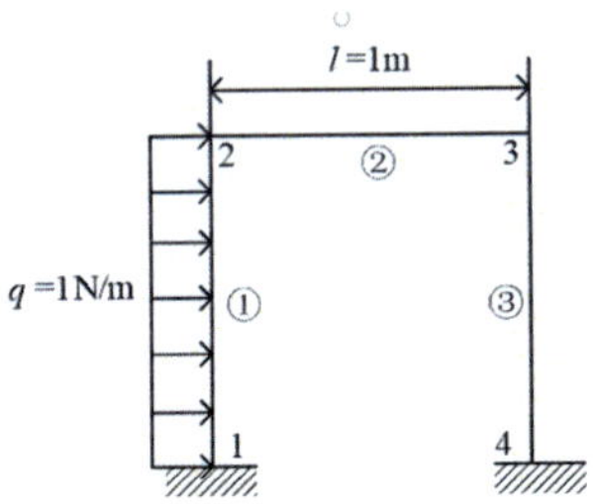

Fig. 5. Structural layout diagram

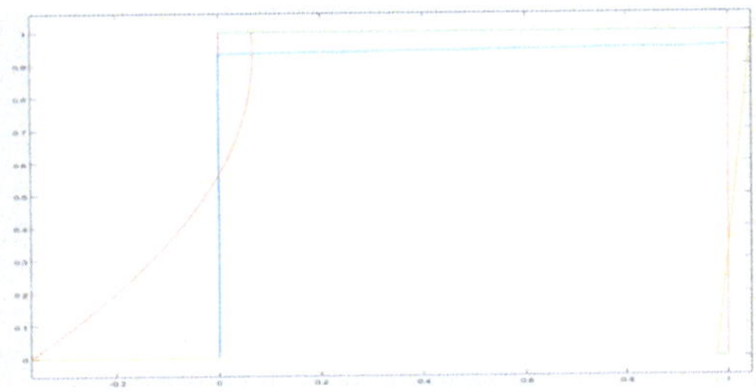

Fig. 6. Bending moment diagram

4.2 Continuous Beam

The continuous beam is composed of 4 elements and 5 nodes. The length of the 5 rods is set as 1m, and the bending stiffness – EI is 1. The load condition is as follows: element 1 and 4 are subjected to uniform load of 1 N/m, element 2 and 3 are subjected to no load; the constraint condition is: node 1 and node 5 are just connected, and the vertical displacements of the other nodes are constrained.

5 Conclusion

There are a few details that need to be noted in the program, such as using a cell matrix to store basic information about the link, so that when calculating the displacement and force of each node, as well as when the information is entered, the coordinates of the two points of each link will be entered, and the program will calculate the length of the link and the Angle in the overall coordinate system.

The program is designed to initially use programming to achieve rigid frame and cantilever beam which aims to solve the node displacement, internal forces and finally draw the bending moment diagram. In a word, based on MATLAB language and the idea of matrix displacement method, the program in the paper to solve the plane bar structure is designed. The program pays attention to generality, and adopts unified analysis mode for rigid frame, continuous beam, truss and composite structure, which is especially advantageous for the analysis of composite structure.

References

1. Yisheng, W., Hai, C., Jinji, G.: Application of array displacement method in bar structure analysis. Guangdong Civil Eng. Archit. **2**(2), 22–23 (2005)
2. Yisheng, W., Hai, C., Jinji, G.: Application of structural matrix displacement method in bar structure analysis. Guangdong Civil Eng. Archit. **2**(2), 22–23 (2005)
3. Logan, D.L., Zhang, R.: A First Course in the Finite Element Method. Publishing House of Electronics Industry, Beijing, pp. 3–4 (2014)
4. Zhiyu, S., Xihong, D.: Aircraft Structural Mechanics, pp. 4–5. National Defense Industry Press, Beijing (2012)
5. Ren-bo, Q.: Finite element programming of planar bar structure based on MATLAB and matrix displacement method. Fujian Archit. **05**, 109–116 (2019)

Brief Analysis of Natural Frequency and Its Corresponding Influences for Old Tacoma Narrows Bridge Using Finite Element Method

Taige Lu[1], Binghong Duan[2], Bofan Wang[3](✉), and Zhe Wang[4]

[1] School of Civil Engineering, The University of Sydney, Sydney, NSW 2006, Australia

[2] School of Civil Engineering, The University of New South Wales, Sydney, NSW 2052, Australia

[3] School of Environmental Science and Engineering, Shandong University, Jinan 250000, China

[4] Huaer Zizhu Academy, Shanghai 200241, China

Abstract. The failure of the 1940 Tacoma Narrow Bridge has become a famous case for the bridge engineering study and structural dynamics courses nowadays. The collapse of the old Tacoma Narrow Bridge was owing to the torsional oscillation caused by the wind at the velocity of 67 km/h, as a resonance occurred when the torsional oscillation met the natural frequency of 1940 Tacoma Narrow Bridge. Therefore, detailed determination of the natural vibration frequency of the old Tacoma Narrow Bridge by applying FEM using will be discussed in this report. For the methodology of research project, the structural behaviour of Tacoma Narrow Bridge is detailed analysed. A model with a scale of 1:1 to the actual bridge is established via ANSYS and its structural behaviour is only focusing on the vibration mode, which will be done by program controlling in modal analysis. By determining the natural frequency of the bridge, the stiffness (k) is directly related to the natural vibration frequency while the mass is inversely relevant to the natural vibration frequency. These features can be manipulated to avoid the phenomenon of resonance and hence prevent the possibility of collapse. Moreover, the aerodynamic structure of the suspended bridges can be developed to avert the resonance caused by torsional flutter. Therefore, this research can assist engineers to address relevant issues and be a simple guideline when designing a suspension bridge.

Keywords: Tacoma narrow bridge · Torsional oscillation · Natural frequency · Finite element method · ANSYS modelling

1 Introduction

With the continuous development of modern society, bridge construction has become more and more critical. Bridges can promote the economic development of the corresponding area and make transportation more convenient. To be specific, a suspension bridge was constructed widely due to its characteristics of holding a longer span. Tacoma bridge constructed in 1940 is a typical example of a suspension bridge. The

P. Mendonça and N. D. Cortiços (Eds.): ICAMC 2021, LNCE 226, pp. 65–73, 2022.
https://doi.org/10.1007/978-3-030-94514-5_8

total height is around 60.35 m, with width of approximately 11.9 m. The total length is around 1810 m with a main span of 853 m [1]. However, it failed and collapsed due to aeroelastic flutter [2].

The failure of the 1940 Tacoma Narrow Bridge has become a famous case for the bridge engineering study and structural dynamics courses nowadays. Von Kármán [3] reported that the collapse of the old Tacoma Narrow Bridge was owing to the torsional oscillation caused by the wind at the velocity of 67 km/h, as a resonance occurred when the torsional oscillation met the natural frequency of 1940 Tacoma Narrow Bridge. In fact, the resonance induced by external forces might happen at all similar suspended bridges. Therefore, the calculation and adjustment of the natural frequency of suspended bridges have become an essential part of avoiding the occurrence of resonance at suspended bridges. However, the determination of the natural vibration frequency of a real-world suspended bridge has been regarded as a complex work even has provided a simplified analytical model of the natural vibration frequency calculation of suspended bridges [2]. Instead of using the analytical method, the numerical method such as finite element analysis was considered a higher efficiency approach to estimate the natural vibration frequency than the traditional analytical method. The finite element analysis was one of the most critical developments in applied mechanics [4], and FEM replaces the governing differential equations of distributed-mass system with a system of linear equations; this also shows how the FEM simplified the calculation of natural-vibration frequency.

In this report, a detailed determination of the natural vibration frequency of the old Tacoma Narrow Bridge by applying FEM using is discussed, and ANSYS will be mainly used to perform the FEA calculation to determine the natural frequency of the old Tacoma Narrow Bridge. Based on the measurement from Farquharson [5], the torsional vibration frequency at 1940 TNB was about 0.2 Hz. Hence, a comparison will be conducted between the final estimation from the modal analysis and harmonic analysis of ANSYS and the calculated value from Farquharson [5].

2 Methodology

For research methodology, the structural behaviour of Tacoma Bridge will be analysed. A model with a scale of 1:1 to the actual bridge will be created using ANSYS and the structural behaviour with only focusing on the vibration mode, which will be done by program controlling in modal analysis. Table 1 below is the summary of the components which formed the bridge.

2.1 Background of the Tacoma Bridge

The original Tacoma Narrow Bridge was firstly opened in July 1940, the state of Washington. The 1940 Tacoma Narrow Bridge was a famous bridge in history, and it was also the 3rd longest suspended bridge in the United States with a length of 1600 km. In 1923, the Tacoma Chamber of Commerce was officially proposed to design and construct a bridge connecting the Tacoma and the Kitsap Peninsula, and Leon Moiseiff, who was one of the Golden Gate Bridge dominated the main designer of the 1940

Tacoma Narrow Bridge. The main design proposal by Leon Moiseiff was to replace the 25-feet deep roadway support truss girders with 8-feet deep plate girders to minimize cost and increase the flexibility of the Tacoma Bridge. Unfortunately, the old Tacoma Narrow Bridge collapsed in November 1940 due to the improper design corresponding with the aerodynamic. Nowadays, the 1940 Tacoma Narrow Bridge accident has become an essential example in these courses, such as dynamic structure and bridge engineering [1].

2.2 Detail of Modelling

The detail procedures of modelling of the Tacoma bridge were listed below:

Using ANSYS to create the bridge deck with length of 1650 m.

Four bridge columns were then be created at length of 400 m, 1250 m and 1650 m. The main span is 850 m in this case.

Corresponding 6 base plates were created at the bottom of the bridge piers.

Main cables and secondary perpendicular cables at one side were created.

For Tacoma Bridge, the structural elements were modeled as bridge deck, columns, cables and bottom plates. Table 1 below is the summary of all the relevant structural components. For the bridge deck, it was composed by beams, girders, and concrete surface. The geometry is a solid plate with a membrane thickness of 0.3m. Another member that is needed to be stated is the pylon. Due to the limitation of the property table in Strand7, the pylon can only be modeled as a rectangular steel hollow section. In addition, the cables were set as cut-off bars for simplicity.

2.3 Modal and Harmonic Analysis of the Model

In the stage of analyzing using FEA method, 2 different analyses were performed using ANSYS. Detail procedures for the analysis were listed in the following.

Modal Analysis

1. Open Workbench, create a model analysis in analysis systems.
2. Input the corresponding geometry into the analysis system and start choosing materials.
3. For materials, bridge decks are chosen to be concrete, primary and secondary cables are structural steel.
4. Each element is chosen to be 5 m when creating the mesh.
5. For the connection of each element, choose surfaces as the connection area and then add the joint option in order to connect cables with the bridge deck.
6. Fix the six bottom plates in 'Analysis'.
7. In 'Solve', chose the deformation and solve.

Harmonic Analysis

1. Open Workbench, create a harmonic analysis in analysis systems.
2. Input the corresponding geometry into the analysis system and start choosing materials.

Table 1. Summary of all relevant structural component for Tacoma Bridge.

Member	Element Type	Properties	Geometry	Dimension
Bridge Deck	Concrete Panel	Modulus 19.36 GPa Density 2392 kg/m3	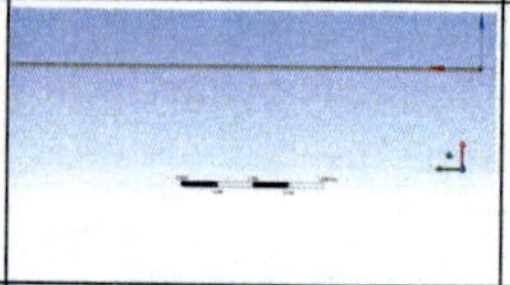	Length: 850 m
Main Cables	Steel Beam	Modulus 200 GPa Density 7850 kg/m3	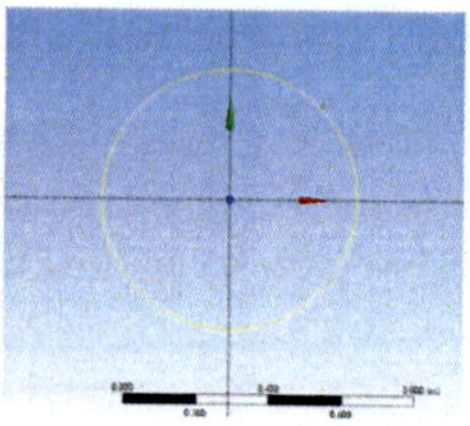	Radius: 0.35 m
Secondary Cables	Steel Beam	Modulus 200 GPa Density 7850 kg/m3	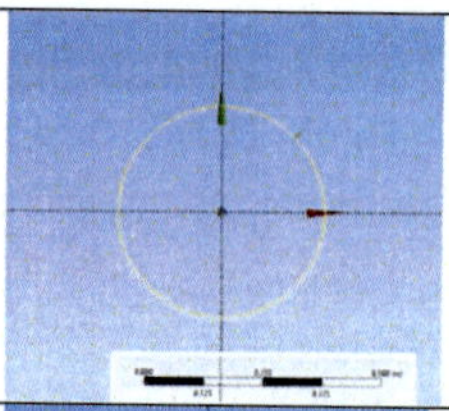	Radius: 0.15 m
Bridge Columns (Primary)	Concrete Panel	Modulus 19.36 GPa Density 2392 kg/m3	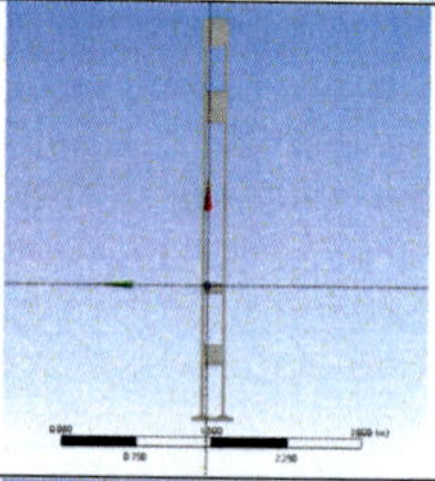	60.35 meters above the deck surface
Bridge Columns (Secondary)	Concrete Panel	Modulus 19.36 GPa Density 2392 kg/m3	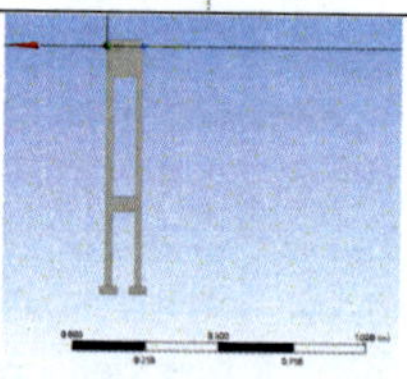	Height: 60 m
Bridge Piers (Bottom Plates)-Primary	Concrete Panel	Modulus 19.36 GPa Density 2392 kg/m3	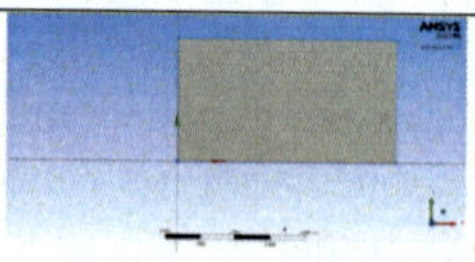	Length * width = 7 * 6
Bridge Piers (Bottom Plates)-Secondary	Concrete Panel	Modulus 19.36 GPa Density 2392 kg/m3	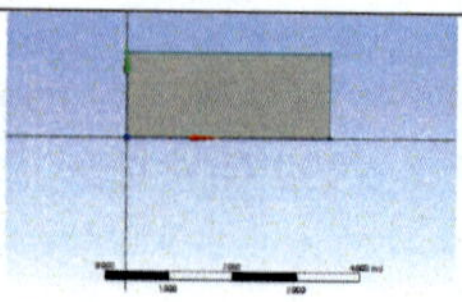	Length * width = 30 * 15

3. For materials, bridge decks are chosen to be concrete, primary and secondary cables are chosen to be structural steel.
4. Each element is chosen to be 5 m when creating the mesh.
5. For the connection of each element, choose surfaces as the connection area and then add the joint option in order to connect cables with the bridge deck.
6. Fix the six bottom plates in 'Analysis'.
7. Apply wind load on the bridge in the horizontal direction, with a total of 1400 kN.
8. In 'Solve', choose the deformation and then add stress and deformation analysis.
9. Finally, click 'solve' to obtain the results.

3 Results

Determination of the natural vibration frequency of Tacoma Narrow Bridge by implying the traditional analytical method is a time-consuming process. Therefore, the utilization of the numerical method is associated with the calculation of the natural vibration frequency. The finite element method is considered as a relatively practical approach for numerical analysis at present. Moreover, the rapid development of computer software also lowers the difficulty of using the FEM. As a result, the natural vibration frequencies of the Old Tacoma Narrow Bridge were estimated by the utilization of ANSYS. The estimated results have included two parts: the modal analysis and the harmonic motion analysis.

3.1 The Modal Analysis

The purpose of performing modal analysis is to demonstrate the natural vibration frequency of the old Tacoma Narrow Bridge at different modes through the finite element method (ANSYS). ANSYS allows users to establish an ideal model structure of the targeted object at the Workbench and computes the required results using the computational finite element method.

The modal analysis of ANSYS has produced 9 modes of the natural vibration frequency as the Table 2 shown:

Figure 1 shows that the linear regression lines of both modal and calculated values of the old Tacoma Narrow Bridge formed a similar trend, and the absolute errors are within an acceptable range. Therefore, it can be concluded that values of natural vibration frequency of the old TNB computed by the finite element method (ANSYS) are highly similar to the calculated values of NVF [5], which was mentioned in the Literature Review section.

3.2 The Harmonic Motion Analysis

The harmonic motion analysis of ANSYS is achieved by applying external forces or accelerations to undertake a simulation of real-world harmonic motion. For this ideal model of the old Tacoma Narrow Bridge, the wind-generated force of 1400 KN at the sideway is regarded as the only external force involved.

Table 2. NVF of the old Tacoma narrow bridge comparison between modal analysis and calculated result from Farquharson [5]

Modes	Modal NVF		Calculated NVF		
	NVF (Hz)	NVP (s)	NVF (cpm)	NVF (Hz)	NVP (s)
1	**0.11072**	9.031791908	7.88	**0.131333**	7.614213
2	**0.12055**	8.295313148	11.63	**0.193833**	5.159071
3	**0.15619**	6.402458544	15.7	**0.261667**	3.821656
4	**0.22727**	4.400052801	16.29	**0.2715**	3.683241
5	**0.23125**	4.324324324	23.6	**0.393333**	2.542373
6	**0.27611**	3.621744957	27.4	**0.456667**	2.189781
7	**0.35917**	2.784196898	31.5	**0.525**	1.904762
8	**0.44211**	2.261880527	35.4	**0.59**	1.694915
9	**0.4475**	2.234636872	39.4	**0.656667**	1.522843

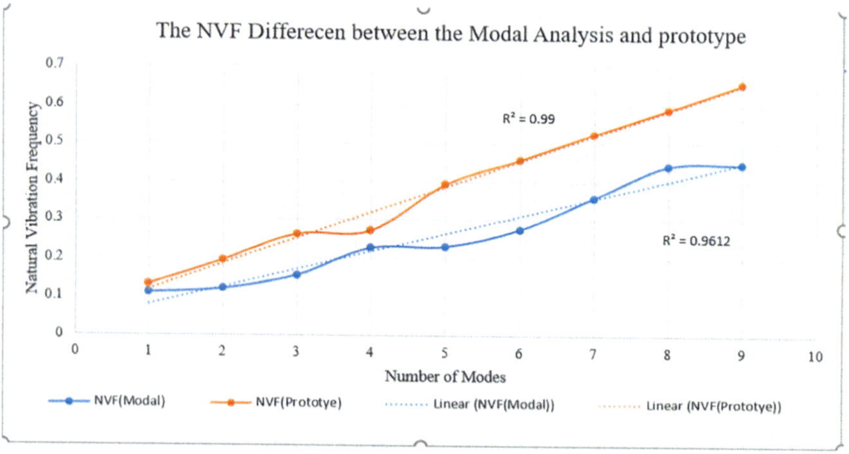

Fig. 1. Linear analysis of the modal result and prototypes results

The harmonic motion analysis aims to determine the maximum amplitude and then obtain its associated frequency for the old Tacoma Narrow Bridge. The phenomenon of resonance is observed at the maximum amplitude point [6]. Hence, its corresponded frequency can be regarded as the most critical role in the failure of the old TNB and Fig. 2 has shown the critical points.

Results of harmonic motion analysis were shown as Table 3. It generally indicated the point where the maximum amplitudes occurred and its corresponding frequency, which is about 0.15 Hz. This indicates the presence of the destructive resonance for this model is at 0.15 Hz, and this also matches with the 3rd mode natural vibration frequency which produced the torsional motion of the bridge.

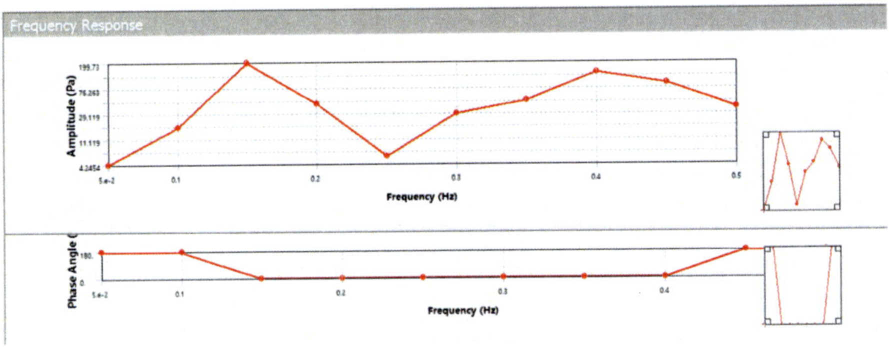

Fig. 2. Results from harmonic modal analysis

Table 3. Values of amplitude and the corresponding frequency

Frequency	Amplitude (pa)
0.05	4.2454
0.1	17.666
0.15	199.73
0.2	42.281
0.25	5.6477
0.3	28.374
0.35	47.404
0.4	136.97

3.3 Comparison

The result from harmonic analysis suggested the most critical natural vibration frequency was around 0.15 Hz, which was slightly lower than the observed data from Farquharson (1949), as shown in Table 4.

Table 4. Difference between numerical and analytical results

Numerical result of natural frequency (ANSYS) [Hz]	Analytical result of natural frequency (Farquharson) [Hz]
0.15	0.21

4 Discussion

During the modelling and the phase of result gathering, errors have certainly occurred. The following are some possibilities that can cause errors for the results displayed in the Table 2.

4.1 Errors in ANSYS Modelling

Firstly, it might be due to the errors in creating the bridge model. The value of structural properties of materials that were put into during modeling might not be 100% the same as the actual materials. In the Finite Element modeling, not all details of the structural elements are considered in the models. Therefore, it may lead to errors occurred and affected the overall structural behavior of the bridge [7]. Also, the dimension of structural elements is a crucial component during the phase of modeling. When the dimensions of different elements were assigned into the model, it might not have the exact same values as the accurate dimensions of the elements [8]. Also, for the bridge deck and bridge column, it should be concrete with composite steel reinforcement. However, in the modeling processes, those were all simply considered as concrete, and this also identified how the errors were induced.

4.2 The Insufficient Numbers of Mesh

The solution from the finite element method is regarded as an approximation. For this case, the accuracy of the solution is highly associated with the number of segments generated in the ANSYS, and due to the limitations of student version's ANSYS, the accuracy of conducting results were largely affected by the insufficient numbers of mesh. Therefore, to increase the amount of meshes could practically minimize the errors induced.

4.3 Inappropriate Parameters and Inputs for Both Modal and Harmonic Analysis

Firstly, the wind load value can be considered as a simple estimation, which controverted to the actual values [9]. Therefore, it can affect the value of harmonic analysis. Also, ANSYS student versions were used as a primary source of modeling. It can cause errors in generating mesh. Thus, errors can occur.

5 Conclusion

In conclusion, the simple estimation of the Old Tacoma Narrows Bridge's natural frequency was performed using the Finite Element Method. To be specific, ANSYS will be used to conduct a modal analysis of the 1:1 model to simulate the bridge's structural performance. Based on the current model and results obtained from modal and harmonic analysis, the numerical solution for the natural frequency is in a reasonable range since it is adequately close to the analytical result. Therefore, this model can furtherly be used to simulate and calculate the natural frequency of other suspension bridges. However, in terms of the validity and reliability of the results produced by the model, errors should be further eliminated. Specifically, dimensions of structural elements and materials error can be minimized by not using ANSYS Student Version.

By determining the natural frequency of the bridge, the stiffness (k) is directly proportional to the natural frequency while the mass is inversely proportional to the natural frequency. These factors can be manipulated to avoid the phenomenon of resonance and hence eliminate the possibility of collapse. Also, the aerodynamic structure of the bridge can be improved to avoid the resonance caused by torsional flutter. Finally, this research can help engineers to solve related problems and be a simple guideline when designing a suspension bridge.

References

1. Tacoma Narrows Bridge history - Bridge - The machine (2021). https://wsdot.wa.gov/TNBhistory/1940-bridge.htm. Accessed 2 Aug 2021
2. Goremikins, V., Rocens, K., Serdjuks, D., Sliseris, J.: Simplified method of determination of natural-vibration frequencies of prestressed suspension bridge. Proc. Eng. **57**, 343–352 (2013)
3. Ammann, O.H., Von Kármán, T., Woodruff, G.B.: The failure of the tacoma narrows bridge. A Report to the Honorable John M. Carmody, Administrator, Federal Works Agency, Washington, DC Board of Engineers (1941)
4. Cook, R.D., et al.: Concepts and Applications of Finite Element Analysis, 4th edn. Wiley, Hoboken (2002)
5. University of Washington. Structural Research Laboratory, & Washington Toll Bridge Authority. Aerodynamic Stability of Suspension Bridges with Special Reference to the Tacoma Narrows Bridge: A Report of an Investigation. University of Washington Press (1949)
6. Franciosi, C., Franciosi, V.: Second order influence line analysis of suspension bridges. Int. J. Mech. Sci. **31**(8), 599–609 (1989)
7. Zheng, X., Yang, D., Yi, T., Li, H.: Development of bridge influence line identification methods based on direct measurement data: a comprehensive review and comparison. Eng. Struct. **198**, 109539 (2019). https://doi.org/10.1016/j.engstruct.2019.109539
8. Zeng, Q., Yang, Y., Dimitrakopoulos, E.: Dynamic response of high-speed vehicles and sustaining curved bridges under conditions of resonance. Eng. Struct. **114**, 61–74 (2016)
9. Clemente, P., Nicolosi, G., Raithel, A.: Preliminary design of very long-span suspension bridges. Eng. Struct. **22**(12), 1699–1706 (2000)

Experimental Study on Prestress Loss of Concrete Beam Under Fatigue Loading

Yongguang Wang[1,2], Fang Liu[1(✉)], Weiguo Wu[1], and Xiying Song[2]

[1] School of Transportation, Wuhan University of Technology, Wuhan, Hubei, China
fang_liu@whut.edu.cn
[2] Zhonggongchengke (Jilin) Engineering Testing Co., Ltd., Jilin, China

Abstract. To understand the prestress loss of prestressed concrete beam under fatigue loading, several post-tension cohesive prestress concrete test model beams were made on the basis of similar theoretic design, the concrete strain, reinforcement stress, and the stress state of prestressed reinforcement were analyzed through the constant-amplitude fatigue test, and the experimental study of the prestress loss was carried out under fatigue loading. The experimental results indicated that the fatigue loading effect has small degree of injury to non-split beam; the prestress loss existed in the prestressed concrete beam under fatigue loading and in three stages in the whole loading process basically, and that the remaining carrying capacity of the prestressed concrete beam has almost no change but increased to a certain extent.

Keywords: Bridge engineering · Prestressed concrete beam · Fatigue loading · Experimental study · Prestress loss

The structure of prestressed concrete beam bridge is one structural style extensively used for bridge at present. The beam sustains fatigue loading such as vehicle load repeatedly for a long time in the bridge operation, and thus the accumulation of fatigue causes crack, excess deflection, and so on of the beam. It is crucial to correct assessment of prestress loss and the remaining carrying capacity of prestressed concrete beam bridge after fatigue loading effect, so as to better maintain and reinforce the bridge. It is considered with regard to the effective prestress of prestressed concrete beam in early studies [1–10] that prestress is unchanged in the constant-amplitude repeated loading process. However, as the fatigue test of prestressed concrete beam is carried out, it is found that the effective prestress and crack bending moment of prestressed reinforcement are reduced with increasing times of repeated loading effect, and that the dynamic creep of concrete around prestressed reinforcement is the main reason for reduction of effective prestress. The carrying capacity of the bridge is also decreased with reduction of effective prestress.

It is comparatively difficult to confirm the effective prestress of prestressed reinforcement under fatigue loading due to the coupling effect of multiple factors on prestressed concrete structure during use. There has not been a method for relatively exact description of stress state of prestressed reinforcement under fatigue loading. Zhou Lv et al. [11] verified the example in literature [12] by utilizing the concrete fatigue strain equation as specified in CEB-FIP 1990 model code [13] and found that

P. Mendonça and N. D. Cortiços (Eds.): ICAMC 2021, LNCE 226, pp. 74–87, 2022.
https://doi.org/10.1007/978-3-030-94514-5_9

the prestress loss caused by concrete fatigue strain was 23.40 MPa which is 5.4% of the original effective prestress, in accordance with the computing method of concrete fatigue strain mentioned in CEB-FIP1990 pattern specifications. Zhang Jiwen et al. [14] found through the repeated load test of prestressed concrete cantilever slab in the bending tension–compression state that the prestress loss of prestressed reinforcement was up to 3%–6% in the whole fatigue loading process; the prestress loss was large in the earlier stage of cyclic loading, and it showed a linear declining trend in the later stage of cyclic loading, with the descent speed slowing down. Zeng Ding et al. [15] found through the constant-amplitude fatigue test that the prestress loss showed a form of exponential function with increase of cycle number in the fatigue loading process relative to the effective prestress after stretch-draw. Kang Jianqiang [16] conducted the relative fatigue test for full prestressed concrete beams and class-A prestressed concrete beams and found that prestress loss existed in the prestressed concrete beam under different degrees of prestress in the repeated loading process, and there was smaller difference in the two kinds of beams in the first 700,000-times loading processes, and difference after 700,000-times loading processes; after 2,000,000-times cyclic loading, the accumulated prestress loss was 79 in the full prestressed concrete beam and 116 in parts of class-A prestressed concrete beams, and the latter was 1.47 times as much as the former.

To deeply understand the remaining carrying capacity and stress performance of prestressed concrete beam after fatigue loading, prestressed concrete beam was selected for the static loading test and the fatigue test after ten thousands of times. In the tests, the destruction test was completed for ST-1 and ST-2 beams with the static loading method, which provided the fatigue loading test with test data; the fatigue test was conducted for FTG beam.

1 Study of Model Test Schemes

The design and manufacture of models include confirmation of structure size, material selection, beam pouring, prestress exertion, etc. Four corresponding test model beams were made by the contraction scale of 1:5 and similarity relation according to typical prestress bridge structures (prefabricated prestressed concrete simply-supported beam, simply supported-continuous box beam, and prestressed concrete simply supported T beam) of highway. Two test beams were used for the static destruction test, and the other two test beams were used for the fatigue test.

1.1 Sectional Dimension

Prototype beam wing plate and web plate are relatively small, and it is difficult for template processing, concrete pouring, and vibrating by the contraction scale of 1:5 completely. As a result, the sectional dimension was adjusted properly with the cross-middle section moment of inertia and the neutral axis as the main control factors under the condition of meeting the production process and reasonable stress of beam body. The section form of the adjusted model beam is shown in Fig. 1.

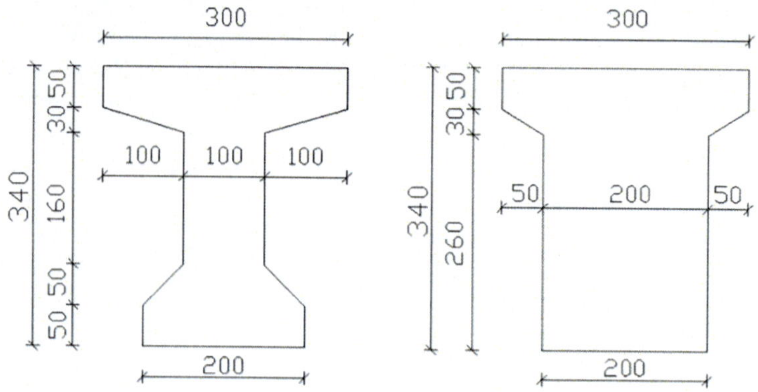

Fig. 1. Sectional dimension of model beam (unit: mm)

1.2 Layout of Prestressed Reinforcement

Two 75 steel strands were confirmed to be used with reference to quantity and layout of prestressed reinforcement of prototype beam, providing that the effective prestress was basically identical. The theoretical calculating value of the controlled stretching stress is confirmed to be 1258.70 MPa under the condition of equal flange stress during pre-applied force stage. The combination of straight line and arc line was used for prestressed reinforcement. See Fig. 2 for details of layout form.

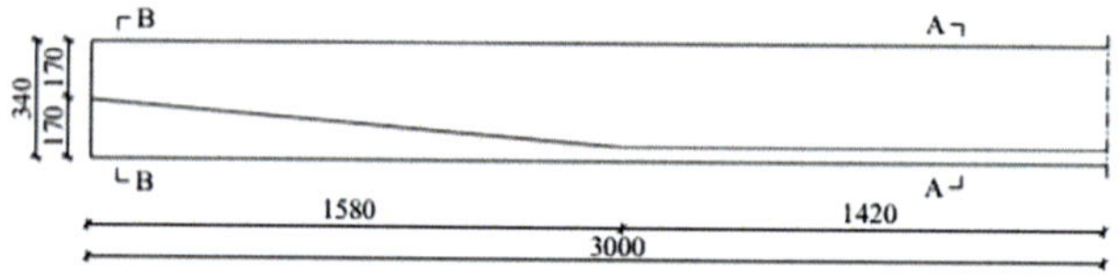

Fig. 2. Layout of prestressed reinforcement of model beam

1.3 Layout of Non-prestressed Reinforcement

Load-bearing bar was arranged with reference to regular reinforcement ratio of prototype beam and regulations of the minimum thickness of concrete protective cover in *Specifications for Design of Highway Reinforced Concrete and Prestressed Concrete Bridge and Culverts* [17], and stirrup and horizontal longitudinal reinforcement were collocated in accordance with construction rules in the Specifications. Load-bearing bars of model beam were HRB400 hot rolled ribbed steel bar with the diameter of 12 mm and HPB300 plain round bar with the diameter of 10 mm. Stirrup and horizontal longitudinal reinforcement WERE HPB235 plain round bar with the diameter of 6.5 mm. The stirrup spacing was 150 mm in the loading point zone, and encrypted to 70 mm near the support, and 100 mm in other zones. The specific layout is seen in Fig. 3.

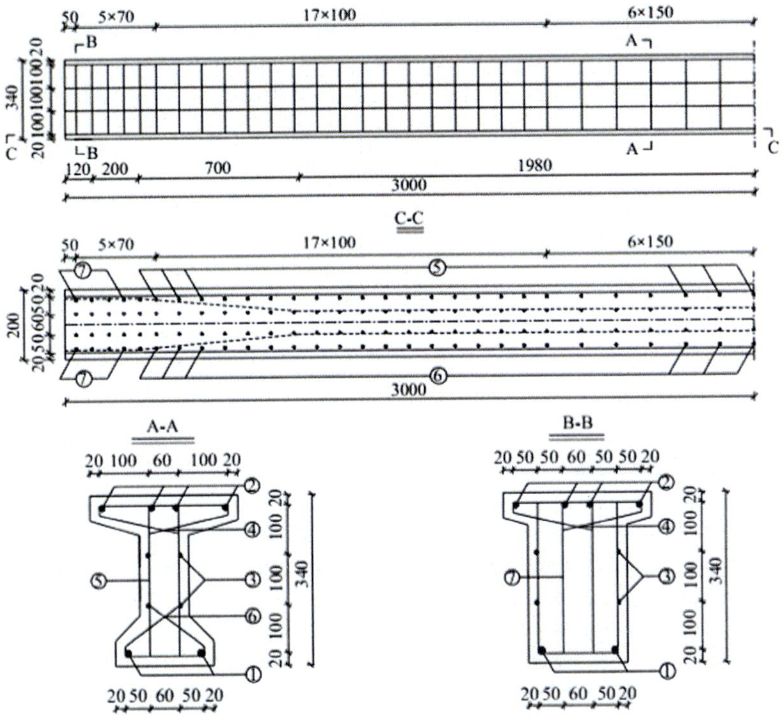

Fig. 3. Layout of regular reinforcement of model beam.

1.4 Concrete Mix Ratio and Mechanical Property

The concrete material ratio of model beams is cement: coal ash: fine aggregate: coarse aggregate: additive: water equal to 363:109:725:1087:5.19:146. Cement is ordinary silicate cement; coal ash is grade I ash; fine aggregate is medium sand; coarse aggregate is limestone macadam; additive is polycarboxylic acid; water is daily drinking water. Six standard test cubes were reserved from each model beam and maintained under the same condition as the beam body. Three test cubes are used for determination of concrete test cube strength during stretch-draw, and the other three test cubes are used for determination of corresponding cube strength before loading. The measured mechanical properties of concrete are shown in Table 1, the measured mechanical properties of reinforcing bars are shown in Table 2.

Table 1. Actually measured mechanical property of concrete

Prefabricated beam no	Test beam no	Stretch-draw cube strength (MPa)
Beam 1	ST-1	65.20
Beam 2	ST-2	73.80
Beam 3	FTG-1	71.24
Beam 4	FTG-2	73.80

Description of serial number: ST and FTG represent the static test and the constant-amplitude fatigue test, and the figures behind ST and FTG represent the specific serial number of fatigue test beam.

Table 2. Actually measured mechanical property of reinforcement

Variety of reinforcement	Nominal diameter (mm)	Yield strength f_y (MPa)	Ultimate strength f_u(MPa)	f_u/f_y
HPB235	6.5	387	463	1.20
HPB300	10	387	516	1.33
HRB400	12	455	619	1.36

2 Loading Schemes

The tests include a static loading test and an alternate loading test, both of which are conducted with the four-point bending mode, with reference to *Standard for Test Method of Concrete Structures* (GB/T 50152-2012) [18]. The third point symmetrical load was achieved through the distribution beam system with the form of simply support. According to the computed results of the maximum deformation theory, a 200 mm jack is used for ST-1 beam, and a pressure sensor was put on the jack to measure the load value. MTS fatigue loading tester was used for loading on ST-2, and the loading value was recorded.

The experimental study of fatigue property was carried out for post-tension cohesive prestress concrete beams with the same reinforcement based on the result of the static loading test for prestressed concrete beam. FTG-1 beam with the upper limit 40 kN of the fatigue loading and FTG-2 beam with the upper limit 60 kN of the fatigue loading were selected and used for the test due to the long period of a fatigue test, and the loading frequency was 3 Hz. The static loading was conducted every other 150,000 times at pause, and dynamic strain data were collected every other 10,000 times. Midspan deflection, regular reinforcement strain, prestressed reinforcement strain, and concrete strain in tension-compression zone were determined, and the crack propagation was observed closely in the loading process.

3 Analysis of Test Results

3.1 Static Loading Test

According to the failure form of the beam body after the test, it can be concluded that the failure of ST-1 beam and ST-2 beam is a typical flexural failure. That is, the failure of the specimen begins with the yield of the tensile reinforcement, and then a failure zone is formed near the concrete collapse point in the compression zone of the pure bending section. See Fig. 4 for details. Summarizing according to the trend of change curve of midspan deflection, reinforcement, prestressed reinforcement and compression concrete during static loading test, it is shown that concrete and tensioned reinforcement can work harmonically and embody better integrity characteristics. The static loading test and numerical calculation results of ST beam are basically consistent. Since the initial strain method is used for application of prestressing, the simulated value of cracking load is relatively large, while the error of the ultimate load value is relatively small, which indicating that the ST beam static loading test data is basically reliable. It can be used to determine the upper and lower limits of fatigue loading for FTG-1 and FTG-2 beams.

Fig. 4. Overall and local failure status of the beam with direct static loading

3.2 Beam Fatigue Behavior Test

No related cracks were observed in the FTG-1 beam and FTG-2 beam during the 2 million cycles of loading. However, the tensile area of the test beam has the phenomenon of continuous growth of concrete strain and continuous reduction of prestressing reinforcement strain. Through static loading failure test after fatigue, it is found that the two beams are still typical bending failure. That is, the failure of the members begin with the cracking of the concrete at the bottom edge, and then the tensile reinforcement yields, and finally the concrete at the compression edge of the pure bending section is crushed. Neither the tensile reinforcement nor the prestressing reinforcement appeared to be broken when it was failure. See Fig. 5 for details.

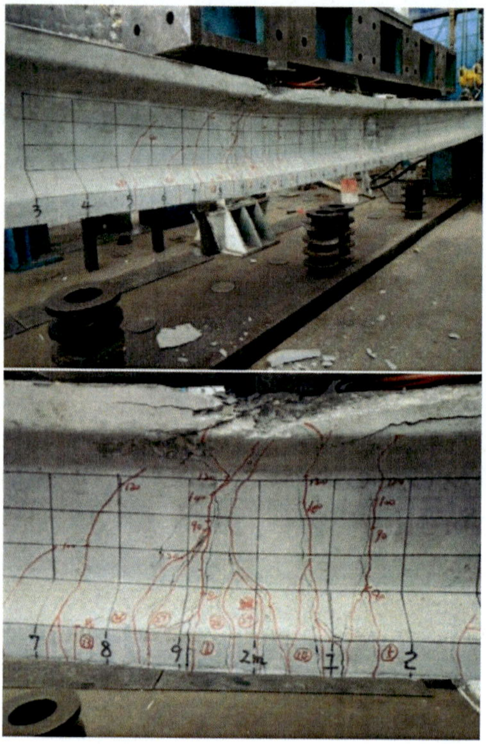

Fig. 5. Overall and local failure status of the beam with static loaded after fatigue loading

According to the test results, summarize the variation curve of strain amplitude of tensile reinforcement, prestressed reinforcement, and tensile and compression zone concrete of FTG-1 beams, FTG-2 beams with the number of cycles. See Fig. 6 and 7 for details.

For FTG-1 and FTG-2 beams, the strain amplitude of ordinary reinforcement, strain amplitude of prestressed reinforcement, strain amplitude of tensile zone concrete and strain amplitude of compression zone concrete basically remain unchanged with the increase of repeated action times. According to the above analysis, the strain amplitudes of prestressed reinforcement and ordinary reinforcement of FTG-1 and FTG-2 beams basically remain unchanged during the fatigue loading process, indicating that FTG-1 and FTG-2 beams are in normal working condition under the upper limit of 40 kN and 60 kN respectively, and the composite materials of the beam body (ordinary reinforcement, prestressed reinforcement and concrete) do not appear obvious damage.

It is found by beam static loading failure test after static failure test and fatigue test that the stress characteristics and failure states of the test beams are basically the same.

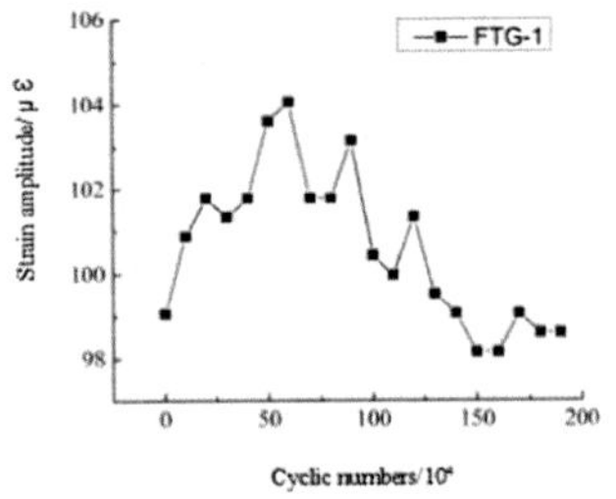

(1) Strain amplitude of tensile reinforcement

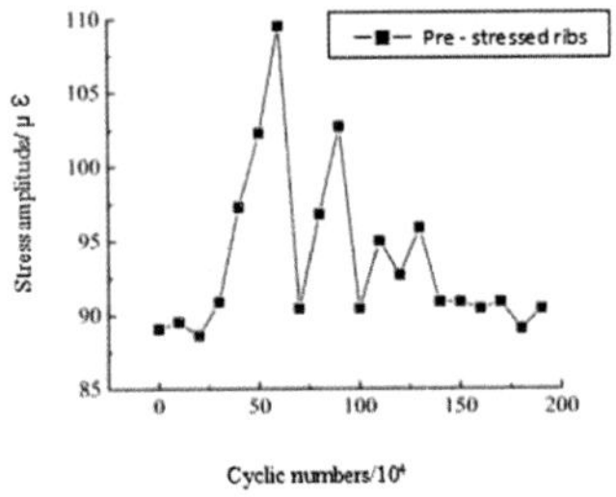

(2) Strain amplitude of prestressed reinforcement

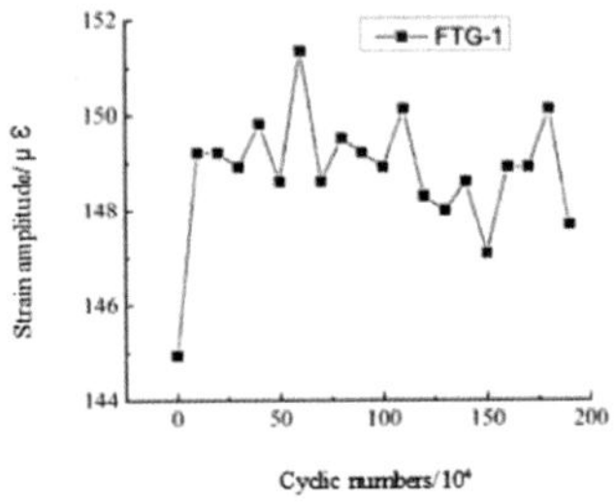

(3) Strain amplitude of tensile zone concrete

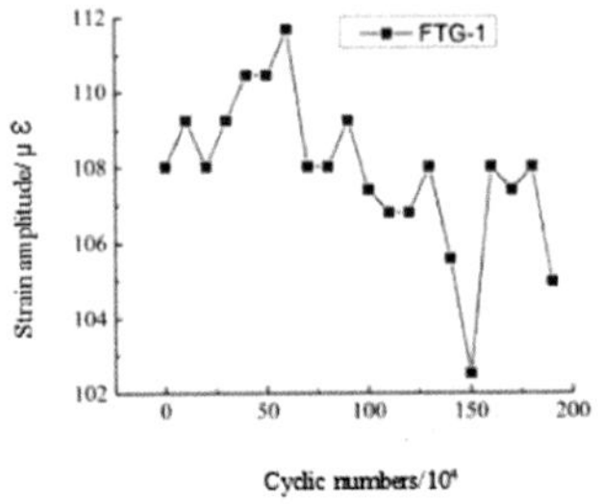

(4) Strain amplitude of compression zone concrete

Fig. 6. Strain amplitude of tensile reinforcement, prestressed reinforcement, and concrete of FTG-1beams varies with the number of cycles

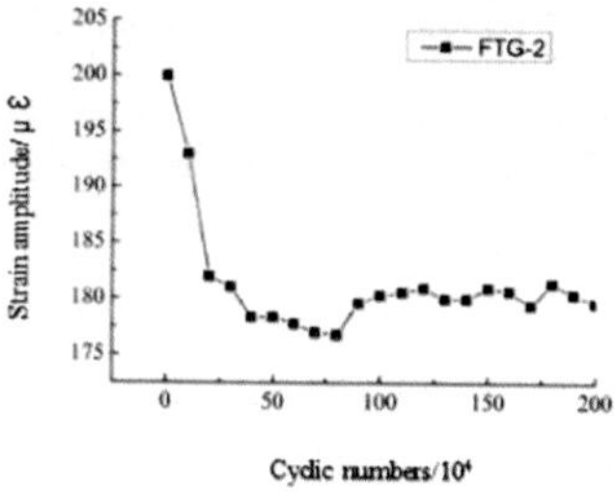

(1) Strain amplitude of tensile reinforcement

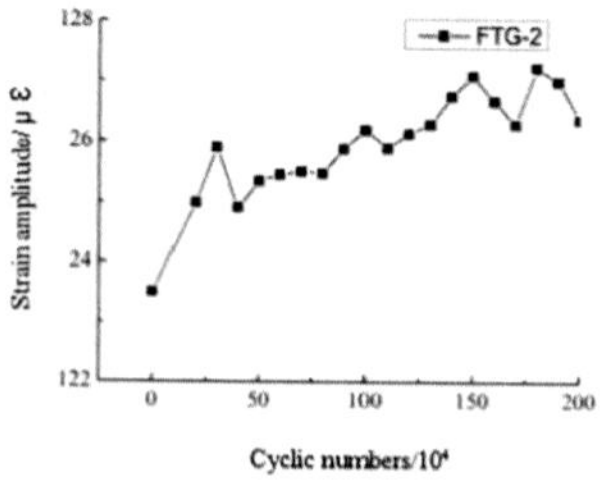

(2) Strain amplitude of prestressed reinfo rcement

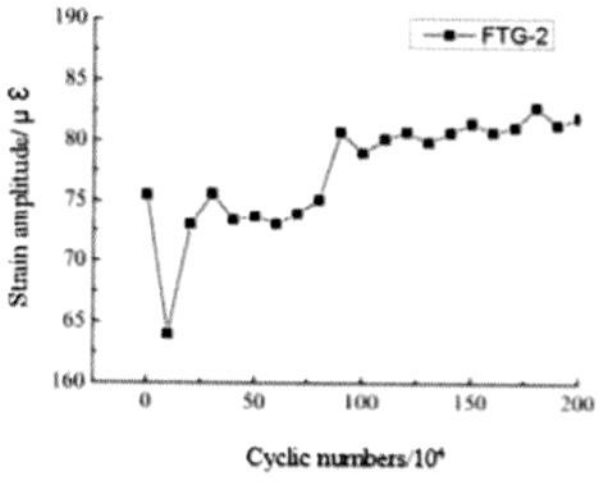

(3) Strain amplitude of tensile zone concrete

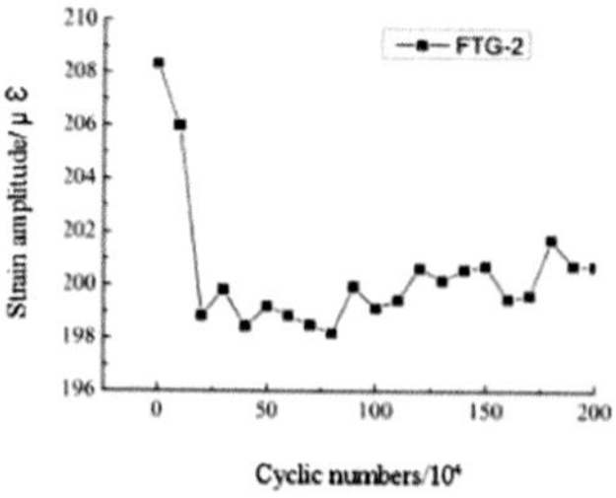

(4) Strain amplitude of compression zone concrete

Fig. 7. Strain amplitude of tensile reinforcement, prestressed reinforcement, and concrete of FTG-1beams varies with the number of cycles

There is no brittle fracture of ordinary reinforcement and prestressed reinforcement during the failure. The ultimate load of FTG-1 beam is 167 kN, and that of FTG-2 beam is 170 kN. Compared with the ST beam, the static loading bearing capacity is improved, and the beam body has a larger corresponding deflection when it is destroyed, showing ductile failure characteristics. During the loading process, the cracks of the four beams are evenly distributed in the pure bending section, and the cracks are basically located near the stirrup. When approaching the ultimate load, the width of the main crack in the pure bending section increases rapidly and extends upward to the upper flange, while the width of other cracks increases relatively slowly. Under the condition that the load remains unchanged, the horizontal cracks of compression zone concrete increase, the compressive strength of concrete is exhausted gradually, and the increase of deflection is accelerated, and when the ultimate load is reached, the concrete in a certain section on both sides of the main crack is crushed to form a triangular failure area.

Figure 8, Fig. 9, Fig. 10 and Fig. 11 respectively show the ST beam static loading and FTG beam static loading after fatigue corresponds to the strain contrast curve of midspan deflection, tensile reinforcement, prestressed reinforcement and compressive concrete.

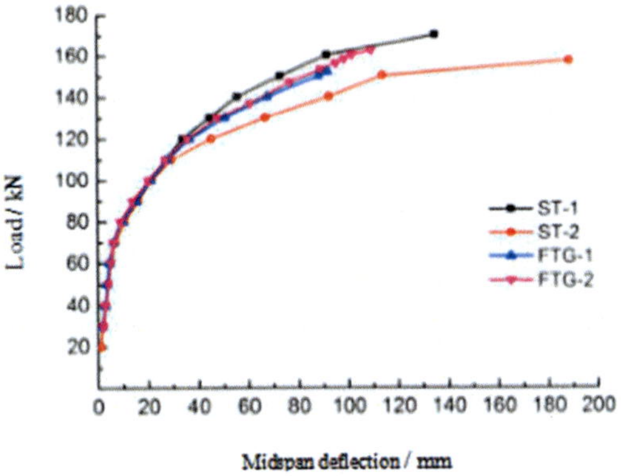

Fig. 8. Strain contrast curve of load - midspan deflection

According to Fig. 8, the midspan deflection of FTG-1 and FTG-2 beams also changes in three stages during static loading after fatigue. It can be seen from FTG-1 and FTG-2 curves that the midspan deflection increment increases sharply under the same load at the later stage of loading. The midspan deflection of 150 kN of FTG-1 beam is 88.20 mm, which is 1/65 of calculated span; the midspan deflection of 166 kN of FTG-1 beam is 135.41 mm, which is 1/43 of calculated span, both beams show the ductile failure characteristics. By comparing the curves of ST-1, ST-2, FTG-1 and FTG-2 curves, we can see that: the deflection curves of the four beams are consistent before 110 kN, and at 120 kN, as the tensile reinforcement of ST-2 beam reaches yield, however, the tensile reinforcement of FTG beam is still in the elastic working stage. There is a deviation between them. When approaching the failure, the deflection value of ST-2 beam is relatively large, but the ultimate load of the four beams is basically the same; the deflection growth rate of The ST-2 beam is deviated from that of the other three beams at the later stage of static loading. The reason is the difference of material composition, casting and curing of model beam. Therefore, the ST-2 curve has a large deviation.

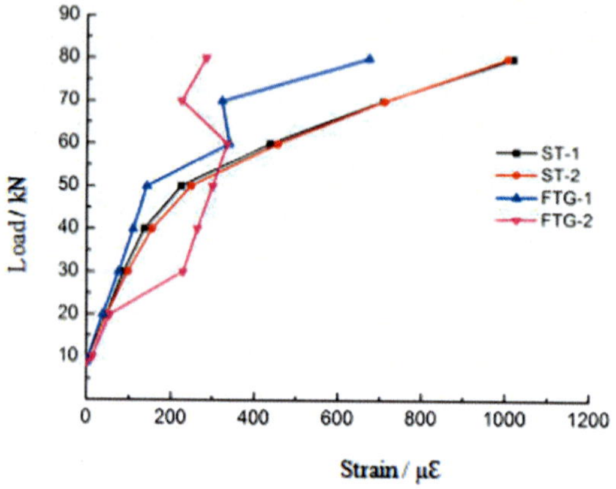

Fig. 9. Strain contrast curve of load - midspan deflection

According to Fig. 9, no complete data of tensile reinforcement was obtained during the static loading of FTG-1 beam after fatigue, so the entire loading process was not analyzed. The strain data of tensile reinforcement retained in FTG-2 beam experiment is only one, with large deviation, and also was not analyzed. Comparison of curve ST-1, ST-2 and FTG-1 and FTG-2 curves, we can see that: the ST-1 and ST-2 curves in the static loading process are consistent, but the FTG curve of static loading after fatigue has a large deviation, which show that the mechanical properties of tensile reinforcement are affected by repeated loading.

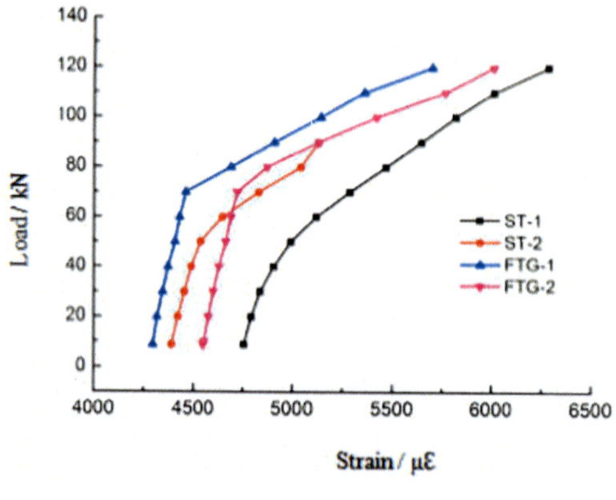

Fig. 10. Strain contrast curve of load – prestressed reinforcement

According to Fig. 10, during the static loading of FTG-1 and FTG-2 after fatigue, the strain of prestressed reinforcement in the pure bending section approximately shows a two-stage variation. Before 68 kN, it was a straight line. At the late loading stage, since the prestressed reinforcement became the main member resisting the external force, and the sudden increase in the strain of the prestressed reinforcement caused the friction between the two steel strands and caused the failure of the strain gauge. No effective analysis of the data after the yield load was carried out. For ST-1 and ST-2, the variation trend is the same within 0–90 kN. By comparing the curves of ST-1, ST-2, FTG-1 and FTG-2 curves, we can see that: the strain at the end of stretch-draw of the four beams is different, so the starting point is different, and the strain after the end of ST-1 stretch-draw is larger. Due to the use of five-times cyclic loading under different load levels in the static loading process, the beam body is damaged to different degrees with the development of cracks in the loading and unloading process. With the accumulation of stiffness damage, the concrete in the tensile zone cracks and withdraw from the work, and the prestressed reinforcement bears more external forces under the same load. Therefore, after the cracking load is reached, the strain growth rate of the prestressed reinforcement of the four beams increases as the load increases.

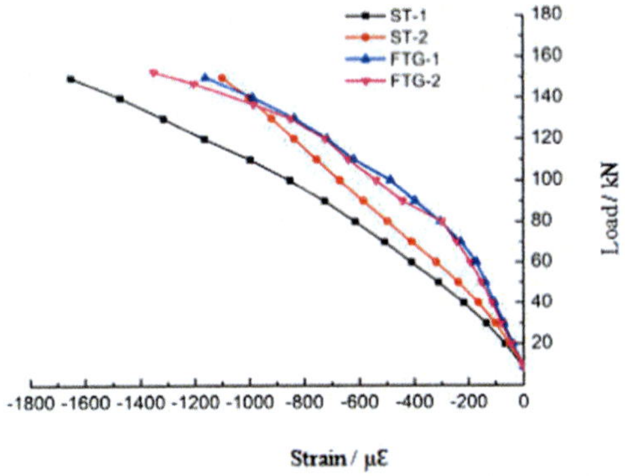

Fig. 11. Strain contrast curve of load – tensile concrete

According to Fig. 11, during the static loading of FTG-1 and FTG-2 after fatigue, the strain of tensile concrete in the pure bending section approximately shows a two-stage variation. By comparing the curves of ST-1, ST-2, FTG-1 and FTG-2 curves, we can see that: FTG-1 and FTG-2 curve have good consistency, after 120 kN, because the cracks of ST beam develop rapidly after entering the yield stage, the ST-1 curve corresponds to relatively large strain data of concrete under the same load; ST-1 curve is the corresponding curve of 0–150 kN in the loading process, due to the use of five-times cyclic loading under different load levels in the static loading process, the beam body is damaged to different degrees with the development of cracks in the loading and unloading process, therefore, the corresponding concrete strain is relatively large under the same load. According to the comparison of the test results of midspan deflection, tensile reinforcement, prestressed reinforcement and concrete in the compression zone during the above different loading processes, it is found that, fatigue loading has little effect on uncracked FTG-1 beams and FTG-2 beams, but after repeated loading, the midspan deflection of the model beam and the concrete strain in the compression zone decreased to a certain extent under the same load.

According to the test results, the ultimate load corresponding to ST beam is 160.9 kN, the ultimate load corresponding to FTG-1 beam is 167 kN, and the ultimate load corresponding to FTG-2 beam is 170 kN. It can be known after comparison that, after 2 million times of repeated loading, the remaining carrying capacity is improved, and the ultimate load of the beam is increased by 3.79% and 5.66% compared with the static loading test.

According to the formula for calculating the flexural capacity of T-shaped flexural members in concrete structure design code, it can be obtained that the ultimate load of FTG-1 beam is 128.72 kN, and the ultimate load test values of ST beam, FTG-1 and FTG-2 beam are respectively 1.25, 1.30 and 1.32 times of the calculated values. It has a certain safety reserve compared with the current specification.

4 Conclusion

In this paper, through the static and fatigue loading tests of the model beam, to carry out correlation analysis about the stress state of the prestressed reinforcement after different cycles times. Referring to the existing calculation methods, the prestress loss of prestressed concrete beams under fatigue loading is discussed, and the following conclusions are drawn:

(1) According to the fatigue loading test of the prestressed concrete model beam with cyclic loading of 2 million times, it is found that the midspan deflection, ordinary reinforcement, prestressed reinforcement and concrete strain are basically unchanged. The fatigue loading has less damage to the uncracked beam.
(2) The strain increment and strain amplitude of prestressed reinforcement remain unchanged under the action of fatigue loading. The prestressed concrete beam has the prestress loss under the fatigue loading, and the prestress loss changes in three stages with the increase of the number of fatigue loading. The prestress loss in the early and late loading stages is relatively large, while the prestress remains unchanged in the middle loading stage. The prestress loss after millions of fatigue loading is about 4%–6% of the effective prestress before loading.
(3) The static loading failure test after fatigue shows that the remaining carrying capacity of prestressed concrete beams has not changed but increased to a certain extent, and it still has a certain safety reserve compared with the current code.

References

1. Guo, Z.H.: Principle of Reinforced Concrete, 3rd edn. Tsinghua University Press (2013)
2. Song, Y.P.: Fatigue Behavior and Design Principle of Concrete Structures. China Machine Press (2006)

3. Huang, H., Wang, W.W., Dai, J.G., et al.: Fatigue behavior of reinforced concrete beams strengthened with externally bonded prestressed CFRP sheets. J. Compos. Constr. **21**(3) (2017)
4. Sun, X.Y., Dai, J.G., Wang, H.L., et al.: Static and fatigue behavior of CFRP-strengthened RC bridge girders subjected to vehicle overloading. Adv. Steel Constr. **11**(3), 359–371 (2015)
5. Editorial Department of China Journal of Highway and Transport. Review on China's bridge engineering research: 2014. J. Highway Transp. **27**(5), 1–96 (2014)
6. Balaguru, P.N.: Analysis of prestressed concrete beams for fatigue loading. PCI J. **26**(3), 70–94 (1981)
7. Hilsdorf, H.K., Kesler, C.E.: Fatigue strength of concrete under varying flexural stresses. J. Am. Concr. Inst. **63**(10), 1059–1076 (1966)
8. Neville, A.M., Whaley, C.P.: Non-elastic deformation of concrete under cyclic compression. Mag. Concr. Res. **25**(84), 145–154 (1973)
9. Feng, X.F., Song, Y.P.: Analysis on normal section fatigue crack resistance for prestressed concreted members. J. Harbin Inst. Technol. **38**(3), 414–416 (2006)
10. Li, J.Z., Xu, Z.W.: Study on the effect of fatigue in 32 m-span PC simply supported T-type beams on existing line due to heavy-haul transport. Railway Stand. Des. **58**(10), 56–61 (2014)
11. Zhou, L., Wu, J.Z.: Discussion on some problems in the analysis of P.C and P.P.C structures. Bridge Constr. **2**, 42–46 (1998)
12. Che, H.M., Shao, H.K., Li, X.P.: Partial Prestressed Concrete: Theory, Design, Engineering Practice. Southwest Jiaotong University Press (1992)
13. CEB-FIP Model Code 1990[S]
14. Zhang, J.W., Zeng, J.C., Zhang, Z.X., et al.: Experimental study on behavior of pre-stressed concrete under flexural tension-compression fatigue loading. Ind. Constr. **39**(8), 72–75 (2009)
15. Zeng, D., Wang, G.L., Xie, J., et al.: Exploratory experiment of fatigue prestress loss of prestressed concrete beam. J. Highway Transp. Res. Dev. **29**(12), 79–83 (2012)
16. Kang, J.Q.: Experimental Study on Carrying Capacity of Prestressed Concrete T Beam. Chang'an University (2010)
17. JTG 3362–2018, Specifications for Design of Highway Reinforced Concrete and Prestressed Concrete Bridge and Culverts
18. GB/T 50152-2012, Standard for Test Method of Concrete Structures

Architectural Design and Building Surveying

Exploration of a Dynamic Building Skin System Responding to Light Environment Information

Zhipeng Guo(✉) and Yuntian Ma

Northwestern Polytechnical University, Xi'an, China

Abstract. Dynamic building skin is an intelligent and transformable skin system, which can improve the climate adaptability of buildings by combining active and passive technologies into the building. The design of dynamic building skin with environmental response strategy is a bottom-up process based on the optimization of the physical performance of the building. With case studies and the development and building of a physical model, the article presents a design method of a dynamic building skin system that responds to light environment information, providing new possibilities for climate-adaptive building design.

Keywords: Dynamic building skin · Light environment information · Climate adaptability

1 Variable Building Skin and Building Environment Response

Since the industrial revolution, the advancement of technology has allowed people to use external equipment such as electric shutters and artificial lights to solve the indoor light environment problems, but this has also caused the relationship between the structure of building and light environment to be gradually ignored. The rise of the concept 'green building' has made people re-recognize the importance of climate-adaptive buildings. Many climate-adaptive building cases have demonstrated how buildings can intelligently take advantage of external conditions to keep themselves in a state of comfort. The skin system is the dominant structural interface of contact and material exchange between the building and the outside world, affecting the indoor environment comfort and energy consumption of the building. Relevant studies have pointed out that nearly 50% of the energy consumption of a building during daily use is directly or indirectly related to the performance of the skin [1].

With the development of parametric design and the reform of digital construction technology, architectural design has undergone major changes in concepts, methods, and processes. New skin design therefore produces more flexible and free facades, with the help of software such as Grasshopper. Even so, most building skins are static, and the development and application of dynamic interactive architectural skins are one of the main methods to make them flexible. Dynamic skin systems take data transmission as the hub, interactive devices as the main body, building construction, architectural

P. Mendonça and N. D. Cortiços (Eds.): ICAMC 2021, LNCE 226, pp. 91–97, 2022.
https://doi.org/10.1007/978-3-030-94514-5_10

aesthetics and other relevant technologies as tools, to achieve an active response to the surrounding environment and adaption to different requirements by sensing external information in real time and controlled variations of geometric shapes, distributions of interactive devices and physical attributions of the skin.

2 The Development of Research into Dynamic Building Skin

With the continuous improvement of digital simulation technology and architectural design methods, climate adaptability has become one of the important considerations in skin design, and more and more relevant cases have emerged. Dynamic building skins can regulate various environmental factors through the intelligent control of skin components, so as to adapt to the ever-changing climate environment and building performance requirements [2]. Its new technical characteristics such as intelligent, dynamic, and adaptive provide new ideas for the design of building skins. The research on its design methods mainly includes: London University College Interactive Architecture Lab in responsive architecture Practice in design led by Ruai Li Green; research into climate-adaptive building interface and proposal of the 'environmental parameterization' skin design method from Architectural Interface Research Group of Delft University of Technology [3]; Jae Wan Park's proposal, design and verification of an interactive skin structure [4]; discussion of the design method of dynamic sunshade skin based on energy consumption and visual comfort from Jie Xiong et al. [5] and Su-Ji Choi et al. [6].

From the perspective of climate adaptability, Roel Loonen et al. classified the variable building skins according to the environmental factors they regulate, and classified them into four categories: thermal environment, light environment, wind environment, and photovoltaic power generation [2]. A skin can focus on one or many of the four categories. In the process of participating in the undergraduate scientific research and innovation project "Intelligent Building Skin Research Based on Unmanned System", we used Grasshopper, Python and Raspberry Pi to design and develop physical models based on light environment adjustment factors, combined with unmanned systems and interactive design technology, attempting to explore an effective scheme for realizing a dynamic building skin system.

3 Design Exploration

3.1 Modular Processing of the Skin

Modularization of the skin is the basis of this design. Since the interactive building skin needs to respond to external factors in real time, the overall freedom of the skin changes determines the complexity of the skin's ability to adapt to the external environment. The modular skin design strategy can greatly increase the application potential of the interactive skin.

The basic unit of the modular skin is the modular skin unit. As the basic part of the system, the units are identical structurally and independent mechanically, achieving high movability and coordination, as well as easy updating and maintenance. The module group can be reorganized according to a certain logic to achieve a specified function. The internal light environment of the building can interact with the outdoor environment by changing the state of the skin modules.

3.2 Model Design of Skin Module Unit

Figure 1, Fig. 2 and Fig. 3 specifically reflect model structure design of skin module unit.

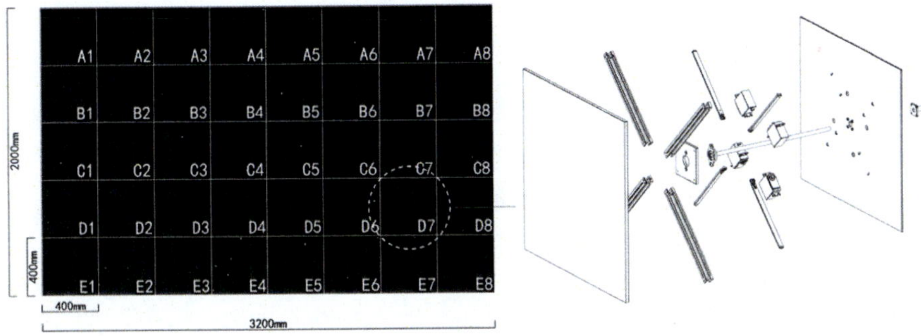

Fig. 1. Smart skin structure and unit

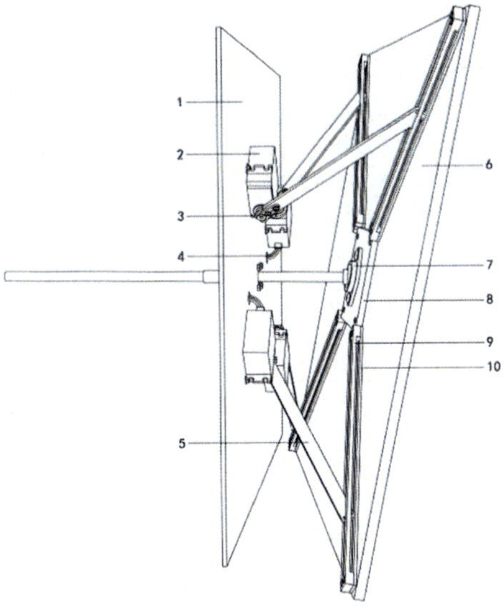

Fig. 2. Movable skin model

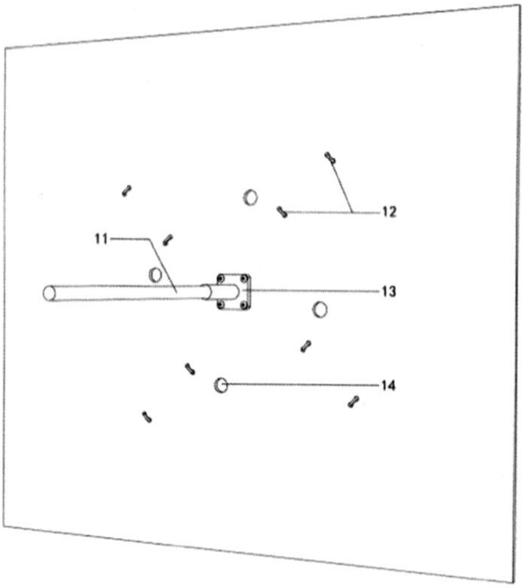

Fig. 3. Model back structure

Component definition: 1, 3 mm acrylic back plate; 2, MG996R steering gear; 3, steering gear arm; 4, steering gear cable; 5, acrylic extension arm parts; 6, 7 mm PVC outer plate; 7, KFL08 spherical bearing; 8, Acrylic bearing base; 9, Acrylic rail parts; 10, Acrylic rail; 11, 8 × 300 mm optical axis; 12, 1.7 mm zip tie; 13, LMK8LUU linear bearing; 14, cable hole.

3.3 Control Program

For the structure, the change of its status mainly depends on the control angle of the four steering gears of each unit. In actual control, for a given status, we need to obtain the angle of the steering gear and the speed of change of each steering gear angle, so that the physical characteristics of the structure are satisfied at every moment.

By analyzing the physical characteristics, we can find that the structure has the following characteristics: the center position only changes in one dimension; the diagonal lines are always perpendicular to each other and intersect at the center. Therefore, in actual control, we convert the four steering gear angles into three parameters, namely: center extension h, diagonal angle 81 and diagonal angle 82. In fact, the structural characteristics of the two diagonal angles are unified, so we will illustrate with an example.

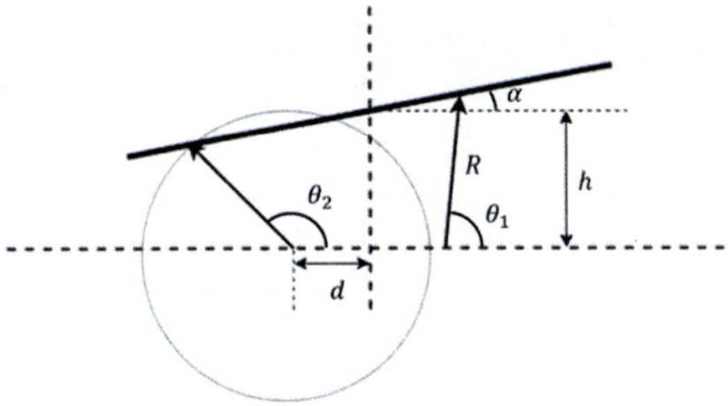

Fig. 4. One-sided schematic

As shown in Fig. 4, when the one-sided parameter h and α is given, the corresponding angle θ_1 and θ_2 can be obtained by solving the following equations.

$$\begin{cases} \tan\alpha(R\cos\theta_1 + d) + h = R\sin\theta_1 & (1) \\ \tan\alpha(R\cos\theta_2 - d) + h = R\sin\theta_2 & (2) \end{cases}$$

4 Model Construction and Experiment

Based on comprehensive design considerations, we built a model of the dynamic building skin system and conducted relevant experiments. The model skin system is composed of 8 horizontal units and 5 vertical units, with each unit being capable of performing individually. The skin can move according to a certain logic to meet the diverse needs of users in the building for natural lighting at different times and locations.

Fig. 5. No lighting mode/Fully closed mode

Fig. 6. Partially lighting mode

Fig. 7. Component arrangement

As shown in Fig. 5 and Fig. 6, the dynamic skin arrangement can set four monomers into a group, forming obvious zoning features in the form of the facade, providing an orderly arrangement of lighting for the interior of the building. The transformation logic of the skin is diverse, which can be changed according to the internal functional layout of different buildings, the outdoor light environment, and the individual needs of users. Figure 7 shows the arrangement of mechanical components.

5 Concluding Remark and Future Research

As the exterior surface of the building, the building skin is one of the components of the urban space. The building facade has been redefined as an interactive space, which can flexibly collect information and respond as the exterior of the building. The main reasons for putting the interaction design of light environment for the building skin into consideration are as follows: 1. The building skin has strong environmental sensitivity; 2. The building skin can greatly affect the indoor light environment of the building; 3. The structure of the building skin can withstand large mechanical movements. The future application potential of interactive architectural skins enables architectural design to adapt more actively to the behavior of users and the needs of the urban environment while considering aesthetic factors.

In the future interactive skin research, the following points ought to be paid more attention: 1. The changes in the light environment needs of building users; 2. The response of the building skin to the changes in the external light environment; 3. The practical building interactive skin structure research; 4. Material research.

The above research will lead to more discussions. The interactive architectural skin will reconstruct the relationship between building indoor and outdoor environment and space. We made an attempt to establish a more organic interaction within the building. In the future, the skin system will be capable of realizing the meaning and value of smart buildings.

References

1. Zhuang, W., Qi, B., Lin, B.: Environmental and Ecological Oriented Architectural Composite Skin Design Strategy. China Construction Industry Press, Beijing (2014)
2. Loonen, R.C., Trčka, M., Cóstola, D., et al.: Climate adaptive building shells: state-of-the-art and future challenges. Renew. Sustain. Energy Rev. **25**, 483–493 (2013)
3. Bowman, R., Schulze, F.: Building a Masterpiece: Milwauke Art Museum. Lund Humphries Publishers, London (2009)
4. Park, J.W., Huang, J., Terzidis, K.: A tectonic approach for integrating kinesis with a building in the design process of interactive skins. J. Asian Architect. Build. Eng. **10**(2), 305–312 (2011)
5. Xiong, J., Tzempelikos, A.: Model-based shading and lighting controls considering visual comfort and energy use. Sol. Energy **134**, 416–428 (2016)
6. Choi, S.J., Lee, D.S., Jo, J.H.: Lighting and cooling energy assessment of multi-purpose control strategies for external movable shading devices by using shaded fraction. Energy Build. **150**, 328–338 (2017)

Housingprime App, A Participatory Design Platform for Housing Customization and Community Self-organization

Zidong Liu(✉)

University College London, London WC16BT, UK
zidong.liu.19@alumni.ucl.ac.uk

Abstract. The traditional bottom-up design model has proved unable to adapt to the complexity and change of cities and buildings, and a participatory design model is gaining increasing attention. The paper presents a participatory design platform, Housingprime, that allows users to design their own houses and community based on mobile interfaces. Housingprime provides the whole process of service from design to construction, committed to meeting the housing crisis among youngsters aged 25 to 44 in megacities. This platform brings a shared housing model in community design, effectively improving spatial efficiency, adaptability and sense of belonging. The concept of participatory architectural design has decades of theoretical history. However, it has not been until the recent popularization of smartphones that it has a technical foundation for extensive practice. The paper first introduces the research background of participatory design and mobile interfaces in architecture practice. Then the paper presents the front-end use flow and back-end algorithmic logic of Housingprime by sharing a design result based on 200 hypothetical users' data. Finally, this article discusses the further development of this community generation system.

Keywords: Participatory design · Mobile application · Housing customization · Community organization

1 Introduction

Worldwide population concentration and migration to megacities have caused a new round of housing crises, posing severe challenges to the existing real estate development and community organization model. Due to the low democracy of the design process, the traditional top-down community planning model has led to low land utilization and further exacerbated the housing crisis [1]. Planners tend to preprogram residents' family structure, surrounding facilities and neighborhood relations and use a preconceived guiding principle to simplify complex relationships between site and city. Once the plan is determined, there is a slight possibility of change. With the increase in housing pressure and the decline of the living quality, society urgently needed a new participatory community design model.

The prototype system presented in this paper, Housingprime, is based on mobile interfaces, providing an interactive platform for people to input their personal

P. Mendonça and N. D. Cortiços (Eds.): ICAMC 2021, LNCE 226, pp. 98–109, 2022.
https://doi.org/10.1007/978-3-030-94514-5_11

information and preview the design result. In this way, customers can design their own houses and participate in the community generation process. Housing-prime is a research project of RC11, B-Pro, the Bartlett, UCL 2020 led by Philippe Morel and Paul Poinet, with members including Zidong Liu, Shiyuan Huang, Yi Li and Huiyu Pan. The research focuses on self-organizing community algorithm design and interactive platform design, examining how residents can participate more in the community design and design their houses.

2 Research Context

2.1 Participatory Design

The idea of a participatory community design model has decades of history. Around the 1960s, with the rise of criticism of modernism, architects tried to explore a novel democratic model of community self-government. Under the influence of cybernetics, a generation of ideal architects with similar thoughts of adaptive architecture emerged around the world, such as Constant's New Babylon (1956–1974), Yona Friedman's Mobile Architecture (1958), Cedric Price's Fun Palaces (1961), Archigram's Plug-in City(1962), Walking City (1964) and Instant City (1968), Metabolism (1970s). These experimental designs often have a decentralized organizational network and movable structural units, making it possible to quickly adapt to the changes of society, environment and people's diverse needs. However, due to the limitations of the technology at the time, these ideal architects could not go far enough into algorithmic programming. Fun Palace may be a rare exception. With the support of Gordon Pask's cybernetics group, Fun Palace moved from architecture to cybernetics [2]. In most other futuristic projects, the operating system is very crude: Cranes move and stack the dwelling units in a prebuilt giant structural frame. After decades of technological development and popularization of computing equipment, the former participatory design based on cybernetics has revived today. In this novel public engagement process, architects are supposed to develop interactive platforms that facilitate participatory design. One of the most notable recent commercial practices is Zaha's real estate development project on Roatan Island in Honduras, where they developed a platform for parametric customization of housing. Before this, there had been many similar attempts to build participatory design interactive platforms [3–8]. The large-scale participatory design will lead to mass housing customization [9, 10].

2.2 Mobile Interfaces

As the most popular human-computer interaction tool, mobile devices provide a large user base for participatory design platforms. The application of mobile devices in participatory design is mainly in two aspects: The collection of environmental data and user information and the real-time visualization of design results. With the rise of GPS technology, environmental data collection is the first practice. Convenient access to location information dramatically accelerated the establishing and updating process of city databases [11, 12]. Later, with the popularity of smartphones, mobile devices also became the most important collectors of user data. Users' active presence on social media and their app usage records have become valuable data resources. At the same time, the design visualization methods of mobile devices have become increasingly rich and vivid. VR and AR technologies bring users a more immersive virtual experience of architecture. This WYSIWYG simulation technology helps to improve users' enthusiasm for participation and enhance users' confidence in investment and purchase [13–15]. The integration of positioning technology and interaction design can be applied to tour guidance, such as the museum AR guidance system based on mobile services [16].

3 Methodology

3.1 Overall Structure

Housingprime is planned to be a whole-process participatory design platform. It ranges from housing unit design to community design and plans to go from conceptual design to mass production optimization in the future. The whole process consists of three steps. First, based on user and city data, the app will help customers complete housing customization and optimize their community design. Second, the app will send design plans to the prefabricated component manufacturers. The last step is the using stage. Clients would receive their new house just like receiving a box from Amazon and share their housing design on the media platform (Fig. 1). The project is currently in the experimental stage, focusing on the conceptual design part. The conceptual design system includes two parts: The backend and the frontend.

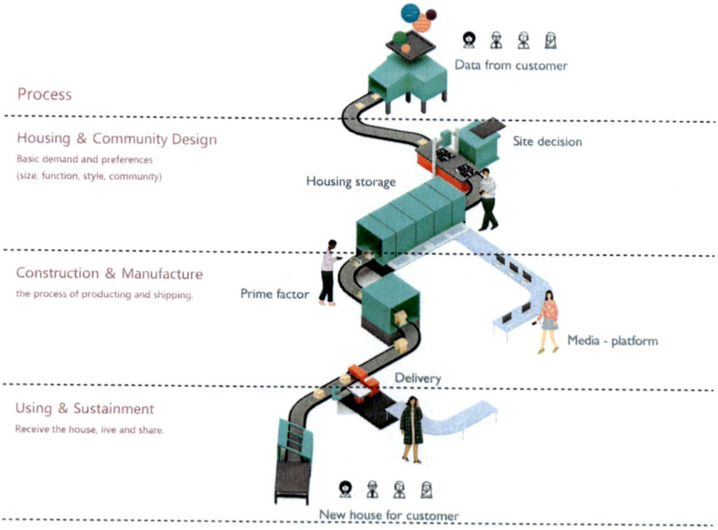

Fig. 1. Working flow of the entire Housingprime system.

3.2 Frontend Design

The frontend is based on mobile devices for account registration, user data collection and displaying design results. Social functions are also considered to be added to the platform. User's participation is thoroughly valued in this process.

Mapping Out Key User Demands. The first step of building an automatic system is to define input parameters. By analyzing user demands, the factors influencing the housing decision can be divided into six dimensions and 52 factors, including dwelling characteristics, surrounding environment and facilities, social identity and so on [17]. To build a generative design system, these influencing factors are regrouped into three quantifiable categories: Housing unit design, community design and site decision (Fig. 2).

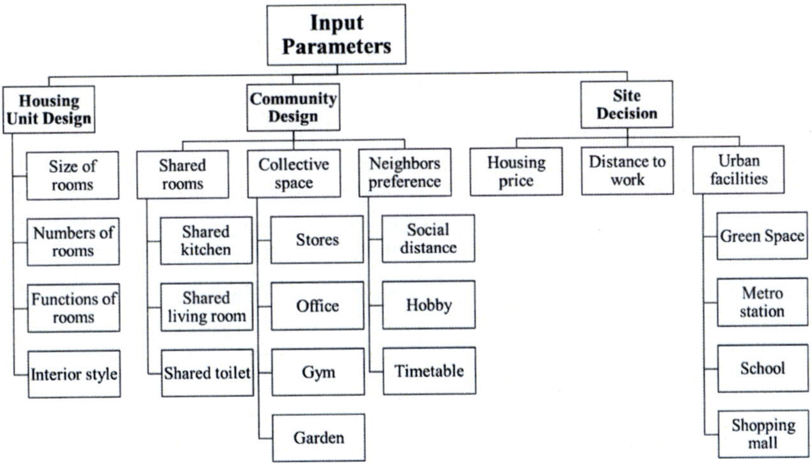

Fig. 2. Classification of user demands in housing decision.

Use Flow. The goal of the frontend design is to allow users to complete data input most and intuitively. Users can get feedback on design results in real-time and modify the results straightforwardly (Fig. 3):

- Initial introduction. On the App introduction pages, customers will get some tips for using this app and then input some basic personal information like the number of households.
- Site selection. The client will first enter the city where he will live. Then the system will directly link to Google Maps and mark available lands on the map. After that, according to the user's preference, the system will automatically evaluate all the lands and give three recommended options.
- Housing design. The housing design is a fully interactive process. Users will first confirm the housing unit size according to their financial budget. The following space design is based on a genetic algorithm idea: The user inputs the initial requirements, and then the system will tentatively generate several design options. According to users' selection, the system will produce new options in the direction

of user preference. After many iterations, the system and users will finally complete a good layout design and interior design.

- Community design. Users can share some housing functions, and those with the same choice will be paired to form a living cluster. The living cluster will find the most suitable location in the community according to the needs of the residents and the surrounding facilities.
- After-sales stage. When the design is completed, the system will upload the design plans and contact the corresponding suppliers. Customers can track the production and shipping process of the product on the app. In addition, the app creates a social community. People can share their evaluation of the product and their design experience.

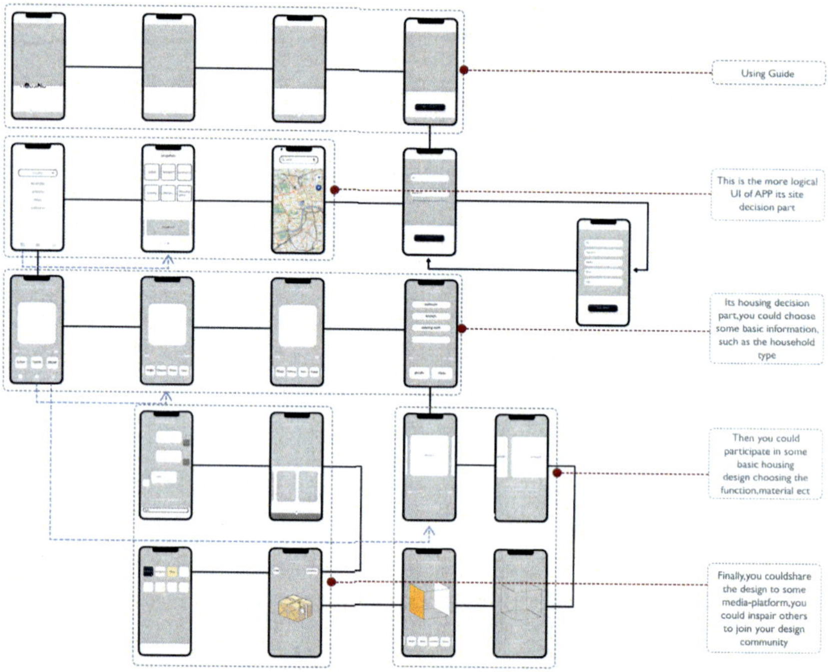

Fig. 3. Use flow of Housingprime App.

Interaction Platform. Based on the use flow chart, we developed the APP user interface via java. The design visualization is realized by OpenGL ES and extended to the web page with the support of Three.js. People can participate in architectural design through mobile devices and personal computers and walk freely in virtual buildings to get an immersive experience (Fig. 4).

Fig. 4. User interfaces of Housingprime App.

3.3 Backend Design

In the backend algorithm design, the generative design process is subdivided into three parts at the backend: site selection, house unit design and community design (Fig. 5). At present, the data processing process mainly runs on the offline platform Grasshopper. In an ideal state, a cloud server will be established as a backend for automatic synchronization.

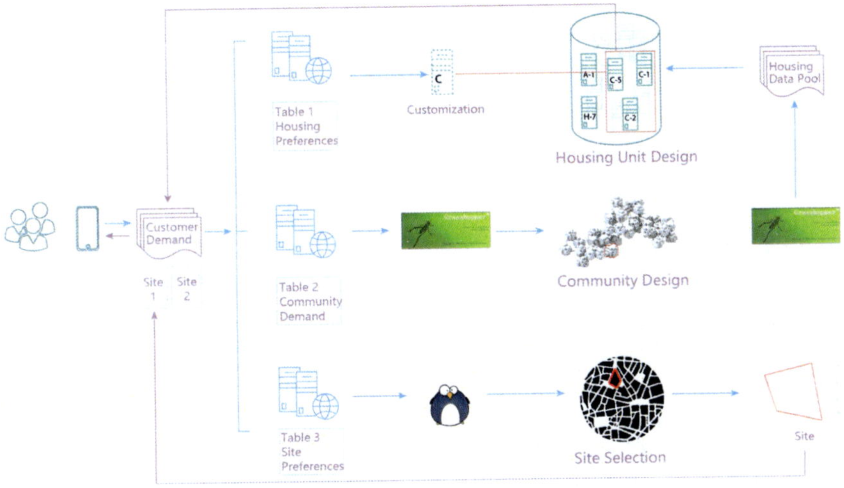

Fig. 5. Overall logic structure of the generative design system in backend.

Site Selection. Site selection is the prerequisite for community generation and the first decision made by clients in housing decisions. Mainly based on GISMO and Python, the algorithm is subdivided into the following steps (Fig. 6):

- Land information collection. By cooperating with real estate companies, the system automatically collects information on the location and size of the available land in the city. In this design experiment, we loaded the London land sale information from the Rightmove website and the facility information from the London Datastore websites through Python.
- Land value quantification. According to Clarence Perry's "neighbourhood unit" concept, 400 m is the most comfortable walking distance. The approach is to draw a circle with a radius of 400 m from the center of each piece of land and count the number of corresponding infrastructures in each circle. The system gives a score for each piece of land based on the different preferences of each user.
- Site geometry information import. After the site is selected, grasshopper will automatically import the city geometric information around the site into the system through GISMO to facilitate the subsequent computational design.

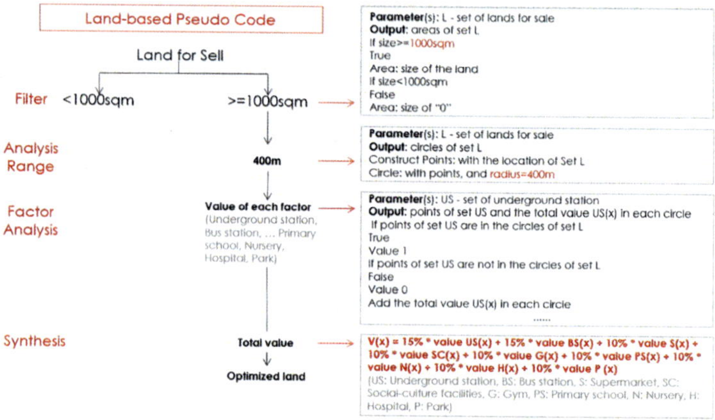

Fig. 6. Algorithmic logic for site selection.

Housing Unit Design. In the housing design, user participation is fully guaranteed. The specific steps of housing customization are as follows (Fig. 7):

- House outline generation. The housing unit size is based on a 5 m × 5 m modulus. First, the user will provide some basic housing information, including the number of households, the required room functions. Based on this information, the system gives the required number of modular units and their room prices.
- Layout automation. The layout automation is always the core step in the parametric design of housing. What we use here is a standard genetic algorithm based on constraints. When the housing's outline and functions are determined, the location of each room is represented by a point. Each functional room has a pre-set area value range. For example, the area of the toilet is 2 to 4 square meters. When two

points are close, they avoid overlapping and change the shape of the room. Some rooms with specific functions have special attributes. For example, the bedroom needs daylighting, so we must set a specific daylighting surface in the building outline which directly intersect with the bedroom edge. Based on these constraints, the system randomly distributes points to generate many solutions. Then through the evaluation of the performance (the degree of compactness of the apartment, the size of the bedroom, etc.), three recommended solutions are generated. Then the user chooses his favourite plan. After several iterations, the optimal solution is obtained.

- Interior design. Interior design is a part to be improved. At present, a kind of folding furniture has been developed, and people can customize the size and function of the furniture according to their needs. The finished furniture plan will be sent to the furniture supplier for manufacturing.

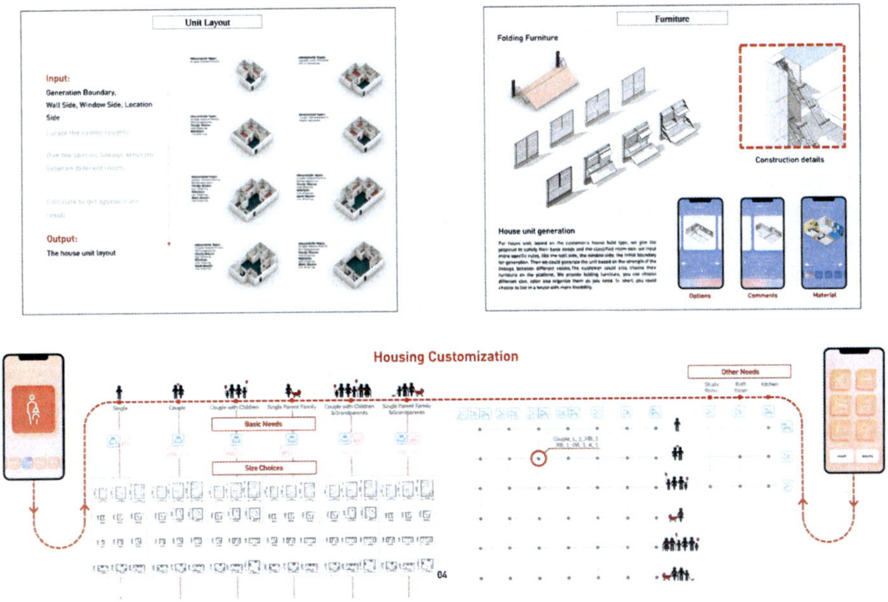

Fig. 7. Algorithmic logic for housing unit design.

Community Design. Community design is the most innovative aspect of backend algorithm design. Because the community scale is significantly larger than the scale of housing units, people have remarkable features of sequence and regularity in community activities. These features inspire algorithm design: Using sequential methods to organize the complex network of the community:

- Living cluster design. The living cluster is the middle level between the house and the community. Residents living in the same cluster share similar living patterns and facility requirements. We use the K-means Clustering algorithm [18] to group residents based on three-dimensional data (shared room, community facilities, house similarity). Due to the flexible layout in the living cluster, we chose particle physics simulation to describe the constraints between the living units for layout generation. By confirming the entrances and exits of each floor, floors can be connected in series to form a complete living cluster with diverse architectural space (Fig. 8).

- Sequencing activity network. The daily activities of residents in the community form a complex topological network. For example, for sports fans, their daily activity is to take a walk in the community park after work, enter the gym for anaerobic exercise, and then go home. We define such a life pattern as a sequence so that the complex network of the entire community is transformed into a collection of intertwined sequences. Based on 200 hypothetical users in 20 living clusters, we simulated 52 activity sequences. These sequences connect a series of public facilities, such as stores, offices, laundries, etc.
- Spatial structure generation. Sequential information is transformed into geometric information based on constraints via physical engine simulation. The central points represent the location of living clusters and service facilities, and the number of squares occupied by points represents their area. Points have repulsive forces against each other to maintain the size of the room. The adjacent elements in a sequence have a strong attraction to keep them stick to each other. Similar algorithms are developed into three dimensions so that the community can grow in the vertical direction. Since the initial position of each point is random and greatly influences the design result, we use genetic algorithms to optimizes the design results based on the rationality of the plan (circulation length, sunshine, etc.).
- Housing unit allocation. After completing the structure and layout design of the community, the system distributes the previously designed housing units to the community structure. At this stage, the customer can see the location of his room in the community and related facilities. In addition, we have developed algorithms for generating modular community facilities like shared kitchens, shops, offices, etc. These service facilities will also be visualized in the design preview (Fig. 9).

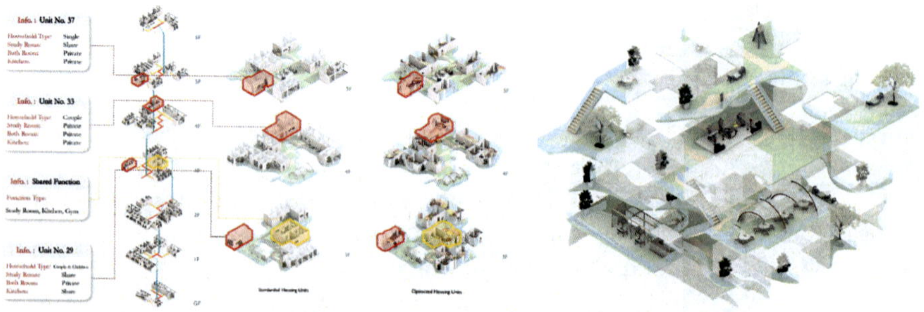

Fig. 8. Floors aggregation of living cluster

Fig. 9. Result of circulation optimization and community design.

4 Conclusion

The paper introduces the front-end use flow and back-end design generation logic of Housingprime, a participatory architectural design platform. Through the establishment of this interactive platform, the research explores how individuals can participate in the management of public affairs, promoting the democratic process of society through a bottom-up community self-organization mechanism. On reflection of modernism, collective intelligence has been proved to be more effective and accurate in solving urban design compared with elite meetings.

As a data-driven design platform, the project explores the potential of architecture automation. Through algorithm design, the system realizes the automation of site selection, housing customization, living cluster design, and community planning step by stepIn the future, architects should no longer be just designers of buildings and cities but developers of interactive platforms. Technological progress makes a wider range of social connections possible in urban design. Based on the concept of digital twins, the system realizes the integration of the virtual world and the real world through continuous data collection and feedback and provides residents with an immersive living experience during the design phase.

Future research will pay more attention to the material realization of virtual design, including the production, transportation, construction technology of building components and the digital community management after construction. At present, our research focuses on the digital design stage. Later Housingprime will develop towards the realization of building lifecycle management. In the future, with the integrated application of cloud computing, big data, IoT and BIM [19], Housingprime will be able achieve its ultimate goal: a whole process participatory design platform.

Acknowledgments. Housingprime is a research project of RC11, B-Pro, the Bartlett, UCL 2020 led by Philippe Morel and Paul Poinet, with members including Zidong Liu, Shiyuan Huang, Yi Li and Huiyu Pan. I would like to thank our tutors Philippe and Paul who constantly gave us inspiration and guidance. I would also like to thank my team members for their company and encouragement.

References

1. Wong, J.F.: Factors affecting open building implementation in high density mass housing design in Hong Kong. Habitat Int. **34**(2), 174–182 (2010)
2. Mathews, S.: The fun palace as virtual architecture. J. Architectural Educ. **59**, 39–48 (2006)
3. Donath, D., Felipe, L., Böhme, G.: Constraint-based design in participatory housing planning. Int. J. Architectural Comput. **6**(1), 97–117 (2008)
4. Elmasry, M.H., Farid, E.M.: User participation and mass customization as key factors in the future residential building. In: The 3th International Conference of the Arab Society for Computer Aided Architectural Design (ASCAAD 2007) Proceedings (Alexandria), pp. 327–342 (2007)
5. Dortheimer, J., Neuman, E., Milo, T.: A novel crowdsourcing-based approach for collaborative architectural design. In: ECAADE 2020 (Berlin), vol. 2, pp. 155–164 (2020)
6. Heckmann, O., Budig, M., Xuereb, C.Z., Cheng, R., Lo, T.T.: User-driven parcellation of high-rise units for future urban habitation - participatory computational design tools for future urban habitation. In: CAADRIA 2020 (Bangkok), vol. 1, pp. 751–760 (2020)
7. Chow, K.L., van Ameijde, J.: Generative housing communities - design of participatory spaces in public housing using network configurational theories. In: CAADRIA 2020 (Bangkok), vol. 1, pp. 283–292 (2020)
8. Ma, C.Y., van Ameijde, J.: Participatory housing: discrete design and construction systems for high-rise housing in Hong Kong. In: The 26th CAADRIA Conference Proceedings (Hong Kong), pp. 271–280 (2021)
9. Tian, L.T., Schnabel, M.A.: Multi user online design environment for mass housing - a collaborative decision-support tool. In: The 34th eCAADe Conference Proceedings (Oulu), p. 7780 (2016)
10. Buš, P.: Large-scale prototyping utilising technologies and participation - on-demand and crowd-driven urban scenarios. In: The 37th eCAADe and 23rd SIGraDi Conference Proceedings (Porto), pp. 847–854 (2019)
11. Brown, A.G., Knight, M.W., Chen, Y., Saeed, G.: City information delivered to Mobile Digital devices - reflection on contemporary potentials and problems. In: The 24th eCAADe Conference Proceedings (Volos), pp. 146–150 (2006)

12. Langenhan, C., Alexander, S., Frank, P., Arne, S., Astrid, T.: Mobile application to collect information about architecture to obtain a collective knowledge base: 'ar:searchbox.app'. In: The 7th International Conference Proceedings of the Arab Society for Computer Aided Architectural Design (ASCAAD 2014 Kingdom of Saudi Arabia), pp. 375–382 (2014)
13. Fukuda, T., Kaga, A., Izumi, H., Terashima, T.: Citizen participatory design method using VR and a blog as a media in the process. Int. J. Archit. Comput. **7**(2), 217–233 (2009)
14. Chernyakova, I., Villeré, M., Casalegno, F., Giusti, L., Schladow, Z.: Civic media platforms and participatory urbanism: a critical reflection. Int. J. Archit. Comput. **10**(2), 253–273 (2012)
15. Loyola, M., Rossi, B., Montiel, C., Daiber, M.: Use of virtual reality in participatory design. In: The 37th eCAADe and 23rd SIGraDi Conference Proceedings (Porto), pp. 449–454 (2019)
16. Chen, Z.R.: The guidance system of gamification and augmented reality in a museum space. In: CAADRIA 2020 (Bangkok), vol. 1, pp. 671–680 (2020)
17. Roy, N., Dubé, R., Després, C., Freitas, A., France, L.: Choosing between staying at home or moving: a systematic review of factors influencing housing decisions among frail older adults PLOS One **13**(1) e0189266 (2018)
18. MacQueen, J.: Some methods for classification and analysis of multivariate observations. In: Proceedings of the Fifth Berkeley Symposium on Mathematical Statistics and Probability, vol. 1, no. 14, pp. 281–297 (1967)
19. Cloud Computing, Big Data: Present and future of integrated applications of BIM, cloud computing, big data and Internet of things. J. Graph. **39**(5), 806 (2018)

The Interdisciplinary Approach to Free-Form Canopies Optimisation in Terms of Geometrical and Structural Logic Design

Anna Stefańska[1(✉)], Małgorzata M. Kurcjusz[2], Marta R. Cygan[2], Justyna A. Buczkowska[2], Katarzyna M. Szmołda[2], and Paulina Morawska[3]

[1] Department of Structure Design, Construction and Technical Infrastructure, Faculty of Architecture, Warsaw University of Technology, Warsaw, Poland
anna.stefanska2@pw.edu.pl
[2] Faculty of Architecture, Warsaw University of Technology, Warsaw, Poland
[3] Faculty of Architecture, Cracow University of Technology, Kraków, Poland

Abstract. The formulation of tasks applies to the transfer of information, analysis, and conclusions on the achievements of architecture based on new solutions in terms of aesthetic, economic, and practical aspects. The investigation process is dedicated to searching for an optimal freeform canopies' geometry of temporary pavilions, in which parametric tools were used. The research aimed to study different spatial construction variants based on four connected double-curved surfaces, considering minimising the total weight while maintaining the strength of the structure. The results present new possibilities to create more efficient and sustainable canopies. The key findings show which geometric variables greatly influence the final performance of the shape of the area of double-curved geometry and its need to be explored for a better understanding of this type of structure. The results show the benefits of implementing structural optimisation at the early design stage, linking to aesthetically attractive and structural logic design.

Keywords: Architectural engineering · Architectural design optimisation · Conceptual architecture · Double-curved shape · Structural logic · Temporary structures

1 Introduction

An interdisciplinary approach to the design of free geometries, following structural logic at an early stage of design, influences the aesthetic expression of contemporary forms. The use of today's advanced tools facilitating the design and manufacturing process thanks to multivariate analyses gives the possibility to optimise the stiffness or deformability of structures and the process of their manufacture or the use of materials. It is crucial because all these elements contribute to minimising the carbon footprint and greenhouse gas emissions [1], which are among the main parameters of sustainable design.

P. Mendonça and N. D. Cortiços (Eds.): ICAMC 2021, LNCE 226, pp. 110–119, 2022.
https://doi.org/10.1007/978-3-030-94514-5_12

The ability to transfer loads on long-span bar structures without additional support is challenging for architects and builders. Modern interdisciplinary design demands research to find solutions to optimise costs, materials, working time, and energy required to carry out a project. Searching for optimal shape through structural, geometric, or fabrication analyses requires multivariate solutions and considerable interdisciplinary knowledge. Modern ADO paradigms using numerical methods of form optimisation, such as dynamic relaxation, force density method, or particle spring systems, allow optimisation of pre-determined forms by the architect [2]. The structure optimisation methods listed above are only elements of the complete topological optimisation that modern architects can use. The search for supporting tools for structural calculations helps designers return to the experimental design, such as the medieval craftsmen, the search for chain curve geometry by Antoni Gaudi or chain surfaces by Frei Otto and Heinz Isler. They all searched experimentally, by trial and error, not only for the most optimal geometries but also investigated the properties of materials such as stone, concrete or membranes. Examples of the search for purely geometrical optimisation of mathematical surfaces, such as the paraboloid using concrete, can be found in the objects designed by Felix Candela or in the lattice roof canopies of Stephan du Chateau. Design according to mathematical principles and geometry is known from antiquity, where the Greeks described a sphere as the most perfect solid [3]. The subsequent search for non-Euclidean geometries gave rise to the analysis of indecomposable solids, working best as shell structures such as a sphere or a saddle, which, thanks to the pre-curvature of the geometry, allow better transmission of the stresses appearing in the form by minimising bending moments.

The interdisciplinary architectural and technological design has its roots in a definition adapted from the field of cybernetics, as all processes: it includes the design of algorithms and information systems. The search for optimal structures also involves using materials following their characteristics, the ability to use today's high-tech digital fabrication machines and computer programs that reduce the problem of man-machine communication. The architects' way of working has not changed; the basic knowledge is how to use the available tools. In line with the Fourth Industrial Revolution, one more factor influencing the way and quality of design is visible - it affects the surrounding environment.

The paper presents the theoretical issues of optimising rod-shaped structures using mathematical models and analysing structures based on structural and geometric optimisation.

1.1 Sustainability in Architectural Design

The growing awareness of the impact (especially of the architectural and construction sector) on the environment and the state of utilisation of the resources generates an increasing discussion on the potential for optimising structures from an early design stage. The search for lightweight systems for material efficiency is interdisciplinary to architectural and structural design [4]. Developing technologies towards sustainable development equated with energy-saving, cost-effective, and environmentally friendly is essential in computer-aided design. The search for materials and technologies corresponding to these issues has led researchers to look to Nature for inspiration. The fundamental parameter of bionic patterns is the ability to evolve for form and self-organisation, searching for an equilibrium of physical forces. Optimisation using FEM or BESO allows infinitely complex calculations; however, many bionic inspirations adapting to environmental variables are statically indeterminate. The search for optimal structures is understood as an investigation of the best systems in the set of acceptable solutions to the projected goal. Objectives encountered in Nature and used in the architectural design include minimising weight, own vibration or manufacturing cost and maximising stiffness. As an example, the chain curve of a spider web structure is often used by designers. By constantly changing its geometry, support points, and surface area; this structure is not an optimal system from the materials engineering or structural statics point of view. But it provides a guideline for the search for a direction of geometry design based on the equilibrium state of forces with the prevailing loads. Another example of Optimal Structures may be the largest cantilevers, i.e. dendriform. Despite their shape following the moment diagram and varying in thickness from the trunk to the highest branches, they only resemble the arboreal structures designed in contemporary architecture. The structural logic of these bionic patterns follows mainly the response to horizontal wind loads, whereas human-designed arboreal structures follow the distribution of vertical loads. Bionics is a source of shape inspiration and primarily learning from various evolutionary algorithms selected over millennia and nature-based technical solutions. The ways used by living organisms to adapt to the conditions and requirements of the environment allow for minimisation of the energy, material and production time used, which in light of the requirements of sustainable development, not to "compromise the ability of future generations to meet their own needs" (WCED 1987) is also becoming a fundamental objective for the construction sector.

1.2 Interdisciplinary Approach to Carbon Footprint Reducing in AEC Sector

The construction sector is responsible for the increased consumption of natural materials and affects the environment by emitting greenhouse gases. The search for rational use of limited materials is turning researchers' attention towards Nature. Nature, through evolution, has followed the path of sustainable development, making effective and efficient use of available raw materials. Due to the high consumption of materials in construction, architects have begun to look at the natural world for aesthetic inspiration together with the patterns and algorithms responsible for shaping organic structures in Nature. The need for changes in the use of materials and thinking about a building in the context of its whole life (cradle-to-cradle or even cradle-to-grave) is enshrined in legal regulations such as the E.U. development strategy 'Europe 2020', the Aarhus Convention, the Helsinki Convention or the Convention on Biological Diversity.

The awareness of the construction aspect and its multithreaded process, which can positively influence the work of geometry while maximising the optimisation of the structure, has the potential to provide greater rationality in the use of material or the weight of the structure itself, which can potentially lead to better performance in the field of the rationality of use of material resources and CO2 emissions already at the production stage. The search for optimal solutions is increasingly borrowing inspiration from bionics as potentially the best way to cope with the development of civilisation, the reduction of environmental contamination and the depletion of non-renewable materials [4]. The search for minimising the consumption of greenhouse gases and carbon dioxide draws attention to the design method and the materials used. The main strategies of designing the following minimising the negative impact on the environment are the selection and use of materials, the way of production, reusing and recycling the material, increasing the lifespan of products and their components, and the possibility of recycling [5]. However, regardless of how the carbon footprint is determined, its minimisation becomes an important boundary parameter during the design process. Interdisciplinary teams combining architects, designers, biologists, and materials engineers consider greenhouse gas emissions when developing concepts for innovative buildings.

2 Background

2.1 Architectural Design Using Multivariate Optimisation

Design is a decision-making process. Contemporary architectural design, using computational tools, is not just about creating a shape using CAD tools, but also formulating algorithms, information systems, controlling the process of creation so that it is possible to understand the functions, techniques of control mechanisms and communication between elements, similar to how living organisms, social or environmental systems work. With such a comprehensive design, decisions are more subtle and precise than analogue design. The modern computer-aided design allows information to be processed and transferred between machine and human, information awareness of design decisions increases. The availability of tools for multivariate optimisation, based on an interdisciplinary environment, changes the way architects work, requiring them to be more aware of the intended effects, the task of boundary conditions and the search for the best, from the point of view of the assumed goal, optimisation of structures in a set of sufficient structures. Generative and parametric design support this process with the ability to design multi-criteria optimisation algorithms, accelerating the ability to decisive designing steps.

Using computational capabilities makes it possible to design complex curvilinear geometries and carry out multi-variant optimisations, strive for typification of solutions and digital fabrication, and use non-standard materials. Multi-variant optimisations are still an extremely complex issue in interdisciplinary design, and research conducted into the possibilities of their implementation on an architectural scale is usually undertaken on the example of small buildings and experimental pavilions [6–9].

The increasing availability of structural calculations in software designed to shape curvilinear geometries makes it possible to use FEM in form-finding. Examples of this include Grasshopper/Karamba or Robot/Dynamo.

2.2 Technological Background for Architectural Optimisation

Analysing ancient Greek temples or Gothic cathedrals, one can see structural logic in shaping architectural orders. The detailed division into massive load-bearing elements and infill was related to the characteristics of the architectural styles of each era. The shaping of architectural forms over the centuries followed the technological possibilities. Structured logical actions representing contemporary experimental architecture often refer to these geometric patterns by imitating the shape or topological specificity [10]. Technical and material developments have influenced the specifics of freeform design. Visible in the 1970s, experimental methods were used to find geometries with optimum distribution of forces. All these shapes were achieved by empirical methods, which today, when developed, enable precise optimisation in the Finite Element Method. Structural optimisation is one of the primary methods of form-finding. Geometric optimisation is another crucial method used in morphogenesis and is used in searches based on mathematical algorithms. These methods use an intuitive choice of parameters and output guidelines, usually preserving good conditions for fabrication [11], and were used in this study.

3 Research Methodology

3.1 Mathematical Bases for Structure Modelling

The search for the optimisation of curvilinear structures in the last century has been mainly related to the exploration of moment-free shells and searches based on the geometric shaping of domes and biconvex surfaces. The search for free-form geometries is evident in reinforced concrete shells, steel or wooden gridshells, and material minimum surfaces. The material properties were already considered in the initial design stage in the types mentioned above, facilitating optimisation processes. Today's implementation of digital technologies allows an interdisciplinary exchange of information, supporting the development of research, fabrication, and use of facilities.

The search for structural forms based on geometrically determinable shapes simplifies fabrication processes, which can be seen in conoids, hypersurfaces or spheres. The apparent interest in the use of hyperbolic parabolas is linked to the achievement of aesthetic and light spatial forms, both in surface coverings (Fig. 1a) and rod structures (Fig. 1b). Nowadays, the search for architectural geometry is a broad branch of interdisciplinary research, especially in fabrication and the associated production and assembly costs. The use of bionic patterns in shaping is not always combined with the possibility of directly implementing them in buildings. The use of simplifications in the design phase, for example, allows the use of flat panels in structural canopies, the unification of nodes and the possibility of using repetitive joints, achieving smooth aesthetic surfaces [12].

Fig. 1. The use of mathematical surfaces in architectural objects, a) Covering the ticket hall of the Warsaw Ochota railway station, 1962, b) The Varberg Observation Tower, 2012.

The topological optimisation and the search for lightweight bar structures through variable round cross-sections allow the minimisation of material and weight. The study of pavilion structures, thanks to the simplified function, allows minimising the boundary parameters necessary in the design [13, 14], often to issues related only to material and structural characteristics and the way of construction of the geometry. The research's inspiration was the two-curvature geometries of mathematical surfaces based on a quadrilateral plan. All geometries consist of saddle surfaces, built by lifting two quadrilateral corners to variable heights and filling the resulting non-flat border with Coons' patch (Fig. 2a). The resulting curves were duplicated so that the whole structure was supported on four symmetrical supports (Fig. 2b, 2c). The resulting figure can change the middle point's height and the symmetrical cantilevers' height independently, analogous to the covering of the L'Oceanogràfic café in Valencia. This shell shape allows bending, and the material used to be minimised, making the execution more economical and sustainable. The form itself can also be easily generated by computer programs [15]. The resulting rod structures with saddle surfaces can also be easily interpolated, and simple lattice bars of repeatable length can be used.

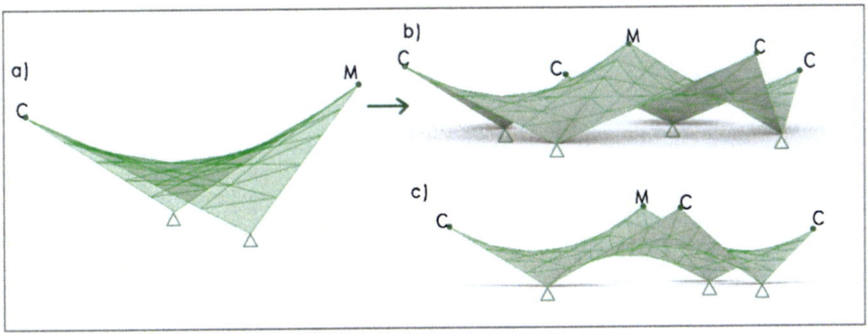

Fig. 2. The geometry formation a) quadrilateral mesh with M and C points elevated to variable heights and the resulting contour filled with Coons' patch, b) duplication of the structure about point M, to create four supports, c) side view

3.2 Case-Study Form-Finding Factors

The study was conducted based on the following assumptions:

- pavilion canopy, based on a square projection of 625 m^2;
- structure based on four supports located at half-length of the sides of the square;
- homogeneous material used: S355 steel with variable tubular section;
- following the functional requirements (the possibility to use it as an exhibition pavilion without additional column) and aesthetic conditions (light and slender form), the height of the offset of the pavilion centre (point M) and the cantilever (points C) was limited to 4.0–8.0 m (Fig. 2);
- Structural calculations, dimensioning of elements were carried out according to E.C. with the combination of loads (dead load, service load, snow and wind load);
- all structures were equally divided into a triangular bar mesh for easy preparation of the flat panels of the cover (Fig. 2a).

The mesh forming the canopy was designed from polycarbonate so that the allowable global deformations were not exceeded in any of the structures studied. In the analyses carried out, the parameters studied were the index of the minimum total weight k_i (unitless) and the total weight related to the total length of all the bars of each variant (kg/m).

A factor k_i defines the proportion of the weight of particular geometries to the first variant, according to the formula:

$$k_i = \frac{k_n}{k_{ref}}$$

where:
k_i - weight factor of the tested structures (unitless)
k_{ref} - weight of variant with a total height of the Middle and Cantilever of 4.0 m [kg]
k_n - weight of structures of the n-variant in kg.

4 Results and Findings

4.1 Reticulated Saddle Surface Canopy Results

The comparison of all the variants in the study (Table 1) shows the tendency that the lightest structures are found among the geometries with centre point heights of M = 6.0 m and M = 7.0 m, with cantilever heights of 4.0–7.0 m. The weight parameter related to the total length of bars (Table 2), with the increasing length of bars as the curvature of the structure increases, shows the preserved trend of the best results in the same group of solutions (Tables 1 and 2).

Table 1. Results of various curvatures adopted to the initial grid, Weight factor of the tested variants k_i (unitless)

	C 4,0	C 5,0	C 6,0	C 7,0	C 8,0
M 4,0	1,00	1,00	0,92	0,93	0,97
M 5,0	0,94	0,91	0,92	1,05	0,94
M 6,0	0,90	0,90	0,91	0,92	1,04
M 7,0	0,92	0,91	0,91	0,91	0,92
M 8,0	0,94	0,92	1,04	0,92	1,05
M 9,0	0,96	0,93	0,93	0,93	0,94

Table 2. Results of various curvatures adopted to the initial grid, weight to the total length of all the bars in the canopy in the variant m_i kg/m

	C 4,0	C 5,0	C 6,0	C 7,0	C 8,0
M 4,0	9,63	9,54	8,71	8,71	8,96
M 5,0	9,00	8,71	8,72	9,79	8,69
M 6,0	8,56	8,55	8,54	8,55	9,56
M 7,0	8,66	8,55	8,47	8,48	8,46
M 8,0	8,72	8,57	9,63	8,49	9,57
M 9,0	8,83	8,56	8,57	8,49	8,50

As a result of the analysis of two-curvature geometries based on saddle surfaces, it is apparent that optimum solutions are achieved in the medium curvature range of the structures. The symmetrical height of points M and C show some slight possibilities of achieving lower total weight than those where the cantilever height is higher than the centre point.

4.2 Findings

Analysis of selected structural and material solutions and fabrication in terms of production speed, material consumption, and the energy consumption is essential in the preliminary design process. The use of boundary conditions to control architectural requirements such as proportionality parameter, size limitation leads to individualised aesthetic solutions. This is particularly interesting from the point of view of prefabrication, which is now more often characterised by the idea of post-Fordism, where the desire for unification does not exclude the creative search for individual solutions.

5 Conclusion and Discission

The practical use of interdisciplinary tools supporting architectural design with unprecedented precision minimises the cost, used materials, and energy to preserve the unique character of objects [16]. This represents a challenge as well as the opportunity for architects and designers, allowing them to return to the possibility of working experimentally, along the lines of the explorations carried out by Frei Otto, among others, but with the possibility of using accurate multivariate optimisations usually calculated in real-time to build a script. Architects' limitations today are understanding multiple processes and multivariate optimisations whose boundary parameters often compete with each other [13], giving conflicting solutions.

References

1. Padilla-Rivera, A., Amor, B., Blanchet, P.: Evaluating the link between low carbon reductions strategies and its performance in the context of climate change: a carbon footprint of a wood-frame residential building in Quebec, Canada. Sustainability **10**(8), 1–20 (2018). https://doi.org/10.3390/su10082715
2. Kilian, A., Ochsendorf, J.: Particle-spring systems for structural form-finding. J. Int. Assoc. Shell Spat. Struct. **46**(148), 77–84 (2005)
3. Tobolczyk, M.: Starożytne wzorce harmonii opartej na prostych bryłach geometrycznych w architekturze współczesnej. **61**(4), 61–71 (2018)
4. Knippers, J., Nickel, K.G., Speck, T. (eds.): Biologically-Inspired Systems Biomimetic Research for Architecture and Building Construction, 1st edn. Springer, Heidelberg (2016)
5. Sipahi, S., Kulözü-Uzunboy, N.: A study on reducing the carbon footprint of architectural buildings based on their materials under the guidance of eco-design strategies. Clean Technol. Environ. Policy **23**(3), 991–1005 (2021). https://doi.org/10.1007/s10098-020-02009-4
6. Meisel, N.A., Watson, N., Bilén, S.G., Duarte, J.P., Nazarian, S.: Design and system considerations for construction-scale concrete additive manufacturing in remote environments via robotic arm deposition. 3D Print. Addit. Manuf. 1–11 (2021). https://doi.org/10.1089/3dp.2020.0335
7. Aldinger, L., Margariti, G., Körner, A., Suzuki, S., Knippers, J.: Tailoring self-formation fabrication and simulation of membrane-actuated stiffness gradient composites. In: Proceedings of the IASS Symposium 2018 Creativity in Structural Design, vol. 141, no. July (2018)
8. Pottmann, H.: Architectural geometry and fabrication-aware design. Nexus Netw. J. **15**(2), 195–208 (2013). https://doi.org/10.1007/s00004-013-0149-5
9. Kormaníková, L., Kormaníková, E., Katunský, D.: Shape design and analysis of adaptive structures. Procedia Eng. **190**, 7–14 (2017). https://doi.org/10.1016/j.proeng.2017.05.300
10. Słyk, J.: Źródła architektury informacyjnej (eng. The origin of information architecture). Oficyna Wydawnicza Politechniki Warszawskiej, Warsaw (2012)
11. Tellier, X.: Morphogenesis of curved structural envelopes under fabrication constraints. Universite Paris-Est Ecole doctorale Science, Ingenierie et Environnement, Paris (2020)
12. Wallner, J., Pottmann, H.: Geometric computing for freeform architecture (2014). https://doi.org/10.1186/2190-5983-1-4
13. Chiandussi, G., Codegone, M., Ferrero, S., Varesio, F.E.: Comparison of multi-objective optimisation methodologies for engineering applications. Comput. Math. Appl. **63**, 912–942 (2012)
14. Chen, Z.-R., Lim, C.-K., Shao, W.-Y.: Comparisons of practice progress of digital design and fabrication in free-form architecture. J. Ind. Prod. Eng. **32**(2), 121–132 (2015). https://doi.org/10.1080/21681015.2015.1023853

15. Sulpizio, C., Fiore, A., Demartino, C., Vanzi, I., Briseghella, B.: Optimal design criteria for form-finding of double-curved surfaces. Procedia Manuf. **44**(2019), 28–35 (2020). https://doi.org/10.1016/j.promfg.2020.02.201
16. Dixit, S., Mandal, S.N., Thanikal, J.V., Saurabh, K.: Evolution of studies in construction productivity: a systematic literature review (2006–2017). Ain Shams Eng. J. **10**(3), 555–564 (2019). https://doi.org/10.1016/j.asej.2018.10.010

Construction Expression of the Pantadome System

Yukun Zhai(✉)

The Chinese University of Hong Kong, Shatin, Hong Kong SAR, China
zhai-yukun@link.cuhk.edu.hk

Abstract. The Pantadome system is a fast and efficient construction method invented by engineer Kawaguchi. In 1990, it was used by the architect Isozaki in the construction of the Palau Sant Jordi for the Barcelona Olympic Games, creating a powerful architectural image. This paper analyzed the construction process and technical characteristics of the Palau Sant Jordi, revealing that the architects enhance the expressiveness of the building by shaping the characteristics of the construction system, and echo the environment of the place.

Keywords: Pantadome · Palau Sant Jordi · Arata Isozaki · Mamoru Kawaguchi · Construction · Expression

1 Introduction

On May 25, 2019, the Pritzker Prize award ceremony was held at the Palace of Versailles in France, and the highest honor in the architectural world was awarded to the Japanese architect Arata Isozaki (1931-). Just 4 days later, the famous Japanese engineer Kawaguchi (1932–2019) passed away suddenly and the master fell. This cannot help but recall the many wonderful collaborations between the two, of which the Palau Sant Jordi (1983–1990, See Fig. 1) at the 1992 Barcelona Games is the most well-known.

Fig. 1. Palau Sant Jordi, Arata Isozaki, 1990.

P. Mendonça and N. D. Cortiços (Eds.): ICAMC 2021, LNCE 226, pp. 120–127, 2022.
https://doi.org/10.1007/978-3-030-94514-5_13

2 Question of Architectural Form

When seeing the exterior image of the Palau Sant Jordi, we may find that its architectural form is very different from the complete and pure volume of a general stadium. The upper and lower parts are composed of three parts, and the roof is cut into four pieces. The plan geometry of the building is also between a square and a circle, and the section is like a dome with "defective" designs. This can't help but make us question: As the Olympic venues that represent the image of the city, why are they adopting such a unique form? What are the considerations for adopting this form?

Architectural form is usually closely related to structure, function, space or culture. Therefore, before we find the answer, we might as well make the following hypothesis. First of all, for large-span stadiums, structural issues are usually given priority. It may be due to the reasons of increasing the structural strength, reducing the deformation and enhancing the stability, the most reasonable form is adopted according to the force calculation. Secondly, the dome may be divided into multiple parts for function and space needs to form a combination of large and small spaces. In addition, it may also be related to construction. Compared with pure geometrical volumes, multi-part forms seem to be easier to construct. Finally, it is purely for the purpose of form, to pursue a certain cultural or symbolic expression. Following section will verify the above hypothesis one by one.

2.1 Verification of Structure Hypothesis

The Palau Sant Jordi adopts a double-layer spatial lattice shell structure. The plane size is 128 × 106 m and the lattice shell height is 21 m. The distance from the ground is 45 m [1]. The reticulated shell structure is composed of 5 parts. The central dome is about 80x60m, supported by 4 surrounding parts. It can be clearly seen from the cross-sectional view (See Fig. 2) that this structural composition is in an "imperfect" arched state.

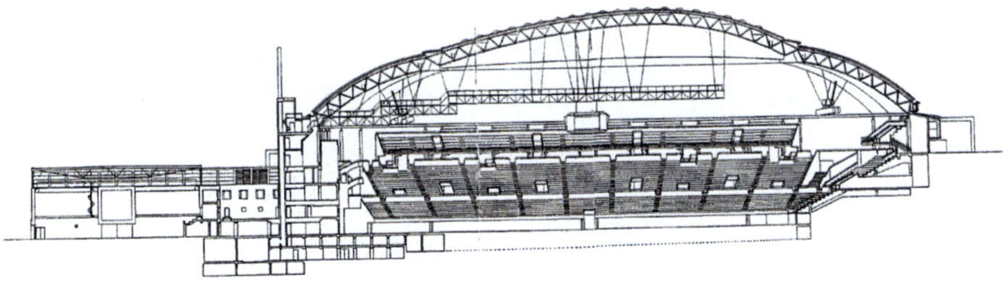

Fig. 2. Section, Palau Sant Jordi.

If try to raise the middle part of the reticulated shell, it can get a complete and continuous arch. Comparing the force of the fulcrum of the reticulated shell in the two states, because the angle between the reticulated shell and the horizontal plane is smaller in the "imperfect" state, it will produce greater lateral thrust, which is not ideal for the force of the supporting point. In addition, if it is a complete and continuous spherical reticulated shell, under load, the force will flow uniformly along the structure. At this time, the structural strength is higher and the stability is better. But when the structure changes locally, the changed part is likely to produce relatively large local stress. Therefore, the structural connection part of the roof reticulated shell of the Palau Sant Jordi is relatively fragile, and the structural form is not completely reasonable. According to the knowledge of graphical statics, the most reasonable arch form should be the overturn of the natural catenary under uniform load, just like the suspension model made by Gaudí in the design of the Sagrada Familia. But the Palau Sant Jordi did not follow this rule. Moreover, the hinged nodes used when connecting the various parts of the dome will also affect the stress and deformation of the structure. Although the hinge point can reduce the damage of uneven settlement to the structure to a certain extent, it reduces the overall rigidity of the structure. When the structure is subjected to external forces, it will deform more.

To sum up, the special form of the Palau Sant Jordi does not bring structural benefits, so its special shape is not considered for structure.

2.2 Analysis of Form and Spatial Factors

When the architect Shin Isozaki participated in the design competition in 1983, the plan was different from the one built in 1990. The roof of the competition proposal is like a ribbon, covering the half of the stand, with a slight bulge in the middle and both ends of the roof, and a depression between the two, forming a soft curve. This curve coincides with the Montjuic hill where the building is located. The completed plan retains the convex curve feature in the middle, and designs a skirt curve at the bottom to coordinate with the site environment (See Fig. 3). Across Montjuic, not far away is the Mediterranean Sea. The stadium echoes the background of mountains and sea with slight bulges and wave-like curves. For an Olympic venue that represents the city, the architect adopted a form that can represent and integrate into the city, so that the building belongs here.

Fig. 3. Sketch, Palau Sant Jordi.

The slightly protruding part of the dome of the building corresponds to the sports field, and the depressed periphery of the dome echoes the auditorium (there is no auditorium on the south side). The internal space is like two domes buckled together in mirror image, the size and function of the space are unified. In the middle dome, there are 10 $\times$ 5 rows of lighting holes in a staggered design. The holes are located just above the structural grid, and the bright middle part becomes the focus of vision. The lighting strips at the junction of the dome structure highlight the area division between the center and the edge (Fig. 4).

Fig. 4. Interior, Palau Sant Jordi.

It can be seen from the above that the special form of the dome echoes the characteristics of the internal space and reflects the characteristics of the natural environment of the site, making the expression of the form seem natural. However, are there other strong reasons why architects can change the mature structural system and adopt new structural forms?

2.3 New Construction Method: Pantadome System

From a construction point of view, large-scale "perfect" geometric shapes are not easy to construct. The early pyramids were in the form of steps or twists because it was difficult to form a perfect pyramid. The dome of the Cathedral of Santa Maria del Fiore consists of eight petals and sits on an octagonal drum base, rather than a pure circular dome. Similarly, the construction of large curved reticulated shells is not easy. A large number of high-altitude constructions make it difficult to control accuracy and reliability. Although, under the existing technical conditions, the construction of space reticulated shells has become relatively easy through the cooperation of scaffolding supports and elevators and other equipment. However, construction efficiency, economy and safety need to be improved.

In the design of the Palau Sant Jordi, the dome is divided into multiple parts, which can effectively reduce the difficulty of construction. Each part can be constructed separately and then assembled. In fact, the surrounding part of the dome is composed of 18 relatively small space grids, all of which are constructed on the ground, which greatly reduces the difficulty. But what followed was a problem: how to assemble these small parts into a large complete dome in the air?

To solve this problem, the engineer Kawaguchi invented a method of connecting these parts to raise the whole: the pantadome system (See Fig. 5). In fact, this is a system that includes both structure and construction methods. The previous decomposition and construction process is also part of this system. Before lifting, put the completed center part on a hydraulic lift. Then, it is hingedly connected to the surrounding part of the 18 fast, and the other end of the surrounding net frame is hinged to the upright on the base. At this time, each part forms a whole linkage mechanism, and the whole mechanism can only move upward in one direction when the hydraulic press is lifted. The dome structure is gradually "unfolded" to form a tall indoor space. After the "unfolding" is over, the 18 parts around the dome need to be connected and fixed together, and the hoop function is used to form an integral spatial reticulated shell structure. At this time, the hydraulic press is removed, and the dome structure can stand alone.

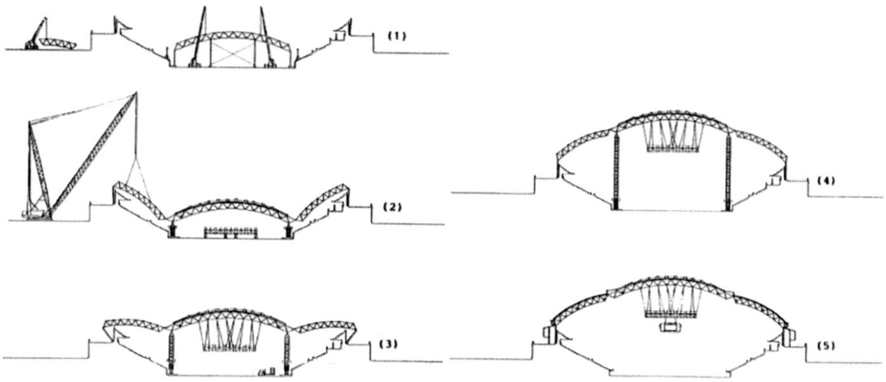

Fig. 5. Construction process, Pantadome system.

From the perspective of the construction process, the application of the Panda dome system has had a great impact on the architectural form, and is even decisive. The architectural form strongly reflects the traces of construction.

3 Question of Architectural Form

Among the possible hypothesis, construction has the greatest influence on form. But is this due to passive reasons of technical limitations, or is it an active choice made by the architect? From the first application of the Pantadome system in 1984, to 2000, there were 6 large buildings constructed by the Panda dome system. They were designed by 4 architects, all of which were made by Kawaguchi as the structural engineer. Among them, only the Palau Sant Jordi clearly shows the unique characteristics of construction in terms of architectural form. Under the same technical conditions, it is not difficult to speculate that this is the intention of the architect to express this form. So what made architects choose to retain the "unfinished" status? How do architects strengthen the expressiveness of shaping construction?

3.1 Expression of Construction

There are six buildings constructed from 1984 to 2000 using the Pantadome structure system: the World Memorial Hall in Kobe, the Singapore Indoor Stadium, the Palau Sant Jordi in Barcelona, Sun Dome in Fukui, Namihaya Dome in Osaka, Nara Centernial Hall [2]. These buildings have various forms and are not restricted by construction methods. The plane changes from a circle to an ellipse to a quadrilateral. The façade height ratio and contour are also very free, some are towering, and some are flat. It can be seen that the form of the Pantadome structure system is not subject to too many constraints and can still be freely shaped. However, compared with the general long-span buildings, most of the Pantadome system buildings have not made a big breakthrough. Only the "imperfect" dome of the Palau Sant Jordi is unique. By retaining the traces of the construction, the architect forms a formal expression of the construction.

The Pantadome system transforms the stable reticulated shell structure into a foldable geometrically unstable body through a series of designs. This system needs to set hinge points along the circumference at least three heights (See Fig. 5), and cut off the transverse connecting rods of these three circumferences. In this way, the reticulated shell will lose the stability of the spatial structure, and can move up and down, fold and unfold. The name of the pantadome comes from the English pantagraph, which refers to a foldable proportional plotter or a pantograph that is raised and lowered by a tram. The name Pantadome vividly expresses its working principle. Since it is a motion mechanism, there will be a variety of intermediate states in the process of motion. Although the final state at the end of the exercise (fully expanded) is the most stable, the intermediate state is more expressive of the movement mechanism. Just as the photographer captures the moment when the athlete is in the air, the architect captures the moment of construction and freezes it.

But there is no doubt that the moment of freezing is designed and practiced. The hinge points divide the dome into three parts: upper, middle, and lower. The upper part is the part that drops the most when folded. It will be suitable for construction on the ground, and the size of this part is designed by controlling the height. For example, in the Namihaya Dome in Osaka, because the upper dome has a smooth surface and small height, which is convenient for construction, the upper part occupies about 3/5 of the total dome. The size of the middle part mainly determines the dropable height of the top. The size of the bottom part needs to meet the movement requirements of the middle part. Because in the process of moving and unfolding, the hinge points of the middle and lower parts will first expand outward and then retract inward, which requires the size of the bottom to meet the requirements of the outer expansion distance. The upper, middle and lower parts are mutually restricted, and the size of each part is related to the height of the dome and the curvature of the curved surface. This makes the Pantadome a whole system, closely related to structure, construction and form.

In the design of the Palau Sant Jordi, the architect used the characteristics of the Pantadome system to create a unique and expressive form. This form is precisely derived from the system itself related to structure and construction. When dividing the folded part, the upper, middle, and lower parts account for a ratio of 4:5:1. The architect deliberately highlights the upper and middle parts and controls them to be approximately equal proportions, while weakening the bottom base. In fact, the 60 vertical columns at the bottom are not part of the space reticulated shell structure, but gently lift the dome as if off the ground. The proper proportion of the dome makes the architectural form not only a powerful expression of construction, but not trivial. If the proportion of the top is too large, the visual focus will be on the dome, losing the capture of the construction; if the proportions of the three parts are the same, the building will be divided too finely, risking becoming a formal game. In addition, the curvature of the roof and the proper degree of expansion also strengthen the expressiveness of the construction form.

3.2 Site-Based Construction

Such a unique form is not only an expression of construction characteristics, but also a response to the construction site. The Palau Sant Jordi is located at the foot of Montjuic mountain, close to the old town of Barcelona. It is a place of extraordinary attraction both in history and topography [3]. It is not easy to build a giant on a site of historical and cultural significance. It is likely to become abrupt or fragmented from the environment. Only buildings integrated into the culture of the place can be accepted by the city.

The streamer-like wave-curved roof in the competition stage has revealed the architect's imagination of the mountains and the sea. Because, crossing the Montjuic mountain to the east from the site is the Mediterranean Sea. In the implementation phase, the architect retained this design intention, but needed to propose specific solutions in terms of construction methods. For a city with a "dome culture", the Pantadome system is an excellent solution, both for construction and implementation, or for responding to the cultural environment. "When the great roof of the Palau Sant Jordi appeared on the slope of the Montjuic mountain, the citizens had the impression that something like a cathedral or a basilica had appeared" [3]. Therefore, this stadium was crowned the patron saint of the city.

The architect is not satisfied with responding to the urban culture with a simple dome, he also allows the building to blend in between the mountains and the sea. "It could have been lifted higher, like any other dome, but it was decided to leave it at its present level. Many invisible efforts were made to give it its flatness. Its line represents the smooth ridge of the small mountains that surround Barcelona like the mountain of Montjuic" [3]. If the Panda dome exactly echoes the mountain, the "kernels" surrounding the dome reflect the sea. Between the mountains and the sea, the Palau Sant Jordi has found the city's belonging through the performance of construction [4].

4 Conclusion

The unique shape of the Palau Sant Jordi is neither a formal or space consideration nor a pursuit of an efficient structural system, but a unique construction method. By capturing the moment of the construction process, the construction is frozen and the construction characteristics are strengthened through design. This kind of performance is not only a rational consideration of construction, but also the response of the site environment. The slight curve of the roof echoes the surrounding hills. The architect unifies the construction and the environment, and echoes the characteristics of the environment by shaping the characteristics of the construction.

References

1. Chilton, J.: Space Grid Structures. Architectural Press, New York (2000)
2. Chen, Z.: Pantadome structure system. Archit. Knowl. (05), 31–32 (2000)
3. Isozaki, A.: A system called architecture: the Palau Sant Jordi. Temes Disseny (07), 107–124 (1999)
4. Isozaki, A., Koshalek, R., Stewart, D.B.: Arata Isozaki: Four Decades of Architecture. Thames and Hudson (1998)

Model Influencing the Success of Performance-Based Design in Construction Project

Sulfiah Dwi Astarini(✉), Christiono Utomo(✉), and M. Arif Rohman

Department of Civil Engineering, Institut Teknologi Sepuluh Nopember, Surabaya, Indonesia
christiono@ce.its.ac.id

Abstract. The design process that involves a multidisciplinary designer is required to work together to produce a design according to performance requirements. Performance Based Building Design (PBBD) can be used in the design process which facilitates the formulation of clear building performance objectives. This is because the PBBD concept emphasizes the concept of how the actual building functions, not the way it is built. The purpose of this study is to examine the effect of perspective designer, attribute designer, interoperability, team management, team support and innovation on successful practice of PBBD. This study uses a questionnaire survey in data collection, and involves expert designers who have been involved in the design process. A 5-point Likert scale is used to assess variables. To find out the correlation of each variable, regression analysis was then developed in the formation of a conceptual model. Based on the results of the study, it was found that there was a positive relationship between the variables of interoperability, team support and innovation on the successful practice of PBBD. Meanwhile, there are two variables that are negatively related, namely, perspective designer and innovation. The findings of this study can be used as a basis for introducing PBBD practices and as a first step in the formulation of a conceptual model of performance-based design in construction projects.

Keywords: Performance design · PBBD · Interoperability · Team support · Innovation · Construction

1 Introduction

The construction design phase has been known as a complex social activity where the activities involve multidisciplinary designers to complete the design work. In general, construction project performance is still based on cost, time and quality performance [1], although researchers such as [2] add performance namely safety, and client satisfaction. However [3] criticized those performance for proving weak and especially if the project objectives were not clearly defined. Mellado et al. [4] also mentioned that in construction projects, the current performance is not focused on the specific design, but rather on the project or the company as a whole. Performance-Based Building Design is a design concept that can be used, where the performance objectives of the design are stated more explicitly. However, current research has not discussed much about PBBD, especially regarding the influence of individual and design team support on the success

P. Mendonça and N. D. Cortiços (Eds.): ICAMC 2021, LNCE 226, pp. 128–137, 2022.
https://doi.org/10.1007/978-3-030-94514-5_14

of its practice. Thus, this study discusses the factors that can influence the practice of PBBD to achieve success.

In the second part, this paper presents a review to describe the performance of the design as well as some of the supporting factors that are considered to have an influence on the success of the design among several researchers. The supporting factors that are formulated and considered to have an effect on the success of performance-based designs are the designer's perspective; designer attributes; interoperability; team management; team support; and design innovation. The research method used is presented in the third section, the research results are discussed in the next section and the conclusions are described in the last section.

2 Conceptual Background

2.1 Performance in Design

Project objectives are related to client satisfaction and end-user needs for the function of the building [5, 6]. PBBD describes how a building operates and is able to support the activities of its users. Support for occupant activities includes various required performance criteria, for example indoor air quality and occupant safety [5]. According to [7] because the performance formulations are interrelated, it is necessary for the designer to consider various aspects of performance simultaneously to translate and ensure that the requirements or needs of clients and end users can be met so as to be able to achieve project objectives. The designer's collaborative and integrated role can facilitate successful practice in a performance-based design.

2.2 Perspectives of Designer

The involvement of multiple multidisciplinary designers is required to complete design work that cannot be completed by a single designer. This is mainly related to the designer's perception of translating the needs of clients and end users of the building [5]. According to [8] in the conceptual design phase, the designer elaborates a lot of ideas in exposing the client's needs to jointly create a solution based on his scientific perception. Adopting more perceptions from the designer is done to obtain the best information and help the designer get a holistic view of the object to be designed. When designers use their experience in completing design tasks, designers with various perspectives can share a common understanding, so that the design process will improve and be able to affect the quality of the final result [9].

2.3 Attribute of Designer

The designer plays an important role during the design phase. Researcher [10] stated that the designer's role is to be responsible for the timely delivery of designs, so that designers are required to produce drawings or design documents quickly [11]. In addition, the designer also ensures that the final product of the building must be able to meet the quality of building performance as described in the design [10]. Supporting this case, designer attributes are needed to encourage designers to achieve optimal performance and contribute to the success of design work [11]. Researcher [10, 12] mentions the designer's attributes, including flexibility in collaborating, the ability to respond quickly to job changes, and the designer's skills, such as problem-solving processes and decision-making efficiency. Meanwhile [13] argues that the designer's attributes are the quality of the designer's work, such as the completeness of drawings and design documents, and the proposed design solution is able to meet technical standards.

2.4 Interoperability

One of the characteristics of the design phase is the exchange of data and information between the various designers involved. The data or information exchanged is useful to minimize ambiguity, so that it can be understood by the designer. Researcher [14] states that design ambiguity can arise if there is loss of data or information during the data exchange process. Usually this happens because of differences in the use of work systems by each designer [14, 15]. In this case, interoperability systems can be used by designers to work together regarding the exchange of data and information during the design process [16]. This system interoperability capability enables heterogeneous designers to interact and communicate with each other using different platforms in a timely manner. Especially if there is a design change that can affect the designs of other designers during design development, so that any changes or modifications made can be immediately understood by other designers that affect their designs [15]. Therefore, [14] mentions that interoperability can improve the quality of design results and work processes more efficiently.

2.5 Team Management

A multidisciplinary team of designers is an important aspect of the design process where team members must adapt assigned tasks and manage work interdependencies. Managing the design team well will contribute to long-term relationships due to the formation of interpersonal relationships between designers, creating a shared culture, and understanding of project goals [17]. According to [18] a good relationship between designers includes finding solutions to any problems found, instead of blaming each other. The project culture can be fostered by getting to know each other from the initial design phase so that each individual can agree on a common goal before starting work on the construction site.

2.6 Team Support

The design process is an activity that uses a lot of knowledge, both theoretical and derived from the designer's years of experience [19]. An important strategy for a successful project is to create a work environment where all members of the various skills and experience who are part of the team can exchange information freely, and can express their point of view to provide practical and innovative solutions [20]. One form of support for the design team is the sharing of knowledge between designers [19].

When designers shape knowledge-sharing behavior during design development, they can directly contribute to subsequent project performance [21, 22] because it minimizes the occurrence of lack of knowledge on similar projects. In addition, team support will also encourage integration, because it has a common goal of project results, and ensures that these results can benefit all parties [18].

2.7 Innovation

Innovation which refers to the process of finding and utilizing new ideas to be implemented into a product [23]. Utilization of ideas in the form of designer knowledge in developing creativity during the design process [24]. Researcher [25] shows that the main driver for implementing innovation lies in the project factor where the integration of all participants is required from the early stages of design. To promote successful innovation practices, this process requires collaboration, and effective coordination among all parties involved, this includes designers, clients, contractors, sub-contractors, etc. In performance-based design, innovative practices are needed because of their flexibility to building code codes so as to be able to introduce new methods in design and construction [26]. Successful innovation practices can have an effect on increasing productivity, client satisfaction, and shortening processing time [24, 25].

Based on each variable explanation from previous research, the proposed model can be seen in Fig. 1 model influencing success of PBBD.

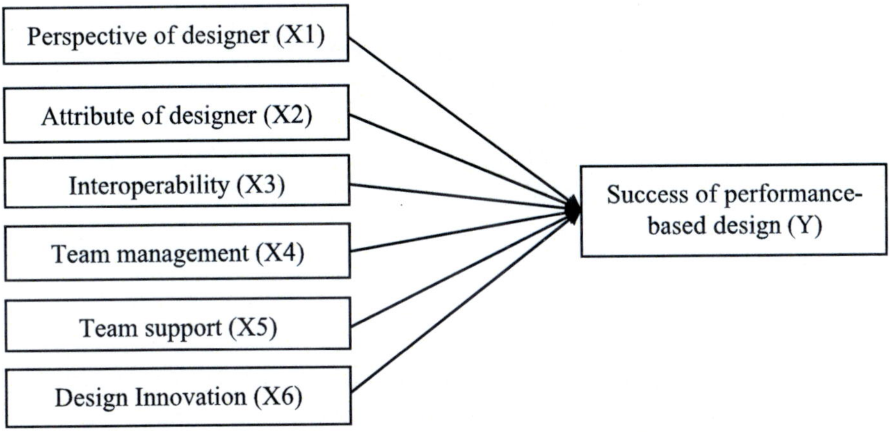

Fig. 1. Model influencing success of performance-based design

From the Fig. 1, six hypotheses are proposed based on the relationship of each independent variable (X) with the dependent variable (Y) which are reviewed as follows:

Hypothesis 1: The designer's perspective has an influence on the success of PBBD practice.

Hypothesis 2: Designer attributes have an influence on the success of PBBD practice.
Hypothesis 3: System interoperability has an influence on the success of PBBD practice.
Hypothesis 4: Team management has an influence on the success of PBBD practice.
Hypothesis 5: Team support has an influence on the success of PBBD practice.
Hypothesis 6: Innovation has an influence on the success of PBBD practice.

3 Method of Reserach

As a first step in conducting a survey, first, develop a research instrument that will be used as research measuring tool, namely a questionnaire. After the questionnaires were ready to be distributed, the samples were selected. The sample selected is expert respondents who are experienced and have been involved in the design of construction projects, Indonesia. The expert designer selected in this study aims to obtain accurate results on the variables reviewed. Since the respondents involved were from research colleagues working in design consulting firms, the links of research questionnaire were sent by email and messaging applications. The questionnaire contains two parts of questions, where the first part is questions regarding the respondent's data, this consists of the respondent's experience in the design process, and the type of building designed by the respondents. The second part contains the main questions regarding the variables under consideration. Respondents who provided feedback on the questionnaire were then collected and analyzed. Analysis of the data used is a regression analysis of the proposed model. The use of regression analysis is intended to examine the relationship of six factors from the independent variables, namely designer perspective, designer attributes, interoperability, team management, team support, and innovation to the dependent variable, namely the success of performance-based design practices. The results of the analysis are then discussed further.

4 Result and Discussion

4.1 Background of Respondents

A total of 58 questionnaires were distributed, and as many as 25 questionnaires were received for further analysis. The response rate of this study was 75.7%. It was found that respondents who were involved had experience less than 10 years were 44%, respondents with experience more than 10 years were 32% and respondents with experience more than 20 years were 24% as presented in Table 1 below.

Table 1. Background of respondent

Category	Contents	Percentage
Respondents of experience in design construction	10 years 11–20 years ≥ 21 years	44% 32% 24%
Type of the building designed by respondent	Residential Shopping center Industrial area Office Others (hospital, airport, hotel)	32% 16% 12% 12% 28%

The table above also show the function of the building designed by respondents. Based on the survey, it was found that function of residential buildings is the most building designed by respondents with a percentage of 32%.

4.2 Regression Analysis

Table 2 shows a summary of the model consisting of the R value, the value of variation of the R Square (R2), the value of adjusted R Square (Adjust R2) and the standard error estimate. Based on the regression analysis, six independent variables namely perspective of designer, attribute designer, interoperability, team management, team support, and innovation contribute 66.5% to the overall success of PBBD practice. Meanwhile, the results obtained also indicate that there is a significant influence between independent factors simultaneously on the dependent factor with a value of F = 8.93; ρ (0.000) shown in Table 3 Anova.

Table 2. Model summary

Model	R	R square	Adjusted R square	Std. error of the estimate
1	0.865	0.749	0.665	0.32229

Notes: Predictors: (Constant), Innovation, Perspective designer, Team support, Interoperability, Attribute designer, Team management. Dependent Variable: Success of PBBD

Table 3. ANOVA

Model		Sum of squares	df	Mean square	F	Sig.
1	Regression	5.570	6	0.928	8.938	0.000[b]
	Residual	1.870	18	0.104		
	Total	7.440	24			

Notes: Dependent Variable: Success of PBBD
Predictors: (Constant), Innovation, Attribute designer, Perspective designer, Interoperability, Team support, Team management

Furthermore, the results of the regression analysis coefficients are shown in Table 4. Based on the table, the results obtained indicate that interoperability, team support and innovation are the main factors that contribute to the success of PBBD practice in the model. The interoperability variable has a significant effect on the success of PBBD practice with a value of $\beta = 0.473$ on $\rho < 0.05$ (0.000), significant influence between the team support variable on the success of PBBD practice with the value of $\beta = 0.229$ on $\rho < 0.05$ (0.021), and innovation variables that have a significant effect on the success of PBBD practice with the value of $\beta = -0.243$ on $\rho < 0.05$ (0.002). The negative beta coefficient on the innovation variable indicates that the designer is lacking in implementing innovation.

Table 4. Coefficients regression

Model		Unstandardized coefficients		Standardized coefficients	t	Sig.
		B	Std. error	Beta		
1	(Constant)	2.145	0.603		3.555	0.002
	Perspective designer	−0.035	0.150	−0.036	−0.236	0.816
	Attribute designer	0.061	0.127	0.086	0.483	0.635
	Interoperability	0.473	0.096	0.670	4.907	0.000
	Team management	0.059	0.168	0.063	0.348	0.732
	Team support	0.229	0.091	0.399	2.524	0.021
	Innovation	−0.243	0.068	−0.463	−3.554	0.002

Dependent Variable: Success of PBBD

This findings confirm the proposed model by testing six independent variables which include perspective designer, attribute designer, interoperability, team management, team support, and innovation on the dependent variable, namely the successful practice of PBBD. Based on the results of the regression analysis, it was found that hypothesis 1 perspective of designer does not have a significant effect on the successful practice of PBBD. The regression coefficient has a negative direction, which means the less of designer gives his perspective in the design process, it will reduce the success of the practice PBBD. The same thing was also found in the hypothesis 2 attributes of designer does not have a significant effect on the success of PBBD. The regression coefficient has a positive direction which indicates the good attributes of designer in terms of work quality, it will increase the successful practice of the PBBD. Furthermore in hypothesis 3, interoperability has a significant effect on the success of PBBD. Hypothesis 4 that is team management is found to have no significant effect on the success of PBBD, while the regression coefficient has a positive direction, meaning that a well-managed team will increase success in the practice of PBBD. Finally, hypothesis 5 and hypothesis 6, namely team support and innovation, each has a significant influence on the success of PBBD practice.

In general, it is known from the six hypotheses tested, there are three hypotheses which accepted and have a significant effect on the success of the PBBD practice, namely hypothesis 3 interoperability and hypothesis 5 team support which has a positive effect, while hypothesis 6 innovation has a negative effect. These findings confirm the research [15] that the ability of interoperability systems can make it easier for designers to communicate with each other, especially with regard to the exchange of data and information on design objects. Since system interoperability removes the barrier of data loss being exchanged due to the use of different systems, this gives the designer an advantage to better understand the content of the information being conveyed. In addition, system interoperability is also supported by the use of the same language between systems, so that it can directly improve performance in the design process.

Subsequent research findings related to team support in the design process are in line with research [20] that a mutually supportive team will create a learning environment. This is because the designers who are members of the team will take advantage of the transfer of knowledge that can be used in future projects. Team support will also remove communication barriers due to hierarchical differences, but still establish respect for all members involved. Having team support in the design success will take advantage of finding solutions to problems, instead of blaming each other [18].

The last, results of research that show innovation contributes to the success of design performance on construction projects that support research results [4]. At the same time, in this study it was also found that generally designers are lacking in implementing innovations. In research [25] stated that one of the emphasis to increase innovation is the use of software/technology that supports the process of forming designer ideas during the design process. Based on this, it is possible that designers have not adopted innovations especially in PBBD practices due to the low use of design software/technology that is able to facilitate the work process or techniques used in the creation of designs. In line with researcher [24] which shows that the designer's reluctance to start is due to his unfamiliarity with the use of software/technology due to difficulties in understanding, resulting in the lack of implementation of innovation in the development of design objects.

5 Conclusion and Recommendation

The purpose of this study is to test the PBBD model that can be used to achieve success in the implementation process. Interoperability, team support, and innovation had a significant correlation to the successful practice of PBBD. Based on these results, interoperability and team support have a positive effect. This means that when a designer uses an interoperability system, it will increase the success of PBBD, as well as a team that supports each other during the design implementation process. Meanwhile, innovation has a negative effect, which means that although innovation is a factor that can increase the success of PBBD practice, this factor is still not adopted by designers in construction projects in Indonesia. Furthermore, the limitations of this study are the limited number of expert respondents involved and only carried out on construction projects in Indonesia, so the evidence produced in this study is not strong enough and the results may not be generalized. However, the model developed in this study can be a starting point for further research in different country contexts, also considering other variables that have not been used in this study.

Acknowledgments. The authors express appreciation for the funding of this research paper through the research grant of "Penelitian Dasar (PD) RistekBRIN 2021".

References

1. Lindhard, S., Jesper, L.: Identifying the key process factors affecting project performance. Eng. Constr. Archit. Manag. **23**(5), 657–673 (2016)
2. Demirkesen, S., Ozorhon, B.: Impact of integration management on construction project management performance. Int. J. Proj. Manag. **35**(8), 1639–1654 (2017)
3. Mellado, F., Lou, E.C.W., Becerra, C.L.C.: Synthesising performance in the construction industry: an analysis of performance indicators to promote project improvement. Eng. Constr. Archit. Manag. **27**(2), 579–608 (2020)
4. Salter, A., Torbett, R.: Innovation and performance in engineering design. Constr. Manag. Econ. **21**(6), 573–580 (2003)
5. Becker, R.: Fundamentals of performance-based building design. Build. Simul. **1**(4), 356–371 (2008). https://doi.org/10.1007/s12273-008-8527-8
6. Astarini, S.D., Utomo, C., Negoro, N.P., Sari, A.F., Rohman, M.A.: The influence of performance-based building design on the strategy of retail property in Indonesia. Sustainability **12**(21), 1–15 (2020)
7. De Wilde, P.: Building Performance Analysis. 1st edn. Wiley Blackwell (2018)
8. Savolainen, J.M., Saari, A., Männistö, A., Kähkonen, K.: Indicators of collaborative design management in construction projects. J. Eng. Design Technol. **16**(4), 674–691 (2018)
9. Kleinsmann, M., Valkenburg, R., Buijs, J.: Why do(n't) actors in collaborative design understand each other? An empirical study towards a better understanding of collaborative design. CoDesign **3**(1), 59–73 (2007)
10. Kärnä, S., Junnonen, J.M.: Designers' performance evaluation in construction projects. Eng. Constr. Archit. Manag. **24**(1), 154–169 (2017)
11. Ling, Y.Y.: Model for predicting performance of architects and engineers. J. Constr. Eng. Manag. **128**(5), 446–455 (2002)
12. McComb, C., Cagan, J., Kotovsky, K.: Rolling with the punches: an examination of team performance in a design task subject to drastic changes. Design Stud. **36**(C), 99–121 (2015)
13. Chow, L.K., Ng, S.T.: Delineating the performance standards of engineering consultants at design stage. Constr. Manag. Econ. **28**(1), 3–11 (2010)
14. Oh, M., Lee, J., Hong, S.W., Jeong, Y.: Integrated system for BIM-based collaborative design. Autom. Constr. **58**, 196–206 (2015)
15. Lai, H., Deng, X., Chang, T.-Y.P.: BIM-Based platform for collaborative building design and project management. J. Comput. Civ. Eng. **33**(3), 05019001 (2019)
16. Pirayesh, N.A., Etienne, A., Kleiner, M., Roucoules, L.: Performance evaluation of collaboration in the design process: using interoperability measurement. Comput. Ind. **72**, 14–26 (2015)
17. Knotten, V., Lædre, O., Hansen, G.K.: Building design management–key success factors. Archit. Eng. Design Manag. **13**(6), 479–493 (2017)
18. Baiden, B.K., Price, A.D.F., Dainty, A.R.J.: The extent of team integration within construction projects. Int. J. Proj. Manag. **24**(1), 13–23 (2006)
19. Ding, Z., Ng, F., Cai, Q.: Personal constructs affecting interpersonal trust and willingness to share knowledge between architects in project design teams. Constr. Manag. Econ. **25**(9), 937–950 (2007)

20. Öztürk, G.B., Arditi, D., Günaydın, H.M., Yitmen, İ.: Organizational learning and performance of architectural design firms in turkey. J. Manag. Eng. **32**(5), 05016015 (2016)
21. Zhikun, D., Fungfai, N.: Knowledge sharing among architects in a project design team: an empirical test of theory of reasoned action in China. Chin. Manag. Stud. **3**(2), 130–142 (2009)
22. Issa, R.R.A., Haddad, J.: Perceptions of the impacts of organizational culture and information technology on knowledge sharing in construction. Constr. Innov. **8**(3), 182–201 (2008)
23. Kline, S.J., Rosenberg, N.: An overview of innovation. Stud. Sci. Innov. Process 173–203 (2010)
24. Elmualim, A., Gilder, J.: BIM: innovation in design management, influence and challenges of implementation. Archit. Eng. Design Manag. **10**(3–4), 183–199 (2014)
25. Ozorhon, B., Oral, K.: Drivers of innovation in construction projects. J. Constr. Eng. Manag. **143**(4), 04016118 (2017)
26. Nwadike, A., Wilkinson, S.: Impacts of innovation in performance-based building code (2020)

Research on the Form and Optimization of Rural Residential Huo-Kang in Northeast China

Jingzhao Feng(✉)

Faculty of Humanities and Arts, Macau University of Science and Technology, Taipa, Macau, China
21098537AT30001@student.must.edu.mo

Abstract. "Huo-Kang" has a history of more than 2,000 years and has many types and internal structures. Its construction form originated from traditional folk construction techniques. Along with the economic revitalization of the countryside, optimizing the heating problem and improving the comfort of the living room in the northern countryside has become one of the measures to update the construction of new rural housing. This study analyzes the internal structure, function and material of the Huo-Kang, which is a common type of the Huo-Kang existing in the rural areas of Northeast China. Optimization measures are proposed in terms of the Huo-Kang's body construction, materials and dimensions to standardize the Huo-Kang construction standards and provide reference materials for future rural Huo-Kang construction.

Keywords: Huo-Kang · Rural area · Form and optimization · Northern area

1 Introduction

The Huo-Kang is a great achievement in the application technology of ancient residents. The residents in northern China invented heating facilities to combat the severe cold weather conditions. Actually, the origin of the Huo-Kang can be traced back to the Qin and Han Dynasties [1, 2]. The Huo-Kang is a facility that can burn fire for heating, while it is an immovable heating device built on the indoor ground [3]. The Huo-Kang can be defined as "a bed in the north, built of earth or masonry, with internal apertures for heating by fire" [4], but the difference lies in the fact that the Huo-Kang is heated to form a heat source body with thermal insulation effect.

The cold winter temperatures and the large temperature difference between day and night in northeastern China have created as severe cold climate, making the Huo-Kang an indispensable indoor facility to fight the cold and keep warm in the Northeast. The Huo-Kang is known as the "fire place" in folklore, and is generally referred to as the "Di-Kang" and "Di Long" in northern China [5]. The Huo-Kang assumes the daily life functions of resident such as resting, meeting and dining, etc. The difficulty and economic cost of Huo-Kang is low, and it is done independently by individuals and the construction materials are easily available [6]. Therefore, the Huo-Kang has gradually

P. Mendonça and N. D. Cortiços (Eds.): ICAMC 2021, LNCE 226, pp. 138–149, 2022.
https://doi.org/10.1007/978-3-030-94514-5_15

become one of the common forms of heating in northern rural areas and is loved by northern residents.

Nowadays, the level of urbanization is accelerating and the economic income of Chinese rural residents continues to grow. The old Huo-Kang in some villages needs to be repaired, renovated and rebuilt work, and the new Huo-Kang is gradually becoming the primary choice for renovation and construction by rural residents. This study proposes optimization recommendations for three common types of the Huo-Kang, the traditional Huo-Kang, the fire-wall Kang and evelated Kang, in terms of insulation, materials and internal structure.

2 Construction Form of Huo-Kang

There are many types of the folk Huo-Kang in rural areas of Northeast China, and there are differences in the forms, materials and techniques of building the Huo-Kang in different regions and ethnic groups. However, the main functions of different types of the Huo-Kang are the same, such as heating against cold, daily rest and recreation. Usually, the Huo-Kang consists of three parts: the stove body, which is used for cooking and provides heat source to the Huo-Kang; the Huo-Kang body, i.e. the Huo-Kang board and the internal flue; and the chimney, i.e. the smoke removal function.

2.1 Surface Structure of Huo-Kang

The surface of the Huo-Kang is an important part of the Huo-Kang between the inner and outer space and its main functions include supporting, storing and transporting heat [7]. The surface of the Huo-Kang is divided into the head, middle and tip of the Huo-Kang, with the head being the higher temperature area. The comfort level of the Huo-Kang depends on the temperature of the Huo-Kang's surface, which is reflected in three aspects: maximum temperature, minimum temperature and local temperature difference.

In the early days, the surface of Huo-Kang was usually made of clay and scraped grass, while in modern times, new materials such as refractory cement, "refractory soil" cement board are used. Different surface materials of the Huo-Kang have different functions, and the surface must have good fire resistance and crack resistance. In addition, in the actual construction process, the height of the Huo-Kang's area is slightly higher than that of the Huo-Kang's area.

2.2 Internal Structure Form of the Huo-Kang's Boby

The Huo-Kang's hole is one of the main structures that guide the flow of smoke and support the Kangban, one of the thermal effect factors of the Huo-Kang. The construction of the Huo-Kang's hole can be divided into flower holes, straight holes, back holes and other holes [7], and straight holes is the most widely used in the rural areas of northeastern villages and towns.

The Huo-Kang's body stores the heat of the flue gas flowing through the Huo-Kang's duct and exchanges heat with the internal high-temperature smoke to release the absorbed heat into the indoor space; the more heat absorbed by the Huo-Kang's body, the more effective the heating capacity is. By optimizing the internal structure of the Huo-Kang, the internal resistance of the Huo-Kang is strengthened or less time, which affects the residence time of the flue gas, and the residence time determines the heat absorbed by the Huo-Kang's body [8].

Construction Style of Flower-Hole Huo-Kang's Body. The traditional construction of flower hole makes the smoke flow disorderly in the internal space of Huo-Kang, which cannot effectively control the specific direction of the smoke flow and may lead to uneven temperature on the surface of Huo-Kang. Therefore, flower holes are not suitable for larger size of Huo-Kang [9]. The new flower hole incorporates other types of Huo-Kang's body form, and by changing the position of the stove and chimney, the flow of smoke is directional and the residence time of smoke in the internal space of Huo-Kang's body is enhanced, as shown in Fig. 1.

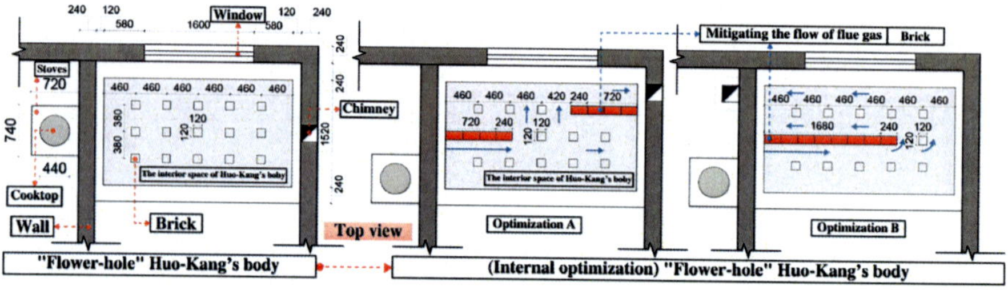

Fig. 1. Construction style and improved method of flower-hole Huo-Kang's body. (Painted by the author)

Construction Style of Straight-Hole Huo-Kang's Body. Straight-hole Huo-Kang's body is one of the widely used Huo-Kang's holes in rural houses in Northeast China. The supporting structure is arranged vertically or horizontally in a regular manner, which has good and stable supporting effect. The improvement of straight hole Huo-Kang's body is to set up a smoke baffle inside the Huo-Kang's body, so that the direction of smoke flow is clear and the residence time of smoke at the end of Huo-Kang is enhanced, as shown in Figs. 2 and 3.

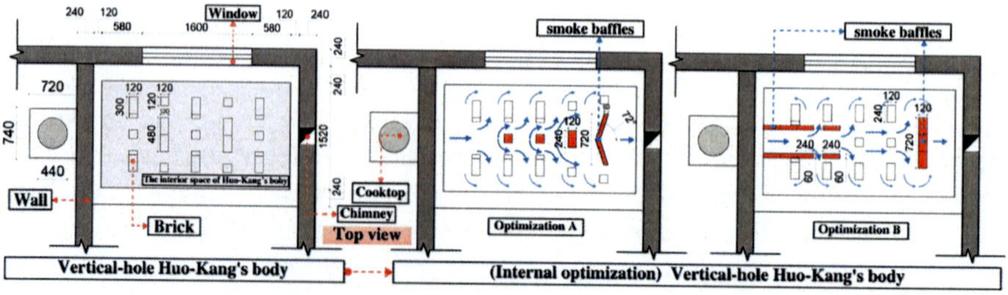

Fig. 2. Construction from and optimized method of vertical-hole Huo-Kang's body. (Painted by the author)

Construction Style of Huo-Kang's Body with Return Hole. Flue gas enters in vertical or parallel direction, concentrates to one main flue and gradually circulates to the rest of the flue, prolonging the flow time of flue gas in Huo-Kang and improving the thermal effect of Huo-Kang's body, as shown in Fig. 4. The construction difficulty of Huo-Kang's body with return hole is low and Huo-Kang's body has good stability. Nevertheless, the resistance of the return-hole Huo-Kang's hole is higher than other forms of Huo-Kang's hole, so it is generally suitable for small-sized Huo-Kang. The way to reduce the resistance is to open holes in the pillar wall to increase the flow rate of flue gas.

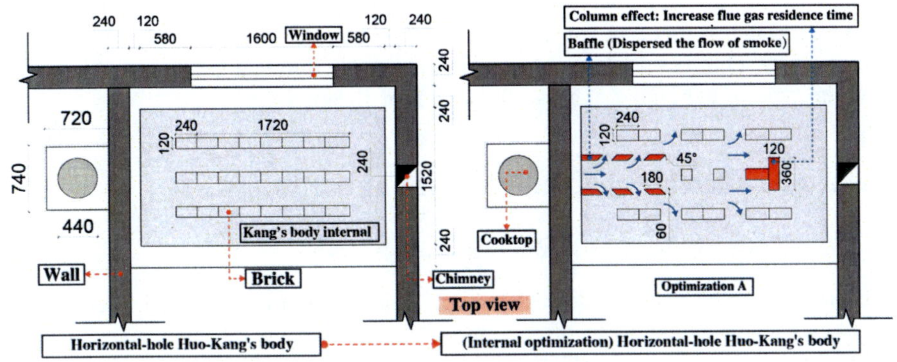

Fig. 3. Construction from and optimized method of horizontal-hole Huo-Kang's body. (Painted by the author)

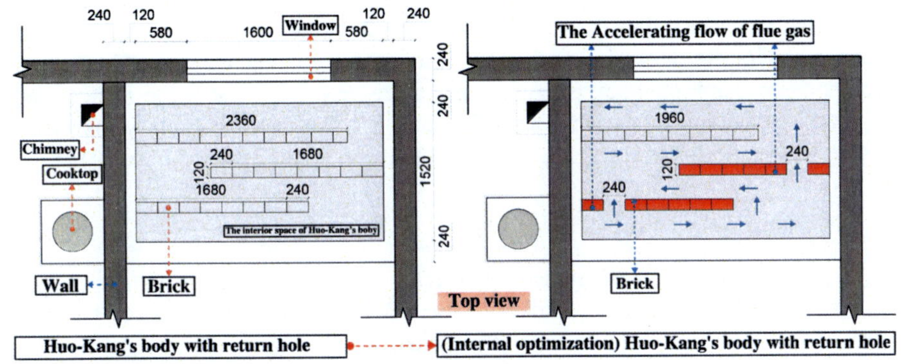

Fig. 4. Construction form and optimized method of Huo-Kang's body with return hole. (Painted by the author)

2.3 Construction Style of Stove

The stove is an important facility for the daily diet of the residents and is one of the sources of heat for Huo-Kang. The common forms of stoves are single stove and double stove. The upper part of the stove is mainly the pot body, while the lower part is the burning area, ash accumulation area and smoke inlet area. The height of the stove hole should be about 30–40 mm lower than the smoke inlet of the Huo-Kang to prevent the occurrence of the prairie smoke and the fire from entering the interior of the antibody when the fuel is burning. The surface of the smoke inlet is treated smoothly to help to reduce the resistance [5].

2.4 Construction Style of Chimney

The main function of a chimney is to discharge smoke, which is exceedingly significant facility for Huo-Kang. Usually, the height of chimney should be more than 500 mm above the roof. The internal diameter of the chimney is about 140 mm–250 mm, and the width of the top and bottom of the chimney should be consistent. The location of the chimney is determined according to the internal structure of the antibody, the type of the Huo-Kang and the layout of indoor space.

3 Factors Affecting the Temperature of the Huo-Kang

3.1 Flow State of Flue Gas in the Huo-Kang's Body

In the Huo-Kang system, the flow of flue gas depends on the extraction force of the chimney. The extraction force of the chimney is influenced by three main factors, height, flue gas temperature and air. Flue gas temperature is affected by the resistance of the bunkder including objects such as smoke outlet, the pillar, baffle, etc. Assuming that the smoke temperature of the Huo-Kang's body is low and the chimney draw is too small, it is not enough to overcome the resistance inside the Huo-Kang, which may cause the Huo-Kang to back up. If the smoke temperature of the Huo-Kang' body is high, the chimney has high extraction power, the flow rate of smoke is fast and the residence time of the smoke inside the Huo-Kang is short, which may result in the loss of temperature. High chimney exhaust temperature may cause the waste of heat and low energy utilization of the Huo-Kang [8], and it is better to keep the smoke temperature between 50–80 °C to ensure the effective extraction of the chimney [10].

3.2 Thermophysical Properties of Fluids

The greater the temperature difference between the flue gas temperature and the surface temperature of the Huo-Kang board, the more effective the heat exchange. The viscosity of flue gas will increase with the increase of flue gas temperature, and the viscosity is the key factor that hinders the flow of the flue gas. The lower specific heat capacity and density of flue gas at higher temperatures reduce the ability of flue gas to carry heat per unit volume and weaken the convection effect [11]. Therefore, the inlet and outlet temperature of the flue gas in the Huo-Kang and to coordinate the heat release of the flue gas with the heat gain of the Huo-Kang.

3.3 Physical Properties of the Huo-Kang' Body

Different types of the Huo-Kang's bodies should have certain heat storage capacity, i.e. thermal inertia of the Huo-Kang's body. Thermal inertia is determined by the physical properties of the Huo-Kang's body itself and the arrangement, size and quantity of the internal pillars of the Huo-Kang's body. The material has good thermal conductivity and heat insulation, and the scientific arrangement of the internal pillars of the Huo-Kang's body is conducive to improving the sustainability of the heat insulation of the Huo-Kang's surface and ensuring the normal exclusion of smoke.

4 Huo-Kang Forms and Improvement Measures

When optimizing different types of the Huo-Kang, the three main principles of the renovation design are: one is economic and technical feasibility; the other is the stability of the Huo-Kang; the third is the temperature distribution on the surface of the Huo-Kang.

Based on the three common forms of the Huo-Kang in Northeast China, it is assumed that the length, height and width of the Huo-Kang are 3000 mm × 2000 mm × 680 mm, the chimney is located inside the outer wall with an inner diameter of 150 mm and extends 100 to 120 mm from the roof, the construction material is standard brick, the construction of the Huo-Kang's surface is 30mm thick stone slab with 20 mm thick yellow clay plaster or clay layer, and the uppermost part is heat-resistant Huo-Kang's paper.

4.1 The Structure and Improvement of Traditional Huo-Kang

In the early days, the structure of traditional Huo-Kang in Northeast China was simple, and 3 to 4 smoke passages were usually reserved in the interior of Huo-Kang to ensure sufficient support strength and space for smoke passage [5].

Firstly, a slope of about 2–5° from the "smoke inlet" to the "chimney entrance" is laid on the bottom layer of the traditional Huo-Kang with loss, but the slope is not too steep. The temperature of the Huo-Kang's area will be reduced and the temperature of the Huo-Kang's surface will be affected. The smoke inlet and the smoke outlet are inclined to facilitate the inflow and outflow of smoke. The inner wall of the chimney will absorb a large amount of smoke for a long time, and the soot will be dislodged from the inner wall of the chimney by gravity after a long period of accumulation, so it is necessary to set up a groove in the smoke outlet [5]. The purpose is to prevent the accumulation of dislodged soot, which leads to blockage phenomenon and is not conducive to the elimination of smoke from the interior of the Huo-Kang.

By changing the internal structure of the Huo-Kang's body, as shown in Fig. 5, the purpose is to disperse the smoke and to increase resistance. Installing pointing smoke baffles in the Kangtou area will reduce the time the smoke stays in the Kangtou, while installing baffles in the smoke outlet area will increase the time the smoke stays in the Huo-Kang.

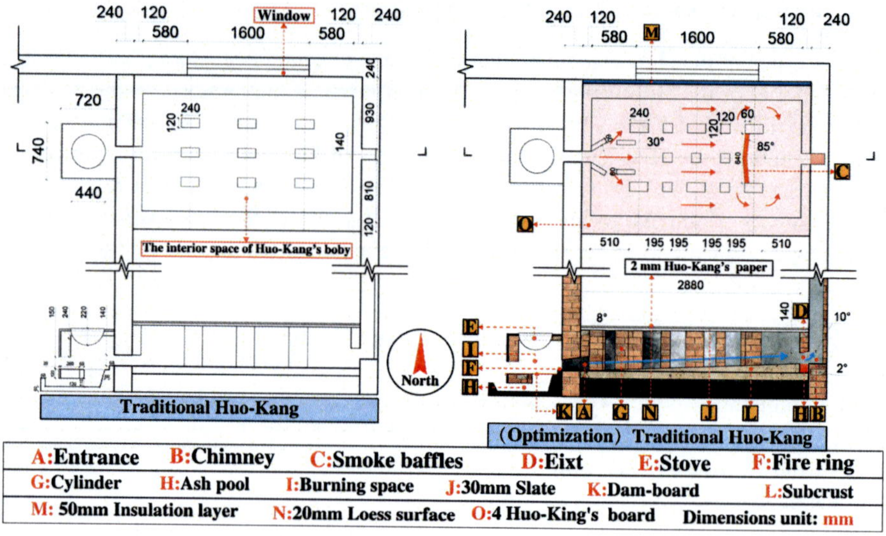

Fig. 5. Construction style and improved method of traditional Huo-Kang. (Painted by the author)

4.2 Construction Form and Improvement Measures of Fire-wall Kang

The fire-wall Kang is decomposed into two independent internal Huo-Kang's bodies, which have the dual thermal performance of the Huo-Kang's body and independent combustion chamber. The single-inlet dual thermal performance of the Huo-Kang's body and independent combustion chamber. The single-inlet smoke mode is transformed into a double-inlet smoke mode, which can meet the heating requirements without auxiliary equipment. The fire-wall Kang has diversified combustion methods, which improves the convenience of heating and heat utilization rate. At the same time, the Huo-Kang's body and the fire wall use a chimney to exclude the smoke, as shown in Fig. 6.

Advantageously: First, the independent combustion space of fire wall and "instant heat" solve the problem that the traditional Huo-Kang cannot increase the temperature in time [12, 13]; second, it improves the comprehensive utilization efficiency of fuel; third, the fire-wall Kang does not need auxiliary facilities [14]. The experiments were conducted on fire-wall Kang and traditional Huo-Kang, and it was proposed that the fire-wall Kang had increased construction cost but reduced fuel costs compared with floor-to-ceiling Huo-Kang [13]. Temperature tests were conducted for the fire-wall Kang (3.0 m × 1.9 m), where the instant heat of the Huo-Kang's surface was improved and the Huo-Kang's surface had the best temperature uniformity when burning under double space [14].

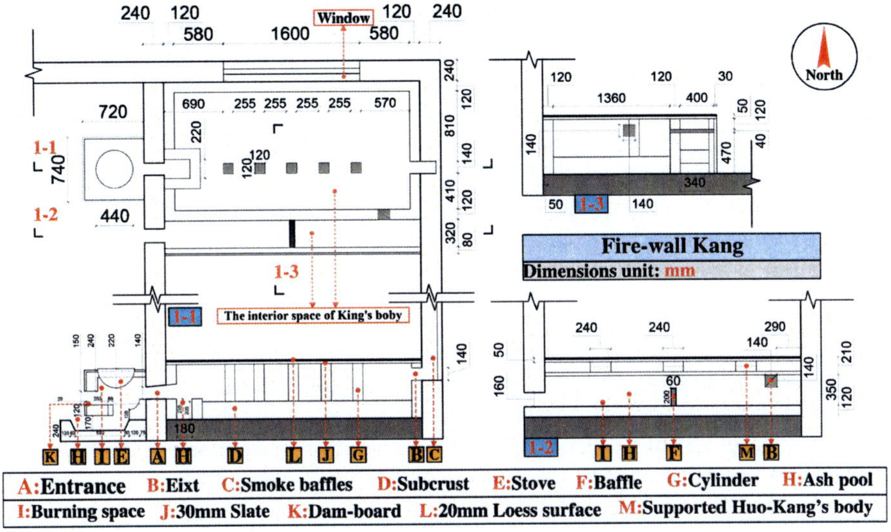

Fig. 6. Construction style of fire-wall Kang. (Painted by the author)

Increase the Setting of the Smoke Baffles Inside the Huo-Kang's Body. The experimental results prove that installing smoke baffles at the smoke outlets of the Huo-Kang's top and fire wall improves the utilization rate of the smoke energy [8]. The width of smoke baffle is 60 mm for two long boards, and the angle design facilitates the flow of smoke. This design increases the residence time of the smoke in the tip area of the Huo-Kang's body, reduces the exhaust temperature and increases the internal temperature of the Huo-Kang's body, as shown in Fig. 7.

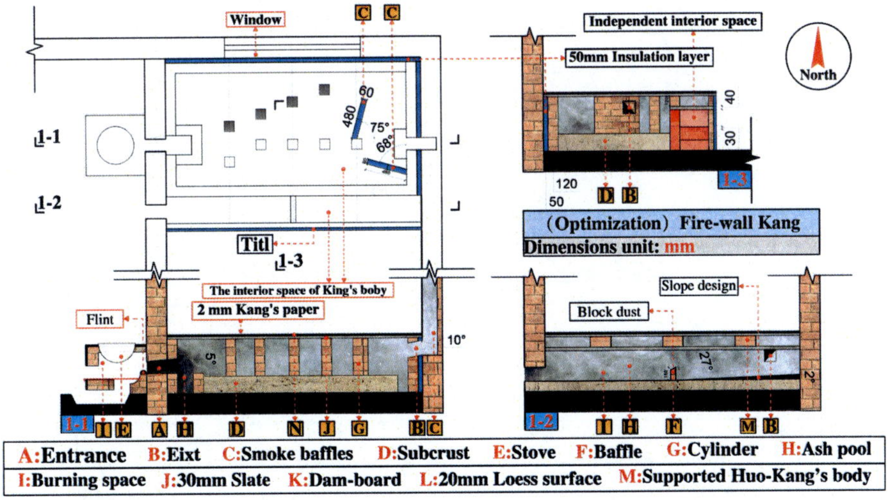

Fig. 7. Improved method of fire-wall Kang. (Painted by the author)

Change and Increase the Internal Columns of the Huo-Kang's Body. One row of supporting columns of the Huo-Kang's body is modified into two rows of supporting columns, forming three trajectories to enhance the smoke dispersion effect. The design form reduces the space of the outer wall of the Huo-Kang's body and increases the internal resistance to further enhance the residence time of the smoke and the temperature of the Huo-Kang's body. By changing the position of the supporting columns at the entrance of the Huo-Kang, the internal smoke distribution is relatively even. In addition, insulation panels are installed on the side of the Huo-Kang's body and the outer wall to enhance the heat preservation effect.

Optimize the Ash-Retaining Stack Inside the Firewall. Fuel can be burned directly in the firewall. In order to prevent excessive ash accumulation in the combustion chamber, ash baffles can be set in the middle area of the firewall to facilitate the cleaning of soot generated after fuel combustion. The ash baffle can slow down the rate of flue gas entry, assuming that that the temperature difference between the front and rear space is large and easy to form backflow. The improvement scheme reduces the ash baffle height and the upper half with slope setting concept [14].

4.3 Structure of Elevated Kang

Elevated Kang is a new type of the Huo-Kang based on the traditional Huo-Kang, also called as "Diao Kang". The bottom of the Huo-Kang's body is hollowed out by columns and separated into two parts, with an air layer of 200 to 300 mm below to increase the heat dissipation surface of the Huo-Kang's body.

Because the space of the hollow Huo-Kang's body is compressed by the modeling, there is less smoke compared with the above two types of the Huo-Kang. The author proposes three ways of improvement.

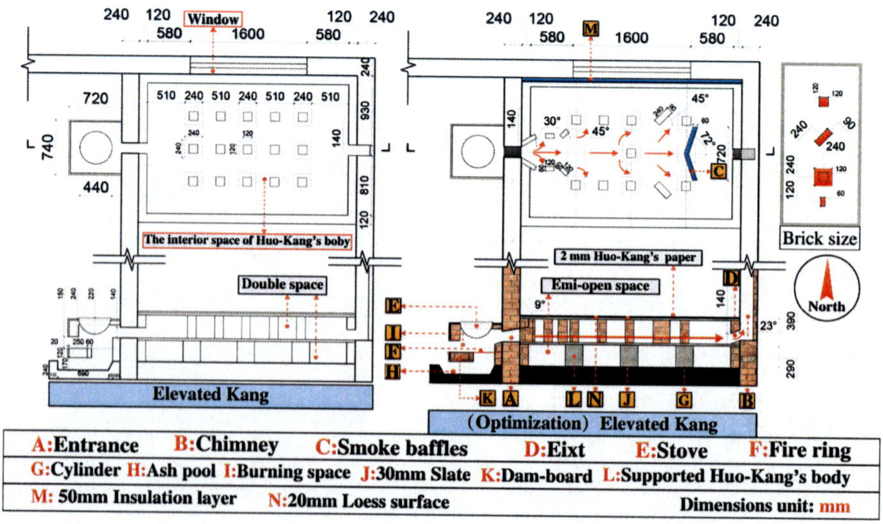

Fig. 8. Construction style and improved method of the elevated Kang. (Painted by the author)

Firstly, the internal column distribution is optimized with the addition of smoke baffles to increase the internal resistance, as shown in Fig. 8.

Secondly, heat-insulating materials are added between the Huo-Kang's body and the outer wall to enhance the effect of heat preservation.

Finally, the ground under the Huo-Kang' body is paved with materials with high heat storage capacity as a way to increase the heat dissipation time and improve the thermal stability of room temperature [13].

Comparing the thermal effects of traditional Huo-Kang with elevated Kang, the elevated Kang is more suitable for heating in rural areas of northeast China, and the indoor temperature difference is smaller [7]. The biggest difference between the two lies in the heating efficiency, which is usually greater for the same size of the elevated Huo-Kang than the traditional type.

4.4 Comparison of the Performance of the Huo-Kang

The above analysis of the internal structure of each type of the Huo-Kang shows that the traditional Huo-Kang is cheaper than fire-wall Kang and elevated Kang, uses less materials, has a simpler internal structure, and has faster heat dissipation [15]; the fire-wall Kang has significantly higher thermal performance than the traditional Huo-Kang and the elevated Huo-Kang, is more flexible and convenient for residents, and has better temperature uniformity on the Huo-Kang's surface. The elevated Huo-Kang has faster heat conduction than the traditional Huo-Kang, and the space between the upper and lower layers enhances the effectiveness of heat dissipation surface, and the heating rate is greater than that of the traditional Huo-Kang, and the instant heat performance is worse than the fire-wall Kang [15].

5 Conclusion

With the development of the times, the existed old Huo-Kang in the Northeastern countryside needs to be re-improved, reinforced and rebuilt. For technological development and material innovation, the optimization of the Huo-Kang is abundant and diverse. Without incorporating modern electronic equipment, this study analyzes the traditional construction methods in five dimensions.

Firstly, the way of improving the interior of the Huo-Kang's body starts from adjusting the column arrangement and the size of the column inside the Huo-Kang's body. The number of internal columns should be scientifically adjusted according to the actual size of the Huo-Kang in the actual construction of the Huo-Kang.

Secondly, smoke baffles can be added to the tip of the Huo-Kang's body or the smoke outlet area to strengthen the space resistance and help maintain the temperature at the Huo-Kang's tip. The high-temperature smoke presents a floating state, and the smoke baffle should be set close to the Huo-Kang's board to disperse the smoke and support it. The number of smoke shields does not set too many smoke baffles, as the smoke will stay for a long time and cause the temperature of the smoke to drop, which is not conducive to the elimination of smoke inside the Huo-Kang's body.

Thirdly, the Huo-Kang's body should be equipped with insulation boards near the outer wall, and the thickness should be adjusted according to the change of local temperature.

Fourthly, the surface of the Huo-Kang is prone to cracking when it is dry for a long time. New materials can be used to avoid such phenomenon, which have high-temperature resistance and good crack resistance.

Fifthly, the details of the Huo-Kang: the surface treatment of the smoke inlet and outlet should be inclined and smooth, with a fire stone and an ash accumulation tank under the outlet. At the same time, good sealing of the Huo-Kang is a key consideration for construction.

Finally, the Huo-Kang has become one of the representative characteristics of rural housing in Northeast China, and the types of fire bed construction are diversified. It is necessary to consider the actual economic factors, the real needs of the occupants and the safety when improving, strengthening and building. At the same time, the future construction of the Huo-Kang should be combined with new construction materials and rationalized with new technologies and innovative ideas.

References

1. Ren, H.G.: Research on Technology Design of Rural Residential Huo-Kang in Northeast China (Master's Thesis). Harbin Institute of Technology, Harbin (2008)
2. Liu, F.: The Study of the Change of the Folk-Culture about Heated Kang of the Northern Han nationality (Master's Thesis). Liaoning University, Shenyang (2013)
3. Song, J.F.: Research on the Origin and Development of Heated Kang in Northeast Aisa (Master's Thesis). Northeast Normal University, Changchun (2013)
4. Xu, C.: Experiment Research on Thermal Properties and the Establishment on Heat Transfer Model of Hot-wall Kang (Master's Thesis). Dalian University of Technology, Dalian (2015)
5. Ren, X.Y., Guan, H.Y.: On ecological consciousness about the design of heated Kang in North China. Ecol. Economy **31**(12), 196–199 (2015)
6. Xu, D., Wang, X.Y., Pei, Y.P.: Analysis of huo-kang and ground huo-kang of rural houses in Liaoning Province. Sichuan Build. Sci. **36**(04), 299–301 (2010)
7. Liu, Y., Gu, Y., Li, Y.: Comparative analysis of energy-saving design between floor-to-ceiling huo-kang and overhead huo-kang in rural residences in Northeast China. J. Anhui Agric. Sci. **42**(14), 4326–4328 (2014)
8. Zhu, J.L., Wang, Z.S., Duan, M.L., Song, B., Ding, Z.W., Fang, Y.N.: Experimental study on structure improvement of hot-wall Kang. Build. Sci. **27**(02), 26–32 (2011)
9. Zhao, Y.: Thermal Performance of Heating System of Hot-wall Kang in Northern Rural Areas (Master's Thesis). Dalian University of Technology, Dalian (2009)
10. Guo, J.Y.: Fire-Saving and Coal-Saving Stove Huo-Kang. China Agriculture Press, Beijing (2003)
11. Duan, M.L., Zhao, Y., Wang, Z.S., Shu, H.W., Zhu, J.L., Shan, H.N.: Research and assessment method of thermal performance of Chinese Kang. Build. Sci. **25**(12), 30–38 (2009)
12. Li, X., Yan, L.A., Wei, R., Zhao, Y.D.: Research on the improvement design of winter heating methods in Northeast Rural Areas. Agric. Technol. **32**(08), 66+68 (2012)

13. Xu, J.M., Xia, N., Xing, K.: Low-energy-consumption huo-kang design for residential buildings in the severe cold Area of Northeast China. Low Temp. Archit. Technol. **38**(01), 11–13 (2016)
14. Zhao, Y., Duan, M.L., Wang, Z.S., Zhu, J.L., Shan, H.N., Shu, H.W.: Experimental study on the thermal performance of hot-wall type Kang. Build. Sci. **26**(02), 29–36 (2010)
15. Xu, D., Wang, X.Y., Pei, Y.P.: Analysis on huo kang and ground kang of rural residential buildings in Liaoning Province. Sichuan Build. Sci. Res. **36**(04), 299–301 (2010)

Local and Derivative —— Research on Architectural Space Design from the Perspective of Nomadic Culture

Jianqiang Yin(✉)

Macau University of Science and Technology, Macao 999078, China

Abstract. In this study, the relationship between the architectural space renewal design and the regional culture is based on the report of "Ma'an Town" in Hohhot, Inner Mongolia. Using the philosophical thought derived locally, through the method of bibliography and field investigation, briefly analyze the characteristics of the architectural space of the provincial hotel in Hohhot. From the perspective of nomadic culture, with culture as the core, through the interpretation of nomadic civilization and nomadic culture, the relationship between nomadic culture and architectural space is considered. In terms of spatial form, the natural outlook, humanistic outlook, and humanistic outlook embodied in nomadic culture is discussed. Daily ethnic characteristic activities, production and lifestyle, religious beliefs, and other aspects just try to solve the design methods of provincial cultural hotel buildings to achieve the purpose of cultural heritage and space regeneration.

Keywords: Locality · Nomadic culture · Hotel space · Building renewal

1 Introduction

1.1 A Subsection Sample

With the rapid development of tourism in Inner Mongolia, the construction and development of regional cultural buildings are of particular importance. As the first artistic hotel with a grassland theme, the Inner Mongolia Hotel has typical regional characteristics and has attracted the attention of a large number of tourists. However, in terms of design, the use of Mongolian elements lacks a holistic concept, the architectural space design decoration is magnificent, and the messy stacking of traditional graphic symbols can no longer meet people's current aesthetic needs and the development trend guided by modern design. Therefore, this article focuses on studying how to integrate regional culture into the architectural space, interpreting the blending of natural geography and humanistic environment, and presenting a hotel building that is fashionable and diverse, diversity in demand, and inherits regional culture. Through an in-depth investigation of the site, the author found the uniqueness of the architectural site and the superiority of the natural terrain, determined the renewal design with the original architectural area as the carrier, and discussed the historical origins of nomadic culture, humanities, and customs, life, and production methods, geographical environment, and other factors Using local-derived philosophical thoughts to analyze

P. Mendonça and N. D. Cortiços (Eds.): ICAMC 2021, LNCE 226, pp. 150–160, 2022.
https://doi.org/10.1007/978-3-030-94514-5_16

the current situation of the resort and the surrounding humanistic environment, trying to activate the architectural space from the following five aspects: 1. The natural view of nomadic culture; 2. Ethnic characteristic activities; 3. Production source of inspiration for life and religious belief; 4. Business operation model; 5. Case studies.

2 The Status Quo and Reflections of the Resorts in "Ma'an Town"

2.1 Project Status

At this stage, Ma'an Town is now in a stagnant state due to construction costs and management and operational problems. Only the racecourse and two yurt dining areas are now in operation. The venue is suitable and fast to reach the urban area, and the transportation is convenient. First, it finds the value of viewing and recreation, which can ensure the comfort and privacy of tourists. It combines the characteristics of the terrain to establish a regional cultural space; Second. The resource allocation can rely on the city's tourism resources and surroundings the supply of the village maintains the form and structure of the site and updates the building functions. The project site is adjacent to Kuisu Village in the east (see Fig. 1). Daqing Mountain in the south, a sport and leisure resort base in the west, and a 70-year-old Daqing main venue in the north. There are eco-parks, horse ranches. Daqingshan forest trails, ten thousand mu grasslands and other attractions near the site. The topography of the site is fairly flat, and the population of the surrounding villages is sparse. Mongolian villagers are mostly gathered here, with strong rural flavor.

Fig. 1. Analysis of the geographical location of Ma'an Town (picture source: drawn by the author)

2.2 Problems with the Project

Judging from the current status of Ma'an Town, the total area of the project is more than 1,000 mu, and the main problems are as follows (see Fig. 2): The project is in a state of stagnation, and many building foundations have been completed, resulting in restrictions on the new functional layout. Weeds grow around the site, and the barren hills are scattered along with the distribution of the terrain. Weeds grow around the site, and the barren hills are scattered along with the distribution of the terrain. Weeds grow around the site, and the barren hills are scattered along with the distribution of the terrain. Weeds grow around the site, and the barren hills are scattered along with the distribution of the terrain.

Fig. 2. Status of Ma'an Town (picture source: drawn by the author)

3 Nomadic Culture Intervenes in Architectural Space Design

3.1 Overview of Nomadic Culture

The spatial state of "Saddle Town" has formed a field spirit characterized by horses, and horse culture is a typical cultural memory of nomads. Therefore, to activate the space method of Maan Town, this article cuts into the nomadic culture, looking for the connection between culture and space. The origin of Chinese civilization has always been based on farming civilization and nomadic civilization, which respectively represent two different living conditions and lifestyles [1]. Nomads are the kind of people who live in nature and are naturally free. They have no fixed residence due to the influence of their ideas, environment, and ideas from ancient times, and their lifestyles are characterized by uncertainty and instability [2]. The nomadic culture brought by the nomads is a core concept of harmony between man and nature and respect for nature. The most representative of nomadic culture is the cultural origins of the Mongolian people. Mongolia has unique geographical elements and living environment, forming a typical nomadic cultural imprint in all aspects.

Living Environment. The nomads are located on the east side of the Mongolian Plateau, the west side of the Kazakh steppe, the south side of the Gar Basin, and the north side of the South Russian grassland [3]. Around the world, nomadic peoples in different regions have developed their characteristics depending on their terrain and vegetation environment. The Mongolian people have lived in the grassland for a long time, living by water and grass.

Ways to Produce. Due to the instability of the living environment, the horse is considered a symbol of nomadic pastures. The nomads adopted a nomadic production method, which promoted the development of harnesses and weapons.

Religious Beliefs. Since ancient times, nomads have believed in a variety of religions, of which shamanism has a far-reaching influence on them, and the religion that continues to this day is mainly Islamic. The development of these religions has had a profound impact on the life, thoughts, and behaviors of the nomads, forming different outlooks on life, values, and the world.

3.2 Architectural Space Design Elements of Theme Culture Hotel

With the continuous development of urbanization, living in a stable state of life is rapidly replacing the nomadic people's unique production and lifestyle of "growth by water and grass", disrupting the primitive natural and humanistic environment, and leading to the gradual disappearance and passing of nomadic culture. The emergence of the themed resort-style cultural hotel is precisely through the spatial continuation of time and the context of culture, distinguishing the composition mode of the general architectural space.

Serving the Crowd-the Main Body of the Cultural Space. Different groups of people, different ages, and different genders form the basis of different cultures. Due to the different life backgrounds of various groups of people, their cultural values inter-act in the space. Through the analysis of the people, the development of space cul-ture is promoted.

Entertainment Activities-time Constitutes a Node. Cultural and entertainment activities with typical characteristics are the key to the formation and recognition of artistic venues. For example, horse racing activities, classic blog projects, folk song bars, archery and other entertainment projects based on the horse culture of the no-mads carry historical and cultural memories and survival methods and are the core force to continue the development of nomadic culture. The crowd will feel the charm of artistic space when they have contact with specific events and history.

Humanistic History-tracing Cultural Media. Mongolian and Yuan culture, Qing-cheng culture, and nomadic culture is all components of Inner Mongolia's regional culture. Each of them has unique characteristics and deep research value, which have produced the diversity of regional cultural development. Through the analysis of cultural factors, integrating culture into the hotel space, highlighting regional cultural characteristics, and stimulating people's sense of identity and belonging to the architectural space is equivalent meaning of the regeneration of the "saddle town".

4 Research on Regeneration Design of Architectural Space in "Ma'an Town"

4.1 Local and Derivative Design Ideas

Nomadic culture has continuously evolved and elaborated in the long river of history. Regional cultural hotels are the main carriers and important places for the manifestation of local characteristics and culture. Over the years, the evolution of productive lifestyles and social developments has seen nomadic culture pass through history. Therefore, the creation of a cultural space must meet two requirements: one is to realize the continuity of culture (locality); the other is to inherit the culture while introducing the old and bring forth the new (derivative), combine the trend of internationalization, and innovate and develop.

Using culture as means to promote the regeneration of cultural space, the regeneration of cultural space, the inheritance and development of culture, the interaction between the two, sustainable development. Resort-style cultural hotels provide people with a comfortable leisure environment, a fully relaxing space form, and functional facilities for entertainment. In the functional layout, regional characteristics and combined topography are considered to reshape the functional cultural space.

Therefore, locality and derivation are a type of ideological concept for the regeneration of Ma'an Town. Only based on the locality, looking forward to receiving the future, seeking integration and common development are the new innovative forms and expression methods for realizing cultural space. While inheriting culture, derive more interesting cultural space.

4.2 The Meaning and Goal of Architectural Space Regeneration

Space gives cultural meanings and creates a connection between people and architecture. The emotional experience of space integrates people's spiritual needs with material space. The significance and goals of the spatial regeneration of Ma'an Town are mainly as follows:

Inherit the tradition, introduce the old and bring forth the new, continue the historical value and significance of nomadic culture, and focus on the combination of natural culture and regional culture.

Create a representative regional cultural resort hotel, with the help of geographical advantages, to organically integrate cultural space with the natural geographical environment.

It seeks to solve the breakthrough in the combination of cultural elements and space in Inner Mongolia, with culture as the core, reprogramming of the original building blocks, and creating a functional layout with regional characteristics according to different population characteristics, cultural characteristics, and activity needs.

Based on the above three aspects, in the spatial renewal of Maan Town, it is necessary to constantly seek the balance between culture, architecture, space, and people, and emphasize the spiritual feeling brought by cultural space to achieve the best performance effect.

4.3 Design Method of Building Space Regeneration

The core concept of nomadic culture is the nomads' worship and dependence on nature, and their cherish and admiration for natural resources; with the development of history, the living environment has led to the formation of the typical characteristics of "living by water and grass"; Promote the production of various living utensils, harnesses, and weapons; the custom culture of religious beliefs has become an indispensable part of nomadic culture. Therefore, in the spatial form of the architectural space, through the analysis of the nomadic people's natural views, values and human environment, production methods, and religious beliefs, combined with the spatial layout of Maan Town, the connection between nomadic culture and space are established.

The natural view of nomadic culture-the integration of man, architecture and the natural environment. With an attitude of advocating nature, exploring the hidden laws of natural construction in architecture, and finding the spiritual existence of people in architecture is a certain blood relationship established by architecture, regional culture, and natural climate [4]. Trying to create a grassland in the city, based on the relationship between site selection and the city, with the help of natural landform advantages, make full use of the supply relationship of surrounding villages, and integrate the building with the environment; increase the depth of the vertical experience function and horizontal layout of the space (see Fig. 3).

Fig. 3. Site profile (picture source: self-drawn by the author)

1. Continuing ethnic characteristic activities-the digging of historical traces is an effective way to pursue ethnic-cultural memories, and the living conditions and daily activities of the predecessors are the carriers of historical traces. To seize the historical memory and inherit the nomadic culture, it is necessary not only to excavate architectural forms but also to retain activities with national characteristics [5]. In the design of the layout of Maan Town, the characteristic items of the racecourse and the racing track have been retained, and a viewing hall, a cultural and creative centre, an ethnic display centre, a blog and an archery sports centre have been implanted to activate the unbuilt functional areas while maintaining the basic functional facilities of the hotel (see Fig. 4).

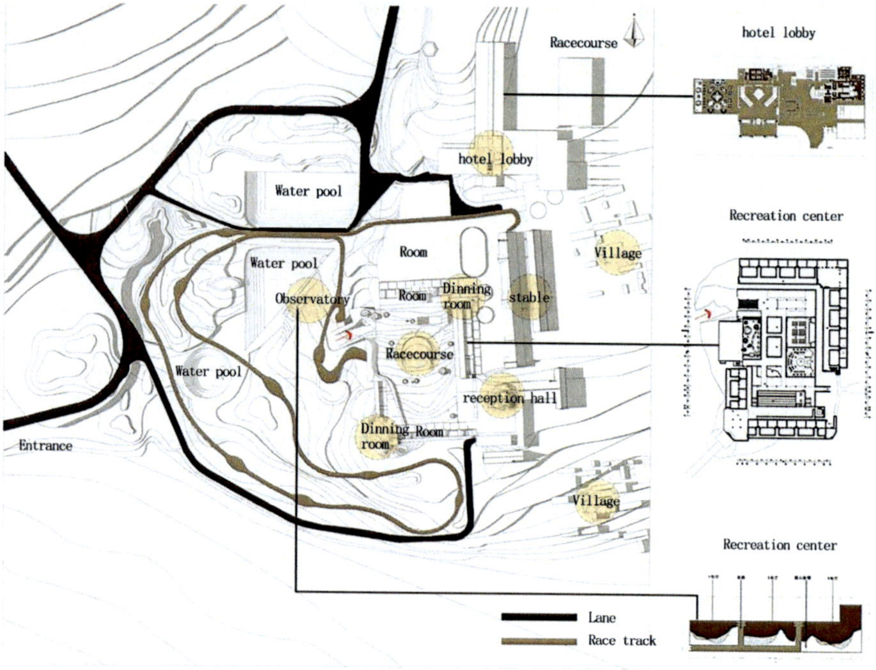

Fig. 4. Site plan analysis diagram (picture source: self-drawn by the author)

2. The source of inspiration for production, life, and religious beliefs-production and lifestyle produce various crafts such as utensils, harnesses, and weapons; various tools are used in the customs of religious beliefs. Looking for traces of the nomads' lives from these materialized tools. For example Stirrups, Turaga, milk teapots, milk buckets, bags, headgear, Sulu ingots, and other utensils all have strong ethnic characteristics. The exquisiteness of these utensils has often reached the highest level of craftsmanship, and the complicated decorative patterns can be used to refine the typical cultural elements by using structural reconstruction design techniques [6] (see Fig. 5).

Fig. 5. Various appliances (picture source: Internet)

In the interior space extrapolation, through the study of local culture and cultural traditions, culture is involved in the space, using the design method of deconstruction and reconstruction, translating into spatial forms, playing the important role of culture in space, and constructing the typical characteristics of regional cultural space [7] (see Fig. 6, 7).

Fig. 6. Hand-drawn scheme sketch (picture source: self-drawn by the author)

Fig. 7. Indoor renderings (picture source: self-drawn by the author)

3. Commercial operation model—Refer to the Disneyland Hotel's integrated business operation model for entertainment, dining, and lodging; the Disney team's management thinking has been running through the design concepts derived from the local area. Every Disney park and hotel in the world will vary according to the regional area. The regional characteristics of the country, the construction of a model suitable for local development. Localization of environmental facilities (Dis-ney Metro Station, Disney Town, a Disney Hotel); localization of service products (localization of tourist souvenirs, localization of park food, service concepts and methods, and localization of human resources). The design concepts and manage-ment measures at Disneyland have formed a complete Disney ecosystem, which has certain enlightenment for the planning and design of saddle town. Many Town is committed to building a regional cultural hotel with the theme of nomadic cul-ture, which needs to be integrated into and integrated with the local; the design of environmental facilities and service supporting products and related experience func-tion areas will start with culture and highlight the regional characteristics [8] (see Fig. 8).

Fig. 8. Outdoor renderings (picture source: self-drawn by the author)

4. Case study—referring to the design concept of Guangzhou Chimelong Hotel and Aman hotel in southwestern Utah. The USA fully demonstrates the integration of cultural and natural geography. Cameroon Hotel is a landmark building of Guangzhou hotel. It fully displays the "wild luxury" of the ecological environment. Through the conceptual design of the four themes of jungle, animals, castles, and Lingnan architecture, it presents a unique safari-style high-end resort hotel. Mangier is located in the tourist resort area of Utah in the southwestern United States. The design of Amangiri is good at combining spatial environment with natural geography, using local raw materials to generate building materials, which have a strong life and cultural atmosphere. The study of two representative cases provided design ideas for the renewal design of Maan Town. In terms of space performance, with the help of regional cultural characteristics, it can achieve a more provincial expression in terms of space material, color matching, and atmosphere creation (see Fig. 9).

Fig. 9. Indoor renderings (picture source: self-drawn by the author)

5 Concluding Remarks

Nomadic civilization and agricultural civilization are both important branches of Chinese civilization. This research explores the contemporary situation of the combination of nomadic culture and space through practical schemes. The design of cultural architectural space must be aware of the natural geographical environment, the local life and production methods, and the cognition of regional culture, humanities and history. With local-derived ideas as the core, it will inevitably make the architectural space show its unique style and distinguish it from the general architectural cultural space [9]. At present, many areas have ancient villages and villages with typical characteristics, but with the development of urbanization, they have not been protected due to various factors, and have even been demolished or abandoned, resulting in a waste of urban folk resources. How to transmit culture through architectural space design? Dissemination of culture? It is a work of long-term exploration. The research on the regeneration of architectural space in Ma'an Town provides a way to regenerate the regional cultural architectural space. It tries to reach a sustainable development path between culture and architecture. At the same time, it is the first regional theme resort in Inner Mongolia. This research hopes to be able to provide a reference for the development of the same type of characteristic resorts in the future.

References

1. Shao, H., Wu, Y.: An analysis of the origin of early nomadic culture. Northern Cult. Relics (01), 28–37 (2020)
2. Wang, M.: Nomadic society and its historical enlightenment. In: Nomadic Culture and Farming Culture-Advanced Forum of Anthropology, vol. 2009, pp. 304–309. Heilongjiang People's Publishing House, Hohhot, Inner Mongolia, China (2009)
3. Ma, G.: Knowledge system of nomadism and sustainable development. J. Qinghai Univ. Natl. (Soc. Sci. Ed.) **43**(04), 36–40 (2017)
4. Zhu, P.: An experiment of natural architecture—design of Shouxian culture and art center. Archit. J. **Z1**, 126–127 (2020)
5. Tang, H., Li, K.: Sacred places and scenic views—the design of Yan'an Baota mountain visitor center under the special geographical context. Archit. J. (08), 86–91 (2020)

6. Yang, T., Yao, S.: Research on interior design techniques based on the concept of localization —taking the design of Xi'an Grand Hyatt Hotel as an example. Furnit. Inter. Decor. (02), 107–109 (2021)
7. Jingtang, H., Jia, H., Guo, W.: From choice to expression—a study on the cultural shaping mode of contemporary cultural architecture. Archit. J. (12), 100–103 (2012)
8. Liu Mingguo, F.: Tracing to the source: on the connotation and essence of regional architecture. Planners **26**(S1), 92–95 (2010)
9. Jingtang, H., Wang, X.: Region, culture, time—ramble on the city and building style with Chinese characteristics. Archit. Pract. (10), 26–35 (2020)

Research on the Optimal Design of Baby Care Room Facilities Based on the Perspective of Public Space

Jingzhao Feng(✉)

Faculty of Humanities and Arts, Macau University of Science and Technology, Taipa, Macau, China
21098537AT30001@student.must.edu.mo

Abstract. With the rapid development of urbanization, the construction and renovation of the public baby care room has been heating up, receiving widespread attention from all walks of life in the city, and becoming the focus of sustainable urban development and livelihood issues. From the perspective of public space, this paper discussed that mother and babies are the main groups, and the facilities become the key entry point through spatial attributes and specific types of people. Therefore, three types of facilities were analyzed from multiple perspectives, namely signs, baby changing station, and entrance doors: user needs, safety, and ergonomics. The proposed optimization strategies provide a reference for the future construction of public baby care room.

Keywords: Baby care room · Facilities · Public space · Optimization

1 Introduction

With the popularization of the concept of breastfeeding, Taiwan, China passed the "Regulations on Breastfeeding in Public Places" in 2010, which emphasizes women's right to breastfeed in public spaces. This regulation highlights the issue of the construction of baby care rooms in public places. With the gradual implementation of three-child policy in China, the number of newborns is expected to continue to grow. In the future, effectively solving the conflict between supply and demand of public space for baby care rooms is a direction that urban construction and development must focus on. According to the city's long-term planning concept, baby care rooms actively constructed in line with sustainable urban development and humanistic care. In addition, modern cities must respect mothers and babies, and protect their rights and interests in public space activities by re-optimizing various facilities in baby care rooms. In the future, the continuously optimization of public facilities and baby care rooms has become a practical problem to be solved in domestic civil engineering projects.

P. Mendonça and N. D. Cortiços (Eds.): ICAMC 2021, LNCE 226, pp. 161–170, 2022.
https://doi.org/10.1007/978-3-030-94514-5_17

2 The Facilities of the Baby Care Room Analysis Status in Urban Public Spaces

According to the results of a survey, mothers of infants aged 0 to 2 years in Shanghai demand for baby care rooms in urban public spaces. More than 80% of mothers have breastfed and changed diapers in public places, and 98% of mothers consider that it is necessary to establish mother-infant facilities in public places [1]. Secondly, in a survey of 69 baby care rooms in five districts of Guangzhou, baby changing station had the highest demand for facilities, and 45.1% of them failed to find a delivery room on time [2]. Another author suggested prioritizing the optimization of baby care room facilities: diaper tables, seats and lockable doors to protect privacy [1].Some provinces and cities in China will develop standards and rules for baby care rooms, such as Shenzhen, Zhejiang and Jiangsu Province. However, many cities have not yet enacted regulations for baby care rooms. This study aims to optimize the use of high frequency facilities in baby care rooms, including signage, infant care counters, and entrance doors. Due to the special characteristics of baby care rooms, the optimized design requires a combination of ergonomics, aesthetics, materials science, optics, and other disciplines.

3 Optimized Design Ideas for Maternal and Infant Space Facilities in Public Spaces

Taken together, maternal and child health facilities need to improve safety, intelligence and aesthetics to provide a comfortable experience for users. Therefore, the premise of optimal design is to first meet the actual needs of users, optimize and improve through a sense of innovation, stimulate user satisfaction, and enhance the impression, goodwill and experience of public areas.

3.1 Optimized Design of Baby Care Room Sign

It is a significant source of information and the logo find can discover the spatial location and identify the type of space. In the baby care room, the optimization of the logo needs to adapted the size of the actual space, including the number, size and color. Designers need to pay attention a wide range of categories, recognition and styles, incorporating both internationally and nationally common signage systems.

Improving the signage system is one of the keys to designing public spaces. The author has designed 19 common signs in baby care rooms, as shown in Fig. 1. Considering the uniformity principle in color matching, one to two primary colors and one secondary color are recommended. The logo color is a combination of dark gray and orange, using a contrasting method of high saturation and low saturation. Meanwhile, the key information of the logo is set to orange, and the warm color with high saturation is called the bold color, which enhances the visual impact and concentration so that users can quickly identify the key information. Logo outline design style adopts a rounded rectangle, and the logo adds a description column according to the actual situation, including introduction, operation methods and notes, so that users can clearly understand the essential functions of each facility.

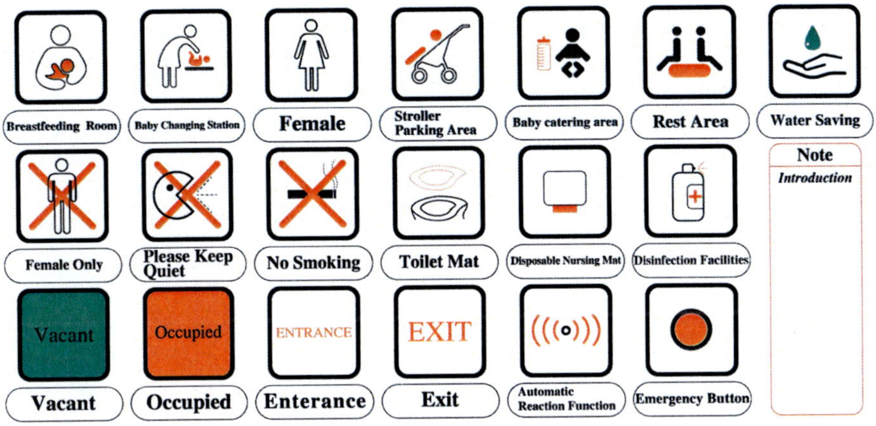

Fig. 1. 19 common signs in baby care room. (Painted by the author)

The logo measures 140 mm × 140 mm × 16 mm square with rounded corners. The material of its external frame is made of 10 mm dark gray rustproof stainless steel. The material is a 120 mm × 120 mm translucent frosted acrylic sheet in the central area. Both materials are stable and durable, moisture and corrosion resistant.

The design of the sign, the black side is designed with ultra-thin LED lights and intelligent sensor device with light slot, different from the traditional sign system. Based on spatial intelligence perception, when a person enters interior space, the backlight around the sign will automatically light up, and the middle area will be slightly brighter. If the person leaves the indoor space, the sign light automatically turns off, as shown in Fig. 2. In addition, babies are more sensitive to bright environments. The sign is not too bright. The author combines special characteristics of babies to enhance the recognition of the sign in a dark environment through the intelligent sensing backlight design.

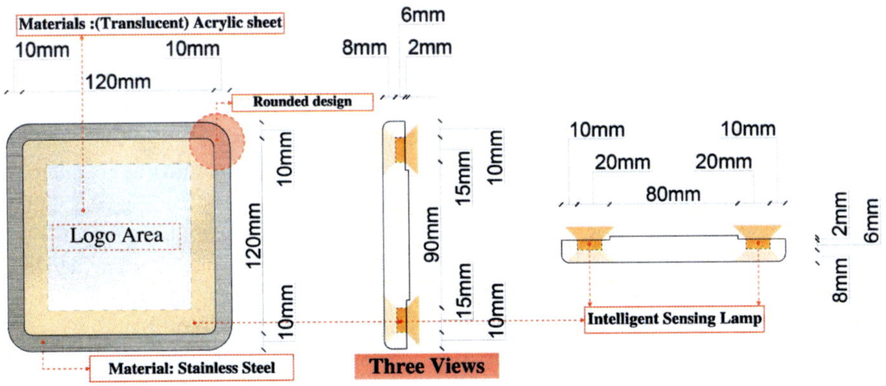

Fig. 2. Three views of the signs. (Painted by the author)

The sign adopts the concept of double-sided design for outdoor space, with the size of 300 mm × 160 mm × 20 mm, as shown in Fig. 3. The outer surface of the sign adopts a combination of stainless steel and acrylic plate, which is highly corrosion resistant, durable and easy to clean. The front and rear of the light fixture are made fully transparent acrylic panels, and the outer surface is made of a circle of embedded LED light strips with an intelligent automatic system. When the outdoor environment is dark, the light automatically turns on. This helps to improve logo recognition in the dark outdoor environment.

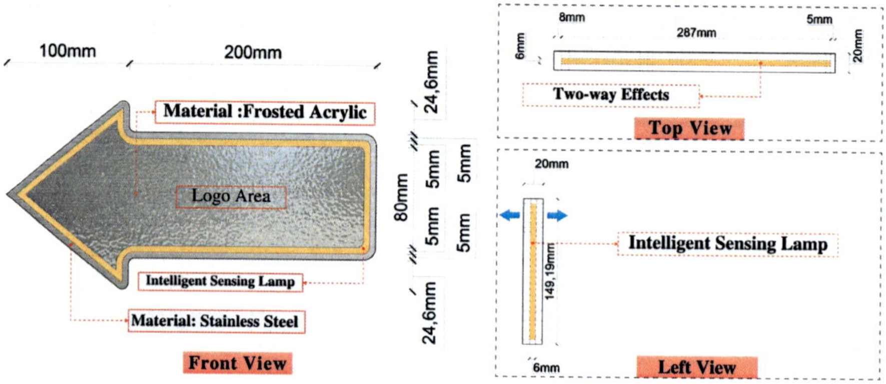

Fig. 3. Three views of outdoor signs. (Painted by the author)

Finally, the information about the baby care room is a cell mobile phone mapping application. Even if the person does not see the indicator in a public space, the person can search for the specific location information about the exact location of the baby care room through the map app.

3.2 Optimized Design of Baby Changing Station

The baby changing station that provides nursing functions, including changing diapers and clothes for babies, and is one of the basic facilities in the baby care room. Therefore, the "Guide to for the Construction of Breastfeeding Facilities in Public Places in Jiangsu Local Standards" and the "Design Regulations for Baby Care Rooms in Public Places in Shenzhen" stipulate that baby changing station must be installed in baby care rooms. Most women use the baby care rooms for breastfeeding and diaper changing [3]. Another author surveyed 81 commercial establishments equipped with sign, a nursing rooms, and a baby changing station in the baby care room [4]. In a survey of 82 places with baby care room in Shanghai, the highest percentage of baby changing stations was found [5]. According to he survey of each city, baby changing stations play a crucial role in baby care rooms. Most of the studies only focus on the number of baby care rooms, installation heights and configuration facilities, and few systematically optimized the baby changing stations. In this study, a set of horizontal recessed folding guardrail baby changing stations was optimized.

The Design Concept of Baby Changing Station. The baby changing station adopts a simple design concept, and the embedded design saves the horizontal space of the facade, maintains the simplicity and integrity of the facade space, and reduces the design of decorative elements and other elements. In addition, this design eliminates the traditional storage box inside the baby care station, reducing the hiding of dirt and facilitating daily cleaning and disinfection. The horizontal folding design is used to close the table when not in use. Rigid brackets and hardware components are made of high-strength metal, with automatic and moderate closing and opening functions, and an overall load-bearing capacity of up to 30 kg. According to the "Reference Standards for Growth and Development of Children Under 7 in China" issued by the National Health and Health Commission of the People's Republic of China, the height of a 2-year-old boy of 33 months is about 84.4–107.2 cm. Girls are lower than boys. The weight of boys is about 10.24–19.89 kg, and the 30 kg weight-bearing design is considerably larger than 19.89 kg to maximize the safety of infants.

Detailed Optimized Design of Baby Changing Station. The length, width and height of the entire baby changing station are 980 mm × 190 mm × 600 mm. The length, width and height of the baby changing station are 910 mm × 510 mm × 36 mm, with a 10mm recess, and the length and width is 850 mm × 450 mm. The bottom of the nursing table extends 25 mm for support, as shown in Fig. 4. Since infants do not have self-protection behavior and awareness, the appearance is designed with a rounded corner concept. Based on the protection concept, the left and right sides of the baby changing station are set with a semi-oval with a height of 84mm, width of 275 mm and 200 mm, and thickness of 14 mm. In front of the baby changing station, there are two semi-oval with a height of 84 mm, width of 195 mm, and thickness of 14 mm, as shown in Fig. 5.

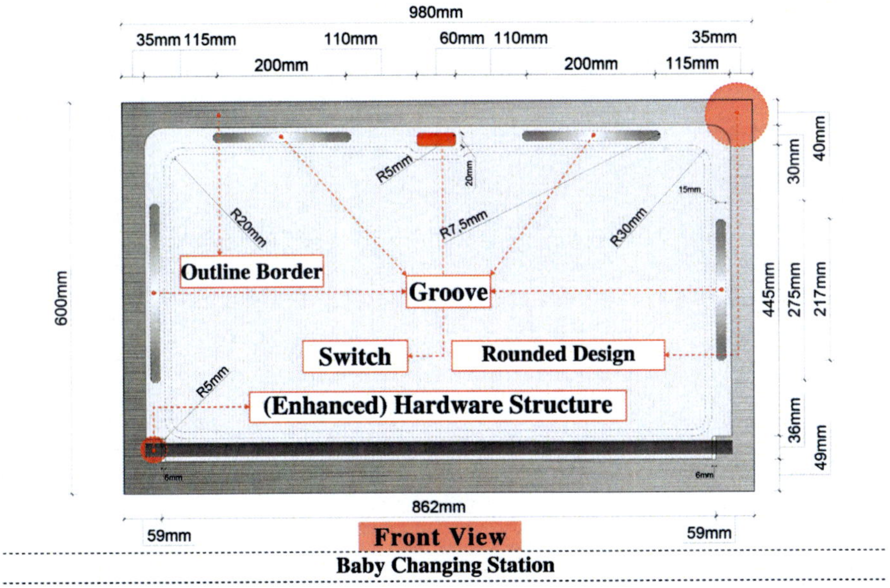

Fig. 4. Front view of baby changing station. (Painted by the author)

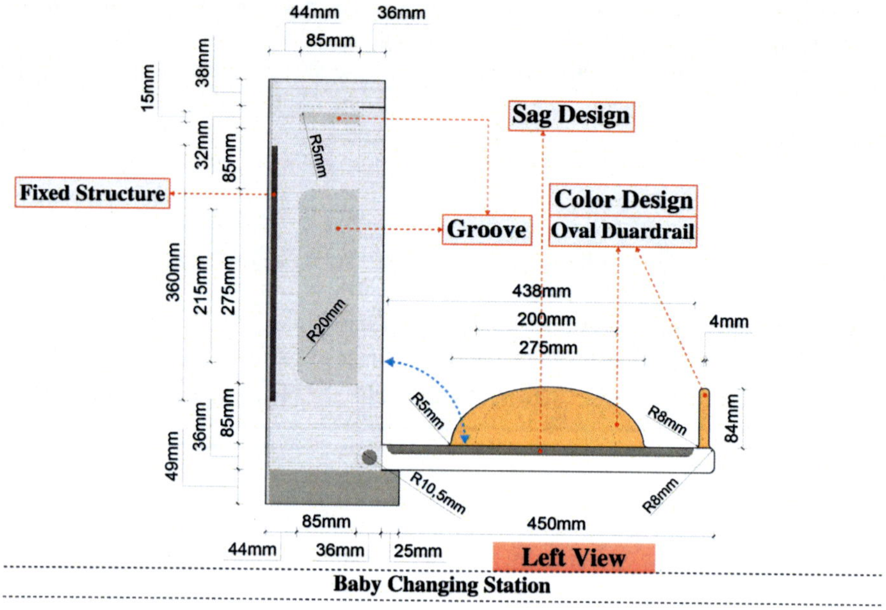

Fig. 5. Left view of the baby changing station. (Painted by the author)

Material and Color of Baby Changing Station. A baby changing station is made of two main materials. The outer facade box is made of 316 stainless steel shell with anti-rust treatment on the surface to ensure a sturdy appearance and good corrosion resistance. The baby changing station chooses "antibacterial high-density polyethylene material," which has good antibacterial properties, good stability, safe and non-toxic, high rigidity and toughness. In the next position, the color is medium to light gray. The gray color has a specific light absorption effect. The three-sided baffle is designed in bright gray to add a vibrant atmosphere, as shown in Fig. 6.

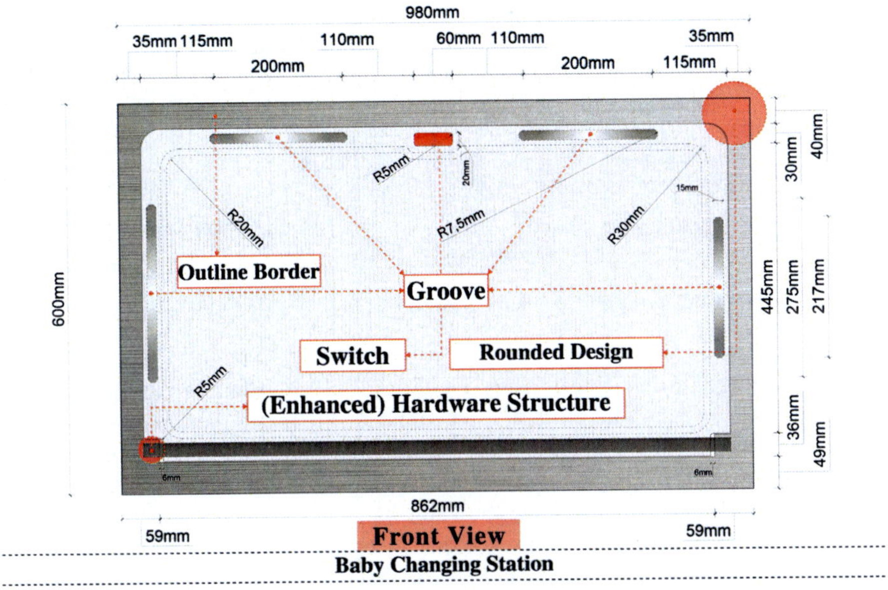

Fig. 6. Top view of the baby changing station. (Painted by the author)

The Installation Height Setting of Baby Changing Station. "Code for Barrier-Free Design (GB50763-2012)" The upper surface of low-level service facilities should be 700 mm–850 mm above the ground. "Jiangsu Provincial Standard (DB32/ T 3375-2018) Guidelines for the Construction of Breastfeeding Facilities in Public Places" and "Shenzhen" Design Regulations for Maternity and Infant Rooms in Public Places (SJG54-2019) "recommend that large, medium and small baby care rooms must be equipped with nursing tables and breastfeeding areas. The local standards of Jiangsu province proposes that the height of the baby changing station should be 700 mm. In addition, some authors suggest that the baby changing station should be installed with a height between 750 mm and 900 mm [6]. Based on Asian operational patterns, other researchers have suggested a relatively comfortable operating height for baby changing station, which is 860 mm [7]. At the same time, baby changing station require enhanced structural the robustness and scientific installation height.

The recommended installation height of baby changing station is about 700 mm–860 mm, which is divided into four height standards of 700 mm, 750 mm, 800 mm and 860 mm. The installation height mainly depends on the actual space area of the baby care room. For this reason, the author proposes that the baby changing stations in the baby care rooms should be 750 mm and 800 mm respectively, and large nursery rooms should have multiple baby-changing stations of different heights. The main principle is to consider women with specific height differences. Secondly, when the other conditions of the baby changing station remain the same, a hidden hook is designed in the middle area at the bottom of the operating floor to hang lighter objects with weights up to the specified range, considering the convenience factor, as shown in Fig. 7.

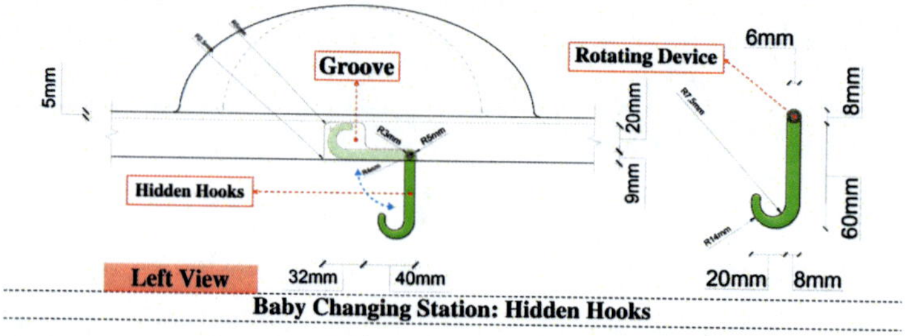

Fig. 7. Hook design of the baby changing station. (Painted by the author)

Baby Changing Station Supporting Facilities Design. Considering the hygiene requirements of the baby care room, it is recommended to install the embedded four-in-one multifunctional box on the facade within 300 mm –400 mm from the baby changing station. The length, width and height of the multifunctional box are 490 mm × 170 mm × 1635 mm, as shown in Fig. 8.

The multifunctional box has a curved corner design and a removable trash can at the bottom. It is equipped with inductive hand sanitizer and disposable paper towels on top. In addition, other modules are divided into disposable pad paper, disposable wet wipes and disinfected supplies. Based on this study, the main original intention of designing the four-in-one was to better protect the safety and health of the baby.

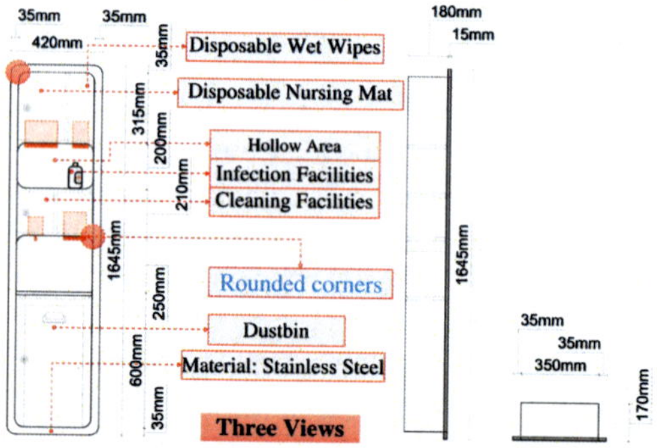

Fig. 8. Embedded four-in-one multifunctional box. (Painted by the author)

4 Optimization of Outside Doors for Baby Care Room

When a person is pushing a stroller or holding a baby with both hands, it can be an inconvenience. Exterior doors are a frequently used facility in buildings. Therefore, the designer has to consider three key factors of durability, stability and safety. At the same time, the design incorporates a new concept of intelligent sensing equipment. If someone approaches the door, it will automatically open and close, thus improving the convenience and experience of entering the space. In addition, when people push strollers, there may be a risk of collisions. Therefore, sliding door should not be made of fragile and easily damaged materials. The author recommends that the center portion of the sliding gate be made of impact-resistant materials to improve safety.

The outer diameter width of a single stroller is about 650 mm and the length is about 1150 mm, and the outer diameter width of the a double stroller is about 815 mm [4]. Regarding the dimensions of conventional strollers, the apparent width of a single door in baby care room is hardly less than 900 mm.

The sliding doors are opened and closed with left and right concealed burial. The smart sensor device replaces the traditional door handle and the motorized track is installed under the sliding door as shown in Fig. 9. The appearance of the sliding door is recommended to be made of high hardness and wear-resistant materials to avoid collision with the stroller. In addition, the sensing device can be used to record the number of people reaching the set number, and activate the restriction command. At this point, the red light comes on and the sliding door is locked. The main function is to prohibit crowding in the interior space.

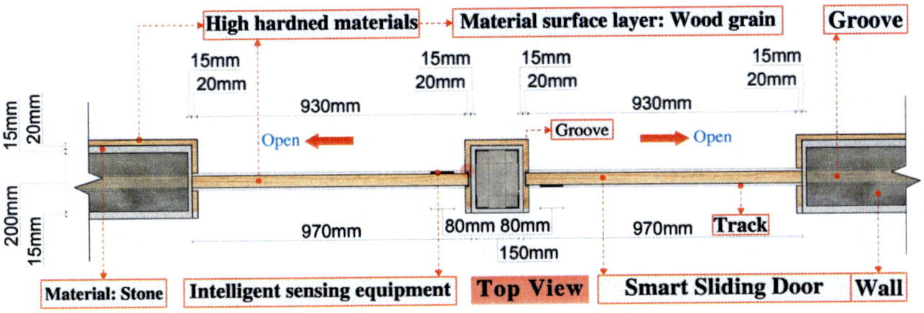

Fig. 9. Optimized design of sliding door for baby care room. (Painted by the author)

5 Conclusion

In a modern cities, baby care rooms are a favorable public space that benefits citizens as the use of space by mothers with children is fundamental. Public baby care rooms have been recently built and optimized to symbolize the modernization of urban civilization to enhance the image of the city.

Firstly, this study proposes the optimal design and improvement measures for baby care rooms with the core principle of safety, functionality, convenience, and aesthetics, with signs, nursing tables and exterior doors as the core principles. In addition, according to the different facilities, the author integrates intelligent modules to enhance the spatial experience, reflecting convenience, automation, and safety.

Secondly, the optimal design of public baby care rooms needs to take into account the actual situation of public space attributes, scale and size. However, the design process meets the needs of users, considering many aspects such as aesthetics, durability, and safety.

Finally, it failed to optimize the materials and the equipment for baby care room on a case-by-case basis. This study hopes to provide a valuable reference for future facilities by optimizing the design of the above three aspects.

References

1. Ma, X.Y., Mu, L., Jiang, H.: Investigation and analysis of mother's use of maternal and child facilities in public places in Shanghai 0–2 Year old infants. Mater. Child Health Care China **34**(13), 3070–3072 (2019)
2. Zhou, S.J., Ke, M.B.: Analyzing the supply and demand characteristics of baby care rooms in public places and exploring development strategies: taking Guangzhou as an example. Stat. Manag. **10**, 59–62 (2017)
3. Li, Y., Zhang, Y.M., Liu, Y.J.: Current problems and implementation advices for urban baby care room—case study of Nanshan district in Shenzhen. Archit. Cult. **05**, 85–88 (2018)
4. Pu, W.J.: A survey of baby care rooms in commercial public space and suggested improvements. Archit. J. **10**, 78–82 (2016)
5. Ma, X.Y., Mu, L., Jiang, H.: Status quo and development dilemma analysis of baby care rooms in public places in Shanghai. Med. Soc. **31**(12), 50–52 (2018)
6. Teng, Q.: The Investigation on the Space Design of the Baby Care Room in Public Places. South China University of Technology, Guangzhou (2017)
7. Jin, S., Huang, S.Y., Lin, L.P.: Design of miniature folding baby care rooms in public space. Packag. Eng. **41**(24), 182–187 (2020)

Surveillance of Anomalous Displacements in Large Buildings Through the Internet of Things: What Can Teach us the Surfside-Miami Events?

Huber Nieto-Chaupis(✉)

Universidad Autónoma del Perú, Panamericana Sur, Km. 16.3 Villa el Salvador, Lima, Perú

Abstract. The recent incidents at Surfside, Miami FL, can be seen as an inherent manifestation of modern structures whose risk were not anticipated in all, so that the periodical assessment of ground layers of particularly very large buildings might be a must henceforth by following local regulations done by specialized analyzers to calculate in a quantitative manner the deformation of structures as well as their exact ageing. In this paper, the idea of the implementation of a dedicated internet of things network to be called Internet of Tall Buildings (ITB in short) that should be running in an unstoppable manner for the sensing of geometrical deviations either at the top or bottom is presented. Thus, geometrical scenarios are proposed in order to measure the growth of angular displacements. A computational algorithm was designed in order to provide a solid idea of the potential advantage of ITB that might be realizable after the events of Surfside-Miami. The algorithm targets to calculate the angular deviations and how it is increasing in time. Once the allowed values are either reached or surpassed then the algorithm triggers all possible alarms that would give enough time to people to make fast decisions as well as to optimize their self-protection actions.

Keywords: Building collapse · Internet of things · Bending

1 Introduction

As witnessed at the end of June (2021) the events of building collapse at Surfside at the area de North Beach, Miami [1], have triggered a plethora of questions that are fairly needed in order to guarantee the public harmony and wellness of habitants, in particular in all those that reside in tall buildings. Certainly, although it is an isolated event, the aftermaths of events have triggered the alarms as well as the re-configuration and re-assessment of both the latest civil engineering and urbanistic technologies [2, 3] and their advanced strategies projected onto the expansion of modern cities. Clearly, one can see here various points that do not necessarily are at the edge of urbanism and civil engineering but also rather linked to those aspects that are totally inside of the territory of telecommunications and their related disciplines. In this way one can state that the

P. Mendonça and N. D. Cortiços (Eds.): ICAMC 2021, LNCE 226, pp. 171–177, 2022.
https://doi.org/10.1007/978-3-030-94514-5_18

continuous surveillance of the urbanistic, security and maintenance of housing might be joined to well-designed program of a robust management of the so-called smart cities by the which are correlate to the wise usage of wireless communications such as the Internet technologies. Therefore, one finds a list of facts that might be well combined each other and that plays a concrete role as to the permanent surveillance at the wellness of housing in coast cities that are closely related to tourism and comfort. These facts seen by the authors can be resumed as follows:

- Full and sincere maintenance,
- Deeply ageing assessment of concrete,
- Random water draining,
- Evolution of weight distribution,
- Impact of seismic events,
- Unstopped surveillance after climate fluctuation [4].

The aggressive apparition of Internet-based technologies suggests in a strong manner its imminent usage on the recent advances of building construction as well as surveillance and variables measurement, all these in the sense of identifying the critic points that can be anticipated to avoid unexpected events.

2 The Anomalous Events

Coast cities that face extreme climate conditions are rather sensitive to undergoing anomalous events in their urbanistic structures in special their tall buildings in the sense of experiencing: strong winds [5], seismic movements [6], water draining by ocean, rivers or sewerage and drinking water networks below the city. In Fig. 1 are sketched up to three possible scenarios of anomalous events in tall buildings: (i) In left side is seen the case of apparent rest in the sense that it is possible overload in the levels of building and roof producing fatigue [7]. Nevertheless, the fulfillment of equilibrium

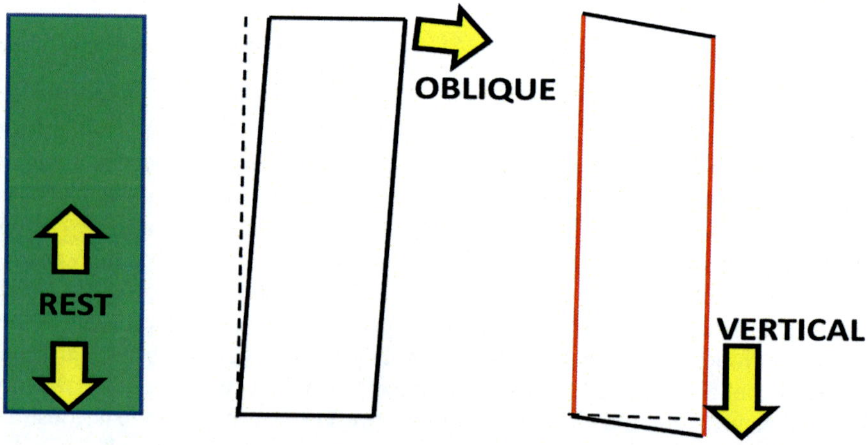

Fig. 1. Possible anomalous and concurrent events in tall buildings.

and the compensation of forces would trigger a scenario of rest that would be enough to satisfy the basic requisites of static with negligible contribution of dynamical effects [8]. (ii) In middle is seen the case that the building can experience deviations by strong winds that can produce small angle inclinations so that one might to expect an oblique position of building that can resist due to a firm ground free of water draining or analogues, (iii) in right side one can see the vertical displacement due to overweight and asymmetric distribution of load by the which the sinking might be imminent.

3 Random Concurrent Events

From Fig. 1 it is possible to expect in a random manner concurrent event in the sense that all scenarios as sketched above can occur simultaneously. It is actually a point that can be well discussed through probabilistic laws more than a sincere analysis done through the Newtonian Dynamics and Static in a mathematical and computational manner. Thus, one can tentatively write down a possible probabilistic formulation of risk given by:

$$P_R = \sum_{n=0}^{N} c_n(t) p_n(t) \tag{1}$$

with $c_n(t)$ time-dependent factor that increases the total risk probability per each one of the different contributions given by $p_n(t)$ running up to N possible scenarios of risk [9]. Clearly Eq. 1 can be perceived as a very basic formulation of a risk against disadvantageous situations coming from the natural facts mainly seismic and humidity causes. For example, consider a coast city and buildings, then one can expect at a certain time a strong climate behavior that can create draining by rains as well as the overflow of oceans that can surpass the security systems in these cases. In addition to this a concurrent activity can be done by the ageing of ground structures produced by silent seismic activity (that might noy be perceived in a first instance) fact that would open new spaces at the range of millimeter level for water accumulation.

4 The Risk Variables Measurement

In Fig. 2, the scenario of an imminent inclination of building produced by winds (for example) is sketched. It would constitute a critic issue at tall building in the sense of rupture at the internal supports that can be progressively under an evolution in a silent manner. An evident variable that can be measured with an acceptable accuracy turns out to be the angular displacement. Although the sketch is showing an exaggerated representation of inclination, one can implement just around the edge of tall buildings a dual device whose end is that of sensing and emitting signals can be done through a emitter and receiver device consisting in a small-power laser. Thus, in left side one can see the case that a detector is receiving a signal from an adjacent building. The role of this device is that of measuring the height of emitter. In right side is the scenario by the which the building can be inclined by a small angle β. Thus, by knowing the distance between two adjacent edges of buildings and the displacement at the arrival of signals, then one can estimate the angle and the position of the gravity center when it is undergoing this anomalous event. Thus, the angle can be estimate from:

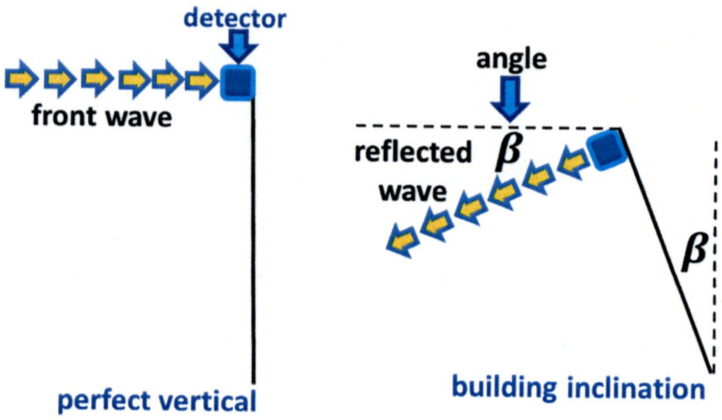

Fig. 2. Inclination of building produced by string winds as commonly is the case of coast cities.

$$\beta = Atan\left(\frac{\Delta z}{L}\right) \quad (2)$$

As shown below in Fig. 3, in left-side it is sketched the inclination can be disadvantageous in the sense that it can produce a net force just initializing a pressure over the structures below. The block called "net force" is composed by all material above that can be roofs, habitants' goods, pools and other type of loads. In right-side one can see a potential sketch of can be expected by the action static of net force. The sketch shows a particular case of vertical displacement of a certain roof but in an oblique manner. In consequence one might expect a new net force the acts onto the all

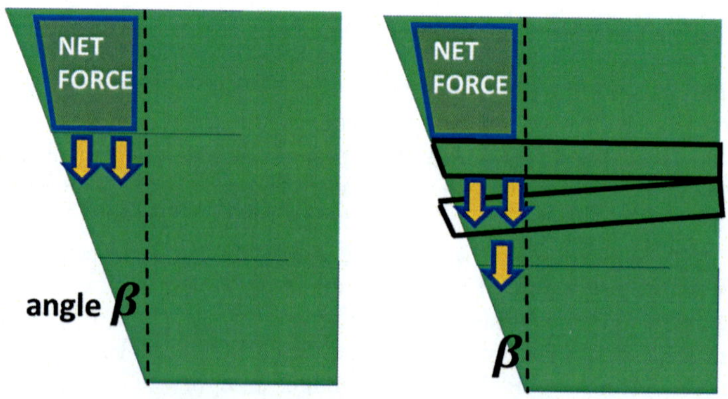

Fig. 3. Apparition of a net force and bending beams over the concrete and all load due the inclination of an angle β.

structures below. This scenario has clear consequences: (i) the deviation of gravity center, (ii) intern of columns and bending construction beams por two vectorial forces: the weight down and gravity. The imminent usage of D'Alembert principle applies because the infinitesimal angular displacement.

5 The Internet of Tall Things (ITT) Implementation

In Fig. 4 it is shown a potential scenario by the which one finds a continuous sensing of angle β by devices deployed on the edges of tops of adjacent buildings. Thus, one can trigger the accumulated signal from an specialized antenna to various centers of signal analysis in order to combine this information and possible consequences by climate impact if any (as well as seismic events). The end of idea of an Internet of Tall Things network (derived from Internet of Things [10–12]) is the permanent measurement of angle β and the sending of alarms in the case of either inclination and sinking and both simultaneously. Since the common loads in large buildings can be of order of tens of tons, one might to expect a substantial deviation from the central axes of building giving rise to the inclination. It should be noted that these prospective devices can also to measure movements by seismic events. It is interesting to note about the duration of any seismic event and its correlation with the amplitude of β. In ither words, one can speculate about its relation in the sense that this can be linear or nonlinear. The continuous measurement of β at time would give rise to a time-dependent curve by which one expects that it exhibits a increasing behavior, this is important since such curve would indicate the evolution of β. Only experience and data can tell us about the shape of curve of $\beta(t)$. The Newton Inertial law would indicate that the angle deviation might be irreversible so that the lack of sensing of angles less than β would be a pure event of imminent alarm by the which habitants can be informed with substantial anticipation to leave buildings. A possible algorithm for triggering alarms is done below:

ALGORITHM

```
DO j = 1, J time slot provided by ITT
 DO q = 1, Q number of buildings
  DO k = 1, K number of devices per building
    INITIALIZE CALIBATION DEVICE
    INITIALIZE ANGLES
    BETA(j,q,k) = ATAN[ΔZ(q,k)/L(q,k)]
    DESV(j) = BETA(j,q,k) + BETA(j-1,q,k)
      IF(j=J) THEN
        DELTA = DESV(j)
         IF (DELTA > REF) THEN
         TRIGGER ALARM
         ENDIF
      ENDIF
   ENDDO
  ENDDO
 ENDDO
```

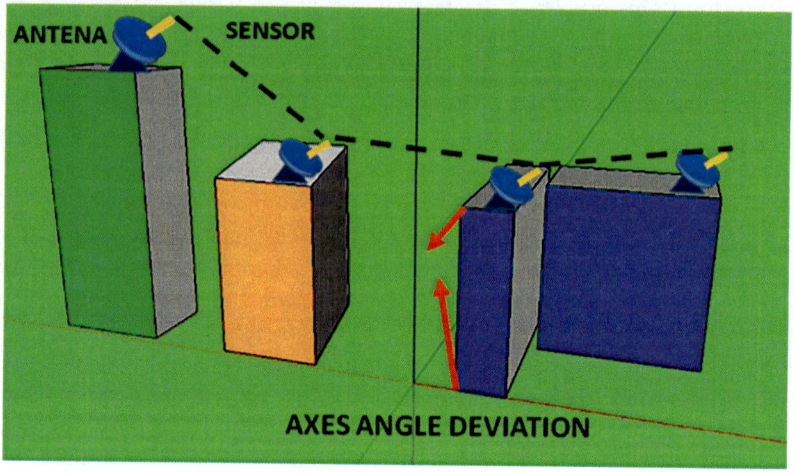

Fig. 4. Possible scenario of implementation of Internet of Tall Things network through the communication of wireless devices at the top of all adjacent buildings.

6 Conclusion

In this paper a proposal of an Internet of tall Things network has been provided under the grounds that tall buildings might to experience angular deviations due to natural causes such as climate and material ageing. The main purpose of this prospective network is that of its permanent measurement of angles by inclination that might appear unexpectedly and that are translated as a kind of pre-collapse [13]. On the other hand, the early identification of these angular variations might be advantageous in the sense that specialized civil engineering might to take preventive actions to restore a firm geometry and exact forces cancellation that would guarantee the wellness residents [14].

References

1. https://www.nytimes.com/2021/06/26/us/miami-building-collapse-investigation.html
2. Li, Q.S., Fang, J.Q., Jeary, A.P., Wong, C.K.: Full scale measurements of wind effects on tall buildings. J. Wind Eng. Ind. Aerodyn. **74–76**, 741–750 (1998)
3. Li, Q.S.: Stability of tall buildings with shear-wall structures. Eng. Struct. **23**(9), 1177–1185 (2001)
4. Ricciardelli, F., David Pizzimenti, A., Mattei, M.: Passive and active mass damper control of the response of tall buildings to wind gustiness. Eng. Struct. **25**(9), 1199–1209 (2003)
5. Li-Teh, L., Chiang, W.-L., Tang, J.-P., Liu, M.-Y., Chen, C.-W.: Active control for a benchmark building under wind excitations. J. Wind Eng. Ind. Aerodyn. **91**(4), 469–493 (2003)
6. Calio, I., Marletta, M., Vinciprova, F.: Seismic response of multi-storey buildings base-isolated by friction devices with restoring properties. Comput. Struct. **81**(28–29), 2589–2599 (2003)

7. Palmeri, A., Ricciardelli, F.: Fatigue analyses of buildings with viscoelastic dampers. J. Wind Eng. Ind. Aerodyn. **94**(5), 377–395 (2006)
8. Kim, Y.-M., You, K.-P., Ko, N.-H.: Across-wind responses of an aeroelastic tapered tall building. J. Wind Eng. Ind. Aerodyn. **96**(8–9), 1307–1319 (2008)
9. Stoica, M., Medina, R.A., McCuen, R.H.: improved probabilistic quantification of drift demands for seismic evaluation. Struct. Saf. **29**(2), 132–145 (2007)
10. Atzori, L., Iera, A., Morabito, G.: The internet of things: a survey. Comput. Netw. **54**(15), 2787–2805 (2010)
11. Fersi, G.: A distributed and flexible architecture for internet of things. Procedia Comput. Sci. **73**, 130–137 (2015)
12. Boyes, H., Hallaq, B., Cunningham, J., Watson, T.: The industrial internet of things (IIoT): an analysis framework. Comput. Ind. **101**, 1–12 (2018)
13. Panahi, S., Zahrai, S.M.: Performance of typical plan concrete buildings under progressive collapse. Structures **31**, 1163–1172 (2021)
14. Yan, Z., Zhao, H., Wang, F., Zhang, X., Cai, S., Mei, X.: Rapid assessment of building collapse based on sequential dynamic fusion of multi-source disaster information from news media. Int. J. Disaster Risk Reduction **51**, 101910 (2020)

Research on the Design of Spatial Environment Health in Primary and Middle Schools in the Post-epidemic Era

Honglei Yu[1,2](✉), Shanshan Zhang[1,2], and Benchen Fu[3,4]

[1] School of Architecture, Harbin Institute of Technology, Harbin 150006, China
[2] Key Laboratory of Cold Region Urban and Rural Human Settlement Environment Science and Technology, Ministry of Industry and Information Technology, Harbin 150006, China
[3] School of Architecture and Urban Planning, Shenzhen University, Shenzhen 518040, China
[4] Shenzhen General Institute of Architectural Design and Research Co., Ltd., Shenzhen 518000, China

Abstract. In the context of the global pandemic of COVID-19, aiming at problems such as restricted teaching activities in primary and middle schools and the impact on the mental health of children and adolescents, through the analysis of the space environmental needs of primary and middle schools in the post-epidemic era, three design strategies are proposed for flexible space design, space information construction, and space security and comfort creation, provide ideas for the construction of primary and middle schools in the future.

Keywords: Post-epidemic era · Primary and middle schools · Spatial flexibility · Spatial informatization · Sense of security and comfort

1 Introduction

The sudden outbreak of COVID-19 brings vast challenges and impacts to all walks of life. Primary and middle schools have been exposed to severe problems and deficiencies in response to public health emergencies. As a temporary measure, the suspension of primary and middle schools interrupts the normal teaching rhythm and brings certain pressure and challenges to parents and society. Therefore, after the epidemic is effectively dealt with and controlled, a full resumption of classes becomes inevitable. Faced with the dense population of primary and middle schools and the poor ability of students to protect themselves, how to provide a safe and healthy space environment for teachers and students returning to campus has become a new topic for campus construction in the post-epidemic era.

P. Mendonça and N. D. Cortiços (Eds.): ICAMC 2021, LNCE 226, pp. 178–187, 2022.
https://doi.org/10.1007/978-3-030-94514-5_19

2 New thinking on the Space Environment of Primary and Middle Schools Under COVID-19

On October 13, 2021, the World Health Organization established a new scientific team to study the traceability of the new crown to deal with future pandemics. "United Nations News" reported that although COVID-19 is the latest virus to cause a pandemic, it will not be the last one, and the world must make better preparations for any future "unknown diseases" [1]. It is foreseeable that after the end of COVID-19, we will not be able to return to the past. The "post-epidemic era" has arrived. The spatial environment of primary and middle schools needs to take daily teaching, epidemic prevention, and control into account, along with a high degree of adaptability to respond to future public health emergencies.

Since the outbreak of COVID-19, relevant government departments have in-depth research and judgment on the form of the epidemic. They have taken practical and effective measures actively to deal with the epidemic. The Ministry of Education and the National Health Commission jointly issued a number of epidemic prevention and control guidelines and technical plans for primary and middle schools (Tab. 1) in view of the conditions such as the learning and living environment in primary and middle schools, the characteristics of students in different age groups. These guidelines and plans aim to clarify the epidemic prevention and control requirements for organizational guarantees and system, personnel control, regarding key areas, environmental sanitation, emergency treatment of suspected infection symptoms, and for overseas teachers and students to return to school. Among them, the following requirements are put forward concerning the space environment: strengthening indoor ventilation and disinfection, avoiding large-scale group activities, controlling time for small meetings, and ensuring that the personnel is separated by more than 1 m, canteens are distributed in time to avoid concentration, and temporary quarantine places are set up [2].

A healthy space environment is an essential guarantee for the regular operation of primary and secondary schools in the "post-epidemic era". From the perspective of users, it has two levels of meaning: The first level of meaning is a space environment that is beneficial to the user's physical health, and the second level of meaning is a space environment that is beneficial to the user's mental health. COVID-19 is highly contagious and has an incubation period, while elementary and middle school students have the characteristics of easy gathering, weak safety awareness, and low immunity. A space environment that can effectively reduce infectiousness is a need for epidemic prevention and control, also the most basic requirement for teachers and students to resume classes. During the epidemic, the Chinese government advocated "suspending classes without stopping school" and actively carried out online teaching with the help of information technology. Although it effectively solves the urgent learning needs of primary and middle school students, factors such as long staying at home, lack of social behavior, and external environmental pressure from the epidemic have led to mental health problems such as increased pressure [3]. Students in the post-epidemic era can return to school. However, there are still potential risks that are threatening to their mental health, such as pressure from the external environment, lack of social trust, as well as the changes in study habits and spatial environment.

Table 1. Epidemic prevention and control guidelines and technical plans jointly issued by the Ministry of Education and the National Health Commission.

Name	Time
Guidelines on COVID-19 Prevention and Control in Primary and Middle Schools	March 12, 2020
Technical Plan for Prevention and Control of COVID-19 in Primary and Middle Schools (Revised Edition)	May 7, 2020
Guiding Opinions on Doing a Good Job in School Physical Education under Normalized Prevention and Control of COVID-19	May 20, 2020
20 Questions about Resuming School and Class under the Normalization of Prevention and Control of COVID-19	June 12, 2020
Technical Plan for Prevention and Control of COVID-19 in Primary and Middle schools in Autumn and Winter (Updated Version)	August 13, 2020
Technical Plan for Prevention and Control of COVID-19 in Primary and Middle schools in the Spring Semester (Third Edition)	February 2 2021

The impact of COVID-19 on the entire society is profound. The design of the space environment of primary and secondary schools should adjust to the physical and psychological health, the safety of teachers and students, and the unknown changes and challenges in the post-epidemic era, in order to realize that the space environment serves people.

3 Primary and Middle School Space Environmental Needs, User's Behavior and Psychology in the Post-epidemic Era

According to the view of environmental behavior, the user's behavior interacts with the space environment and influences each other. On the one hand, the space environment can guide or restrict user behavior; on the other hand, users can also modify the space environment to meet their own behavioral psychological needs. In other words, space and behavior are embodied in two aspects. The space influences the behavior of users, and users' behaviour demands create space. Based on the coupling relationship between the user's behavior psychology and the space environment, this paper analyzes the space environment needs of primary and middle schools in the "post-epidemic era" by studying the user's behavior psychology. It is embodied in three aspects, the upgrade of management behavior, the change of teaching behavior, and the attention to mental health that have caused new demands for the space environment.

3.1 Upgrade of Management Behavior

Compared with before the epidemic, the standards of school management, operation and maintenance in primary and secondary schools have been upgraded. First, it is essential to strengthen school access management, and to increase epidemic prevention measures such as temperature measurement, registration, and code scanning. In addition, there should be space to respond to emergency isolation. When parents pick up and drop off students, they should avoid the crowd and crossing flow lines. Therefore, reasonable design of flow line and space are necessary. Secondly, gathering should be avoided in school public spaces. Specific measures include ensuring a safe distance of more than 1 m, advocating extracurricular activities in different areas, and controlling the number of elevator passengers. Finally, to ensure the daily disinfection and natural ventilation of the campus, schools without natural ventilation should regulate the use of air conditioners and take full advantage of the information platform to form information sharing among multiple departments such as educational administration, student management, logistics security, and campus health. The original space and environment of campus cannot meet the upgrade of management behavior, which brings challenges to the future development of primary and secondary schools.

3.2 Change of Teaching Behavior

The epidemic has boosted the development of distance teaching, and teaching behavior has undergone tremendous changes. When COVID-19 was severe, the school's teaching venues were moved, and the schools were asked to close. The classrooms were idled as the realistic teaching space that has lasted for many years, and the network space has been transformed from the original auxiliary method into an indispensable educational method. It is regarded as the only safe teaching space because it can avoid gathering and human contact. The rapid development of information technology has created conditions for distance teaching. During COVID-19, teachers and students across the country can resume classes through applications such as "Tencent Conference" and "DingTalk". The distance teaching has been transformed from the one-way teaching of MOOC and flipped classrooms to interactive teaching [4].

3.3 Attention to Mental Health

As the epidemic takes longer and longer, students' mental health problems are more likely to occur. Elementary and middle school students are in a key stage of brain development and hormonal changes. They are more susceptible to the influence of external environment, social behaviors, and emotional factors. Primary and middle school students become more anxious due to the changes in the way of learning and their lifestyles caused by the epidemic situation.It is more likely for these children to have psychological problems such as depression and anxiety. The students who are not in good relationship with their families may psychologically suffer more from the long-term quarantine at home. In April 2020, it was shown by a survey of 3,239 students from high school or with higher education in the United States that 57% of students thought their mental health got worse, while 18% said that their mental health was sharply worsened due to the new crown epidemic [5]. It is the responsibility of teachers and parents to pay attention to students' mental health. How to benefit the psychological problems of primary and secondary school students through space design is the focus of the future development of primary and secondary schools.

4 A Healthy Design Strategy for the Space Environment of Primary and Middle Schools Adapted to the Post-epidemic Era

The research on the psychological changes of users' behavior in the post-epidemic era has brought about widespread attention about the space environmental needs of primary and middle schools. Based on this, we propose a healthy design strategy for the space environment of primary and middle schools through flexible space design, smart space construction, space security and comfort creation.

4.1 Flexible Space Design

The flexible design of primary and secondary schools is an significant measure and means that can ensure the health of the space environment and respondence to public health emergencies. In a broad sense, flexible design covers multiple levels, including the flexible planning of the entire campus and the campus buildings.

The flexible planning of the campus needs to start from two aspects, improving the student average index and layout. In the early design stage, striving for a reasonable per capita floor area index can prevent primary and secondary schools from being stretched and lacking room for maneuver when they need more campus safe use of places under particular circumstances [6]. The flexible layout of campus planning should consider different needs of the campus under normal and epidemic conditions. The design strategy is as follows. 1) Both public sharing and group division coexist. At present, the design of primary and secondary schools pays more and more attention to the creation of public shared spaces to promote communication between students and improve the efficiency of public facilities. In certain circumstances, these advantages become disadvantages. The way of grouping and partitioning can help with the start of school by grade and time, independent relatively, to achieve the purpose of effective isolation. 2) Relatively independent areas should be reserved. These areas include space for temperature measurement, registration, code scanning, and emergency isolation in the pre-school area, as well as areas for epidemic prevention and control and logistics personnel under special conditions. 3) Close attention should be paid to streamline design. The overall flow planning of the campus regarding the convenience under regular conditions and the epidemic situation should be considered. In particular, the flow lines for logistics and the picking up need to be designed independently. It is important to emphasize that if the system design for picking up and receiving students includes an underground garage, it is recommended to set up a sunken courtyard in the waiting area for parents and students to achieve natural ventilation.

The flexible design of campus buildings needs to be considered comprehensively with foresight and variability. The foresight of design emphasizes that various potential risks that may be dealt with. The risks should be considered in the early stage of project planning when risk classification and functional zoning need to be carried out. The variability of design refers to the provision of a space that can be easily expand and transform during emergency and future reconstruction. The design strategies are as follows. 1) The adaptability of libraries and relevant public places should be improved.

In response to the epidemic, the existing public spaces such as libraries and conference rooms are only used in specific time sessions or even out of service, resulting in a decline in resource utilization rate. Designing modularized functional spaces and fast-convertible indoor spaces can separate the regions by age groups, which can increase the utilization rate of public spaces. 2) The corridor can be widened. A corridor with sufficient width is conducive to ventilation. It can be also used as space for activities and display under normal conditions. Under special circumstances, the furniture can be dismantled and transformed into a single traffic space to meet the requirements of personnel separation for epidemic prevention and control (Fig. 1). The gymnasium can attribute of a non-medical epidemic prevention building. domparing with other types of public facilities, the gymnasium is a diversified flexible development space that has the advantages of rapid transformation comparing with other types of public facilities. If necessary, it can be transformed into a "square cabin hospital" in response to emergent public health incidents [7].

(a) (b) (c)

Fig. 1. Corridor with increased width.

4.2 Smart Space Construction

The construction of spatial informatization is mostly used in two levels including intelligent management and control, and informatization teaching.

Intelligent management and operation refer to the use of information-based intelligent platforms, which can form real-time management and monitoring of the space environment, in order to ensure the health of the indoor space environment. The information-based intelligent platform can interact with information such as natural ventilation, noise isolation, light, heat, humidity, and sound environment in real time, creating a comfortable and healthy indoor physical environment. At the same time, this mutual feedback mechanism can play a significant role in the occurrence of health emergencies. Students' body temperature and physical condition can be grasped in time, and the epidemic can be detected as soon as possible. By monitoring data such as the air exchange rate of the fresh air system of the air conditioner, the air quality in the indoor space can be effectively improved and the probability of virus transmission can be reduced. Linking the detection data with the control system, such as linking the infrared body temperature monitor with the medical office alarm system, can respond to emergencies in a timely manner and buy time for subsequent prevention and isolation.

The current application of information technology in the field of education has stimulated people's infinite imagination of the future of schools. With the outbreak of

COVID-19, various types of schools at all levels in different regions are actively implementing distance education. Relevant research reports pointed out that although distance learning has a positive impact on students' academic performance, their stress and anxiety have been increased, which makes it difficult for them to concentrate. This shows that the obstacle to comprehensive distance teaching is not only a challenge in technical support and teaching methods, but also a social and emotional challenge brought about by loneliness and social distance [8]. Therefore, the informatization and intelligentization of teaching space in primary and secondary schools can both give play to the advantages of information technology in the field of education, and avoid the disadvantages of distance teaching. The intelligent teaching space breaks the traditional teacher-centered space model in which teachers teach at the podium and students' desks and chairs are neatly arranged under the podium. Different types of tables and chairs can be used to form different spaces for discussion, communication, and interaction. The solidified platform of the teacher's identity is liberated. The role of teachers transforms from traditional knowledge imparters to guides of learning (Fig. 2). Holographic projection and mobile digital terminals (such as mobile phones, tablet computers) can weaken the boundaries of the teaching space. These enables teachers and students to carry out teaching activities at any time and place. Furniture, bookshelves and foldable, sliding lightweight partitions can separate the size and shape of the space, that can make the space flexible and changeable (Fig. 3–4).

(a) (b) (c)

Fig. 2. Three combinations of the same set of tables and chairs.

Fig. 3. Furniture divide the space.

Fig. 4. Bookshelf divide space.

4.3 Space Security and Comfort Creation

A healthy space environment is the guarantee for normal teaching activities in elementary and middle schools in the post-epidemic era. In the face of the anxiety and pressure brought by the epidemic, the mental health problems of elementary and middle school students have emerged. In the space design of primary and middle schools, we must not only pay attention to the safety and health of the space environment itself, but also the creation of a sense of security and comfort that the space environment brings to users. Including environmental protection to enhance the sense of security, comfortable environment.

Partitions and compartments are effective environmental protection measures that can provide a sense of security. For example, the use of translucent partitions on the dining room tables plays a protective role by blocking the spread of droplets during dining (Fig. 5); The use of translucent partitions on the self-study table can achieve the purpose of protection. It can also improve the privacy of the learning space and reduce external interference during learning (Fig. 6). The single-use "learning warehouse" has become a common module unit in primary and secondary schools in the post-epidemic era (Fig. 7). The modular unit can become an intelligent informal learning unit can provide safe protection and privacy, integrate lighting, ventilation, and information equipment (Fig. 8). Semi-open compartments for 1–4 people are mostly located in the public area. They can effectively divide the area, avoid crowded people in the public area, and provide a safe and private space for students' informal learning (Fig. 9).

Increasing green plants, the use of colors, and natural lighting are effective strategies to improve the environmental comfort. Green plants can optimize the air environment, provide a healthy and comfortable space environment for primary and middle school students. It is also very helpful for protecting eyesight of the students. Through the color matching in the indoor space environment, the indoor environment can be relaxing, comfortable, and harmonious. So, primary and middle school students can feel the changes brought about by colors, which produce a sense of relaxation and comfort psychologically, and alleviate their psychological pressure (Fig. 10). The light environment will influence people psychologically. Good natural lighting can improve the comfort of the space environment. It also plays a significant impact on the myopia protection of primary and middle school students.

Fig. 5. Dining table partition.

Fig. 6. Study table partition.

(a) (b)

Fig. 7. "Learning Warehouse".

Fig. 8. Equipment integration in the learning warehouse.

(a) (b)

Fig. 9. Semi-open cubicle.

(a) (b) (c)

Fig. 10. Application of colors.

5 Concluding

A space environment conducive to the physical and mental health of teachers and students in primary and middle schools is a prerequisite for normal teaching in primary and middle schools. Faced with the new challenges and opportunities brought by the epidemic to the construction of primary and middle schools, it is urgent to redesign or transform the space environment. This article proposes a healthy design strategy for the primary and middle school space environment, including flexible space design, smart space construction, space security and comfort creation, in order to benefit the future development of our country's primary and middle school buildings.

Acknowledgments. The authors would like to thank Shi Sun and Jing An for their assistance. This paper was part of project named "Teaching modular system and design integration research of primary and middle school driven by information technology", which was supported by National Natural Science Foundation of China (Grant no. 5187082147).

References

1. UN News Homepage. https://news.un.org/en/story/2021/10/1102992
2. Ma, J.: Place Management, volume 4 of Guidelines on COVID-19 Prevention and Control in Primary and Middle Schools, People's Medical Publishing (2020)
3. Jiang, J., Cai, Q.: Prevention, control and relief of mental health problems of primary and secondary school students in the :post-epidemic period". Moral Educ. China **24**, 36–39 (2020)
4. He, L.: "Presence" and "Regulation": a study of live teaching situation and behaviors. Contemp. Educ. Sci. **08**, 28–33 (2020)
5. China Education News Homepage. https://baijiahao.baidu.com/s?id=1708399063264190994&wfr=spider&for=pc
6. Wu, Z.: Sanitation, epidemic prevention and safe use of campus buildings in elementary and secondary schools. School Adm. **04**, 46–48 (2020)
7. Li, Y.: Research on flexible architectural design combining peace and epidemic_renovation design of Fangcang shelter hospital in Lianjiang County. Fujian Architect. Constr. **05**, 28–31 (2021)
8. Lemay, D.J., Bazelais, P., Doleck, T.: Transition to online learning during the COVID-19 pandemic. Comput. Human Behav. Rep. **04**, 100130 (2021)

Building Materials and Structures

Comparing the Emerging Walling Materials to the Burnt Clay Brick

Ankita Gupta[1] and Piyali Debnath[2(✉)]

[1] Colliers International, Mumbai, India
[2] Inspireli, Praha, Czech Republic
piyalidebnath163@gmail.com

Abstract. Wall is one of the most important elements in architecture, as it provides an enclosure and defines the building elevation. It occupies a major part of the exposed building façade and plays a key role in optimizing energy, noise reduction, firefighting etc. In this regard, the selection of suitable walling material is very important. India has seen a steady upheaval in construction practices. Several walling materials have emerged like AAC block, fly ash brick, precast concrete wall etc. that possess unique benefits. However, the traditional burnt clay brick continues to be the predominant walling material in India. One of the major reasons is the unavailability of studies that provide an extensive idea about the selection of walling material, which fits the functional requirements. This research focuses on comparing the burnt clay brick with emerging walling materials - AAC block, fly ash brick, precast concrete wall, to find out an alternative in terms of cost and function.

Keywords: Emerging walling materials · Burnt clay brick · AAC block · Fly ash brick · Precast concrete wall

1 Introduction

Walls play an important role in providing strength, fire safety, thermal comfort, acoustic comfort and protecting the interior from weathering action. In architecture, walls are considered as the predominant factor in defining space and enclosure, and designing the elevation. The walls in a building can be either load-bearing or non-load-bearing. The main functional requirements of a partition wall is to provide stability to bear self-weight, eccentric vertical loads and out-of-plane horizontal loads; The other requirements include good acoustic performance; thermal and water resistance; flexibility to accommodate the services; quality and aesthetic considerations; the flexibility for changing special requirements; less width to maximize the usable carpet areas and an optimal cost. The selection of walling materials in itself is a complex and iterative process to accommodate all the desired functional requirements. The varying unit of measurements for different functional parameters make the selection process even more difficult. In the booming construction industry of India, there are many emerging walling materials like AAC block, fly-ash bricks and precast concrete wall. Despite the availability of a variety of emerging walling materials, the traditional burnt clay bricks

P. Debnath—Inspireli, Praha, Czech Republic (Affiliated digitally).

P. Mendonça and N. D. Cortiços (Eds.): ICAMC 2021, LNCE 226, pp. 191–203, 2022.
https://doi.org/10.1007/978-3-030-94514-5_20

have been the predominant material of choice. As per the latest research, the production of brick has increased at 4% per year, consuming 24 million tons of coal [1].

A vast pool of research is available on the improvement of performance of walling materials like burnt clay brick, AAC block [2–4]. However, there is little research on comparing burnt clay brick with AAC blocks, based on their performance parameters like compressive strength, density, fire resistance, thermal conductivity etc. [5–7]. A study is available on cost analysis between AAC block, fly ash brick and burnt clay brick to arrive at an economic material [8]. But there is no data available on finding alternatives to burnt clay brick or comparing AAC blocks, fly ash bricks, precast concrete wall and burnt clay bricks based on their physical properties and site-specific parameters.

The aim of this research is to prepare a consolidated comparison between emerging walling materials - fly ash brick, AAC block, precast concrete wall and the traditional burnt clay brick - based on their performance parameters and derive possible alternatives to the burnt clay brick.

2 Methodology

A comprehensive comparative analysis is done on the physical parameters through literature study and site-specific parameters through interview, primary site survey (by taking direct measurement) and by collecting and collating online data. Based on the analyzed data, all the walling materials will be ranked by using the Weighted Score Method to find out possible alternatives to the burnt clay brick.

2.1 Weighted Score Method [9, 10]

Weighted Score Method, is one of the most used multi-criteria decision-making tools, where the alternative with the highest weighted score will provide the best possible solution. In this research, this method has been used in the following way-

- The criteria j (where j = 1, 2, 3…n) needs to be identified against which the alternative i (where i = 1, 2, 3…n) will be scored. In this paper the alternatives are AAC block, fly ash brick, precast concrete wall and burnt clay brick; and Criteria are the physical and site-dependent parameters.
- The weight W_j for each criterion j will be determined by using RII (Refer Sect. 2.2).
- S_{ij} is the score of each walling material against each criterion j which is obtained by using Sect. 2.3.
- The score S_i for each walling material against each criterion j will be calculated by using the Eq. (1).

$$S_i = W_j S_{ij} \tag{1}$$

- The total weighted score (S_T) for each walling material will be evaluated by using the Eq. (2).

$$S_T = \sum_{i=1}^{j} S_i \tag{2}$$

2.2 RII (Relative Importance Index)

Relative Importance Index [11] is used to assign weight to the each criterion by using the Eq. (3).

$$\text{RII} = W_j = \sum W\,(AN^{-1}) \tag{3}$$

Where, W represents weightage given to each factor by the respondents, which ranges from 1 to 5, A is the highest ranking available (i.e. 5 in this case) and N is the total number of respondents [11]. The value of RII lies between 0 and 1 [12].

2.3 Procedure to Calculate Score (S_{ij}) for Each Walling Material

As the units of measurements of different parameters or criteria differ, it is necessary to normalize the dataset on a scale of 0 to 1 [13] for the ease of evaluation. Here the Normalization process has been carried out based on the following systems:

- Higher value means higher performance.
- Higher value means lower performance.

Higher Value Means Higher Performance. There are certain criteria where a higher value means that the particular material has performed well with respect to its pertinent driver. These factors include strength, fire resistance, rate of installation or productivity, plastering etc. To normalize the values of these factors, the following steps have been followed

- *Level 1*- Assuming the rates of installation or productivity of burnt clay brick, fly ash brick, AAC block, and precast concrete wall are X_A, X_B, X_C, X_D respectively.
- *Level 2*- The total value (T) has been calculated by using the Eq. (4).

$$T = X_A + X_B + X_C + X_D \tag{4}$$

- *Level 3*- To normalize each value received at the level 1 (X_A, X_B, X_C, X_D) has been divided by the total value (T) obtained at the level 2 by using the Eq. (5) –

$$\text{N} = \frac{X_A}{T} \text{ or } \frac{X_B}{T} \text{ or } \frac{X_C}{T} \text{ or } \frac{X_D}{T} \left(\text{where, } \frac{X_A}{T} + \frac{X_B}{T} + \frac{X_C}{T} + \frac{X_D}{T} = 1\right) \tag{5}$$

Higher Value Means Lower Performance. There are some criteria, whose higher value means that particular material has performed poorly with respect to the pertinent criterion. These criteria include wastage, mortar thickness, plaster thickness, dampness etc. In such cases, a higher value means the cost implication is higher or more resources are needed to complete that particular activity. To normalize the values of these factors following steps have been followed –

- *Level 1*- Assuming wastage of burnt clay brick, fly ash brick, AAC block, precast concrete wall are X_A, X_B, X_C, X_D respectively.

- *Level 2*- The total value (T) has been calculated by using the Eq. (4).
- *Level 3*- The Normalized values (N) have been evaluated by using the Eq. (5). These normalized values represent, the materials possessing higher values have a higher amount of wastage.
- *Level 4*- To assign lower value X' to each material, the Normalized value received at the level 3 of each material has been subtracted from the total value 1 as described in the Eq. (6). So, the Modified value or Transformed value of each material will be-

$$X' = \left(1 - \frac{X_A}{T}\right) \text{or} \left(1 - \frac{X_B}{T}\right) \text{or} \left(1 - \frac{X_C}{T}\right) \text{or} \left(1 - \frac{X_D}{T}\right) \left(\text{where, } \frac{X_A}{T} + \frac{X_B}{T} + \frac{X_C}{T} + \frac{X_D}{T} = 1\right) \tag{6}$$

- *Level 5*- The modified total value (t) has been evaluated by using the Eq. (7)

$$t = X'_A + X'_B + X'_C + X'_D \tag{7}$$

- *Level 6*- The Modified normalized value N', has been calculated for each material by dividing the value received at the level 4 (X'_A, X'_B, X'_C, X'_D) by the total value (t) obtained from the level 5 [Refer Eq. (8)]-

$$N' = \frac{X'_A}{t} \text{ or } \frac{X'_B}{t} \text{ or } \frac{X'_C}{t} \text{ or } \frac{X'_D}{t} \left(\text{where, } \frac{X'_A}{t} + \frac{X'_B}{t} + \frac{X'_C}{t} + \frac{X'_D}{t} = 1\right) \tag{8}$$

3 Analysis

All walling materials have been analyzed based on the selected physical parameters & site-dependent parameters. The values of the different physical parameters (compressive strength, water absorption, fire resistance, thermal conductivity & STC rating) of different walling materials have been obtained directly from the relevant codes. Among the site-dependent parameters, few parameters (Internal plaster thickness, external plaster thickness and mortar thickness) have been measured directly at the site and few parameters, (wastage, productivity & rate of plastering) which vary over a period of time, depending on the climate condition, availability of labour, availability of tools etc. have been quantified by interviewing 56 industry professionals. Project managers, managers, contractors, supervisors, senior-supervisors, civil engineers etc. have been interviewed, who are directly related and solely responsible for construction. Since the ease of construction and price of block vary from region to region, the data has been collected either from e-commerce websites or from Delhi Schedule of Rates (DSR) 2018.

3.1 Comparative Analysis Between Physical Parameters of Different Walling Materials

In this section the 4 walling materials are compared based on the following 5 physical parameters which play a significant role in their selection.

Compressive Strength. It is the capacity of a material to resist breaking under a compressive load (which tends to reduce size). Amongst the 4 selected walling materials precast concrete wall possesses the maximum compressive strength. However, the minimum required strength is 10 N/mm^2 [14], but depending on the percentage of steel, type of aggregate, grades of cement a higher strength can be achieved. In case of fly ash brick up to 30 N/mm^2 [15] and burnt clay brick up to 35 N/mm^2 [16] of compressive strength can be achieved. But mostly in superstructure 1st class fly ash brick & burnt clay brick are used (Compressive strength 10 N/mm^2). On the contrary, though the maximum attainable strength in AAC block is up to 7 N/mm^2 [17], but in construction mostly 3–5 N/mm^2 strength is used. Based on the usage in the superstructure, if we compare the 4 walling materials precast concrete wall has the maximum compressive strength and AAC block has the least compressive strength.

Water Absorption. This indicates the water tightness or resistance to dampness. The lower the percentage of water absorption, the lower will be possibility of dampness. The average water absorption of burnt clay brick is 14.56% [18], whereas that of a fly ash brick is 9% (range- 6–12%) [19]. Compared to burnt clay brick & fly ash brick, AAC block possesses a higher range of water absorption 21–33% [20]. Precast concrete wall possesses the least water absorption of 4.45–5.75% [21].

Fire Resistance. Fire resistance of a wall ensures a certain period of time where the material does not fail structurally or allow the transmission of heat. An optimal fire resistance ensures that a building can be evacuated safely during an event of a fire. While a burnt clay brick wall of 200mm thickness has fire resistance of 2 h, an AAC brick wall of the same thickness offers fire resistance capacity of 7 h [5]. Fly ash brick wall of 200mm thickness possesses 4 h of fire resistance [22]. Precast concrete wall of 168mm offers fire resistance of 4 h [23].

Thermal Conductivity. The higher the value of thermal conductivity, the higher will be the rate of heat transfer through that particular walling material. After arranging all the walling materials in descending order based on thermal conductivity, we get precast concrete wall (1.58 $Wm^{-1}K^{-1}$) > burnt clay brick (0.98 $Wm^{-1}K^{-1}$) > fly ash brick (0.9 $Wm^{-1}K^{-1}$) > AAC block (0.18 $Wm^{-1}K^{-1}$) [24]. Due to the lowest thermal conductivity AAC block wall possesses the lowest heat transfer rate, resulting in the coolest environment inside.

Sound Transmission Class (STC) Rating. A material with a higher STC rating indicates that the material is more efficient in reducing sound transmission in comparison to other materials. While the precast concrete wall of 200mm thickness has the highest STC rating of 58, a fly ash brick wall of similar thickness possesses the lowest STC rating of 42 [25, 26]. AAC block wall & burnt clay brick wall of similar thickness have STC rating of 45 & 52 respectively [27, 28].

3.2 Comparative Analysis Between Site-Dependent Parameters of Different Walling Materials

The Site-dependent parameters include wastage, productivity, rate of plastering, plaster thickness, mortar thickness and cost which play an important role during the selection of materials as they have a direct impact on project cost and timeline.

Wastage. Wastage of material during wall erection can be broadly divided into two parts i.e. wastage of blocks and wastage of plaster.

The wastage of blocks mainly occurs due to the fragility of materials, mishandling, unavailability of proper cutting & shaping tools etc. Amongst the 4 walling materials fly ash brick possesses maximum amount of wastage and precast concrete wall exhibits the least amount of wastage. If all walling materials are arranged in descending order based on average total wastage, the result is: AAC block (12.78%) > fly ash brick (10.25%) > burnt clay brick (5.95%) > precast concrete wall (0.37%).

The wastage of plaster depends upon the bonding between walling material and plaster, and the quality of plaster used. The higher the bonding between walling material and plaster, the less will be the wastage. Wastage is maximum in case of fly ash brick wall due to poor bonding with plaster. AAC block possesses minimum wastage due to its rough and porous surface which results in good bonding with plaster. Due to the factory-made smooth surface of the precast concrete wall, generally it doesn't require plaster. After applying putty on precast concrete wall surface, paint can be directly applied to its surface. The average wastage of plaster for different walling materials in descending order: Fly ash brick (10.25%) > burnt clay brick (5.10%) > AAC block (1.67%).

Rate of Installation or Productivity. It mainly depends on the heaviness of blocks, working capacity of labour, climate, availability of modern equipment (hoist, automatic block cutter etc.). Since, precast concrete wall is manufactured in standard sizes and its installation is completely mechanized, it possesses maximum productivity of 2.438 m^3/hr. As burnt clay brick is heavy it has least rate of installation 0.106 m^3/hr. After arranging the average productivity in descending order, we get: Precast concrete wall (2.438 m^3/hr.) > AAC block (0.514 m^3/hr.) > fly ash brick (0.155 m^3/hr.) > burnt clay brick (0.106 m^3/hr.).

Plastering. Area of plastering done per hour depends on the bonding between blocks & plaster, working capacity of labour, climate, and availability of modern plastering equipment. Where the undulating surface of fly ash brick & burnt clay brick wall requires more quantity of plaster to make an even surface (resulting in lower productivity), AAC block surface requires less quantity of plaster due to its modular & larger size. In case of precast concrete wall plastering is normally not required due to its smooth surface. Painting is done after applying 2 coats of primer or putty. If all the walling materials are arranged in descending order based on the average rate of plastering, it has been found: AAC block (2.853 m^2/day) > burnt clay brick (1.61 m^2/day) > fly ash brick (1.583 m^2/day).

Plaster Thickness & Mortar Thickness. Plaster thickness: The thickness of plaster depends on the evenness of the wall. The higher the evenness of the surface, the less will be the thickness of the plaster. Walling materials like fly ash brick, burnt clay bricks etc. which have an uneven surface due to the smaller size of bricks, as compared to AAC block, result in higher thickness of plaster. The average thicknesses of internal plaster (13.8 mm) and external plaster (11.7 mm) are least in case of AAC block wall, due to larger block size & regular surface. Average thicknesses of internal and external plasters in case of different walling materials in descending order are: Burnt clay brick ≈ fly ash brick > AAC block (Refer Fig. 1).

Mortar thickness: Mortar thickness (horizontal:16.1mm and vertical:14.96mm) is minimum in case of AAC blocks wall due to their modular surface, larger size and lightweight nature. When JK bond is used as mortar the thickness reduces even further (horizontal: 7.85mm and vertical: 7.55mm). Burnt clay brick (horizontal: 20mm and vertical: 19.78mm) and fly ash brick (horizontal: 19.4mm and vertical: 18.4mm) walls possess maximum mortar thickness due to their heavy weight, smaller size and uneven surface. Precast concrete wall doesn't require mortar, as it's made of solid concrete. Average thicknesses of vertical and horizontal mortar in case of different walling materials in descending order are: Fly ash brick wall ≈ Burnt clay brick wall > AAC block wall (Refer Fig. 1).

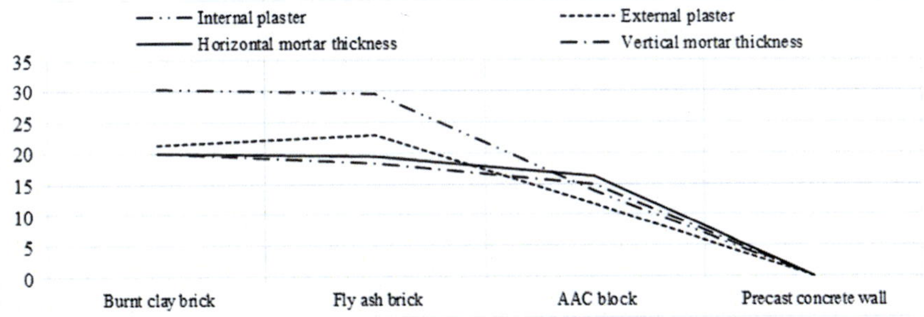

Fig. 1. Comparison between internal plaster, external plaster, vertical mortar thickness & horizontal mortar thickness of the selected walling materials.

Ease of Construction. While choosing walling materials, the construction method is integral to project cost, time and quality. Where the materials like precast concrete wall require heavy equipment for installation, the materials like AAC block, fly ash brick & burnt clay brick don't rely on heavy machinery for installation. If space is a major constraint at any construction site, a precast concrete wall becomes very challenging to install.

To quantify the ease of construction, 2 factors i.e. the hiring charges of labor and equipment have been considered. Required data for this analysis has been taken from the DSR (Delhi Schedule of Rates) 2018 [29] and conducted survey. As we can see, in the Fig. 2 that the precast concrete wall possesses the highest labour and equipment charges as it requires mobile crane for installation [Skilled labour(3), Semi-skilled labour(3), mobile crane(1)]. Other walling materials have considerably low charges as they require only 1 mason and 2 helpers and don't require any heavy machinery.

Price of Block. Price is one of the most important factors that governs the selection of the materials. Prices from several online sellers have been referred for each walling material and based on the average price the following comparison is done. The Fig. 2 depicts that the precast concrete wall is the costliest material while AAC block is the cheapest material amongst the 4 selected walling materials.

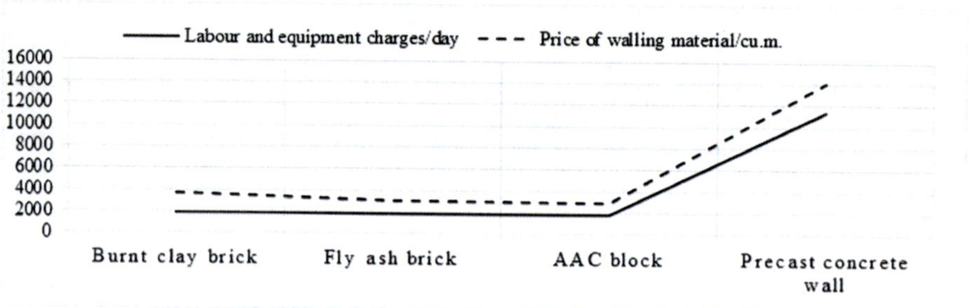

Fig. 2. Comparative analysis between the ease of construction (labour & equipment charges) and price for different walling materials.

4 Result and Discussion

In this section, the derived values of the different parameters have been used to rate the 4 walling materials through the Weighted Score Method (As per Sect. 2.1). The different physical parameters and site-dependent parameters have been used as criteria. To weigh each criterion (W_j) a survey has been conducted among the 41 experts. All the respondents are asked to rate the different criteria on a scale of 1 to 5. The surveyed data has been analyzed by using RII method to find out the weights of different criteria (W_j). To get the values of S_{ij} for different walling materials against each criterion, the values derived from the Sect. 3, have been converted on a scale of 0–1 as described in Table 2. Then the weighted marks (S_i) for different walling materials have been calculated by multiplying W_j (weight for each criterion) with S_{ij} (score of each walling material against each criterion). Finally, all the weighted marks have been added together to obtain the Total weighted marks (S_T) for a particular walling material as shown in Table 1.

If we arrange all the walling materials in descending order based on the derived total weighted marks (S_T), we obtain - Precast concrete wall > AAC block > Fly ash brick > Burnt clay brick (As shown in Table 1).

Table 1. Calculation of Total weighted marks (S_T)

Criteria (j)	Weight for each criterion (W_j)	Score of each walling material (S_{ij})				Weighted marks ($S_i = W_j\ S_{ij}$)			
		Burnt clay brick (A)	Fly ash brick (B)	AAC block (C)	Precast concrete wall (D)	Burnt clay brick (A)	Fly ash brick (B)	AAC block (C)	Precast concrete wall (D)
Physical parameters									
Compressive strength	0.8	0.286	0.286	0.143	0.286	0.229	0.229	0.114	0.229
Water absorption	0.805	0.244	0.280	0.173	0.304	0.196	0.225	0.139	0.244
Fire resistance	0.8	0.112	0.227	0.391	0.270	0.089	0.182	0.313	0.216
Thermal conductivity	0.810	0.244	0.251	0.317	0.189	0.197	0.203	0.257	0.153
STC rating	0.702	0.264	0.213	0.228	0.294	0.185	0.150	0.160	0.207
Site-dependent parameters									
Wastage of blocks	0.654	0.266	0.217	0.188	0.329	0.174	0.142	0.123	0.215
Wastage of plaster	0.6	0.233	0.133	0.301	0.333	0.140	0.080	0.180	0.200
Speed of installation of blocks	0.659	0.033	0.048	0.16	0.759	0.022	0.032	0.105	0.500
Speed of plastering	0.571	0.133	0.131	0.236	0.500	0.076	0.075	0.135	0.285
Plaster thickness (Internal)	0.590	0.196	0.200	0.27	0.333	0.116	0.118	0.160	0.197
Plaster thickness (External)	0.590	0.207	0.197	0.263	0.333	0.122	0.116	0.155	0.197
Mortar thickness (Horizontal)	0.561	0.213	0.217	0.237	0.333	0.120	0.122	0.133	0.187
Mortar thickness (Vertical)	0.561	0.209	0.218	0.239	0.333	0.117	0.122	0.134	0.187
Ease of construction	0.751	0.297	0.297	0.297	0.109	0.223	0.223	0.223	0.082
Price of blocks	0.820	0.282	0.290	0.293	0.136	0.231	0.238	0.240	0.111
Total weighted marks (S_T)						2.237	2.255	2.572	3.210

Table 2. Calculation of S_{ij} of different walling materials against each parameter (Scale of 0–1)

Criteria	Components	Burnt clay brick (A)	Fly ash brick (B)	AAC block (C)	Precast concrete wall (D)	TOTAL (T, t) A+B+C+D
Physical parameters						
Compressive strength	Actual value (X)	10	10	5	10	Total (T) = 45
	Normalized value (XT^{-1})	0.286	0.286	0.143	0.286	
Water absorption	Actual value (X)	0.15	0.09	0.27	0.05	Total (T) = 0.56
	Normalized value (XT^{-1})	0.268	0.161	0.482	0.089	
	Transformed value [$X' = 1 - (XT^{-1})$]	0.732	0.839	0.518	0.911	Total (t) = 3
	Modified normalized value ($X't^{-1}$)	0.244	0.280	0.173	0.304	
Fire resistance	Actual value (X)	0.59	1.2	2.07	1.43	Total (T) = 5.29
	Normalized value (XT^{-1})	0.112	0.227	0.391	0.270	
Thermal conductivity	Actual value (X)	0.98	0.9	0.18	1.58	Total (T) = 3.64
	Normalized value (XT^{-1})	0.269	0.247	0.049	0.434	
	Transformed value [$X' = 1 - (XT^{-1})$]	0.731	0.753	0.951	0.566	Total (t) = 3
	Modified normalized value ($X't^{-1}$)	0.244	0.251	0.317	0.189	
STC rating	Actual value (X)	52.00	42.0	45.00	58.00	Total (T) = 197
	Normalized value (XT^{-1})	0.264	0.213	0.228	0.294	

(*continued*)

Table 2. (*continued*)

Criteria	Components	Burnt clay brick (A)	Fly ash brick (B)	AAC block (C)	Precast concrete wall (D)	TOTAL (T, t) A+B+C+D
Site-dependent parameters						
Wastage of blocks	Actual value (X)	0.060	0.103	0.128	0.004	Total (T) = 0.23
	Normalized value (XT^{-1})	0.203	0.349	0.435	0.013	
	Transformed value [$X' = 1 - (XT^{-1})$]	0.797	0.651	0.565	0.987	Total (t) = 3
	Modified normalized value ($X't^{-1}$)	0.266	0.217	0.188	0.329	
Wastage of plaster	Actual value (X)	0.051	0.1025	0.0167	0	Total (T) = 0.17
	Normalized value (XT^{-1})	0.300	0.602	0.098	0.000	
	Transformed value [$X' = 1 - (XT^{-1})$]	0.700	0.398	0.902	1.000	Total (t) = 3
	Modified normalized value ($X't^{-1}$)	0.233	0.133	0.301	0.333	
Speed of installation of blocks	Actual value (X)	0.106	0.1546	0.514	2.4375	Total (T) = 3.21
	Normalized value (XT^{-1})	0.033	0.048	0.160	0.759	
Speed of plastering	Actual value (X)	1.61	1.58	2.85	0	Total (T) = 6.05
	Normalized value (XT^{-1})	0.266	0.262	0.472	0.000	
	Transformed value (X')	0.266	0.262	0.472	1	Total (t) = 2
	Modified normalized value ($X't^{-1}$)	0.133	0.131	0.236	0.500	
Plaster thickness (Internal)	Actual value (X)	30.06	29.3	13.8	0	Total (T) = 73.16
	Normalized value (XT^{-1})	0.411	0.400	0.189	0.000	
	Transformed value [$X' = 1 - (XT^{-1})$]	0.589	0.600	0.811	1	Total (t) = 3
	Modified normalized value ($X't^{-1}$)	0.196	0.200	0.270	0.333	
Plaster thickness (External)	Actual value (X)	21.12	22.82	11.7	0	Total (T) = 55.64
	Normalized value (XT^{-1})	0.380	0.410	0.210	0.000	
	Transformed value [$X' = 1 - (XT^{-1})$]	0.620	0.590	0.790	1	Total (t) = 3
	Modified normalized value ($X't^{-1}$)	0.207	0.197	0.263	0.333	
Mortar thickness (Horizontal)	Actual value (X)	20	19.4	16.13	0	Total (T) = 55.53
	Normalized value (XT^{-1})	0.360	0.349	0.290	0.000	
	Transformed value [$X' = 1 - (XT^{-1})$]	0.640	0.651	0.710	1	Total (t) = 3
	Modified normalized value ($X't^{-1}$)	0.213	0.217	0.237	0.333	
Mortar thickness (Vertical)	Actual value (X)	19.78	18.4	14.96	0	Total (T) = 53.14
	Normalized value (XT^{-1})	0.372	0.346	0.282	0.000	
	Transformed value [$X' = 1 - (XT^{-1})$]	0.628	0.654	0.718	1	Total (t) = 3
	Modified normalized value ($X't^{-1}$)	0.209	0.218	0.239	0.333	
Ease of construction	Actual value (X)	1854	1854	1854	11388	Total(T) = 16950
	Normalized value (XT^{-1})	0.109	0.109	0.109	0.672	
	Transformed value [$X' = 1 - (XT^{-1})$]	0.891	0.891	0.891	0.328	Total (t) = 3
	Modified normalized value ($X't^{-1}$)	0.297	0.297	0.297	0.109	
Price of blocks	Actual value (X)	3670.34	3077.62	2885	14010	Total(T) = 23643
	Normalized value (XT^{-1})	0.155	0.130	0.122	0.593	
	Transformed value [$X' = 1 - (XT^{-1})$]	0.845	0.870	0.878	0.407	Total (t) = 3
	Modified normalized value ($X't^{-1}$)	0.282	0.290	0.293	0.136	

5 Conclusion

The analysis shows that the precast concrete wall is the highest scorer amongst all walling materials. It possesses high strength, least water absorption, highest STC rating, least wastage, fastest speed of installation and good quality- making it the best alternative walling material to the burnt clay brick. Moreover, due to its smooth surface plaster is not required before painting, which saves on construction time and cost. Its drawbacks include a substantially higher price point as compared to other walling materials and the need of heavy equipment during installation. However, its positive attributes surpass these identified drawbacks.

The highest fire resistance, least thermal conductivity, higher STC rating, faster speed of installation, least mortar and plaster thickness requirement, result in the AAC block being the second best alternative material to the burnt clay brick. Due to its modular size and porous surface, the bond between plaster and AAC block is strong which results in minimal plaster thickness. It is not only lighter in weight but also the cheapest amongst the 4 walling materials. Its drawbacks include: maximum wastage of block, least strength and highest water absorption.

Fly ash brick is the third alternative solution to the burnt clay brick. It possesses less water absorption, higher fire resistance, higher thermal conductivity, and higher speed of installation than the burnt clay brick. However, it poses issues like huge wastage, difficulty in bonding with plaster, uneven surface and requirement of increased amounts of mortar & plaster. If these issues can be addressed, fly ash brick has the potential to become one of the most cost-effective alternatives to the burnt clay brick.

On the other hand, the burnt clay brick has good strength, moderate thermal & acoustic performance, a reasonable cost of construction and moderate wastage which justifies its present use as a popular walling material.

References

1. Development Alternatives: Enabling policies in the Indian brick sector – current status and future trends. p 10. Shakti Sustainable Energy Foundation, New Delhi (2012). https://shaktifoundation.in/wp-content/uploads/2017/06/SDA-2012-Enabling-Policies-in-The-Indian.pdf
2. Khaliq, W., Bashir, M.F.: High temperature mechanical and material properties of burnt masonry bricks. Mater. Struct. **49**(12), 5195–5208 (2016). https://doi.org/10.1617/s11527-016-0854-0
3. Kumar, S.: A perspective study on fly ash–lime–gypsum bricks and hollow blocks for low cost housing development. Constr. Build. Mater. **16**, 519–525 (2002). https://doi.org/10.1016/s0950-0618(02)00034-x
4. Naganathan, S., Mohamed, A.Y., Mustapha, K.N.: Performance of bricks made using fly ash and bottom ash. Constr. Build. Mater. **96**, 576–580 (2015). https://doi.org/10.1016/j.conbuildmat.2015.08.068
5. Shukla, R.: Burnt clay bricks versus autoclaved aerated concrete blocks. Int. J. Eng. Res. Technol. **3**, 575–580 (2014). https://www.ijert.org/burnt-clay-bricks-versus-autoclaved-aerated-concrete-blocks

6. Jain, U., Jain, M., Mandaokar, S.: Comparative study of AAC blocks and clay brick and costing. Int. J. Res. Eng. Sci. Manag. **1** 539–543 (2018). http://www.ijresm.com/Vol_1_2018/Vol1_Iss9_September18/IJRESM_19_129.pdf
7. Duwadi, I.N.: Comparative study on autoclaved aerated concrete block and conventional burnt clay brick. In: KEC Conference, Kantipur Engineering College, Lalitpur, Nepal, pp. 269–271 (2019). http://kec.edu.np/wp-content/uploads/2020/01/Paper_48.pdf
8. Khan, A.A., Shrivastava, S.: Cost analysis of building construction by using autoclaved aerated concrete block (AAC), fly ash brick and clay brick as masonry component. Int. J. Adv. Res. Ideas Innov. Technol. **4** 778–783 (2018). https://www.ijariit.com/manuscripts/v4i5/V4I5-1481.pdf
9. Mazda, C.: Weighted Score And Topsis SlideShare (2009). https://www.slideshare.net/CardinaleMazda/weighted-score-and-topsis
10. Weighted Scoring n.d. ProductPlan. https://www.productplan.com/glossary/weighted-scoring/
11. Gündüz, M., Nielsen, Y., Özdemir, M.: Quantification of delay factors using the relative importance index method for construction projects in Turkey. J. Manag. Eng. **29**, 135 (2013). https://doi.org/10.1061/(asce)me.1943-5479.0000129
12. El-Sawalhi, N.I., Hammad, S.: Factors affecting stakeholder management in construction projects in the Gaza Strip. Int. J. Constr. Manag. **15**, 160 (2015). https://doi.org/10.1080/15623599.2015.1035626
13. Normalized function, normalized data and normalization n.d. calculushowto.com. https://www.calculushowto.com/types-of-functions/normalized-function-data-normalization/
14. Standards India 1986 IS 11447 - 1985: Code of practice for construction with large panel prefabricates, Bureau of Indian Standards, New Delhi, p. 13
15. Standards India 2002 IS 12894 - 2002: Pulverized fuel ash-lime bricks-specification, 1st revision, Bureau of Indian Standards, New Delhi, p. 2
16. Standards India (Reaffirmed in 2007) IS 1077 - 1992: Common burnt clay building bricks - specification, 5th revision, Bureau of Indian Standards, New Delhi, p. 1
17. Standards India (Reaffirmed in 2005) IS 2185 (Part-3) - 1984: Autoclaved cellular (aerated) concrete blocks, first version, Bureau of Indian Standards, New Delhi, p. 7
18. Labaran, Y.H., Musa, N.M., Musa, A.A., Garba, L.: Comparative performance analysis of fired clay bricks and quarry dust sandcrete bricks. Int. Res. J. Eng. Technol. **7,** 1730 (2020). https://www.irjet.net/archives/V7/i5/IRJET-V7I5334.pdf
19. Kumar, R., Patyal, V., Lallotra, B., Ashish, D.K.: Study of properties of light weight fly ash brick. In: National Conference on Advances in Engineering and Technology, International Journal of Engineering Research and Applications, p. 50 (2014). https://www.ijera.com/special_issue/AET_Mar_2014/CE/Version%20%201/J4953.pdf
20. Raj, A., Borsaikia, A.C., Dixit, U.S.: Evaluation of mechanical properties of autoclaved aerated concrete (AAC) block and its masonry. J. Inst. Eng. (India) Ser. A, **101,** 324 (2020). https://doi.org/10.1007/s40030-020-00437-5
21. Alsayed, S., Amjad, M.: Strength, water absorption and porosity of concrete incorporating natural and crushed aggregate. J. King Saud University – Eng. Sci. **8**(1), 113 (1996). https://doi.org/10.1016/S1018-3639(18)30642-1
22. Dolphin maxlite construction block n.d. dolphinmaxlite.in. https://www.dolphinmaxlite.in/fly-ash-block-and-brick.html
23. International Building Code, p. 146. International Code Council, Inc., Illinois (2009)
24. Standards India 2017 Energy Conservation Building Code for Residential Building (Part I: Building envelope)
25. Noise transmission n.d. concretenz.org.nz. https://concretenz.org.nz/page/noise

26. Dolphin maxlite construction block n.d. dolphinmaxlite.in. https://www.dolphinmaxlite.in/fly-ash-block-and-brick.html
27. Autoclaved Aerated Concrete n.d. cement.org. https://www.cement.org/cement-concrete/paving/buildings-structures/concrete-homes/building-systems-for-every-need/autoclaved-aerated-concrete
28. The Brick Industry Association, Technical notes on brick construction gobrick.com (2000). https://www.gobrick.com/docs/default-source/read-research-documents/technicalnotes/5a-sound-insulation—clay-masonry-walls.pdf?sfvrsn=0
29. Standards India 2018, Delhi Schedule of Rates, CPWD, New Delhi, pp. 4–7, vol. 1 (2018)

Review of UHPC Reinforcement Technology in Reinforcement Concrete Beams

Yuyang Zhao(✉)

Chang'an Dublin International College of Transportation, Chang'an University, Xi'an 710016, China
2018902705@chd.ehu.cn

Abstract. After years of operation, normal concrete (NC) beams usually deteriorate, which will adversely affect the performance and safety of the whole bridge. Ultra-high performance concrete (UHPC) is suitable for beam reinforcement due to its far superior mechanical properties and durability to normal concrete, and is used in practical reinforcement projects. This paper discusses the definition and composition of UHPC, the mechanical properties of UHPC and the application of UHPC in the reinforcement of normal concrete beams. It is found that different reinforcement methods and interface treatment methods can improve the bending strength of NC beams. UHPC is a kind of reinforcement material with ultra-high performance. The purpose of this paper is to provide some reference for similar studies on strengthening RC beams with UHPC.

Keywords: Ultra high performance concrete · Normal concrete beams · Retrofitting · Interface treatment · Bending strength

1 Introduction

In today's world, many countries have invested a lot of money in infrastructure construction. Bridges are the typical representatives of infrastructure, and most of them are made up of reinforced concrete structures. Beams are the important components of bridges. After years of operation, due to significant increase in external load, harsh environment, the freeze-thaw cycle and so forth, the beams usually deteriorate, which reduces the working performance of bridges and even poses a threat to the safety of bridges [1]. Therefore, bridge reinforcement technology has been widely concerned by scholars in the past few decades. Ultra-high performance concrete (UHPC) is suitable for beam reinforcement because of its far superior mechanical properties and durability than normal concrete (NC), and has been used in the continuous research of scholars in beam reinforcement. This paper is to discuss the definition and composition of UHPC, mechanical properties of UHPC and the application of UHPC in the reinforcement of normal concrete beams.

P. Mendonça and N. D. Cortiços (Eds.): ICAMC 2021, LNCE 226, pp. 204–212, 2022.
https://doi.org/10.1007/978-3-030-94514-5_21

2 Definition and Composition of UHPC

2.1 Definition of UHPC

Ultra-high performance concrete (UHPC) is one of the important breakthroughs in the field of concrete technology in the 21st century. Although it has not yet formed an international unified definition, in general, it is a new cement-based composite material based on the maximum density theory, which has the characteristics of high strength, high density, high ductility and high durability. Since the 1970s, UHPC has been systematically and deeply studied in European countries. According to the material mix ratio and production process, reactive powder concrete (RPC), CARDIFRC, ultra-high performance fiber reinforced concrete (UHPFRC), strain hardening cementitious composites (SHCC) and other scientific research achievements have been gradually born [2–7]. Larrard et al. first proposed the concept of ultra-high-performance concrete, and UHPC gradually became the general term of this kind of ultra-high performance cement-based composites [20].

2.2 Composition of UHPC

Design mixture of UHPC. UHPC is mainly composed of Portland cement, silica fume, water, superplasticizer, steel fiber and fine sand [8]. Through the study of Bajaber et al., the optimum mix proportions of these materials in 1 cubic meter UHPC are shown in Fig. 1. Different from normal concrete, UHPC ultra-high performance is achieved by adding steel fiber and removing coarse aggregate to improve the uniformity of mix proportion [9]. The dense matrix with very low permeability is obtained by the uniform distribution of fine particles, and the addition of steel fiber is also helpful to improve the ductility of the material.

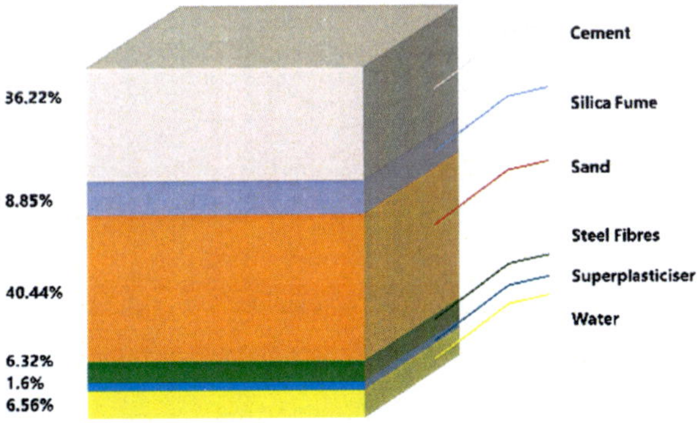

Fig. 1. The optimum mix proportions of UHPC [10]

Steel Fiber Content. Because the concrete material is brittle, concrete without reinforcement is easy to fracture. Therefore, steel fiber is used to strengthen UHPC matrix and improve its mechanical properties. Adding steel fiber size can be divided into long fiber (13mm and above), short fiber (6mm and below), and usually long fiber is better than short fiber [11].

The volume fraction of steel fiber has a great influence on the mechanical properties of UHPC. In a certain range, the higher the steel fiber content, the better the mechanical properties of UHPC. However, when the steel fiber content exceeds a certain value, the increase of steel fiber content is unfavorable to some properties of UHPC. Ju et al. found that: 1. When the steel fiber content is between 1.0% and 3.5%, the compressive strength of UHPC increases with the increase of steel fiber content. When the steel fiber content exceeds 3.5%, the compressive strength of UHPC decreases. 2. The greater the steel fiber content, the better the ductility and fracture modulus of UHPC.

3 Mechanical Properties of UHPC

Compared with normal concrete, UHPC has better tensile, flexural and compressive properties, which will be discussed in detail below.

3.1 Compression Properties of UHPC

Generally, the compressive strength of UHPC is greater than 120 MPa. Compared with normal concrete, the compressive strength of UHPC is usually 3 – 4 times. This is mainly due to the low water-cement ratio of UHPC, uniform distribution of fine aggregate. Figure 2 shows the comparison of stress-strain curves of UHPC, steel fiber reinforced concrete (SFRC) and NC under compression [12].

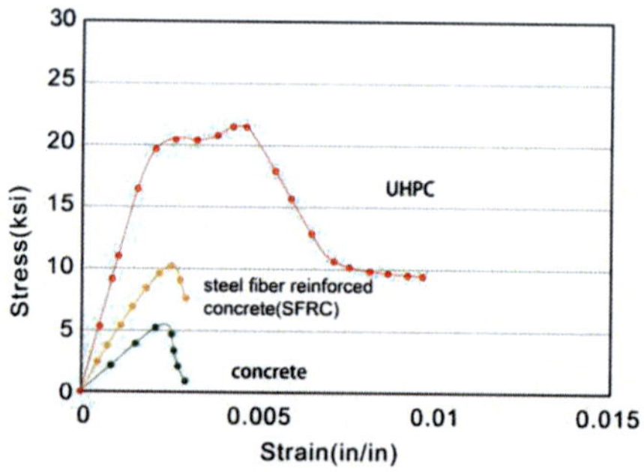

Fig. 2. The stress-strain curve based on cylinder test of the specimens [12]

3.2 Tensile Properties of UHPC

The ultra-high strength characteristics of UHPC can improve the tensile strength of cement-based materials by leaps and bounds. More importantly, UHPC can effectively use the strength of steel fiber and its close bonding with cement paste to achieve tensile 'strain hardening' behavior (similar to the 'yield' of steel), and has large deformation capacity. Figure 3 shows the uniaxial tensile test results of UHPC specimens, conventional steel fiber reinforced concrete (SFRC) and normal concrete. Compared with other cementitious materials, the tensile strength and strain hardening properties of UHPC are significantly improved [13].

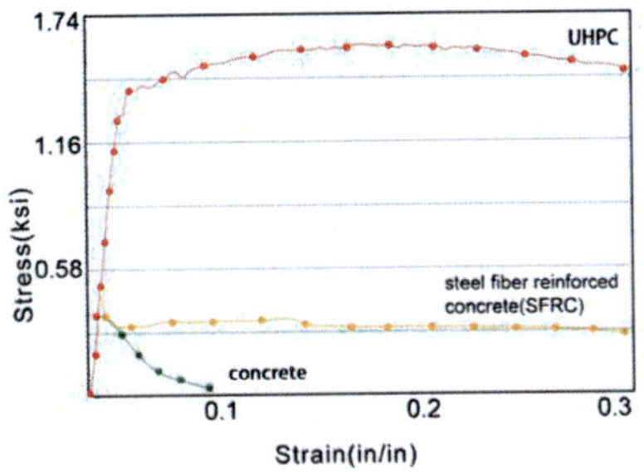

Fig. 3. Uniaxial tensile behavior of UHPC, SFRC and normal concrete [13].

3.3 Flexural Strength of UHPC

UHPC can greatly improve the strength utilization efficiency of steel in concrete, and form a new steel-concrete composite with more coordinated concrete, steel fiber and steel bar, and realize the lightweight, high strength and high toughness of concrete structures. Richard and Cheyrezy (1995) pointed out in the study that the ductility of UHPFC is about 250 times that of conventional concrete. UHPFC under bending load can be divided into three stages: (1) linear elastic behavior continues until the first cracking of the material, (2) displacement hardening stage continues until the maximum load (3) flexural softening stage after the maximum load [1].

4 Application of UHPC in Retrofitting of Normal Concrete Beams

4.1 Bending Strength of UHPC-NC Beam

The beam structure is mainly subjected to bending moment and shear force. The decrease of its flexural capacity is often accompanied by excessive mid-span deflection. Improving its flexural capacity and inhibiting its further deflection are important contents of its reinforcement.

There are two main ways to reinforce normal concrete beams with UHPC, using U-shaped hoop reinforcement and pouring UHPC layer. For U-shaped hoop method, in 2003, Alaee et al. using CARDIFRC material found that the reinforcement plate with a thickness of only 16 mm could improve the flexural capacity of the reinforced member by 66%, and the deflection was only 11% of that of the unreinforced member[17]. In 2017, Al-Osta and others obtained that the U-shaped hoop reinforcement not only improves the flexural capacity of the component by 89%, but also reduces the midspan deflection by 80% [18].

For the method of pouring UHPC layer, in 2016, Safdar et al. used UHPFRC to replace the normal concrete at the top and bottom of the NC beam, respectively, and found that it could improve the flexural capacity of the beam by 23% [21]. In 2017, Al-Osta et al. poured UHPC layer at the bottom of the beam, and found that it could improve the flexural capacity by 16%, and the mid-span deflection was 80% of that without reinforcement.

Based on the above results, it can be found that both methods can effectively improve the flexural capacity of the beam and reduce the midspan deflection of the beam. As shown in Fig. 4, since UHPC is added at the bottom of the NC beam, UHPC contributes a part of the tensile strength in the tensile region of the beam, thereby improving the flexural capacity of the beam. U-shaped hoop reinforcement can also improve the shear capacity of the strengthened beam, but it has the disadvantage of inconvenient construction. Pouring UHPC layer is relatively simple, but there is no UHPC layer on the web, it cannot provide shear resistance. In addition, in the application process, these two methods may lead to micro cracks in UHPC layer, thereby increasing the permeability of UHPC layer and reducing the bending performance. The reason for micro cracks may be that the thickness of UHPC layer is too small, the autogenous shrinkage of UHPC is too large, or the old NC layer restricts the shrinkage of the new poured UHPC layer.

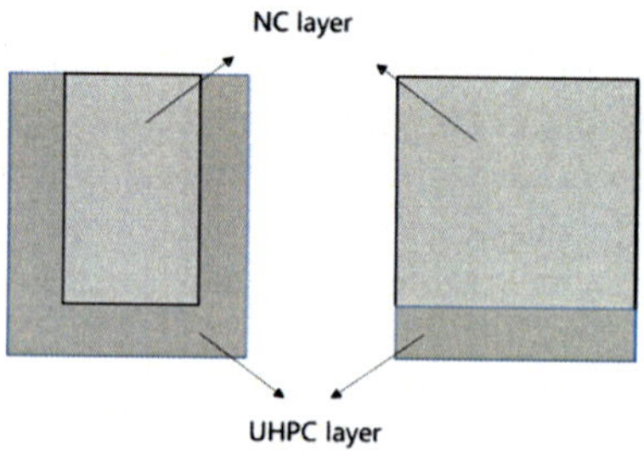

Fig. 4. U-shaped hoop reinforcement (left) and pouring UHPC layer (right)

4.2 UHPC-NC Interface Treatment

It should be noted that the interface treatment also affects the bending performance of the beam. Since the stress of UHPC-NC beam is essentially a composite beam, the longitudinal parts of the two beams are bonded together, and there is a shear transfer between them. If the shear strength is not enough, the two parts will be separated and slide mutually. The interface treatment between UHPC and NC will directly affect the bonding performance, and determine whether UHPC layer and NC layer can be subjected to common force, thereby affecting the reinforcement effect.

UHPC-NC interface bonding is mainly divided into tensile and shear parts, usually by using common interface agents (such as epoxy resin adhesive, polymer cement mortar, etc.) to enhance the tensile strength of the interface. For the enhancement of interface shear strength, it can be divided into two ways in general, surface chisel type and shear key type. It can be found that:

For surface chisel type, through comparative experiments, Tayeh et al. (2013) found that compared with the group without rough interface treatment, the interfacial bonding strength after wire brushing treatment was 1.6 times that of it, the interfacial bonding strength after drilling holes treatment was 1.4 times that of it, and the interfacial bonding strength after sand blasting treatment was 2.0 times that of it [1]. Jang et al. (2017) found that compared with pouring UHPC layer directly on the NC layer, the interfacial bonding strength after high pressure jet drilling was 3.0 times that of UHPC layer [19].

Through the above research, we can find that the bonding strength of UHPC-NC interface after chiseling is improved. However, due to the excessive impact force, the traditional chiseling treatment will cause micro cracks on the original concrete structure and weaken the shear strength of the interface. Therefore, new methods such as sandblasting method and high-pressure jet method have been proposed. These methods can remove the weak layer on the surface of concrete and the roughness of the surface of concrete without imposing excessive impact on the surface of concrete, ensuring the surface strength of the original structure, and the interfacial bonding strength is higher. However, due to the high cost of equipment and the difficulty of implementation, these methods still need to be developed.

For shear keys, there are two most common ways:

(1) Reinforcement planting, as shown in Fig. 5, relying on the strong shear performance of steel bars to maintain interface bonding. Zhang et al. (2020) found that the interfacial bond strength after planting reinforcement was about 1.5–2.5 times that of the ordinary chiseling treatment. (2) Cutting groove, as shown in Fig. 6, forming UHPC, NC alternative tooth structure at the interface. Zhang et al. (2020) found that compared with ordinary chiseling treatment, the interfacial bond strength after grooving treatment was about 1.1–1.8 times [23].

Either way, the bonding strength of UHPC-NC interface is restricted by the shear strength of normal concrete. Compared with other methods, the reinforcement planting method has better failure ductility. In practical engineering, we should flexibly combine multiple methods such as interfacial agent, chiseling and planting reinforcement according to the needs of practical engineering to obtain stable interfacial bonding performance.

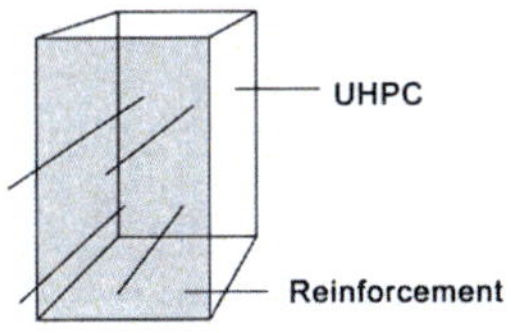

Fig. 5. Reinforcement planting

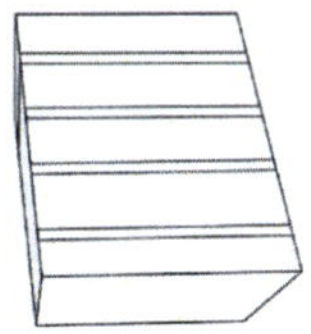

Fig. 6. Cutting groove [22]

4.3 Case Study

In recent years, UHPC has been applied to practical RC beam reinforcement projects in some countries. The Chillon viaducts in Switzerland are two parallel highway post-tensioned concrete bridges built in the late 1960 s [14]. After the long-term operation, the concrete of the beams of these bridges shows signs of early stages of the alkali-aggregate reaction (AAR), which refers to the chemical reaction between the components of the aggregate material and the alkali in the concrete under certain conditions, resulting in expansion, cracking and even destruction of the concrete structure, thus threatening the safety of the bridges [15]. In 2014, to ensure the structural safety of future traffic demand, Swiss authorities decided to use the method of pouring UHPC layers at the top of the beams to reinforce the Chillon viaducts, which are shown in Fig. 7 and 8. The purpose is to improve the bending strength and stiffness of the beams, and provide waterproof measures to protect the structure from the infiltration of water and chloride ions, so as to improve durability. The thickness of beam is 180 mm, and the thickness of UHPC layer is 40 mm. After reinforcement, the shear strength of the beams increased by 40%, and the ultimate flexural capacity increased by 20%. In addition, the deflection measurements of the long-term monitoring system installed since 1986 show that the UHPFRC layer causes a significant decrease in the mid-span deflection of the beams [16].

Fig.7. Chillon Viaducts along Lake Geneva

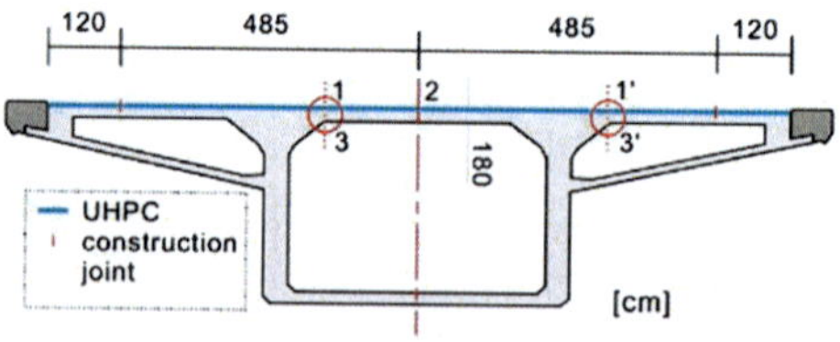

Fig. 8. Geometry of the beams cross-section [16]

5 Conclusions

UHPC has become an ultra-high performance material due to the addition of steel fibers, low water-binder ratio, low porosity, and optimized mixing ratio. Compared with normal concrete, UHPC has better compressive strength, tensile strength and flexural strength. Both U-type hoop reinforcement and UHPC layer pouring can improve the flexural capacity of NC beam and restrain its mid-span deflection. Surface chiseling can improve the interfacial bonding strength, and more chiseling methods suitable for practical engineering need to be developed. Shear keys can improve interfacial bonding strength. However, the effect of these two methods is subject to the shear strength of normal concrete. In practical engineering, we should flexibly choose different methods according to needs to obtain stable interface bonding performance.

For the retrofitting of reinforced concrete beams with UHPC materials, UHPC is a good material due to its good mechanical properties and durability, which can greatly improve the performance of the repaired components and ensure the safety of the entire bridge structure. However, due to its relatively expensive price and the need for strict site pouring and maintenance conditions, the use of this material for construction has brought some difficulties. Therefore, at present, UHPC is difficult to be widely used in practical engineering and replace normal concrete. Researchers still need to deepen the research on the requirements of the material itself for construction conditions and on-site construction technology, so as to make this material as much as possible to be used in engineering construction in the future.

Acknowledgement. Many people have made valuable contributions directly or indirectly to my research. Here I would like to thank my tutors and the senior for their instructive suggestions and valuable comments on the writing of this paper. Without their valuable help and generous encouragement, this paper will not be completed. In addition, I also thank my parents for their support and endless love. My heart is filled with gratitude to all those who have helped me.

References

1. Tayeh, B.A., Bakar, B.H.A., Johari, M.A.M., Voo, Y.L.: Utilization of Ultra-High Performance Fibre Concrete (UHPFC) for rehabilitation –a review. Procedia Eng. **54**, 525–538 (2013)
2. Basha, A., Fayed, S., Elsamak, G.: Flexural behavior of cracked RC beams retrofitted with strain hardening cementitious composites. KSCE J. Civ. Eng. **23**(6), 2644–2656 (2019). https://doi.org/10.1007/s12205-019-1874-4
3. Farhat, F.A., Nicolaides, D., Kanellopoulos, A., Karihaloo, B.L.: High performance fibre-reinforced cementitious composite (CARDIFRC) – performance and application to retrofitting. Eng. Fract. Mech. **74**(1), 151–167 (2007)
4. Khalil, A.E., Etman, E., Atta, A., Essam, M.: Behavior of RC beams strengthened with strain hardening cementitious composites (SHCC) subjected to monotonic and repeated loads. Eng. Struct. **140**, 151–163 (2017)
5. Kim, Y.Y., Lee, B.Y., Bang, J., Han, B., Feo, L., Cho, C.: Flexural performance of reinforced concrete beams strengthened with strain-hardening cementitious composite and high strength reinforcing steel bar, Composites. Part B, Eng. **56**, 512–519 (2014)
6. Lampropoulos, A.P., Paschalis, S.A., Tsioulou, O.T., Dritsos, S.E.: Strengthening of reinforced concrete beams using Ultra High Performance Fibre Reinforced Concrete (UHPFRC). Eng. Struct. **106**, 370–384 (2016)
7. Tanarslan, H.M.: Flexural strengthening of RC beams with prefabricated ultra high performance fibre reinforced concrete laminates. Eng. Struct. **151**, 337–348 (2017)

8. Richard, P., Cheyrezy, M.: Composition of reactive powder concretes. Cem. Concr. Res. **25** (7), 1501–1511 (1995)
9. Graybeal, B.A.: Characterization of the behavior of ultra-high performance concrete. Doctoral Thesis, University of Maryland (2005)
10. Bajaber, M.A., Hakeem, I.Y.: UHPC evolution, development, and utilization in construction: a review. J. Market. Res. **10**, 1058–1074 (2021)
11. Ju, Y.Z., Wang, D.H., Jiang, F.: Experimental investigation on stress-strain curves of reactive powder concrete under uniaxial compression. AMR **261–263**, 192–196 (2011)
12. Shafieifar, M., Farzad, M., Azizinamini, A.: Experimental and numerical study on mechanical properties of Ultra High Performance Concrete (UHPC). Constr. Build. Mater. **156**, 402–411 (2017)
13. Habel, K., Denarié, E., Brühwiler, E.: Experimental investigation of composite ultra-high-performance fiber-reinforced concrete and conventional concrete members. ACI Struct. J. **104**(1), 93–101 (2007)
14. Bastien-Masse, M., Brühwiler, E.: Strengthening the Chillon viaducts deck slabs with reinforced UHPFRC. In: IABSE Conference-Structural Engineering: Providing Solutions to Global Challenges, Geneva (2015)
15. Fournier, B., Bérubé, M.: Alkali-aggregate reaction in concrete: a review of basic concepts and engineering implications. Can. J. Civ. Eng. **27**(2), 167–191 (2000)
16. Brühwiler, E.: UHPFRC technology to enhance the performance of existing concrete bridges. Struct. Infrastruct. Eng. **16**(1), 94–105 (2019)
17. Alaee, F.J., Karihaloo, B.L.: Retrofitting of reinforced concrete beams with CARDIFRC. J. Compos. Constr. **7**(3), 174–186 (2003)
18. Al-Osta, M.A., Isa, M.N., Baluch, M.H., Rahman, M.K.: Flexural behavior of reinforced concrete beams strengthened with ultra-high performance fiber reinforced concrete. Constr. Build. Mater. **134**, 279–296 (2017)
19. Jang, H., Lee, H., Cho, K., Kim, J.: Experimental study on shear performance of plain construction joints integrated with Ultra-High Performance Concrete (UHPC). Constr. Build. Mater. **152**, 16 (2017)
20. Larrard, F., Sedran, T.: Optimization of ultra-high-performance concrete by the use of a packing model. Cem. Concr. Res. **24**(6), 997–1009 (1994)
21. Safdar, M., Matsumoto, T., Kakuma, K.: Flexural behavior of reinforced concrete beams repaired with Ultra-High Performance Fiber Reinforced Concrete (UHPFRC). Compos. Struct. **157**, 448–460 (2016)
22. Wang, X.W.: Research on the interfacial shear behavior of UHPC and reinforced concrete", Master's Thesis, Hunan University (2016)
23. Zhang, Y., Zhu, P., Wang, X., Wu, J.: Shear properties of the interface between ultra-high performance concrete and normal strength concrete. Constr. Build. Mater. **248**, 118455 (2020)

Artificial Neural Network Combined with Grey Wolf Optimizer for Period Determination of Light-Frame Wood Buildings

M. Nikoo and G. Hafeez(✉)

Department of Building, Civil and Environmental Engineering, Concordia University, Montréal, Canada
ghazanfarah.hafeez@concordia.ca

Abstract. The paper investigates the accuracy and competitiveness of the Grey Wolf Optimizer (GWO) algorithm employing the Feed Forward (FF) neural network. The algorithm is implemented on the period database of 47 light-frame wood buildings samples of different dimensions and geometry. Five physical characteristics were selected as input variables to determine the fundamental period of light-frame wood buildings. The structure of the Artificial Neural Network (ANN) was optimized using the GWO algorithm. The model's accuracy was evaluated by comparing the results with the Multiple Linear Regression (MLR) model and the model available in the National Building Code of Canada (NBCC, 2015). The results indicated the accuracy of the GWO algorithm optimized in ANN over other models for predicting the period of light-frame wood buildings.

Keywords: Fundamental period · Wood buildings · Grey Wolf · Light frame building · Seismic design

1 Introduction

In North America, the light-frame wood building represents a gross share of the construction industry. The dynamic response of light-frame wood buildings under strong ground motion is described by their dynamic properties such as fundamental natural period, mode shapes, and damping ratios. The fundamental period of vibration, (T), of a building, is an essential parameter, as it determines the spectral accelerations for Uniform Hazards Spectra, (UHS), for calculating the base shear by the equivalent static force procedure. The fundamental period of a structure is governed by the strength and stiffness characteristics of the building, hence influenced by many parameters, including building height, (h), number of stories, (N), dimensions of the shear wall, etc., [1]. The available models in building codes and literature are mostly height dependent and do not account for these parameters, leading to the period's non-accurate prediction. The current study is motivated to provide a reliable and convenient method for computing the fundamental period of light-frame wood buildings, which accounts for various physical parameters of buildings, including the building height.

P. Mendonça and N. D. Cortiços (Eds.): ICAMC 2021, LNCE 226, pp. 213–222, 2022.
https://doi.org/10.1007/978-3-030-94514-5_22

Lately, a growing interest has been seen in the application of Artificial Neural Network (ANN) in the domain of wood engineering and sciences [2–5]. This study presents an ANN model for predicting the fundamental period of light-frame wood buildings, implementing the GWO algorithm.

2 Fundamental Period Equations

The National Building Code of Canada [6] provides an empirical equation as a function of building height, (h), shown as Eq. (1), to estimate the fundamental period of wood shear wall buildings. This equation does not distinguish between shear walls of different materials, such as concrete, steel, and wood, and is only the function of building height (m).

$$T_a = 0.05 \times h^{0.75} \tag{1}$$

The measured data used for the derivation of Eq. (1) were based on reinforced concrete shear wall buildings, and a large scatter was observed in the collected data when the measured period was related to only the height of the structures [7]. Several researchers evaluated the applicability of this period equation on wood-based [8, 9] buildings, and some improved methods of period estimate were suggested.

3 Introducing Models

3.1 Artificial Neural Network

An artificial neural network (ANN), a data processor, consists of artificial neurons that collectively solve a specific problem. The system is created based on human brain structure and meant to act intelligently in a complex scenario when basic computational approaches do not provide a feasible solution [10]. Artificial Neural Network (ANN) is trained through experience with appropriate learning exemplars and collects information by detecting the patterns in data and relationships between the input and output variables. The ability of ANN to achieve partial tasks and rough outcomes characterizes it from other data-driven models [10]. The neural structure involving hundreds of artificial units connected with coefficients (weights) is typically arranged in three separate layers, namely the input layer, hidden layer, and output layer. The use of a single hidden layer with several nodes is highly recommended for accurately predicting the input and output relationships [11]. The current study also benefits from this recommendation and uses one hidden layer for data analyses. The weighted connections in each layer are the ones that link each neuron to several other nodes. The weights are trained in the ANN so that the target parameters are closest to the output values. The ANN procedure typically consists of three steps, including training, validation and testing. The training step aims at minimizing error, and the validation autonomously builds the model. Finally, the testing stage estimates the algorithm's accuracy [10].

3.2 Grey Wolf Optimizer

Grey Wolf Optimizer (GWO) is the Meta-heuristic optimization technique instigated by the grey wolves' leadership hierarchy and hunting mechanism. Grey wolves live in packs with 5–12 members on average and have a stringent socially dominant hierarchy [12]. Their hunting mechanism comprises three steps, chasing and approaching the prey, circling until it surrenders, finally attacking. They are categorized into four different types α, β, δ and ω, with α and ω being the most and the least dominant characters in the pack [13]. The α generally imposes the decisions on the pack and ought not to be the strongest member, rather the one with the most effective managerial skills in the pack. Next comes β wolve in the grey wolf hierarchy, responsible for assisting α, considered to be the potential candidate for succession after α [12]. GWO algorithm follows the same social behaviours by considering α and δ as the best and least fit solutions, while ω follows the rest. Figure 1 depicts their social hierarchy and hunting approach considered for designing the GWO algorithm. As seen in Fig. 2, a grey wolf can relocate within the space surrounding the prey in any random location, hence modeling algorithm α, β, and δ evaluate the position of the prey.The other wolves relocate their positions randomly around the prey [12].

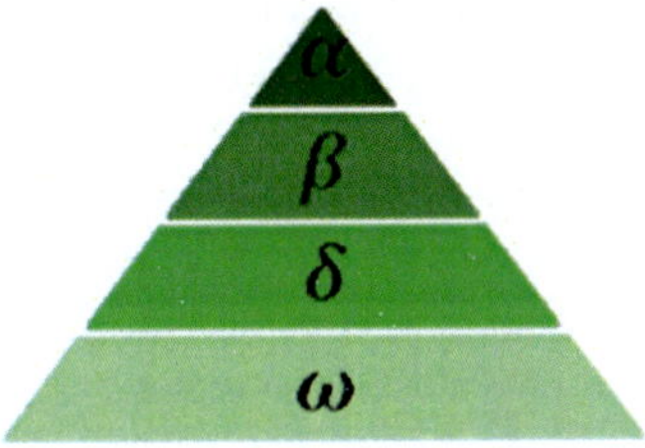

Fig. 1. Hierarchy of the grey wolf optimizer (dominance drops top-bottom) [12].

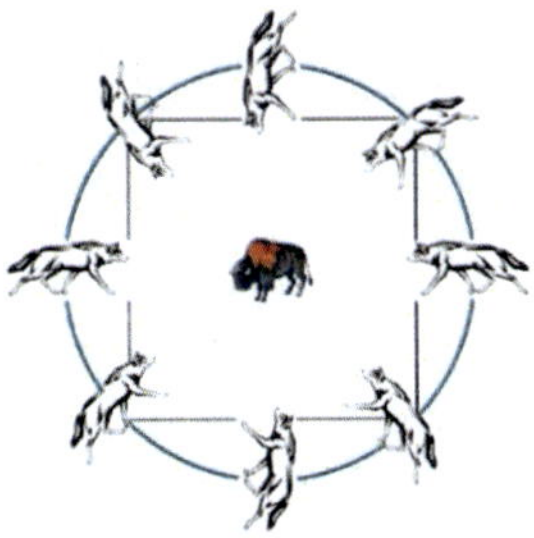

Fig. 2. 2D position vectors and their probable subsequent locations [12].

3.3 Multiple Linear Regression

Engineering problems dealing with the large dataset often require the assessment of relationships between the variables. The regression models are often employed for approximating the correlations between input and output variables as a procedure of determining the best model fitting to the dataset. These models are mostly fitted with the least square method. The linear regression classified into simple and multiple linear regressions [14, 15] is applied to the dataset for obtaining output parameters, using linear predictor functions [16]. If the linear correlation is desired between a predictor, also called the independent variable, and a dependent variable, simple linear regression (SLR) is functional. However, if the predictors are more and the dependent variable stays the same (one), Multiple Linear Regression (MLR) serves the objective appropriately and weighs the correlation between two or more input variables by fitting a linear equation to the desired data [17, 18].

4 Methods and Materials

4.1 Dataset

The database of the fundamental period used for this investigation is derived from the literature on dynamic characteristics of light-frame wood buildings [8, 19, 20]. The authors investigated the dynamic properties of 47 light-frame wood buildings of various heights, plans, and geometry located at different locations in Canada. Buildings were measured using low amplitude ambient vibration methods, and the buildings' dynamic properties were identified by performing analyses of the measured records in the frequency domain.

Feed Forward (FF) neural networks were used to precisely determine the fundamental period of light-frame wood buildings, considering the building's physical properties such as the number of stories, area, height, maximum length, and maximum width as input parameters. The statistical characteristics related to the parameters are shown in Table 1.

Table 1. Statistical characteristics of the collected data [8, 19, 20].

Statistical Index	Unit	Type	Max	Min	Average	STD
Stories	Number	Input	6.00	1.00	3.70	1.20
Area	m^2	Input	4359.80	61.00	1040.00	1069.50
Height	m	Input	20.20	3.00	13.10	4.10
Length	m	Input	184.00	8.50	43.60	32.40
Width	m	Input	80.00	6.70	24.20	16.60
Period	s	Output	0.50	0.10	0.30	0.10

4.2 Empirical Model Development Using ANN and GWO

For creating a reliable model, the input parameters need to be chosen carefully. The input variables should be selected in such a way that the model performs efficiently among variables. In this study, area, height, length, width and number of stories were considered as input variables for predicting the fundamental period of wood buildings.

The data was distributed into training and testing groups, a recommended method when the samples are less than 100. The aggregate of the samples was limited to 47 buildings; hence most of the data, 80% (38 samples), was contributed to network training while the rest, 20% (9 samples), was used for network performance testing. In the ANN model, the number of hidden layers and neurons depends on the problem type and impacts the network performance. In this study, the input layer of each trained ANN carries five neurons, namely, area, height, length, width and number of stories. At the same time, the output layer holds only one neuron, a maximum period. A standard heuristic method for the total number of neural cells of an ANN is provided in Eq. (3) [21], where N_H is the number of hidden layer nodes, and N_I is the number of input parameters. Following Eq. (2), there was a maximum of 11 nodes in hidden layers in the ANN model.

$$N_H \leq min(2N_I + 1) \tag{2}$$

The performance of the various topologies provided in Table 2 was assessed using three statistical indices, coefficient of determination (R^2), Variance Account Factor (VAF) and Root Mean Square (RMSE), shown in Eqs. (3) to (5), respectively. The models were optimized by implementing GWO Algorithm on the trained ANN for reducing the error and adjusting the weights and biases. Properties of the GWO are presented in Table 3.

$$R^2 = 1 - \frac{Sum\,Squared\,Regression\,(SSR)}{Total\,Sum\,of\,Squares\,(TSS)} \tag{3}$$

$$VAF = \left[1 - \frac{var(O_i - P_i)}{var(O_i)}\right] \times 100 \tag{4}$$

$$RMSE = \left[\frac{1}{n}\sum_{i=1}^{n}(P_i - O_i)^2\right]^{\frac{1}{2}} \tag{5}$$

Table 2. Different topologies used in ANN models.

No	Hidden Layer 1	Hidden Layer 2	Hidden activation functions	Output activation functions	No	Hidden Layer 1	Hidden Layer 2	Hidden activation functions	Output activation functions
1	6	5	TANSIG	PURELIN	6	5	5	POSLIN	PURELIN
2	6	4	TANSIG	PURELIN	7	5	4	LOGSIG	PURELIN
3	6	3	TANSIG	TANSIG	8	5	3	PURELIN	PURELIN
4	6	2	POSLIN	PURELIN	9	5	2	POSLIN	PURELIN
5	6	1	LOGSIG	PURELIN	10	5	1	LOGSIG	PURELIN

Table 3. Grey Wolf Optimized (GWO) parameters.

Parameter	Value	Parameter	Value
Max generations	300	Search agents	10

4.3 Empirical Model Evaluation

Ten different ANN models were created with two hidden layers, and their weights were modelled and optimized using the GWO algorithm. The best-performed models based on RMSE and VAF values were then selected for predicting the fundamental period of light-frame wood buildings. Table 4 provides RMSE and VAF values for trained and tested samples. Table 4 displays the lowest RMSE index and the highest VAF percentage, in both training and testing phases, of the GWO-ANN with two layers (2L) network (5–5), showing its preciseness over other GWO-optimized models.

In order to visually illustrate the performance of GWO-ANN 2L (5–5), the predicted and experimental values of the empirical model for the training and testing data are displayed in Fig. 3. The model's predicted values closer to $y = x$ line show the desired accuracy of the model.

Table 4. Statistics of the best two ANN models combined with GWO algorithm for a fundamental period estimate.

Topology	Train		Test	
	RMSE	VAF%	RMSE	VAF%
GWO-ANN 2L (5–5)	0.022	91	0.024	89
GWO-ANN 2L (5–3)	0.032	83	0.028	73

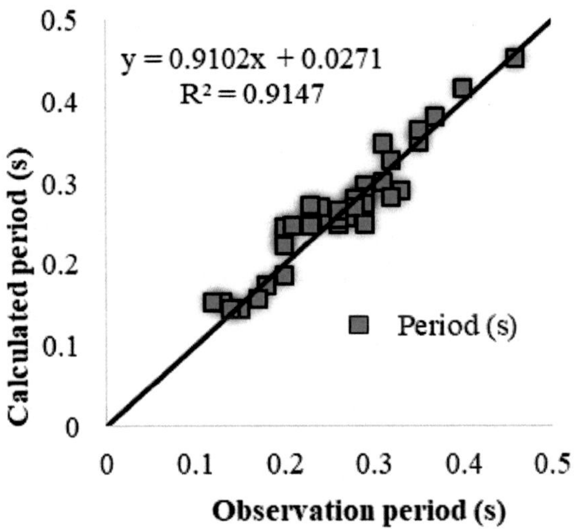

(a)

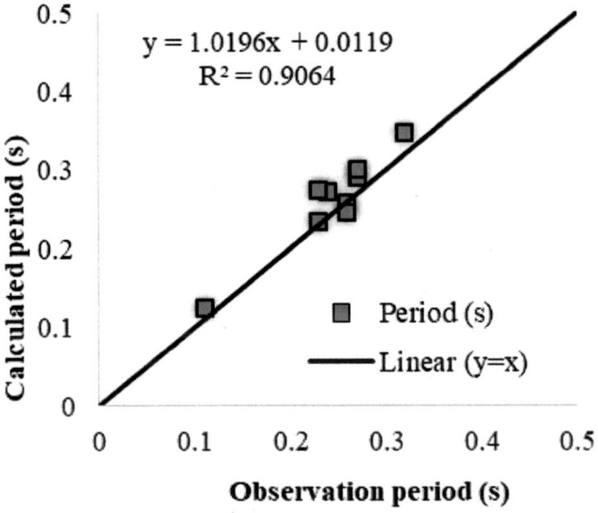

(b)

Fig. 3. Experimental versus predicted values of fundamental period for GWO-ANN (5–5) model, (a) Training data, (b) Testing data.

4.4 Best Performed Model

MLR Models. Several MLR models, developed using physical characteristics of wood buildings, were examined for predicting the fundamental period of these buildings. The best MLR model capable of accurately predicting the fundamental period of wood buildings is shown in Eq. (6). The related statistical indices of MLR model analysis are provided in Table 5.

$$y = 0.101 + 0.0207 \times x_1 + 0.000021 \times x_2 + 0.00825 \times x_3 - 0.000031 \times x_4 - 0.00198 \times x_5 \tag{6}$$

Where, y is period, x_1: Stories, x_2: Area, x_3: Height, x_4: Length, x_5: Width.

Comparison of Results Between Various Models. The best-performed regression models, MLR and GWO-ANN 2L(5–5) are compared with the building code [6]. This comparison is based on statistical indicators and evaluated the models with the least error. The values of statistical indicators, VAF, RMSE, R^2 used for the comparison are provided in Table 5. The table also shows the straight-line slope when the fundamental experimental period of the buildings was plotted against their predicted counterparts. The table shows that the GWO-ANN model yields maximum VAF and minimum RMSE values with the most significant R^2 when compared to the other evaluated models. Comparison of GWO-ANN with NBCC and MLR models is displayed in Figs. 4(a) and 4(b) respectively. The results indicate that the ANN optimized using the GWO algorithm offers much higher flexibility and accuracy for period estimate of wood buildings.

Table 5. Statistics of different models.

Numbers	Models	RMSE	VAF%	R^2	$y = ax + b$
1	GWO-ANN 2L (5–5)	0.022	91	0.908	y = 0.9157x + 0.0278
2	NBCC	0.096	15	0.445	y = 0.805x + 0.1329
3	MLR	0.047	57	0.570	y = 0.5685x + 0.1112

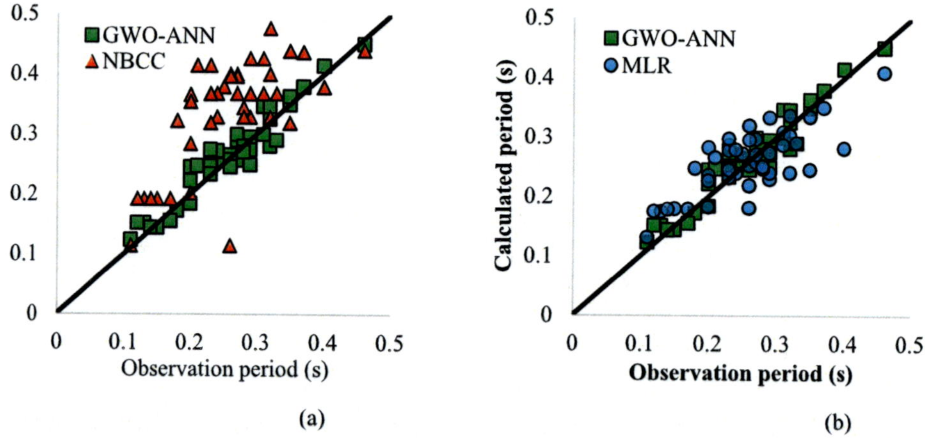

Fig. 4. Experimental versus predicted values of the fundamental period, (a) Comparison with NBCC, (b) Comparison with MLR model.

5 Conclusion

The paper presents an Artificial Neural Network (ANN) model combined with the Grey Wolf Optimizer algorithm (GWO-ANN), developed for estimating the fundamental period of light-frame wood buildings. The accuracy of the model for predicting the period of wood buildings was evaluated by comparing the results with the Multiple Linear Regression (MLR) and NBCC models. The outcomes of the research are presented next.

i. The statistical records of VAF and RMSE from the GWO-ANN model for the training (0.022 and 91%) and testing (0.024 and 89%) data set approved the reliability of the GWO-ANN model when comparing with the other informational models. MLR model showed the better potential of estimating the fundamental period with the estimated R^2 value as 0.57, compared to the NBCC model (R^2 = 0.44).
ii. The study concluded a relatively larger deviation of the NBCC model for estimating the fundamental period of light-frame wood shear wall buildings compared to the informative models. The result confirmed that estimating the building period as a function of only building height is insufficient to represent its dynamic behaviour. As a function of area, height, length, width and number of stories, the proposed model can precisely predict the fundamental period of light-frame wood buildings.

References

1. Asteris, P., Repapis, C., Cavaleri, L., Sarhosis, V., Athanasopoulou, A.: On the fundamental period of infilled RC frame buildings. Struct Eng. Mech. **54**, 1175–1200 (2015)
2. Tiryaki, S., Aydın, A.: An artificial neural network model for predicting compression strength of heat treated woods and comparison with a multiple linear regression model. Constr. Build. Mater. **62**, 102–108 (2014)
3. Fathi, H., Nasir, V., Kazemirad, S.: Prediction of the mechanical properties of wood using guided wave propagation and machine learning. Constr. Build. Mater. **262**, 120848 (2020)
4. Asteris, P.G., Nikoo, M.: Artificial bee colony-based neural network for the prediction of the fundamental period of infilled frame structures. Neural Comput. Appl. **31**, 4837–4847 (2019). https://doi.org/10.1007/s00521-018-03965-1
5. Bardak, S., Tiryaki, S., Nemli, G., Aydın, A.: Investigation and neural network prediction of wood bonding quality based on pressing conditions. Int. J. Adhes. Adhes. **68**, 115–123 (2016)
6. NRC/IRC 2015 National Building Code of Canada (Ottawa, Ontario: National Research Council of Canada, Institute for Research in Construction)
7. Saatcioglu, M., Humar, J.: Dynamic analysis of buildings for earthquake-resistant design. Can J. Civ. Eng. **30**, 338–359 (2003)
8. Hafeez, G., Doudak, G., McClure, G.: Establishing the fundamental period of light-frame wood buildings on the basis of ambient vibration tests. Can. J. Civ. Eng. **45**, 752–765 (2018)
9. Camelo, V.: Dynamic Characteristics of Woodframe Structures (2002)

10. Khademi, F., Akbari, M., Nikoo, M.: Displacement determination of concrete reinforcement building using data-driven models. Int. J. Sustain. Built. Environ. **6**, 400–411 (2017)
11. Deshpande, N., Londhe, S., Kulkarni, S.: Modeling compressive strength of recycled aggregate concrete by artificial neural network, model tree and Non-linear regression. Int. J. Sustain. Built. Environ. **3**, 187–198 (2014)
12. Mirjalili, S., Mirjalili, S.M., Lewis, A.: Grey Wolf optimizer. Adv. Eng. Softw. **69**, 46–61 (2014)
13. Mech, L.D.: Alpha status, dominance, and division of labor in wolf packs. Can. J. Zool. **77**, 1196–1203 (1999)
14. Nikoo, M., Torabian Moghadam, F., Sadowski, Ł: Prediction of concrete compressive strength by evolutionary artificial neural networks. Adv Mater. Sci. Eng. **2015**, 1–9 (2015)
15. Khademi, F., Akbari, M., Nikoo, M.: Displacement determination of concrete reinforcement building using data-driven models. Int J. Sustain. Built. Environ. **6**, 400–411 (2017)
16. Khademi, F., Akbari, M., Jamal, S.M., Nikoo, M.: Multiple linear regression, artificial neural network, and fuzzy logic prediction of 28 days compressive strength of concrete. Front Struct. Civ. Eng. **11**, 90–99 (2017). https://doi.org/10.1007/s11709-016-0363-9
17. Khademi, F., Behfarnia, K.: Evaluation of concrete compressive strength using artificial neural network and multiple linear regression models. IUST **6**, 423–432 (2016)
18. Sadrmomtazi, A., Sobhani, J., Mirgozar, M.A.: Modeling compressive strength of EPS lightweight concrete using regression, neural network and ANFIS. Constr. Build. Mater. **42**, 205–216 (2013)
19. Hafeez, G., Doudak, G., McClure, G.: Dynamic characteristics of light-frame wood buildings. Can. J. Civ. Eng. **46**, 1–12 (2018)
20. Hafeez, G., Mustafa, A., Doudak, G., McClure, G.: Predicting the fundamental period of light-frame wood buildings. J. Perform. Constr. Facil. **28**, A4014004 (2014)
21. Bowden, G.J., Dandy, G.C., Maier, H.R.: Input determination for neural network models in water resources applications. Part 1—background and methodology. J. Hydrol. **301**, 75–92 (2005)

Effect of Hollow Body Shape on the Internal Stress Distribution of Bubble Concrete

Xiangdong Yan(✉) and Pei-Shan Chen

Department of Civil and Architectural Engineering, Kyushu Institute of Technology, Kitakyushu 804-8550, Japan
yan.xiangdong760@mail.kyutech.jp

Abstract. Bubble concrete is a lightweight achieved by mixing high-strength hollow bodies into concrete. Bubble concrete is different from biaxial voided slabs, it can reduce the density of concrete while ensuring the strength. As the existence of the hollow body will greatly affect the stress distribution inside the concrete, in order to improve the strength of the bubble concrete, this study conducted a nonlinear elastoplastic analysis with four different hollow body models. The optimal hollow body model is determined by comparing the stress distribution states inside the concrete. Based on the analytical results, as the capsule model inherits the good force transfer mechanism of the sphere, it can well reduce the lateral expansion of the hemisphere and further reduce the amount of concrete. Therefore, the capsule model is the best hollow body model for the bubble concrete in terms of stress distribution.

Keywords: Bubble concrete · Lightweight aggregate concrete · Mechanical simulation · Elastoplastic analysis · Steel material

1 Introduction

Lightweight concrete materials can reduce the weight of structures and mitigate the effects of seismic loads, which has been widely embraced by scholars over the world. However, the porous structure of lightweight concrete makes it difficult to carry large loads. Therefore, the current applications of lightweight concrete structures in the world are mainly focused on insulated walls, partitions wall and environmentally friendly pavements [1–4].

In previous research, author Pei-Shan Chen was inspired by hollow bird bones and invented a new type of lightweight concrete material called bubble concrete [5, 6]. It is achieved by mixing high-strength hollow bodies into concrete, as shown in Fig. 1. This new type of lightweight concrete material is different from the biaxial voided slab used in Europe [7–9]. The hollow body inside the concrete is not only used to create cavities but also to transmit internal stress. Therefore, bubble concrete can reduce the weight of the concrete while ensuring its strength.

P. Mendonça and N. D. Cortiços (Eds.): ICAMC 2021, LNCE 226, pp. 223–229, 2022.
https://doi.org/10.1007/978-3-030-94514-5_23

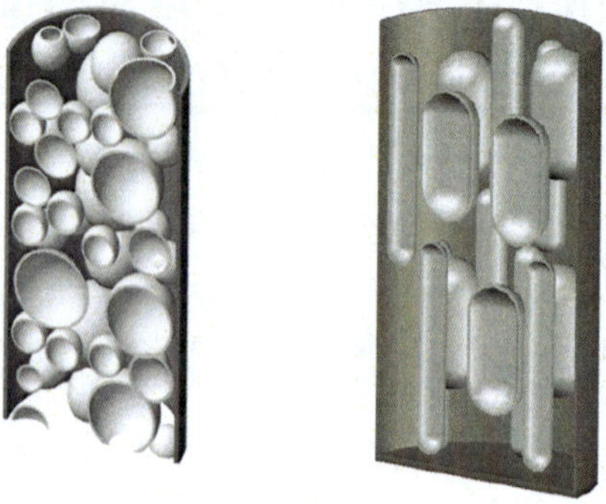

Fig. 1. Bubble concrete contains high-strength hollow bodies.

In previous studies, it was found that the presence of hollow bodies affects the stress distribution inside the concrete significantly [10, 11]. Moreover, the premature failure of concrete is the main reason for the reduction of concrete strength. Therefore, how to improve the stress distribution state inside the concrete is the key to increase its strength. In this study, the nonlinear elastoplastic analysis of the model was performed to compare the stress distribution inside the concrete under the use of four different hollow bodies. Finally, find the most suitable hollow body model for the bubble concrete.

1.1 Analytical Model and Method

In this study, ADINA 9.3 was used for elastic-plastic nonlinear analysis. The concrete strength was 40 MP and the hollow body was steel with a strength of SS400 (yield point = 235 MP). The stress-strain curves of concrete and model are shown in Fig. 2.

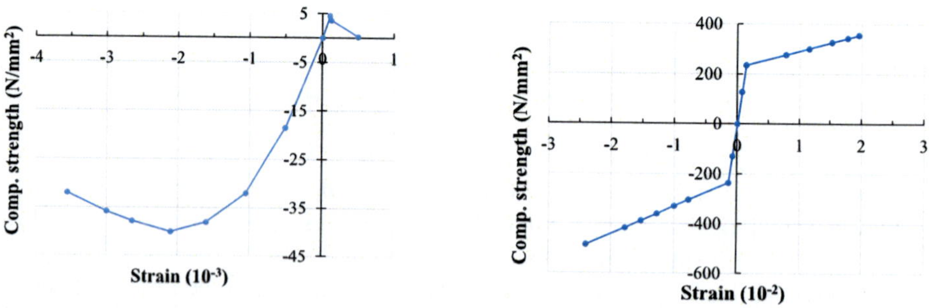

(**a**) Material curve of concrete specimen. (**b**) Material curve of steel sphere.

Fig. 2. The dimension and material curves.

The concrete model dimensions are referred to the Japanese standard concrete specimen, which is a cylindrical model with 150 mm diameter and 300 mm height [12]. The concrete and four hollow body models are shown in Fig. 3, where the thickness of all the hollow body models is 2 mm. The hollow body is placed in the center of the concrete, and then a uniform load is applied at the top, and the fixed at the bottom. The load is increased uniformly in 20 steps starting from 0 until 40 MP. In order to explore the stress distribution around the hollow body, the stress variation diagrams of the top, bottom, left and right directions of the hollow body were counted separately.

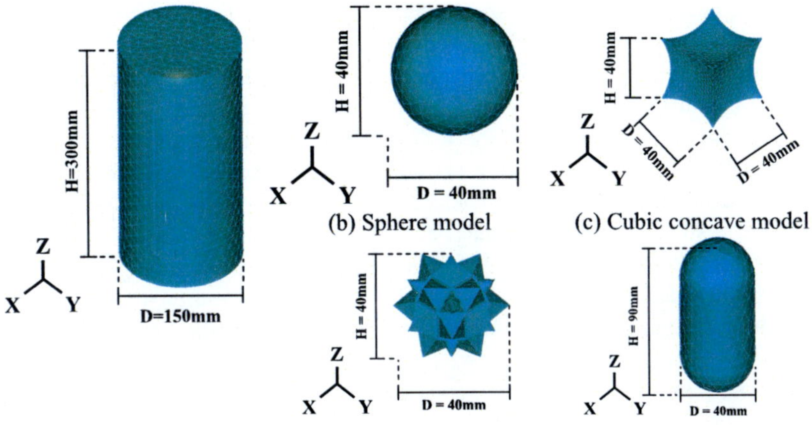

(a) Concrete models (d) Origami model (d) Capsule models

Fig. 3. The specific dimensions of the concrete and hollow body models.

2 Results and Discussion

The stress distribution inside the bubble concrete with the four types hollow body models are shown in Fig. 4. Following the analysis results, when the load is applied to the top of the concrete, the hollow body undergoes vertical compression and lateral expansion deformation. The hollow body needs to transmit the corresponding force downward, so the stress on the hollow body is much greater than the concrete. In addition, the difference in the shape of the hollow body changes its deformation direction, so the stress distribution inside the concrete will change correspondingly.

When a load is applied to the top of the concrete, the sphere model will undergo vertical compression and oblique expansion. As the hollow body and concrete hold different stiffness, the concrete and sphere model are peeled off from each other, and squeezed the surrounding concrete in oblique direction. Therefore, the concrete stress on the vertical direction is decreased, and the concrete stress in the oblique direction of the sphere is increased. With the increase of the load, the concrete in the oblique direction reaches the failure load so that the concrete specimen fails.

To suppress the deformation of the hollow body in the oblique direction, we came up with the cubic concave model. Compared with the spherical model, the concave body model can restrain the expansion and deformation of the hollow body effectively. However, due to the shape mechanism of the concave model which increased the cross-section area, so the concave model is subjected to a larger load, which make it suffer a larger compressive deformation. The deformation of the concave body lets it unable to continue to bear the vertical load. Therefore, the concrete stress on the upper and lower sides of the concave body is decreased, while increase in the lateral direction. Moreover, the shape mechanism of the concave body makes its edges prone to damage after being loaded. Therefore, the strength of concrete with concave model is lower than when using sphere model. In addition, it is difficult to fix the concave body so that the front side faces upward in experiment. When the oblique side of the concave model is upward, which will cause the hollow body to deform rapidly.

The origami model can suppress the hollow body to generate vertical compression and lateral expansion at the same time, so that the concrete stress around the hollow body is evenly distributed. Since the hollow body model needs to transmit vertical stress, the cross-sectional area of the origami model is smaller than sphere and cubic concave model due to its hollow characteristics. Therefore, the origami model needs to carry a greater stress thereby will breaks earlier. In addition, after the tip of the origami model is subjected to a vertical load, a large bending moment will be generated to the origami model, making it prone to damage. Consequently, the strength performance of the bubble concrete with the origami model is worse than the sphere or the cubic concave model are used.

The capsule model is based on the sphere model and extends the middle segment of the model. The capsule model inherits the good stress transfer mechanism of the sphere model and ensures the cross-sectional area so that it can effectively transfer the vertical load. Moreover, the cylindrical part of the capsule model in turn reduces the oblique deformation of the hemisphere and makes the stress distribution inside the concrete more uniform. Therefore, compared with the sphere model, the capsule model can guarantee the strength of the concrete and further reduce the amount of concrete. Among the four types of bubble concrete, capsule model is the best model for bubble concrete in terms of strength performance.

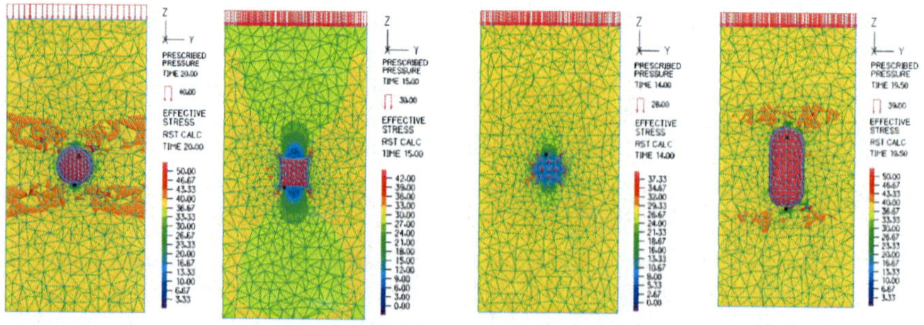

(a) Sphere model (b) Cubic concave model (c) Origami model (d) Capsule model

Fig. 4. The internal stress diagrams.

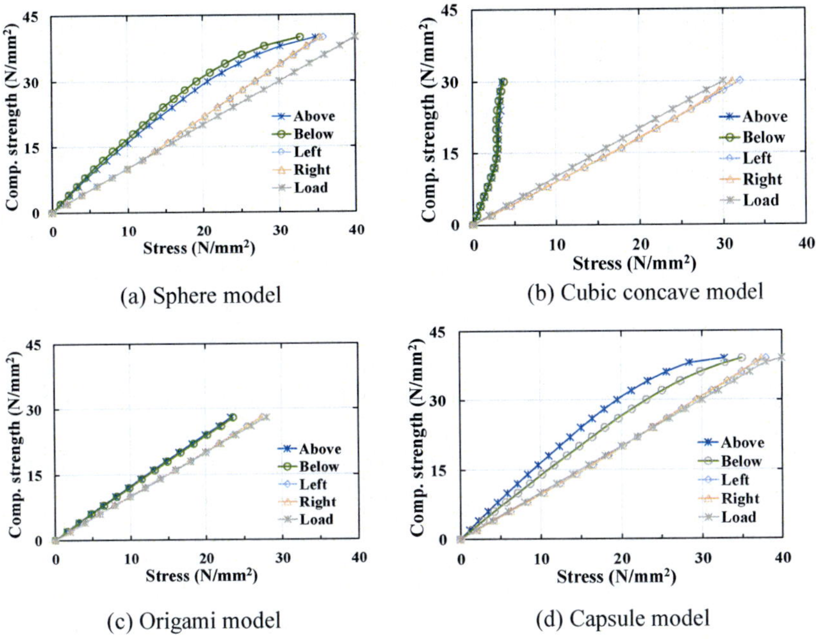

Fig. 5. Concrete stress curve around the model.

In order to compare the impact of the hollow body deformation on the stress distribution of the concrete directly, we made statistics on the horizontal and vertical concrete stress around the hollow body. The concrete stresses at a distance of 5 mm from the hollow body in the four directions (Above, below, left and right) and plotted them in the concrete stress curve, as shown in Fig. 5. The vertical axis represents the load of the concrete, and the horizontal axis represents the effective stress experienced by the concrete.

According to the analytical results, the sphere model is deformed after being loaded, so that the vertical concrete stress is less than the load, while the lateral concrete stress is close to the load. The good force transmission mechanism of the sphere model makes the stress distribution inside the concrete more uniform, and the concrete strength does not decrease significantly. The cubic concave model quickly deforms after being loaded and cannot continue to transmit the load, so the vertical concrete stress is reduced. In addition, the stripping of the cubic concave model from the concrete reduces the effective cross-sectional area to transfer the stress, so the lateral concrete stress is greater than the load. The origami model reduces the vertical and lateral deformation and improves the stress distribution inside the concrete effectively. However, as the tip of the origami model will generate a large bending moment after being loaded, the origami model will be damaged prematurely. Therefore, the strength of bubble concrete with origami model has been reduced obviously. The capsule model inherits the advantages of good stress distribution of the sphere model and reduces the impact of lateral deformation on the surrounding concrete stress. Therefore, for bubble concrete with the capsule model further reduced the amount of concrete while maintaining high concrete strength.

3 Conclusion

In this paper, the effect of four types of hollow body models on the stress distribution of bubble concrete is studied. By comparing the non-linear analytical results of the four types of bubble concrete, the stress mechanism and the cause of bubble concrete was explored. The hollow body inside the concrete uses its material strength and structure to transfer the load downward, so the hollow body will carry the force of the corresponding cross-sectional area. When the shape of the hollow body is different, its force mechanism also changes accordingly.

First, the sphere model undergoes vertical compression and lateral expansion n after being loaded. The stress on the upper and lower sides of concrete is reduced, while the concrete stress in the oblique direction is increased. As the load increases, the concrete stress in the oblique direction reaches the failure load and then concrete fails. The concave shape of the cubic concave model can reduce the lateral deformation effectively. However, as the top of the concave model bears a larger force-receiving area, which let the concave model deform quickly, and then make the concave model unable to transmit load anymore. Therefore, the effective cross-sectional area that can transmit the load is reduced and the lateral concrete stresses are increased. The origami model suppresses the horizontal and vertical deformation and makes the stress distribution inside the concrete more uniform. However, the effective cross-sectional area of the origami model to bear the vertical load is small, which makes it subject to greater stress. In addition, its protruding tip caused it to be subjected to a greater bending moment, which cause it to be damaged early. Because the capsule model inherits the good force transmission mechanism of the sphere model, it can evenly distribute the stress inside the concrete. Moreover, the cylindrical structure also reduces the lateral deformation of the hemisphere and reduces the amount of concrete. Therefore, among the four types of hollow body models, the capsule model is the best model for the bubble concrete in terms of stress distribution.

References

1. Haque, M.N., Al-Khaiat, H., Kayali, O.: Strength and durability of lightweight concrete. Cem. Concr. Compos. **26**, 307–314 (2004)https://doi.org/10.1016/S0958-9465(02)00141-5
2. Min, H.Z., Odd, E.G.: Mechanical properties of high-strength lightweight concrete. Mater. J. **88**, 240–247 (1991)
3. Kan, A., Demirboğa, R.: A novel material for lightweight concrete production. Cem. Concr. Compos. **31**, 489–495 (2009). https://doi.org/10.1016/j.cemconcomp.2009.05.002
4. Lo, T.Y., Tang, W.C., Cui, H.Z.: The effects of aggregate properties on lightweight concrete. Build. Environ. **42**, 3025–3029 (2007). https://doi.org/10.1016/j.buildenv.2005.06.031

5. Chen, P.-S.: A study report on light weight concrete mixed with high strength hollow bubbles. In: Proceeding of AIJ Tohoku Chapter Architectural Research Meeting (Kouzoukei); Architectural Institute of Japan: Tokyo, Japan, no.71, pp. 57–62 (2008)
6. Chen, P.-S.: Bubble Concrete. Japan Patent Office, JP2008308395, 25 December 2008. https://patentfield.com/patents/JP2007181500A. Accessed 25 Dec 2008
7. Andersen, O., Waag, U., Schneider, L., Stephani, G., Kieback, B.: Novel metallic hollow sphere structures. Adv. Eng. Mater. **4**, 192–195 (2000)
8. Lim, T.-J., Smith, B., McDowell, D.L.: Behavior of a random hollow sphere metal foam. Acta Mater. **50**, 2867–2879 (2002). https://doi.org/10.1016/S1359-6454(02)00111-8
9. Schnellenbach-Held, M., Pfeffer, K.: Punching behavior of Biaxial hollow slabs. Cem. Concr. Compos. **24**, 551–556 (2002)https://doi.org/10.1016/S0958-9465(01)00071-3
10. Chen, P.-S., Tsukinaga, Y.: Basic research on the development of light weight concrete mixed with hollow spheres. Soc. Mater. Sci. Jpn. **64**, 711–717 (2015). https://doi.org/10.2472/jsms.64.711
11. Yan, X., Chen, P.-S., Al-Fakih, A., Liu, B., Mohammed, B.S., Jin, J.: Experiments and mechanical simulation on bubble concrete: studies on the effects of shape and position of hollow bodies mixed in concrete. Crystals, **11**, 858 (2021) https://doi.org/10.3390/cryst11080858
12. Japanese Industrial Standard (JIS). JIS A 1108 (2018). https://webdesk.jsa.or.jp/preview/pre_jis_a_01108_000_000_2018_j_ed10_ch.pdf. Accessed 28 June 2021

The Whale Optimization Algorithm Based ANN for Predicting the Fundamental Period of Light-Frame Wood Buildings

M. Nikoo and G. Hafeez(✉)

Department of Building, Civil and Environmental Engineering, Concordia University, Montréal, Canada
ghazanfarah.hafeez@concordia.ca

Abstract. The paper introduces the convenient and reliable method for the period estimate of wood buildings employing Artificial Neural Network (ANN) combined with Whale Optimization Algorithm (WOA). The algorithmic model developed using Feed Forward (FF) networks is based on the physical parameters of 47 light-frame wood buildings whose dynamic properties were measured using the ambient vibration testing method. The proposed model produces a better period estimate than the model available in the National Building Code of Canada (NBCC, 2015).

Keywords: Period formula · Wood buildings · Whale optimization · Feed forward · Neural network

1 Introduction

Building's vibration response is an essential source of information in engineering design and construction to improve the user's living comfort and prevent the building from malfunctioning under extreme events. Estimation of seismic loads depends on the fundamental period of the structure, and inadequately estimated design load could lead to unsafe design. Building codes around the world provide period approximation formulae of the buildings of different materials and seismic force-resisting systems as the function of building geometry.

National Building Code of Canada [1] provides a simple formula for calculating the fundamental period of buildings with the shear wall as a function of building height, $T_a = 0.05 \times h^{0.75}$ where, h is meters (m). The equation is derived from the recorded motion of concrete buildings measured during different seismic events in California [2]. It has been recognized that the equation does not fit the measured period data appropriately when plotted against the building height [3].

Artificial Neural Networks (ANNs) have become a popular and practical tool for classification and prediction in many engineering disciplines in the last few decades. ANN is a machine learning model that mimics brain network has become competitive to conventional regression and statistical models due to its convenience and effectiveness [4]. In this study, the fundamental period of light-frame wood buildings is determined using artificial neural networks optimized using the WOA.

P. Mendonça and N. D. Cortiços (Eds.): ICAMC 2021, LNCE 226, pp. 230–236, 2022.
https://doi.org/10.1007/978-3-030-94514-5_24

2 Whale Optimization Algorithm

Whales are mainly considered predators. The similarities between a human and a whale's brain composition make this creature accountable for its intelligence and foraging. The Whale Optimization Algorithm (WOA) is a swarm intelligence algorithm developed inspired by the foraging behaviour of humpback whales. The algorithm is used for continuous optimization problems with the proven better performance comparing other available algorithmic techniques [4]. The use of improved candidate solutions in each step of optimization differs from the WOA with other algorithms [5]. Figure 1 mimics the schematic of bubble-net foraging behavior of whales. The algorithm follows the foraging behavior of humpback whales, bubble-net hunting, searching, encircling, and finally attacking prey.

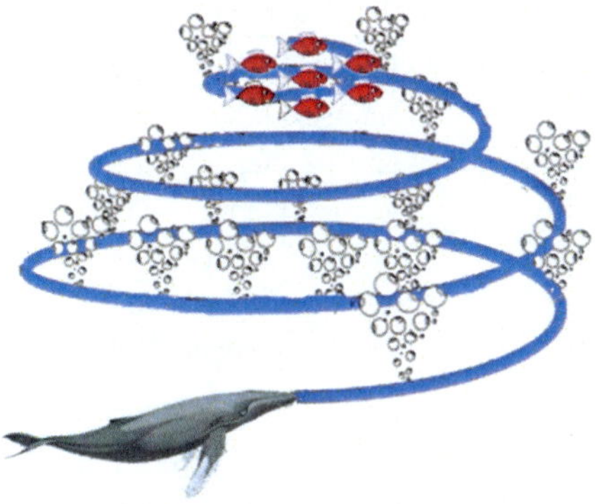

Fig. 1. Schematic diagram of whale bubble-net foraging [5].

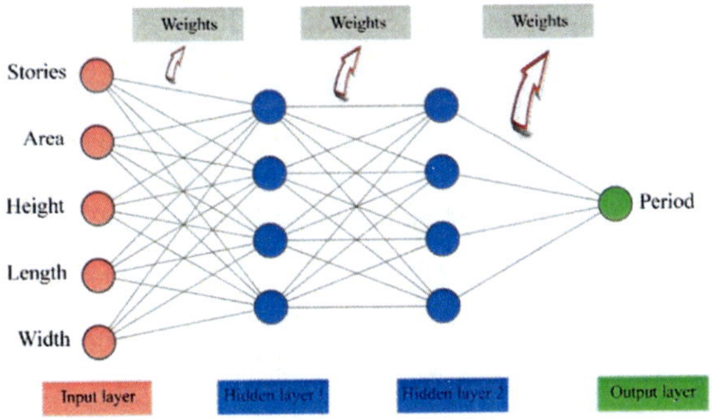

Fig. 2. Architecture of an artificial neural network with topology 5–4-4–1.

3 Research Methodology

This study used fundamental period data set of finished buildings derived from the comprehensive database established by measuring forty-seven light-frame wood buildings of different heights and geometry located in Canada. The buildings were measured using the Ambient Vibration Testing (AVT) method, and the data were analyzed in Frequency Domain Decompositions (FDD) to yield the dynamic properties (fundamental frequency, associated mode shape and damping ratio) of the measured buildings [6].

Building area, height, length, width, and the number of stories were considered as input variables for determining the fundamental period of wood buildings as an output parameter. General characteristics of buildings are provided below in Table 1.

Table 1. Statistical characteristics of dataset [6–8].

Statistical Index	Unit	Type	Max	Min
Stories	Number	Input	6	1
Area	m^2	Input	4359.8	60.97
Height	m	Input	20.2	3
Length	m	Input	184	8.5
Width	m	Input	80	6.7
Period	s	Output	0.46	0.11

The period dataset comprised of 47 samples were randomly divided into two sets. 38 samples were used for network training, and the remaining 9 samples were used for network performance testing. A network is trained by altering the weights of the connections between the nodes. These weights can be randomly chosen or individually chosen. The trained ANN is Feed-Forward (FF) network, with five neurons in the input layer and one in the output layer. A common heuristic method for the total number of neural cells of an ANN is given in Eq. (1) [9].

$$N_H \leq min(2N_I + 1) \tag{1}$$

N_H: Number of hidden layer nodes, N_I: Number of inputs. The performance of the model was evaluated using the coefficient of determination (R^2), Average Absolute Error (AAE), and Mean Absolute Error (MAE) shown in Eqs. (2) to (4) respectively [10, 11].

$$R^2 = 1 - \frac{Sum\ Squared\ Regression\ (SSR)}{Total\ Sum\ of\ Squares\ (TSS)} \tag{2}$$

$$MAE = \frac{1}{n}\sum_{i=1}^{n}|P_i - O_i| \tag{3}$$

$$AAE = \frac{\sum_{i=1}^{n} |O_i - P_i|}{n} \times 100 \tag{4}$$

Employing Eq. (1), the maximum number of nodes in the hidden layers were calculated as 11. the WOA was used for minimizing the error and, adjusting the weights and biases in the neural network. A typical representation of ANN is provided in Fig. 2.

Various topologies of neural network and the properties of the Whale Optimization Algorithm are presented on Tables 2 and 3 respectively.

Table 2. Different topologies used in ANN models.

No	Hidden Layer 1	Hidden Layer 2	Hidden Activation	Output Activation	No	Hidden Layer 1	Hidden Layer 2	Hidden Activation	Output Activation
1	3	3	PURELIN	PURELIN	9	5	3	PURELIN	PURELIN
2	3	4	POSLIN	PURELIN	10	5	4	POSLIN	PURELIN
3	3	5	TANSIG	PURELIN	11	5	5	TANSIG	PURELIN
4	3	6	LOGSIG	PURELIN	12	5	6	LOGSIG	PURELIN
5	4	3	PURELIN	PURELIN	13	6	3	PURELIN	PURELIN
6	4	4	POSLIN	PURELIN	14	6	4	POSLIN	PURELIN
7	4	5	TANSIG	PURELIN	15	6	5	TANSIG	PURELIN
8	4	6	LOGSIG	PURELIN	16	6	6	LOGSIG	PURELIN

Table 3. WOA parameters.

Parameter	Value	Parameter	Value	Parameter	Value
Lower and Upper limit of the optimization parameters (L_b, U_b)	−0.2, 0.2	Search Agents (N)	25	Decrease linear (a)	2
Coefficient vectors (C)	+ 0.6	Loop counter (t)	0	Coefficient vectors (A)	−0.8

Sixteen different ANN architectured with two hidden layers and their weights were optimized using the WOA for network modeling. The top three models were selected based on the results of MAE, AAE and R^2, shown in Table 4.

The WOA-ANN 2L (4–4) network, optimized using WOA, was found to be the most accurate model as it yields the lowest MAE and AAE values while the highest value of R^2 in both training and testing stages.

Figures 3(a) and 3(b) graphically present the performance and accuracy of the WOA-ANN two layers (2L) topology (4–4) model. The plotted period data closely follow the y = x line.

Table 4. Statistics of top three artificial neural networks combined With WAO for period estimate.

Topology	Train			Test		
	MAE	AAE	R^2	MAE	AAE	R^2
WOA-ANN 2L(4–4)	0.022	0.090	0.8747	0.023	0.079	0.8598
WOA-ANN 2L(6–5)	0.025	0.105	0.838	0.026	0.096	0.748
WOA-ANN 2L(6–4)	0.026	0.108	0.8015	0.024	0.089	0.8147

4 Model Validation

Model accuracy was evaluated by comparing the statistical indices of the WOA-ANN 2L (4–4) model to that of the NBCC model. The comparison between the models is based on all statistical indicators and evaluates the models according to the lowest error rate. Comparative results of the statistical MAE, AAE, and the analytical, R^2, indices are provided in Table 5, and the predictions of fundamental period using models the WOA-ANN 2L (4–4) model and NBCC model is presented in Fig. 4.

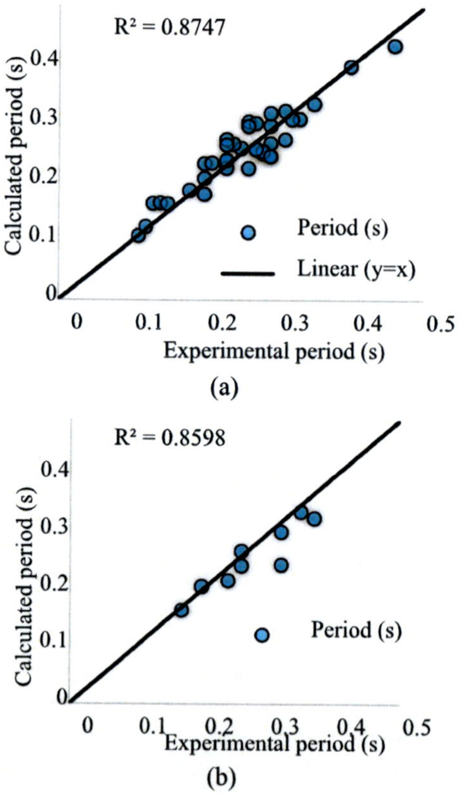

Fig. 3. Experimental versus calculated values of fundamental period for WOA-ANN 2L(4–4) model (a) Training data, (b) Testing data.

Table 5. Statistics of different models in all datasets.

Model	MAE	AAE	R^2	Model	MAE	AAE	R^2
WOA-ANN 2L(4–4)	0.022	0.088	0.861	NBCC	0.092	0.384	0.445

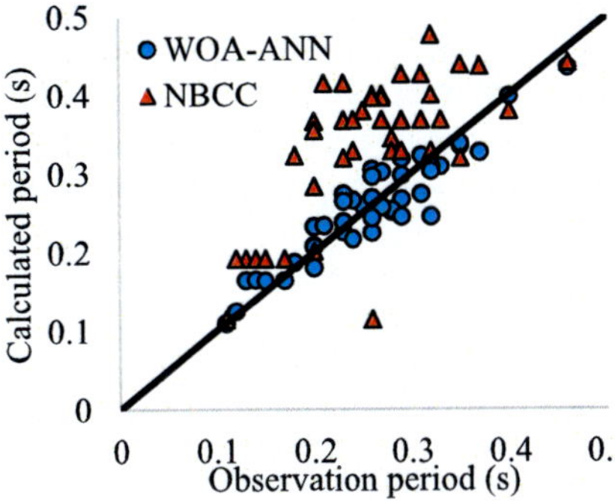

Fig. 4. Experimental versus calculated values of the fundamental period.

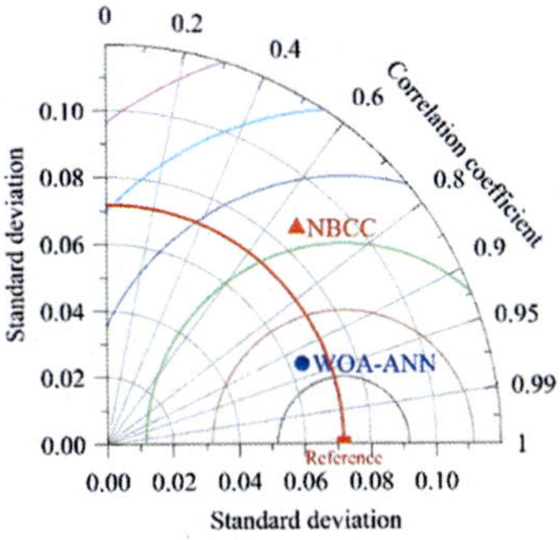

Fig. 5. Taylor diagram visualization of model performance in terms of period predictions.

Taylor diagram, shown in Fig. 5, graphically represents the root-mean-square difference (RMS), correlation coefficient, and standard deviation indices of WOA-ANN and NBCC models. The results shown in Figs. 4 and 5 confirm the higher flexibility and accuracy of the ANN model optimized using the Whale Optimization Algorithm.

5 Conclusions

Sixteen ANN models were developed using the dataset of fundamental period of light-frame wood buildings available in the literature [6, 7]. It was found that the ANN model optimized using the WAO with the topology (4–4) has better prediction capability than the other ANN and NBCC models. The MAE and AAE statistical indices of the WOA-ANN model are 0.022 and 0.090 for training data and 0.023 and 0.079 for the test data, indicating the model's high accuracy.

References

1. NRC/IRC: National Building Code of Canada, Ottawa, Ontario: National Research Council of Canada, Institute for Research in Construction (2015)
2. Alto, P.: Applied Technology Council (ATC 3–06)-Tentative Provisions for the Development of Seismic Regulations for Buildings, California (1978)
3. Saatcioglu, M., Humar, J.: Dynamic analysis of buildings for earthquake-resistant design. Can. J. Civ. Eng. **30**, 338–359 (2003)
4. Dave, V.S., Dutta, K.: Neural network based models for software effort estimation: a review. Artif. Intell. Rev. **42**, 295–307 (2014). https://doi.org/10.1007/s10462-012-9339-x

5. Mirjalili, S., Lewis, A.: The whale optimization algorithm. Adv. Eng. Softw. **95**, 51–67 (2016)
6. Hafeez, G., Doudak, G., McClure, G.: Dynamic characteristics of light-frame wood buildings. Can. J. Civ. Eng. **46**, 1–12 (2018)
7. Hafeez, G., Doudak, G., McClure, G.: Establishing the fundamental period of light-frame wood buildings on the basis of ambient vibration tests. Can. J. Civ. Eng. **45**, 752–765 (2018)
8. Hafeez, G., Mustafa, A., Doudak, G., McClure, G.: Predicting the fundamental period of light-frame wood buildings. J. Perform. Constr. Facil. **28**, A4014004 (2014)
9. Bowden, G.J., Dandy, G.C., Maier, H.R.: Input determination for neural network models in water resources applications. Part 1—background and methodology. J. Hydrol. **301**, 75–92 (2005)
10. Li, J., Heap, A.D.: A Review of Spatial Interpolation Methods for Environmental Scientists. Australian Government, Canberra (2008)
11. Sadowski, Ł, Nikoo, M., Nikoo, M.: Hybrid metaheuristic-neural assessment of the adhesion in existing cement composites. Coatings **7**, 1–12 (2017)

Remediation of Heavy Metals in Coal Mine Brownfield Soil Using Porous Chitosan Beads-Supported $MnFe_2O_4$ Nanoparticles

Hongxia Li[1,2], Ying Liu[1,2], Xinglan Cui[1,2(✉)], Juan Zhong[1,2], Rongzhen Jin[1,2], Xuetao Yuan[1,2], Qidong Zhang[1,2], and Hao Fu[3]

[1] National Engineering Laboratory of Biohydrometallurgy, GRINM Resources and Environmental Technology Corporation Limited, Beijing 101407, China
{rddly, cuixinglan, yuanxt}@grinm.com, kimyeongjin@icloud.com,
[2] GRINM Group Corporation Limited, Beijing 100088, China
[3] CECEP DADI (Shandong) Environmental Technology Corporation Limited, Zibo 255314, China
fuhao@dadi-ep.com

Abstract. Heavy metals (As, Cd, Cu and Pb) co-contamination of coal mine brownfield soil has caused disease risks to human health. But remediation of As anion and Cd cation co-contamination is difficult due to their opposite chemical behaviors to pH and redox potential. Aiming at the low-concentration of As, Cd, Cu and Pb in coal mine brownfield soil, a novel porous nanocomposite was synthesized using crosslinked chitosan and polyethylene glycol (PEG) beads to support $MnFe_2O_4$ nanoparticles (CPM). Characteristics of CPM showed porous structure, well dispersed $MnFe_2O_4$ and plenty of hydroxyl and amino groups (–OH, –NH_2). The addition of CPM in soil would increase the soil pH, but would not cause alkaline pollution to the soil system. Meanwhile, CPM could reduce the percentages of exchangeable and/or reducible fractions and increase the percentages of residual fractions of heavy metals in coal mine brownfield soil.

Keywords: Remediation · Heavy Metal · Coal mine brownfield · Chitosan · Nanoparticles

1 Introduction

China is the largest coal producer and consumer in the world [1]. Simultaneously, there has accumulated 5 billion tons of coal gangue in China, releasing SO_4^{2-}, Fe ions and toxic heavy metals (such as As, Cd, Cu and Pb) through weathering, leaching, biological reaction and spontaneous combustion [2]. Heavy metals would pose threat to the water and soil quality and ecology of the surrounding environment. Ecological restoration was usually adopted for land reclamation in coal mining areas, but heavy metal pollution was easily ignored due to their low concentrations.

Chitosan is a natural cationic flocculant, which can be precipitated by forming chelates with Cu, Hg and other metal ions [3]. However, adsorption of heavy metals by chitosan still has some inherent defects, such as low specific surface area, weak

P. Mendonça and N. D. Cortiços (Eds.): ICAMC 2021, LNCE 226, pp. 237–243, 2022.
https://doi.org/10.1007/978-3-030-94514-5_25

chemical and mechanical properties, and weak adsorption capacity for As anion, resulting in difficulties in the actual use process. Many scholars have tried lots of physical or chemical methods, such as complex, mixing, crosslinking, forming internal hydrogen bonds and grafting polymerization, to obtain the improved chitosan materials [4].

Nanomaterials have been used more and more for soil remediation due to their high specific surface area and excellent adsorption performance. $MnFe_2O_4$ nanomaterial combined with the oxidation property of manganese oxide and the high adsorption property of iron oxide has a good adsorption property to As, which can make up for the weak adsorption property of chitosan to As [5]. However, $MnFe_2$04 nanoparticles have the characteristics of agglomeration of nanometer ions, and can quickly aggregate to form micron or millimeter aggregates, thus losing their unique advantages of high specific surface area and soil migration ability. At the same time, nanoparticles are easy to be absorbed by plants and other plants, resulting in secondary pollution.

Therefore, combining the advantages of chitosan and $MnFe_2O_4$, we attempted to synthesize a composite remediation material by crosslinking chitosan and supporting $MnFe_2O_4$ nanoparticles, to overcome the defects of the two while removing heavy metal cations and anions. It is expected to be a novel and efficient material for remediation of co-contaminated soil.

2 Experimental Materials and Methods

2.1 Materials

Analytical reagent of low molecular weight chitosan (75%–85% degree of deacetylation), polyethylene glycol (PEG, Mw = 2000), dimethylsulfoxide, chloroform, acetic anhydride, methylene chloride, anhydrous ethyl ether, CH_3COOH, NaOH, $NaBH_4$, $FeCl_3.6H_2O$ and $Mn(NO_3)_2.4H_2O$ were used in this study. The contaminated soil was collected from coal mine brownfield, with As, Cd, Cu and Pb contents of 28.31 mg/kg, 0.75 mg/kg, 162.0 mg/kg and 58.9 mg/kg, respectively [6].

2.2 Synthesis of CPM

The CPM were prepared in three successive steps. (1) $MnFe_2O_4$ nanoparticles were prepared by coprecipitation of $FeCl_3.6H_2O$ and $Mn(NO_3)_2.4H_2O$. (2) PEG was oxidized by acetic anhydride in dimethyl sulfoxide and chloroform solution, and then precipitated by anhydrous diethyl ether and methylene chloride for two or three times to obtain PEG-dialdehyde. (3) Chitosan was dissolved in acetic acid solution, and mixed with PEG-dialdehyde and $MnFe_2O_4$ nanoparticles. The pH of the mixed solution was adjusted to 6.5 with 4 mol/L NaOH solution, and $NaBH_4$ solution was added drop by drop, and the reaction time was 18 h. Finally, the solution was precipitated drop by drop with 2 mol/L NaOH to obtain CPM.

Different proportions (0, 1%, 5%, 10%, and 15%) of CPM was added to 100 g of contaminated soil in 250 mL culture flasks, respectively. Deionized water was irregularly added to the culture bottle to maintain 50% of field capacity. After 1, 3, 5, 7, 9,

11, 13, 15, 20, and 30 days of incubation, the soil samples with 3 repetitions were collected and dried. Speciation of heavy metals in soil were extracted by BCR method and analyzed.

3 Results and Discussion

3.1 Characterization of CPM

Figure 1 showed the FTIR spectrum corresponding to PEG, PEG-dialdehyde, chitosan and CPM. Comparing the FTIR spectrum corresponding to PEG and PEG-dialdehyde, the appearance of peak at 1738 cm–1 in PEG-dialdehyde can be assigned to C = O group, indicating successful oxidation of PEG to PEG-dialdehyde. The FTIR spectrum of chitosan showed amide I and amide II bands, corresponding to 1661 and 1599^{cm-1}, respectively [7]. The characteristic peaks of CPM at 3367, 2876, 1661, 1568 and 1080 $^{cm-1}$ demonstrated the overlap of O−H and N−H of chitosan, alkyl C−H of PEG and chitosan, C = O in $-NHCOCH_3$ of amide I, the overlap of C−N and N−H of amide II, and C−O−C of PEG. When comparing FTIR spectrum of chitosan and CPM, two important observations were noted that slight shift of the peak at 1599^{cm-1} (amide II) in chitosan to 1568 $^{cm-1}$ in CPM and dramatic increase of absorbance of peak at 1568^{cm-1} in CPM, indicating more C–N base formed in $-NHCH_2-$. The new C–N base was formed between $-NH_2$ of chitosan and –CHO of PEG. Because N–H group content of chitosan was nearly constant before and after crosslink reaction. The disappear of C = O group in PEG-dialdehyde when forming CPM also indicated the successful reaction between PEG and chitosan. Therefore, the FTIR results demonstrated the crosslink of chitosan and PEG.

The typical outer and inner surface morphology of CPM were shown in Fig. 2a and 2b. The CPM demonstrated an interconnected porous network structure. Figure 2c was a representative SEM image of CPM with corresponding EDS elemental mapping for C (Fig. 2d), O (Fig. 2e), N (Fig. 2f), Mn (Fig. 2g) and Fe (Fig. 2h). It could be seen that C, N and O atoms were discrete and corresponding to orifice structure, indicating a hierarchical frame structure. In contrast, Mn and Fe elements presented the well disperse of $MnFe_2O_4$ nanoparticles on the polymer frame. These results indicated the high potential of CPM regarding to permeability and adsorption.

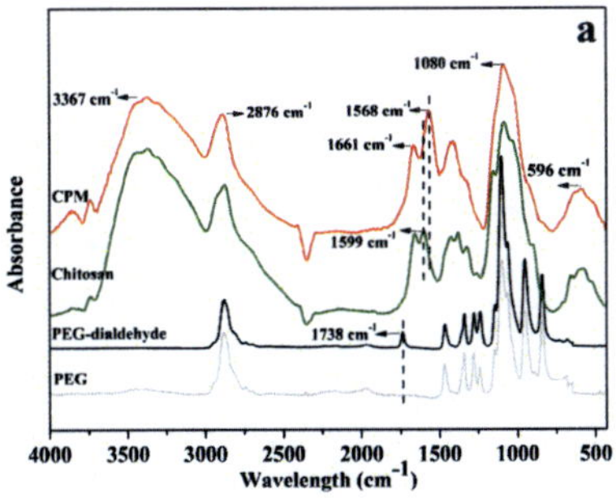

Fig. 1. FTIR spectrum of PEG, PEG-dialdehyde, chitosan and CPM.

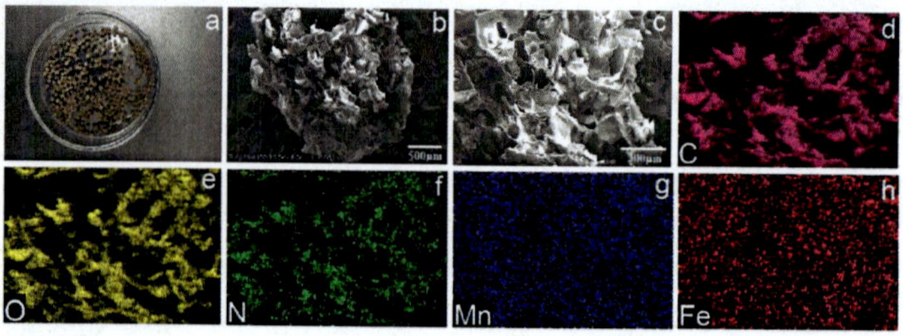

Fig. 2. SEM images (a, b and c) of outer surface and inner surface morphology of CPM and EDS distribution maps of C (d), O (e), N (f), Mn (g) and Fe (h).

3.2 The Influence of CPM on Physical and Chemical Properties of Soil

The change characteristics of soil pH after adding different proportions of CPM were shown in Fig. 3a. It could be seen that the untreated contaminated soil was weakly acidic, with a pH of 6.62. CPM with different proportions could significantly increase the soil pH, which was between 6.74 and 7.97, with an average of 7.31. As the addition of CPM increased, the pH of the contaminated soil increased. After remediation of 30 days, the average pH of 10% and 15% addition were 7.49 and 7.94, respectively. The increase in soil pH could enhance the buffering capacity to Ca^{2+}, and ultimately affect the existence of heavy metals in soil, for example, increasing the precipitation or adsorption of heavy metal cations. Although CPM could increase the soil pH, the influence scope was limited. After the maximum soil pH was reached, the soil pH was basically at a stable stage with the extension of the remediation period. This was because that CPM was a stable crosslinked polymer. The pH range of the soil system suitable for the survival of microorganisms was from 6.5 to 9.5 [8], which requirement was conformed after remediation.

Figure 3b showed the influence of CPM on the cation exchange capacity (CEC) of contaminated soil. Different proportions of CPM all slightly reduced the soil CEC. With the increasing proportion of CPM, the reduction of soil CEC was gradually strengthened. Overall, the impact of CPM on the soil CEC was slightly fluctuating. The influence process of CPM on CEC was mainly through physical or chemical adsorption of heavy metal ions in soil. CPM could form stable coordination compounds with cations such as Ca^{2+} and Mg^{2+} in soil through the $-NH_2$ and -OH groups. The coordination adsorption had a strong binding force, and the process was also irreversible. So that the total amount of cations in soil used for ion exchange was reduced, resulting in a decrease in soil CEC. However, Zhang et al. found that humic acid copolymers could increase soil CEC, which presented an opposite trend to this experiment [9]. This was due to that humic acid copolymers adsorbed heavy metal cations through electrostatic force and van der Waals force, which was physical adsorption and reversible process.

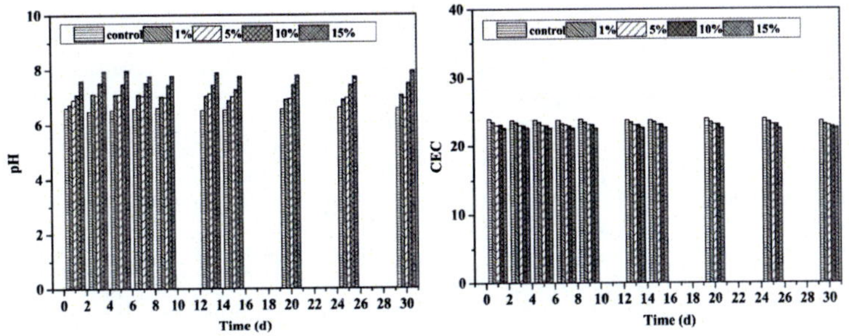

Fig. 3. The influence of CPM on (a) soil pH and (b) CEC

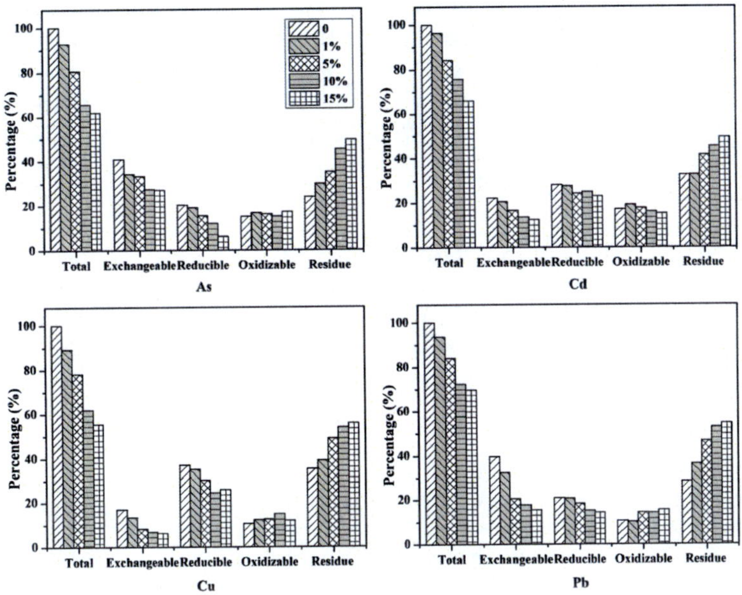

Fig. 4. The influence of CPM on fractions of heavy metals in soil

3.3 Influence of CPM on Total and Fractions of Heavy Metals in Soil

The focus on remediation of heavy metals in soil was the reduction of the total amount of heavy metals as well as the change of heavy metal factions [10]. As shown in Fig. 4, the removal efficiencies (REs) of heavy metals were the lowest at 1% addition of CPM. With the increasing addition of CPM, the total amount of heavy metals in soil showed a rapid downward trend, and then gradually slowed down. Although 15% addition of CPM could achieve the best REs, it was not significantly different from that of 10% addition of CPM. Meanwhile, 10% addition of CPM reduced the contents of As, Cd, Cu and Pb to 18.54, 0.57, 100.24, and 42.67 mg/kg, respectively, which was lower than the second level of farmland soil quality standards. On the whole, there were similar trends in the changes of heavy metals fractions. Although the main fractions of heavy metals in untreated soil were different, the percentages of exchangeable or reducible fractions of heavy metals were all reduced, and the percentages of residue fractions were significantly increased.

4 Conclusion

In this study, a novel granular porous CPM was successfully synthesized using crosslinked chitosan and PEG beads to support $MnFe_2O_4$ nanoparticles and used in As, Cd, Cu and Pb co-contaminated coal mine brownfield soil. The CPM possessed abundant functional groups which thus enhanced its applying feasibility in co-contaminated soil. The addition of CPM in soil could increase the soil pH, and it tended to be stable after reaching the maximum pH. The increased pH was between $6.74 - 7.97$, which would not cause alkaline pollution to the soil system. The addition of CPM could slightly reduce the soil CEC, which was related to the coordination of CPM and cations in soil. When 10% CPM was added, the contents of heavy metals could meet the soil quality standard. With the increasing addition of CPM, the percentages of exchangeable and/or reducible fractions decreased, while the percentages of residue fractions increased.

Acknowledgments. The project was funded by the National Key Research and Development Project (No. 2019YFC1805900), the National Key Research and Development Project (No. 2020YFC1807700), the Youth Fund Project of GRINM (No. 12008), and the Open Foundation of State Key Laboratory of Mineral Processing (No. BGRIMM-KJSKL-2020-07).

References

1. Li, W., et al.: Addressing the CO2 emissions of the world's largest coal producer and consumer: lessons from the Haishiwan Coalfield, China. Energy **80**, 400–413 (2015)
2. Schippers, A., et al.: Geomicrobiological and geochemical investigation of a pyrrhotite-containing mine waste tailings dam near Selebi-Phikwe in Botswana. J. Geochem. Explor. **92**(2), 151–158 (2007)
3. Xia, T.T., et al.: Efficiently remove of Cs(I) by metals hexacyanoferrate modified magnetic Fe3O4-chitosan nanoparticles. Chem. Phys. Lett. **746**, 137293 (2020)
4. Dev, V.V., et al.: Zwitterion-chitosan bed for the simultaneous immobilization of Zn(II), Cd (II), Pb(II) and Cu(II) from multi-metal aqueous systems. J. Clean. Prod. **255**, 120309 (2020)
5. Ma, S.L., et al.: Highly efficient antibacterial and Pb(II) removal effects of Ag-CoFe2O4-GO Nanocomposite. ACS Appl. Mater. Inter. **7**(19), 10576–10586 (2015)
6. Li, H.X., et al.: Distribution of heavy metals and metalloids in bulk and particle size fractions of soils from coal-mine brownfield and implications on human health. Chemosphere **172**, 505–515 (2017)
7. Rafat, M., et al.: PEG-stabilized carbodiimide crosslinked collagen-chitosan hydrogels for corneal tissue engineering. Biomaterials **29**(29), 3960–3972 (2008)
8. Wang, Y.L., et al.: Effect of sterilization on cadmium immobilization and bacterial community in alkaline soil remediated by mercapto-palygorskite. Environ. Pollut. **273**, 116446 (2021)

9. Zhang, H.W., Long, M.J., Zeng, F.S.: Study of amelioration to lateritic red soil by copolymer of humic acids grafted vinyl monomer. Res. Soil Water Conserv. **5**(17), 296–302 (2001)
10. Cui, J.H., et al.: Oxidation and removal of As(III) from soil using novel magnetic nanocomposite derived from biomass waste. Environ. Sci. Nano **6**(2), 478–488 (2019)

Experimental Study and Numerical Simulation for Temperature Distribution of Tropical Concrete with External Self-Limiting Temperature

Yingna Wang, Lin Liu(✉), Jialun Zhang, and Lixin Li

School of Civil Engineering, Shenyang Jianzhu University, Shenyang, China

Abstract. The traditional winter concrete curing method has some problems in the field construction of high-rise concrete and mass concrete, such as inconvenient space conversion, uneven temperature field and construction safety. In order to solve practical problems on site and improve project quality, a new winter curing method for concrete was proposed in this paper. The curing method was applied to the concrete by an external self-limiting heating belt, the law of concrete temperature distribution is obtained. At the same time, MIDAS software was used to simulate the temperature field. The simulation results show that the fitting effect of the model is basically consistent with the experimental results.

Keywords: Winter curing of concrete · Self-limiting heating zone · Air layer · Numerical simulation · MIDAS

The most important construction measure in winter is to provide reasonable curing temperature to ensure the normal development of concrete hydration and accelerate concrete hardening. Traditional winter concrete curing methods include furnace heating, steam heating, electric heating, etc. These methods have some problems such as inconvenient space conversion, uneven temperature field, construction safety and so on in the field construction of high-rise concrete and mass concrete [1, 2]. In order to solve practical problems on site and improve project quality, a new winter curing method for concrete was proposed in this paper. The curing method was applied to the concrete by an external self-limiting heating belt, and the temperature distribution characteristics and rules of the components were studied.

1 Self-limiting Heating Zone

The self-limiting heating strip is a high-resistance structure composed of PTC high polymer material, and the conductive polymer composite material is used as the strip electric heater of heating element. It has the characteristics of automatic adjustment of output power and automatic adjustment of its own surface temperature based on the temperature change of the system being heated. At low temperature, the heating can be started quickly according to its structural advantages, and the temperature will be

P. Mendonça and N. D. Cortiços (Eds.): ICAMC 2021, LNCE 226, pp. 244–254, 2022.
https://doi.org/10.1007/978-3-030-94514-5_26

automatically restricted when it reaches a certain temperature. It can improve the process, improve the quality, shorten the construction period, and save energy and protect the environment.

2 Temperature Distribution Test in Self-limiting Heating Zone

2.1 Experiment

In view of the existing studies [3–9], the temperature distribution characteristics of the concrete components heated by self-limiting heating belt for curing were studied, and the influence of different heating belt arrangements on the temperature field and strength of the components in the process of component curing was analyzed. A total of four groups of component tests were carried out, including the influence of different heating zone layout and different insulation structure on the maintenance of components, especially the influence of air interlayer on the temperature field. In the end, the influence of different insulation structure on concrete curing is measured by comparing the temperature uniformity and component strength of the components with different self-limiting temperature and tropical arrangement.

2.2 Experimental Study

Model Design and Manufacture. Four cuboid components were poured on site, each side length being 0.4 m × 0.4 m × 0.6 m, as shown in Fig. 1 and Fig. 2. The Numbers are specimen 1, specimen 2, specimen 3 and specimen 4. The structure and parameters of the protective layer for heating and curing are shown in Table 1. The arrangement of the heating strip is as follows: the steel concrete inside, the concrete surface, the formwork outside and the heating strip is not used. Dkt-j low temperature limiting (self-limiting temperature ≤ 70 °C) self-limiting heating belt was selected for the experiment. The test component was placed outdoors with an average temperature of –15 °C. Place the temperature sensor inside the test component and the temperature acquisition equipment indoors. The arrangement model of test components plus tropics is shown in Fig. 3.

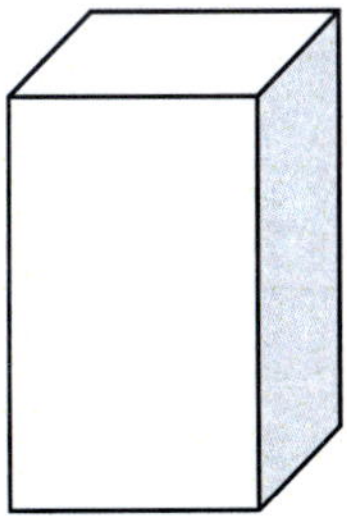

Fig. 1. Concrete test specimen

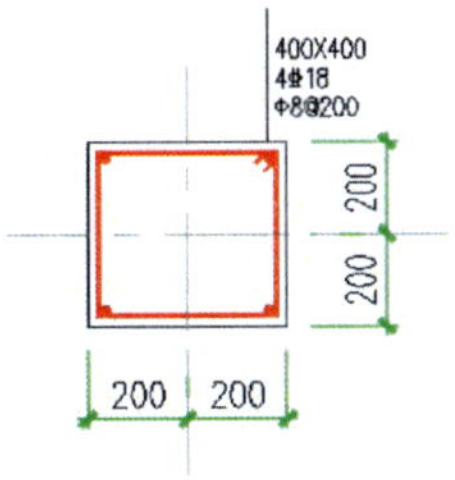

Fig. 2. Concrete specimen reinforcement layout

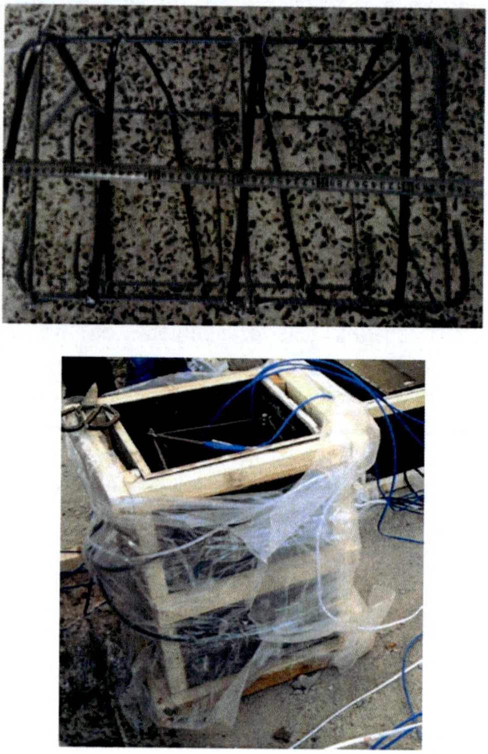

Fig. 3. Electric belt layout picture

The arrangement position of heating belt and the composition of protective layer are shown in Table 1.

Table 1. Heating belt placement and composition of the protective layer

Number	Thermal insulation layer thickness mm	Self limiting temperature plus tropical length m	Thermal insulation layer	Decorate position of self limiting heating
1#	36	4	Template quilt	Concrete interior
2#	36	4	Template quilt	Concrete surface
3#	56	4	Template air quilt	Outside surface of formwork
4#	36	0	Template quilt	–

Temperature Monitoring of Model Concrete Structure. Temperature measurement points mainly measure the temperature change of concrete components under the combined action of hydration and heating zone. The measuring points are the temperature difference cooling rate between the center and the surface of the component.

A total of 15 temperature measuring points were set for the four groups of components, 4 for each of the three components in the heating belt, 2 for the unheated components, and 1 for the ambient temperature. In order to record and analyze the temperature change rule of concrete under the combined action of cement hydration and external heating, four test areas are divided according to the different locations of the test points. The arrangement of temperature measurement points is shown in Fig. 4.

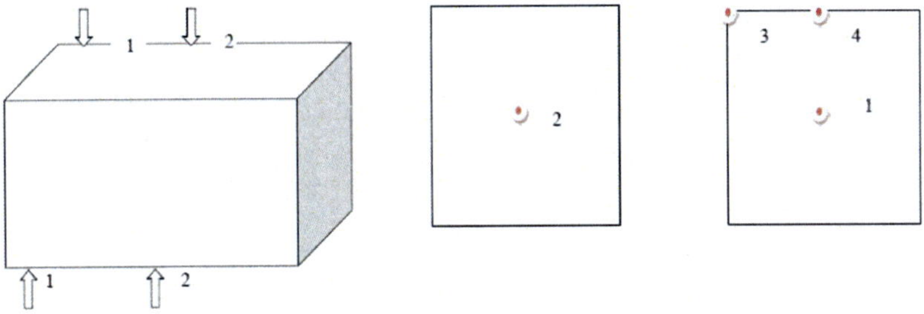

Fig. 4. Concrete specimen temperature measuring point layout

For example, C12, C stands for temperature measurement, 1 stands for no. 1 component, and 2 stands for measuring points with no. 2 inside no. 1 component. Where T represents the ambient temperature, and the Numbers of all measurement points are summarized in Table 2.

Table 2. Component internal temperature measurent point number

Content of measurement	Sensor category	Number of measuring points
Temperature	Temperature sensor	C11, C12, C13, C14, C21, C22, C23, C24, C31, C32, C33, C34, C41, C42, T

2.3 Experiment Results

When the components are heated and maintained, the three components are concreted and electrified at the same time. The heating time is 3 days. The temperature data of each temperature measurement point are extracted, and then the temperature changes at different measurement points of the same component are plotted into a curve. The temperature changes at different measuring points of the same component are analyzed and the temperature changes at the same location of different components are compared. The strength of concrete was measured by the rebound method, and the strength was measured when the heating belt was cut off. The compressive strength of the curing specimen in 3 days, 7 days, 28 days and 90 days was measured, and the strength development curve of each component was drawn.

Comparison and Analysis of Temperature Variation Curves of each Specimen in the Heating Belt. The temperature change curve of each measuring point inside the concrete of specimen 1 is shown in Fig. 5. By comparing the temperature change curves of each temperature measurement point, it can be seen that the temperature change trend of each temperature measurement point of the same component is basically the same. The peak temperature is reached after heating for about 20 h, and the peak value of all temperature measurement points is between 40 °C and 50 °C. The cooling rate of all temperature measurement points is approximately 0.6–0.8 °C/h.

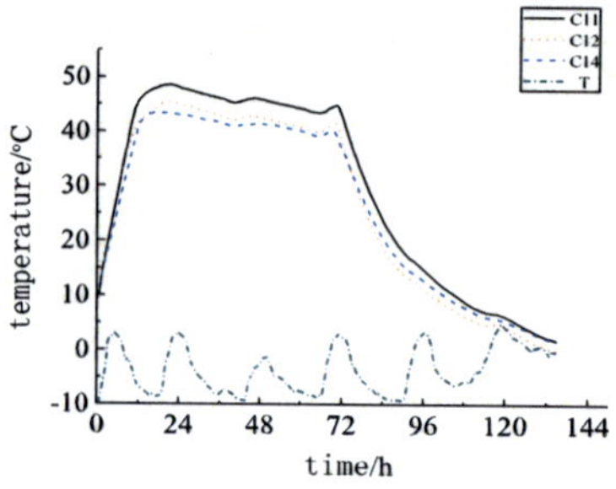

Fig. 5. Specimen 1

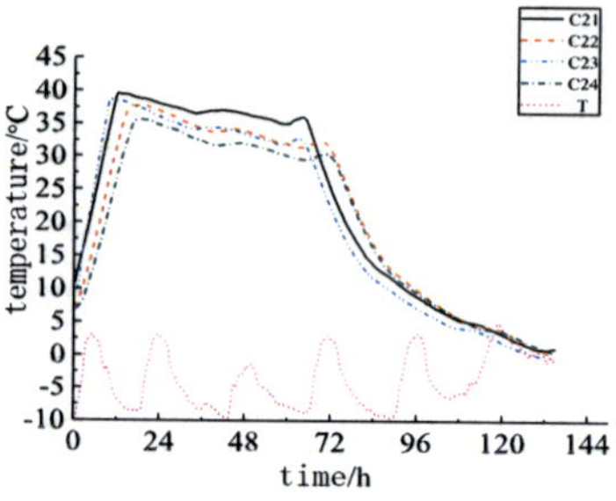

Fig. 6. Specimen 2

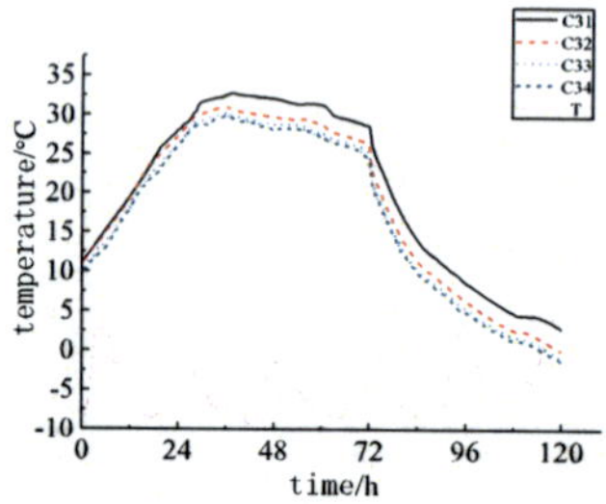

Fig. 7. Specimen 3

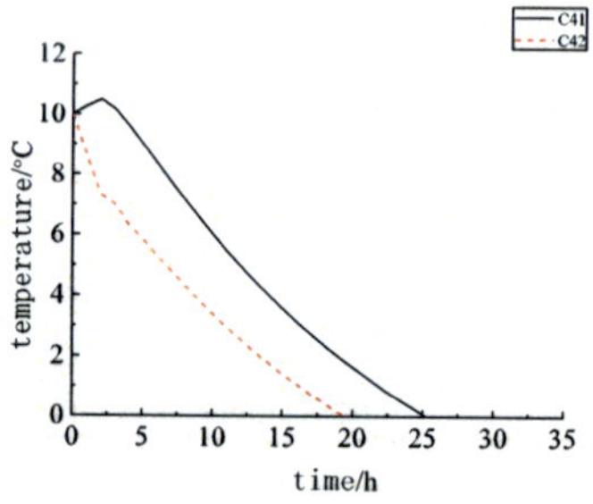

Fig. 8. Specimen 4

The temperature change curve of each measuring point inside the concrete of specimen 2 is shown in Fig. 6. By drawing the temperature change curve of each temperature measurement point and the temperature change curve of all temperature measurement points of the same component, it can be known that the temperature change trend of each temperature measurement point of the same component is consistent. The peak temperature is reached at about 10–20 h of heating. Due to the influence of the arrangement of heating belt on the heating rate of each point, the time to reach the peak is inconsistent. The peak value of all temperature measurement points is between 35 °C and 40 °C, and the cooling rate of temperature measurement points is roughly between 0.5 °C and 0.6 °C/h.

The temperature change curve of each measuring point inside the concrete of specimen 3 is shown in Fig. 7. By drawing the temperature change curve of each temperature measurement point and the temperature change curve of all temperature measurement points of the same component, it can be known that the temperature change trend of each temperature measurement point of the same component is consistent. The peak temperature is reached after 38 h of heating, and the peak value of all temperature measurement points is between 29–32 °C. The cooling rate of temperature measurement points is about 0.6 °C/h.

As shown in Fig. 8, it can be seen that only the temperature at the measuring point of the central point rises within 3 h after pouring, while the temperature at the measuring point 5cm below the central point of the upper end face keeps falling. Due to the absence of external heating conditions, the temperature of the component is low, and the heat generated by hydration is less than the heat lost, resulting in a constant drop in temperature.

Comparative Analysis of Temperature Curves at the Same Position of each Specimen. As shown in Fig. 9 and 10, three component center under the center point temperature and the upper end face of 5 cm change trend is consistent, including 1 specimen with specimen temperature rise of 2 3 rise faster, the main reason is that specimen 1 and 2 is heated directly on concrete specimens for heating, heating zone of the quantity of heat of heat lost to the natural environment is relatively small, lead to hydration relatively fast, both together, result in relatively high temperature.

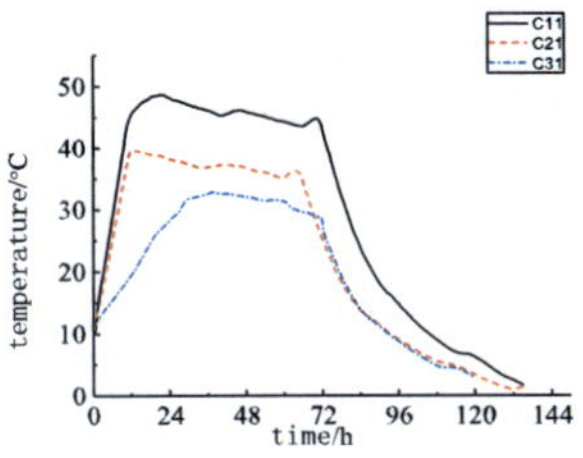

Fig. 9. Temperature change curve at the center of three components

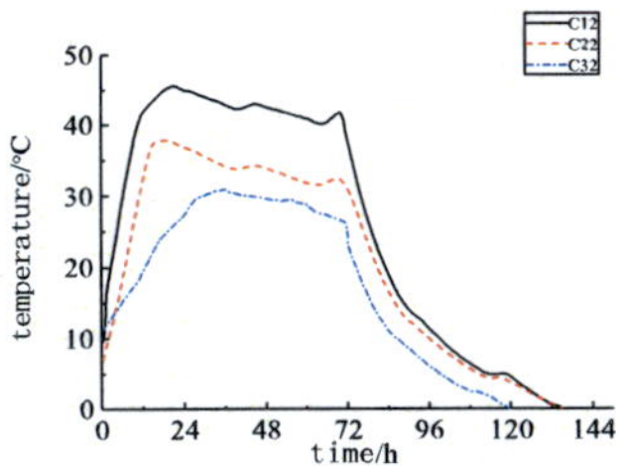

Fig. 10. Temperature change curve of 5cm below the center point of the top end surface of three components

2.4 Strength Development Analysis of Each Specimen

All the components were heated and maintained for 3 days, and the compressive strength of each test component at 3, 7, 28 and 90 days after pouring was measured with the springback instrument (see Table 3) and the strength development curve over time (see Fig. 11).

Table 3. The rebounder measured the strength after different pouring time

Time (d)	3d	7d	28d	90d
artifacts 1	32.3MP	34.5MP	35.7MP	36MP
artifacts 2	32.1MP	33.4MP	34.6MP	35.2MP
artifacts 3	31.7MP	32.9MP	34.8MP	35.6MP

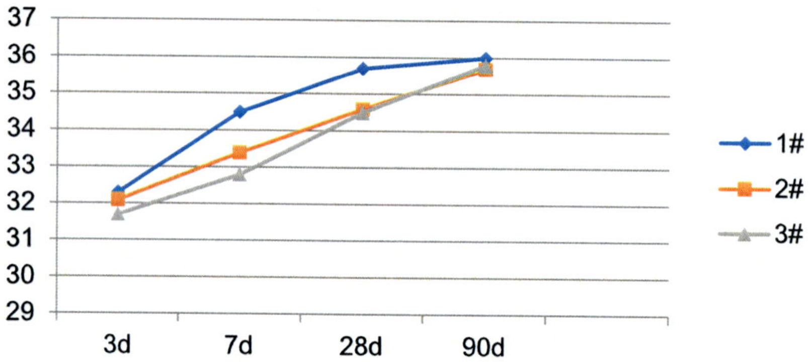

Fig. 11. Relationship of test components between strength and maintenance time

It can be seen from the figure above that all components reach the design strength after three days of heating and curing, and the strength increases slowly after 90 days. At the time of three days, the strength of no. 1 component inside the reinforced concrete is the strongest, while the strength of no. 3 component arranged outside the formwork of the reinforced concrete is relatively weak. At the time of 90 days, the strength value of all components is basically the same. The compressive strength data of the three components indicate that the heating strip, as a heat source, can provide a suitable curing temperature for concrete in winter. The final strength of the template outside the heating strip is basically the same as that inside the concrete and on the concrete surface.

2.5 Analysis of Uniformity of Temperature Field of Each Specimen

The uniformity of the temperature field of all test components was compared, and the temperature difference was obtained by subtracting the temperature at the highest measuring point (central point) and the lowest measuring point (5 cm below the lateral central point) of the same component. The temperature difference of the component over time is plotted as a curve, as shown in Fig. 12 below:

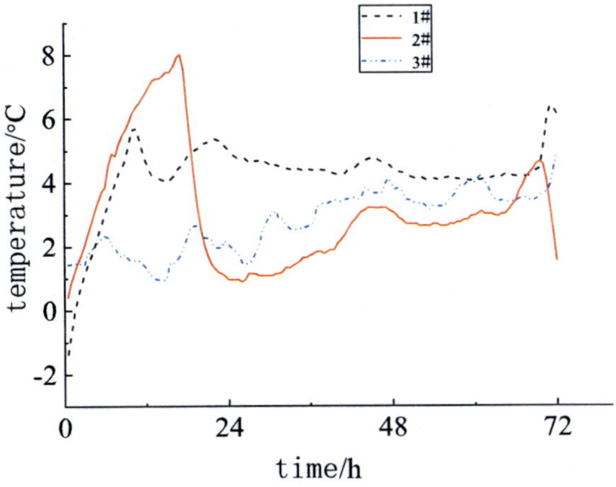

Fig. 12. Relationship between maximum temperature difference of temperature field and age of test component

It can be seen from the temperature difference that the temperature difference between the no. 1 member inside the concrete and the no. 2 member placed on the concrete surface of the heating belt is very unstable, and the temperature difference value is relatively large. The temperature difference is relatively small and the value of temperature difference is relatively stable. This indicates that the temperature field formed by the external curing of the tropical formwork is more uniform and the curing effect is better than the other two schemes.

3 Finite Element Simulation Using Midas Software

3.1 Calculation Basis and Basic Parameters

The temperature after concrete pouring is related to the temperature rise of hydration heat of cement, pouring temperature and pouring schedule of concrete, external air temperature, surface protection and other factors.

1) Mechanical properties of concrete

Specimen concrete: The specimen is made of C30 concrete, and its mechanical properties are shown in Table 4.

Table 4. Mechanical properties of concrete

28 d strength	Elasticity modulus	Poisson ratio	Density
30 N/mm^2	3.0 × 104 N/mm^2	0.2	2400 kg/m^3

2) Thermal properties of concrete

The thermal conductivity, specific heat and mix ratio of raw materials are shown in Table 5 and 6.

Table 5. Thermal properties summary table of raw materials

Name of the material	Cement	Sand	Stone	Water
Heat conductivity coefficient{KJ/(m.h.°C)}	4.446	11.129	14.528	2.16
Specific heat c{KJ/(kg.°C)}	0.456	0.699	0.749	4.187

Table 6. C30 concrete mix ratio

Name of the material	Cement	Sand	Stone	Water
Weight {kg}	461	512	1252	175
Mix proportion	1	1.11	2.72	0.38

Table 7. CONCRETE thermal performance table

Name of the material	Heat conductivity coefficient {KJ/(m.h.° C)}	Specific heat c{KJ/(kg.° C)}
C30concrete	10.89	0.96

Coefficient of expansion: = 1×10^{-5}

According to the weight percentage weighting method in Table 2 and 3, the thermal performance parameters of C30 concrete are shown in Table 7.

Adiabatic temperature rise: According to Table 3 of Research and Development of Hydration Heat of Cement Concrete [31], the hydration heat coefficient of ordinary Portland cement is found. According to the hydration heat of cement, the calculation formula of adiabatic temperature rise of C30 condensate is as follows:

Where: 0 is the adiabatic temperature rise of cement; Q0 is cement hydration heat; W is the amount of cement; K is the reduction coefficient; F is mixed material (0 here); C is specific heat of concrete; is the density of concrete.

According to the actual situation, the specimen was poured in December, and the average temperature in December was −10 °C. Concrete casting temperature: casting temperature is 10 °C.

3) Convection boundary conditions and the six faces of the convection coefficient specimen are set as convection boundary, and the convection coefficient is set as 10 \,kcal/m2.h. [C].
4) Electric tropical parameters: initial temperature of 65 °C, width of 0.012 m (Figs. 13,14).

3.2 Finite Element Results and Analysis

1) At the heating curing stage of the same measuring point, the simulated and tested peak point temperature, the temperature in the stationary stage, and the time needed to reach the peak point are basically the same.

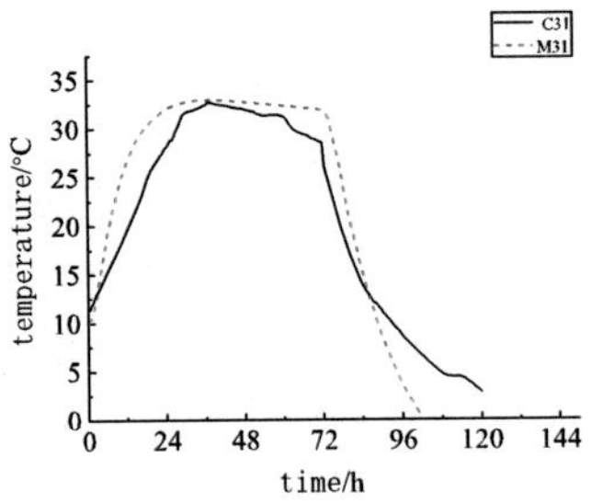

Fig. 13. Simulation and experiment center point temperature curve

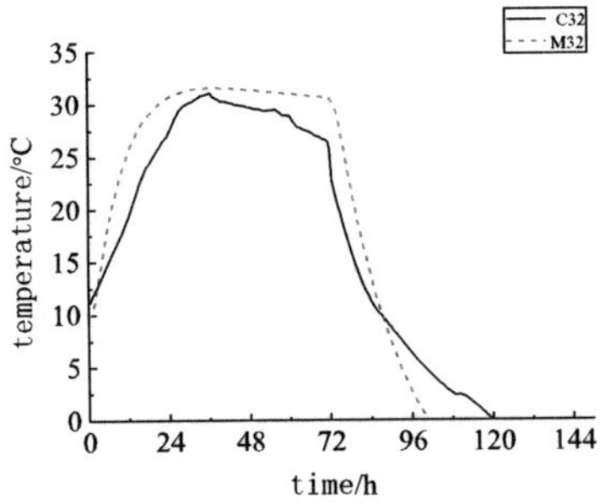

Fig. 14. Simulation and experiment center point of the top surface 5cm measuring point temperature curve

2) At the stop heating stage, the temperature drop rate of the simulated measuring point is basically the same as that of the test measuring point, and the temperature drop rate of the simulated component is faster than that of the test component. When simulating the curing process of specimens, the external environment temperature was set as constant –10 °C, and the natural environment temperature changed periodically. In the heating stage, the external environment temperature had little influence on the test due to the joint action of heating belt and hydration.

4 Results

Through the experiment, the following conclusions are obtained:

A. Under the condition of electrified heating and maintenance, after 3 days of heating, the three groups of heated components all reach the critical strength required by the specification, and all components basically reach the same compressive strength in 90 days.
B. The air interlayer and heating zone work together to improve the environmental temperature of concrete maintenance. Compared with the case where heating zone directly heats concrete, the temperature of the temperature field rises slowly, the

stable stage is more stable, and the temperature drops slowly in the descending stage.

C. The layout of the internal surface of the reinforced concrete and the reinforced formwork in the heating zone, which takes up the volume of concrete, affects the bite force of concrete and steel bars and reduces the bearing capacity of the members, and the safety of the reinforced formwork in the process of casting and vibration is lower than that of the reinforced formwork outside.

D. Through the analysis of test and simulation data, it can be seen the finite element software MIDAS can accurately simulate the temperature change rule of components in the heating curing stage. The heating zone is arranged outside the template, and the area of the component with lower temperature is obviously less.

Based on the above analysis and test results, it can be concluded that it is feasible to adopt the combined action of the self-limiting heating zone and the air layer in the winter concrete curing.

References

1. Bernander, S.: Assessment of thermal cracking in hardening concrete. J. Struct. Eng. (1994)
2. Embor, M., Bernander, S.: Assessment of risk the male racking in harding concrete. J. Struct. Eng. **120**(10), 2893–2911 (1994)
3. Liancai, L.: New technology and scheme selection of winter concrete construction in alpine region of Xizang. Master's Thesis of Sichuan University, pp. 1–2 (2003)
4. Zaiguang, W.: Research on concrete construction methods in cold areas in winter. Master's thesis of Harbin Engineering University, pp. 3–5 (2011)
5. Bian, C.C., Sun, D.S.: Structure and principle of common self-limiting heating strips and precautions for use. Electr. Wire Cable **6**(6), 60–63 (2010)
6. Xiong, Z.Q.: Electric heating construction technology of concrete in winter. Eng. Technol. 77–84 (2013)
7. Guo, J.B., Liu, L., Wang, Q.: Application self-regulating heating cable curing of concrete in winter, vol. 3489, pp. 1531–1535 (2014)
8. Lu, Y.Y.: Analysis and practice of electric heating method for winter construction of concrete. Constr. Technol. **35**(8), 60–61 (2006)
9. Feng, N.: Application of maintenance method of built-in electric heating circuit in winter concrete of Baltic pearl project. Build. Constr. **32**(7), 687–689 (2010)

Experimental Study on Unconfined Compressive Strength of Nano-Clay Modified Lime Soil

Lin Zhou, Weiqing Zhang, and Ping Jiang(✉)

School of Civil Engineering, Shaoxing University, Shaoxing, Zhejiang Province, China
jiangping@usx.edu.cn

Abstract. In order to study the influence of nano-clay on the unconfined strength characteristics and energy dissipation of lime soil, the unconfined compressive strength test of lime soil was carried out to analyze the unconfined compressive strength, elastic modulus and strain energy density. On this basis, the functional expressions of unconfined compressive strength, strain energy density and nano-clay content were established. The results show that the unconfined compressive strength, elastic modulus and strain energy density at different ages all increase with the increase of nano-clay content. The unconfined compressive strength and elastic modulus have a linear function relationship with the content of nano-clay, and the strain energy density has an exponential function relationship with the content of nano-clay. With the increase of the content of nano-clay, the energy dissipation of lime soil is also increasing.

Keywords: Nano-clay · Lime soil · Unconfined compressive strength · Elastic modulus · Strain energy density

1 Introduction

Lime soil is a material composed of lime and soil mixed in proper proportion, mixed and hardened [1]. Lime soil is widely used in highway subgrade engineering due to its advantages of convenient material acquisition, economy and environmental protection [2]. However, as the requirements for highway grades are getting higher and higher, the requirements for the strength grade of roadbed fillers are also getting higher and higher. Therefore, it is difficult to meet the design requirements of highway subgrade by using lime stabilized soil as subgrade filler, and it needs to be modified.

With the development of materials science, nano-clay began to be used as a stable material in the field of civil engineering. Because the free surface of nano-clay has a large number of atoms and molecules, it can improve the surface activity of the soil from a physical and chemical point of view, and promote the generation of pozzolanic reactions in the soil [3–6]. Mohamadzadeh [7] performed liquid-plastic limit and direct shear tests on clay samples by changing the mass percentage of nano-clay (0.5%, 1% and 2%). The results show that the addition of nano-clay reduces the plasticity index and improves the shear strength. Arabani [8] studied the use of nano-clays to improve the mechanical properties of cement stabilized sand by using the unconfined

P. Mendonça and N. D. Cortiços (Eds.): ICAMC 2021, LNCE 226, pp. 255–262, 2022.
https://doi.org/10.1007/978-3-030-94514-5_27

compressive strength (UCS) test and the California bearing ratio (CBR) test. The results show that nano-clay improves its strength properties. Zhang [9, 10] showed through unconfined compressive strength tests that: adding nano-clay to cement can improve its compressive strength. Yang [11] studied the shear characteristics of nano-clay modified cement soil, and the research results showed that adding nano-clay can improve the shear strength, cohesion and friction angle of cement soil. The above studies are all studying the effect of nano-clay on the mechanical properties of plain soil and cement soil, and the study of the effect of nano-clay on the unconfined compressive strength and energy dissipation of lime soil is less.

Therefore, this experiment took the clay of a construction site in Shaoxing, China as the research object, mixed with lime and different amounts of nano-clay (0%, 2%, 4%, 6%, 8%). Through the unconfined compressive strength test, the unconfined compressive strength, elastic modulus, and strain energy density of lime soil of different ages under different nano-clay content are studied. Analyze the evolution law of nano-clay on the strength, stiffness and energy dissipation of lime soil, and provide test basis for nano-clay lime soil as roadbed filler.

2 Test Materials and Methods

2.1 Test Materials

The soil used in the experiment is the clay of a construction foundation pit at a construction site in Shaoxing City, Zhejiang Province, China, with a depth of 5 m. The soil has a natural moisture content of 33.8%, a weight of 19.2 kN/m^3, a void ratio of 0.74, a liquid limit of 41.60%, a plastic limit of 24.30%, and a plasticity index of 17.30. The lime used in the experiment is quicklime powder produced by Jiangxi Xinyu Liangliang Trading Co., Ltd. The main component is CaO, and the content of CaO is 90%. The nano-clay used in the experiment was produced by Hubei Jinxi Montmorillonite Technology Co., Ltd., in the form of light yellow powder.

2.2 Sample Production and Methods

In the experiment, lime soil was prepared according to the mass ratio of dry soil: lime: water = 1:0.06:0.175. On this basis, nano-clay was mixed with 2%, 4%, 6% and 8% of dry soil mass respectively.

According to the Standard for geotechnical testing method GBT50123-2019, the undisturbed soil is dried, crushed and passed through a 2 mm sieve. Add water, lime and nano-clay in sequence according to the mixing ratio to prepare the required soil sample, and seal it for 24 h. After standing still, the soil sample is made into a sample with a size of φ50 mm × h50 mm by static compaction method. Then demould the sample, wrap it in plastic wrap, and cure it in a standard curing room (the curing temperature is (20 ± 2) °C, humidity is 95%) for 7 days and 28 days. 1d before reaching the age, put the sample in water and soak for 24 h, take out the sample and wipe the water dry with absorbent paper for unconfined compressive strength test. The test instrument is a fully automatic unconfined compressive strength meter, and the stress-strain curve can be automatically collected through a computer.

3 Test Results and Discussion

3.1 Unconfined Compressive Strength

The unconfined compressive strength test of lime soil mixed with different nano-clay content was carried out, and the loading rate was 1mm/min. The unconfined compressive strength of nano-clay lime soil at different ages is shown in Table 1.

Table 1. Unconfined compressive strength of nano-clay lime soil (kPa)

Content of nano-clay (%)	Curing age (d)	
	7	*28*
0	1239	1950
2	1425	2420
4	1635	2540
6	1947	2909
8	2055	2956

It can be seen from Table 1 that the unconfined compressive strength of lime soil at different ages increases with the increase of nano-clay content. On the one hand, nano-clay can fill the gap between lime soil particles to make its microstructure more dense, on the other hand, it can promote the pozzolanic reaction between lime and soil, and more complete formation of more cementitious substances such as calcium silicate hydrate C-S-H and calcium aluminate hydrate C-A-H. When the curing age is 7 days, the unconfined compressive strength of lime soil increases most when the content of nano-clay is from 4% to 6%, and the improvement range of unconfined compressive strength of lime soil is the smallest when the content of nano-clay is from 6% to 8%. When the curing age is 28 days, when the content of nano-clay is from 0% to 2%, the improvement of unconfined compressive strength of lime soil is the most obvious, and the improvement of unconfined strength of lime soil is not obvious when the content is 6% to 8%. When the content of nano-clay is the same, the unconfined compressive strength of lime soil increases with the increase of age. The main reason is that with the increase of age, more calcium silicate hydrate, calcium aluminate hydrate and Ettringite are produced inside, which makes the internal structure more compact.

The relationship between unconfined compressive strength and nano-clay content of nano-clay lime soil at different ages in Table 1 is fitted, and the results are shown in Fig. 1.

According to Fig. 1, the fitting formulas for the relationship between unconfined compressive strength of nano-clay lime soil and the content of nano-clay at different ages are as follows:

$$q_u = 106.3x + 1239 \quad R^2 = 0.98\,(7d) \tag{1}$$

$$q_u = 142.6x + 1950 \quad R^2 = 0.91\,(28d) \tag{2}$$

In the formula: q_u is the unconfined compressive strength (kPa); x is the content of nano-clay (%).

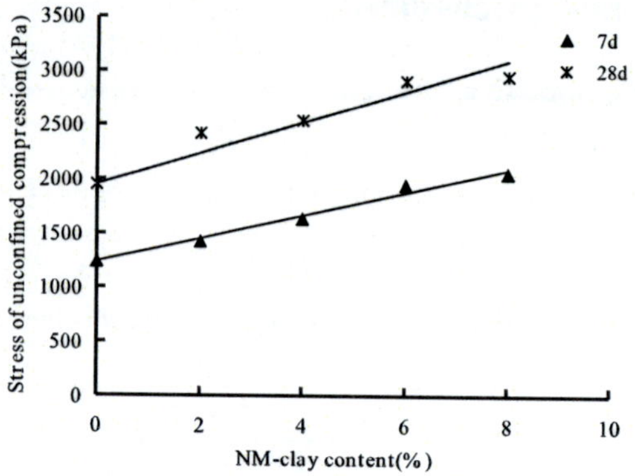

Fig. 1. Relationship between unconfined compressive strength and content of nano-clay lime soil

3.2 Elastic Modulus

The elastic modulus is the ratio of stress to strain in the linear elastic range when a load is slowly applied, and it is an important parameter to measure the stiffness of a material. In this paper, the unconfined test data is analyzed and processed to obtain the elastic modulus value of nano-clay lime soil. Select a straight line segment on the stress-strain curve of unconfined compressive strength, and use a straight line to fit the stress-strain data. The slope of the straight line is the elastic modulus. Take the 7d data of lime soil without nano-clay as an example, as shown in Fig. 2. According to this method, the elastic modulus of nano-clay lime soil at different ages can be obtained. The results are shown in Table 2.

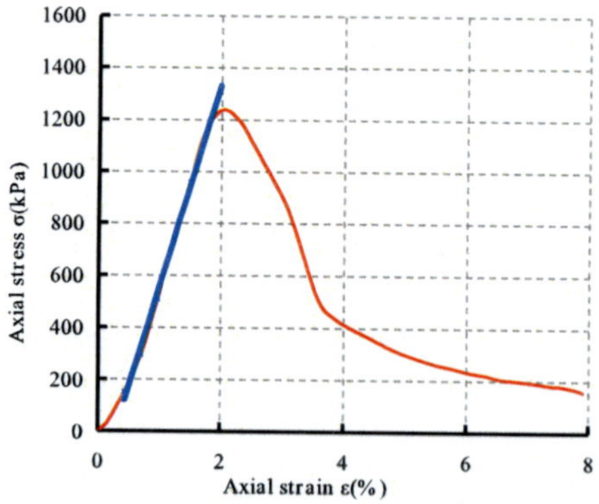

Fig. 2. Schematic diagram of elastic modulus value

It can be seen from Table 2 that the elastic modulus of lime soil at different ages increases with the addition of nano-clay, indicating that the addition of nano-clay can improve the resistance of lime soil to deformation. When the curing age is 7 days, the elastic modulus of lime soil mixed with 2%, 4%, 6%, and 8% nano-clay is increased by 19 Mpa, 37 Mpa, 44 Mpa, 53 Mpa, respectively, than that of lime soil without nano-clay; When the age is 28 days, the elastic modulus of lime soil mixed with 2%, 4%, 6%, and 8% nano-clay increases by 13 Mpa, 17 Mpa, 27 Mpa, 42 Mpa, respectively. The elastic modulus of nano-clay lime soil increases with the increase of age, but the effect of nano-clay on the elastic modulus of lime soil decreases with the increase of age.

Table 2. Elastic modulus value of nano-clay lime soil (MPa)

Content of nano-clay (%)	Curing age (d)	
	7	*28*
0	80	146
2	89	159
4	117	163
6	122	173
8	133	188

According to Table 2, the relationship between elastic modulus of nano-clay lime soil and nano-clay content at different ages is fitted, and the results are shown in Fig. 5.

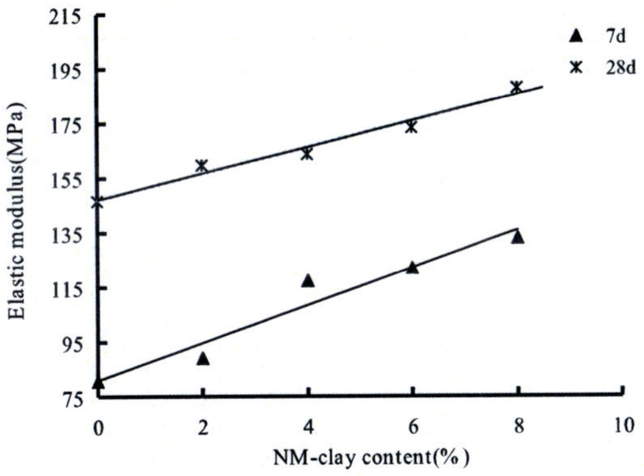

Fig. 3. Relationship between elastic modulus and content of nano-clay lime soil

From Fig. 3, the fitting formulas for the relationship between the elastic modulus of nano-clay lime soil and the content of nano-clay at different ages are:

$$E = 6.877x + 80.9\ R^2 = 0.94(7d) \tag{3}$$

$$E = 4.615x + 147.1\ R^2 = 0.93(28d) \tag{4}$$

Where E is the elastic modulus (Mpa); x is the content of nano-clay(%).

3.3 Test Process

In the unconfined compressive strength test, the formation of internal cracks in lime soil is actually an energy consumption process. According to the material damage criterion, the material has experienced the initiation of cracks, the subcritical state and the final failure. Assuming that the damage of the material corresponds to the energy per unit volume, it can be expressed by the strain energy density dw/dv. The expression of strain energy density is as follows:

$$\frac{dw}{dv} = \int_0^{\varepsilon_0} \sigma_{ij} d\varepsilon_{ij} \tag{5}$$

In the formula: σ_{ij} represents the stress component; ε_{ij} represents the strain component; ε_0 represents the strain value corresponding to the maximum stress.

The strain energy density is obtained by analyzing the stress-strain curve of the non-side compressive strength test of nano-clay lime soil. The value obtained from the stress-strain integration is the area of the shaded area in the figure below. Take the stress-strain curve of lime soil without nano-clay at 7 days as an example for analysis. The strain energy density is the shaded area in the figure, as shown in Fig. 4. From this calculation, the strain energy density of nano-clay lime soil is shown in Table 3.

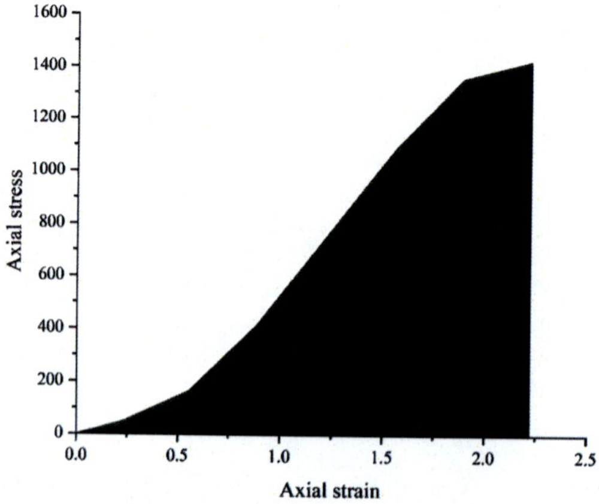

Fig. 4. Integration of stress-strain curve of 7-d age lime soil without nano-clay soil

Table 3. Strain energy density of nano-clay lime soil(kPa)

Content of nano-clay(%)	Curing age(d)	
	7	8
0	11.9	16.5
2	15.1	22.5
4	15.2	23.4
6	17.6	27.2
8	23.5	32.6

It can be seen from Table 3 that the strain energy density of nano-clay lime soils at different ages increases with the increase of the content of nano-clay, and the change law is almost consistent with the change law of unconfined compressive strength and elastic modulus. With the increase of nano-clay, the strain energy density can reflect the law of internal structure damage of cement soil to a certain extent. At the same age, the greater the strain energy density, the greater the load-bearing capacity required when the specimen fails. Due to the effect of nano-clay, the overall value shows an upward and downward trend. This shows that with the increase of nano-clay, the lime soil needs to be applied with a larger external load to induce the generation of cracks. Therefore, the expansion of cracks after compression needs to absorb energy from the outside, and the energy dissipation is greater.

According to Table 3, the relationship between the strain energy density of nano-clay lime soil and the content of nano-clay is fitted, and the results are shown in Fig. 5.

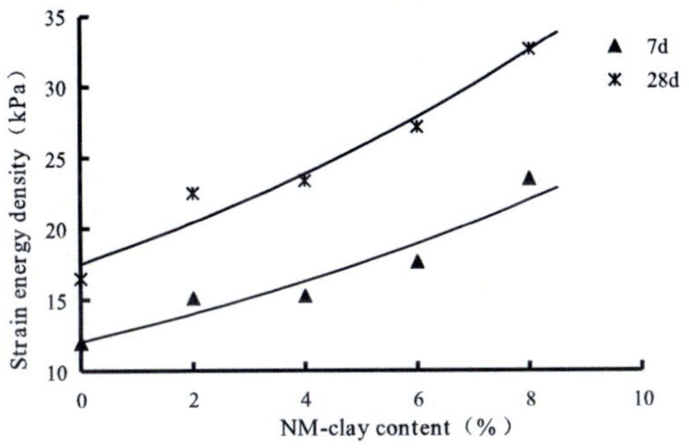

Fig. 5. Relationship between strain energy density and content of nano-clay lime soil

According to Fig. 5, the fitting formulas for the relationship between strain energy density and nano-clay content of nano-clay lime soil at different ages are as follows:

$$\frac{dw}{dv} = 12.029e^{0.0754x} \quad R^2 = 0.92 \quad (7d) \tag{6}$$

$$\frac{dw}{dv} = 17.482e^{0.0776x} \quad R^2 = 0.95 \quad (28d) \tag{7}$$

Where dw/dv is the strain energy density; x is the content of nano-clay.

4 Conclusions

1. The unconfined compressive strength and elastic modulus of lime soil at different ages increase with the increase of nano-clay content, indicating that nano-clay can not only increase the strength of lime soil but also increase the stiffness of lime soil.
2. The relationship between unconfined compressive strength, elastic modulus, strain energy density and nano-clay content was established. The unconfined compressive strength and elastic modulus have a linear function relationship with the nano-clay content, and the strain energy density has an exponential function relationship with the nano-clay content.
3. Through the analysis of the strain energy density of nano-clay lime soil, it is obtained that the strain energy density increases with the increase of the content of nano-clay at different ages. With the increase of nano-clay, the energy required for the destruction of lime soil gradually increases, and the energy dissipation is also higher.

References

1. Gong X.N.: Foundation treatment technology and development prospect (2014)
2. Festugato, L., Diambra, A.: Modelling tensile/compressive strength ratio of fibre reinforced cemented soils. Geotext. Geomembr. **46**(2), 155–165 (2018)
3. Changizi, F., Haddad, A.: Strength properties of soft clay treated with mixture of nano-SiO2 and recycled polyester fiber. J. Rock Mech. Geotech. Eng. **7**(4), 367–378 (2015)
4. Choobbasti, A.J., Vafaei, A., Kutanaei, S.S.: Static and cyclic triaxial behavior of cemented sand with nanosilica. J. Mater. Civ. Eng **30**, 04018269 (2015)
5. Edalati, S.A., Bayat, M., Pakar, I.: A novel approximate solution for nonlinear problems of vibratory systems. Struct. Eng. Mech **57**, 1039–1049 (2016)
6. Anvari, S.M., Shooshpasha, I., Kutanae, S.S.: Effect of granulated rubber on shear strength of fine-grained sand. J. Roc. Mech. Geotech. Eng. **9**, 936–944 (2017)
7. Mohamadzadeh, S.A., Arabani, M., Haghi, A.M.: Effect of nanoclay additive on the geotechnical properties of silty sand. In: Proceedings of the 4th International Conference on Geotechnical Engineering and Soil Mechanics (2010)
8. Arabani, M., Hagh, A.K., Mohamadzadeh, S.A.: Use of nanoclay for improvement the microstructure and mechanical properties of soil stabilized by cement. In: Proceedings of the 4th International Conference on Nanostructures (ICNS4) (2012)
9. Zhang, M.H., Yang, J., Liu, Y.J.: Effect of nano materials on early strength of low content cement soil. Chin. Foreign Highways **35**(03), 239–242 (2015)
10. Zhang, M.H., Liu, Y.J., Yang, J.: Experimental study on unconfined compressive strength of cement soil mixed with nano materials. Constr. Technol. **44**(15), 78–81 (2015)
11. Yang, J.: Shear properties of nano cement soil. Northeast Forestry University (2015)

Model of Concrete Chloride-Ion Diffusion Coefficient in Chinese Salt Lake

Liming Zhang[1(✉)], Jia Li[2], and Hongfa Yu[3]

[1] School of Civil and Architectural Engineering, Nanchang Institute of Technology, Nanchang, China
[2] Library, Nanchang Institute of Technology, Nanchang, China
[3] Department of Civil Engineering, Nanjing University of Aeronautics and Astronautics, Nanjing, China

Abstract. Early destruction of concrete structures is common in salt lake environment in China, So far, there is no model which can predict concrete service life in Chinese salt lake environment. Immersion in Chinese salt-lake solution was adopted to periodically test the concrete chloride-ion diffusion coefficient (CCDC). The results is that CCDC gradually decreased with increasing time and chloride-ion concentration. A calculation model for CCDC was established. Combining the prediction results and the measured data reported in this paper, the effectiveness and applicability of the established CCDC calculation model were compared and verified. The results of analysis provide important boundary conditions.

Keywords: Salt-lake environment · Chloride-ion concentration · Concrete chloride-ion diffusion coefficient · Multi-factor coupling model

1 Introduction

Salt solutions chloride-ion concentration in China [1] is more than 10 times that of the marine environment. Therefore, in salt-lake environments, as shown in Fig. 1. the problem of insufficient durability of concrete structure generally [2, 3] leads to the lack of durability of concrete structure, which is the main factor causing the difference between the internal and external chloride-ion concentration caused by chloride-ion diffusion [4, 5] and steel corrosion in concrete structures [6]. Thus, there is an urgent need to adopt the corresponding technical means to achieve more than 50 years of service life for concrete structures. The current method of estimating the service life of concrete structures in salt-lake environments is mainly based on Fick's second law of diffusion [7], therefore, the concrete chloride-ion diffusion coefficient is an important parameter in predicting the service life of concrete structures.

A multi-factor model of determining the CCDC in a salt-lake environment has not been published, to the best of our knowledge, in the literature. In view of this, analyses

P. Mendonça and N. D. Cortiços (Eds.): ICAMC 2021, LNCE 226, pp. 263–271, 2022.
https://doi.org/10.1007/978-3-030-94514-5_28

Fig. 1. Chaerhan Salt Lake, Qinghai, China

of actual salt-lake brine immersion test data from the Qinghai, Inner Mongolia, Xinjiang, and Tibet regions have been carried out, and the concentration of chloride-ion solution, water-to-cement ratio, and the influence of the exposure time of surface chloride-ion concentration discussed. A salt-lake environment was built using experimental data from a concrete chloride-ion diffusion coefficient calculation model using multiple factors. An experiment comparing empirical model calculation values with measured data was conducted and the results compared with the corrected values, thus verifying the effectiveness and applicability of the model.

2 Experimental Procedure

The concrete test was divided into two test groups as shown in Table 1. The tests used a mix of conch cement (P.O. 42. 5), medium-grade natural river sand, stone produced in Nanchang Anyi with a (5–16) mm continuous gradation of gravel, tap water, and carboxylic acid as a superplasticizer, with a water-reduction rate of >20%. The chemical compositions of the salt solution are shown in Table 2.

Table 1. Concrete test groups and mix ratios

Group	Factors	Sample no.	Cl^- concentration (%)	Water to cement ratio	Material content/(kg.m^{-3})				
					Cement	Stone	Sand	Water	Superplasticizer
1	Salt-lake solution	XZ	9.2	0.55	369	1103	735	203	0
		NMG	10.8						
		XJ	13.3						
		QH	20.4						
2	W/C	S50	20.4	0.50	398	1362	690	196	0
		S45		0.45	407	1328	715	183	
		S40		0.40	435	1085	665	174	
		S35		0.35	452	1131	679	158	
		S30		0.30	500	1116	657	150	4.07
		S28		0.25	544	1062	708	136	5.85

The test steps are as follows:

(1) The sizes of all concrete block specimens were 100 mm × 100 mm × 100 mm and were cured, on average, for 28d. After removing the specimens from molds, specimens designated Group 1 were soaked in Xinjiang, Inner Mongolia, Qinghai, and Tibet experimental salt solutions, while designated Group 2 specimens were soaked in a Qinghai salt-lake experimental solution. The exposure times were t = 28, 56, 90, 270, and 520d.

Table 2. Chemical compositions of various salt-lake brines

Salt lake	Chemical compositions (g/L)							
	Na^+	Mg^{2+}	K^+	Ca^{2+}	Cl^-	SO_4^{2-}	CO_3^{2-}	HCO_3^-
Xinjiang	84.23	9.85	1.87	0.72	133.43	33.36	2.04	3.87
Qinghai	68.36	35.13	5.98	4.24	204.21	22.29	0.17	0.13
Inner Mongolia	97.17	3.96	2.64	0.13	108.64	36.44	25.38	4.59
Xizang	61.61	4.66	6.65	0.16	92.29	27.95	1.192	0.95

As described in the literature [12, 13] the three-dimensional was derived based on Fick's second law applied to the chloride-ion-diffusion theory model:

$$C_{f}{=}C_s+\sum_{m=1,3,5}^{\infty}\sum_{m=1,3,5}^{\infty}\sum_{m=1,3,5}^{\infty}\frac{64}{mnp\pi^3}(C_0-C_s)\sin\left(\frac{m\pi}{L_1}x\right)\sin\left(\frac{n\pi}{L_2}y\right)\sin\left(\frac{p\pi}{L_3}x\right)\exp\left[-D_t t\left(\frac{m^2\pi^2}{L_1^2}+\frac{n^2\pi^2}{L_2^2}+\frac{p^2\pi^2}{L_3^2}\right)\right] \tag{1}$$

(2) The specimens were removed after different exposure times and powder was collected from the two sides of the specimen by drilling. The drilling equipment comprised a small drill with a drill diameter of 6 mm; the drilling position is shown in Fig. 2. In the figure, labels I–IV show the four powder collection points. The distances from a collection point to the two sides of the specimen are d_1 = 20 mm and d_2 = 20 mm. Eight holes were drilled on each test piece. The sampling depths were selected as 0–5, 5–10, 10–15, 15–20, and 20–25 mm, and the powder collected from each concrete sample weighed approximately 5 g. A sieve with a pore size of 0.15 mm was used to remove coarse particles.
(3) The concrete chloride-ion content determination method used for different exposure times can be found in Zhang et al.'s study [14].

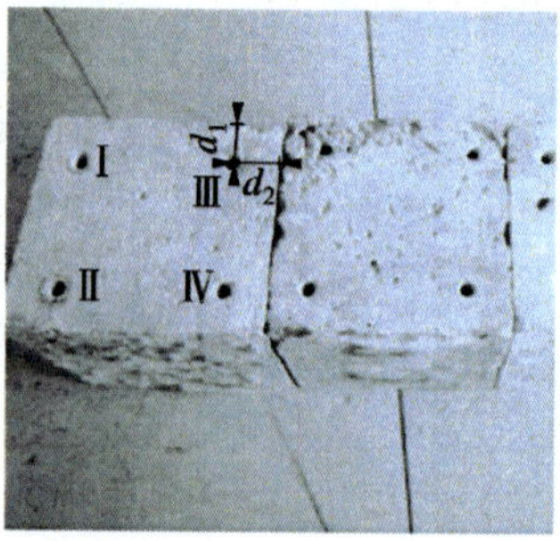

Fig. 2. Sample drilling position.

3 Results and Discussion

3.1 Relationship Between Solution Concentration, Water to Cement Ratio(W/C) and Chloride-Ion Diffusion Coefficient

As shown in Fig. 3(a), the chloride-ion diffusion coefficient of concrete in each group increases with increasing chloride-ion concentration in the salt-lake solution. In addition, the growth rate is faster when the chloride-ion concentration is less than 10.8%, and then gradually slows down and gradually increases until it reaches a stable state. In the same salt-lake solution, the longer the corrosion time, the smaller the chloride ion diffusion coefficient of the concrete. The linear function, power function, exponential function, polynomials, and other functions of MatLab software (MathWorks, USA) were used to perform regression analysis on the data plotted in Fig. 2. Considering the boundary value (i.e., the chloride-ion diffusion coefficient when the corrosion time is zero is also zero), the results show that the chloride-ion diffusion coefficient on the concrete surface has a significant power-The correlation coefficients R for exposure times of 28, 56, 90, 270, and 520d are 0.953, 0.951, 0.942, 0.958, and 0.961, respectively. Therefore, CCDC can be expressed by an exponential equation:

$$D_t = \alpha C_{cl}^{\beta} \tag{2}$$

Where α and β are the fitting parameters as shown in Table 3. D_t is the chloride-ion diffusion coefficient and C_{Cl} is the chloride-ion concentration in salt-lake solution in percent.

As shown in Fig. 3(b), the chloride-ion diffusion coefficient of the concrete surface increases with increasing water-to-cement ratio. The same regression analysis described above was performed on the data plotted in Fig. 3 in MatLab. Considering the boundary value (i.e., the chloride-ion diffusion coefficient when the water-to-cement ratio is zero is also zero), the results show that CCDC has a significant exponential function relationship with the water-cement ratio. For exposure times of 28, 56, 90, 270, and 520 d, the correlation coefficients R2 are 0.973, 0.970, 0.965, 0.967, and

0.936, respectively. Therefore, CCDC can be expressed by the following power–function equation that follows the law in Song et al.'s study [15].

$$D_t = 10^{n + bR_{W/B}} \tag{3}$$

Where n and b are fitting parameters with values shown in Table 3, D_t, the concrete chloride-ion diffusion coefficient, and $R_{W/B}$, the concrete water-to-binder ratio.

Many studies [16, 17] have shown that the time-varying model of the concrete chloride-ion diffusion time-varying model o can more reasonably describe the time-varying characteristics of CCDC. Values for concrete chloride-ion diffusion time dependence coefficient m are shown in Table 3. As shown in the table, the time-varying coefficient m decreases as the chloride-ion concentration and water-to-binder ratio in the salt-lake solution increases. The m values in Table 3 were fitted using a spreadsheet m was exponentially related to the chloride-ion concentration and water-to-binder ratio in the salt-lake solution. The correlation coefficients were both higher than 0.90, namely, 0.957 and 0.989. The functional relationship is established, as is the relationship between m and the solution's chloride-ion concentration and water-to-binder ratio:

$$D_t = D_{28}\left(\frac{28}{t}\right)^m \tag{4}$$

Where D_{28} is the chloride-ion diffusion coefficient of the concrete exposed for 28d, D_t, the chloride-ion diffusion coefficient of concrete, t, the exposure time (in d), and m, the time-varying coefficient:

$$m = \frac{0.133}{R_{W/B}C_{Cl}^{0.2}} \tag{5}$$

Where C_{Cl} is the chloride ion concentration (in %) in the salt-lake solution and $R_{W/B}$ is the concrete water-to-binder ratio.

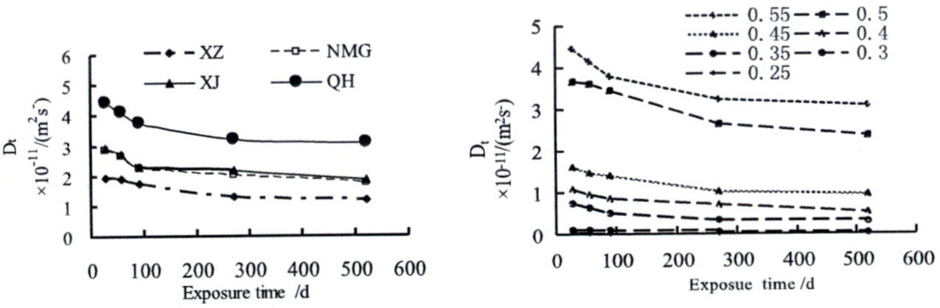

(a) Different salt lake. (b) Different W/C.

Fig. 3. Relationship between concentration, W/C, and Dt.

Table 3. Test data-fitting parameters

Exposure time (d)	Fitting parameter		R	Fitting parameter		R	Cl− (%)	m	R	$R_{W/B}$	m	R
	α	β		N	b					0.55	0.164	0.934
28	3E −12	0.937	0.953	−13.8	6.763	0.973	9.2	0.183	0.987	0.45	0.195	0.969
56	2E −12	0.894	0.951	−14.5	8.175	0.970	10.8	0.166	0.948	0.4	0.235	0.967
90	1E −12	0.865	0.942	−14.8	8.475	0.965	13.3	0.149	0.969	0.35	0.334	0.983
270	8E −13	0.801	0.958	−15.3	8.838	0.967	20.4	0.132	0.945	0.3	0.501	0.914
520	6E −13	0.757	0.961	−15.6	9.115	0.969	-	-	-	0.25	0.596	0.952

3.2 Establishment of the Model for Chloride-Ion Diffusion Coefficient

The Life365 model [17] was applied to calculate the chloride-ion diffusion coefficient and water-to-binder ratio using the following formula:

$$D_t = 10^{\left(-12.06 + 2.40\,R_{W/B}\right)}\left(\frac{28}{t}\right)^{0.2} \tag{6}$$

Where t is the exposure time (in d) and $R_{W/B}$ is the water-to-binder ratio. Inserting Eqs. (4) and (5) into Eq. (6), we obtain the model proposed:

$$D_t = 1.732C_{Cl}^{0.468}10^{\left(-12.93 + 2.682\,R_{W/B}\right)}\left(\frac{28}{t}\right)^{\frac{0.133}{R_{W/B}C_{Cl}^{0.2}}} \tag{7}$$

Where t is the exposure time (in d), $R_{W/B}$, the water-to-binder ratio, and C_{Cl}, the chloride-ion concentration (in %) in the solution.

3.3 Establishment and Verification of the Multi-factor Model for the Chloride-Ion Diffusion Coefficient

Using the test data in this paper, the results of various calculation models for verifying the chloride-ion diffusion coefficient of concrete, including the Life365 model and a multi-factor chloride-ion diffusion coefficient model are shown in Fig. 4. It can be seen from the figure that fore-mentioned two models both reflect that the chloride-ion coefficient of concrete decreases with increasing exposure time. In general, the results obtained by multi-factor chloride-ion diffusion coefficient model are closest to the measured values, while the Life365 prediction results deviate significantly from the measured values. Through further comparison and analysis of the difference between the value predicted by the Life365 model and the model proposed in this paper and the field-measured values, it can be seen from Fig. 4. that the value predicted by the Life365

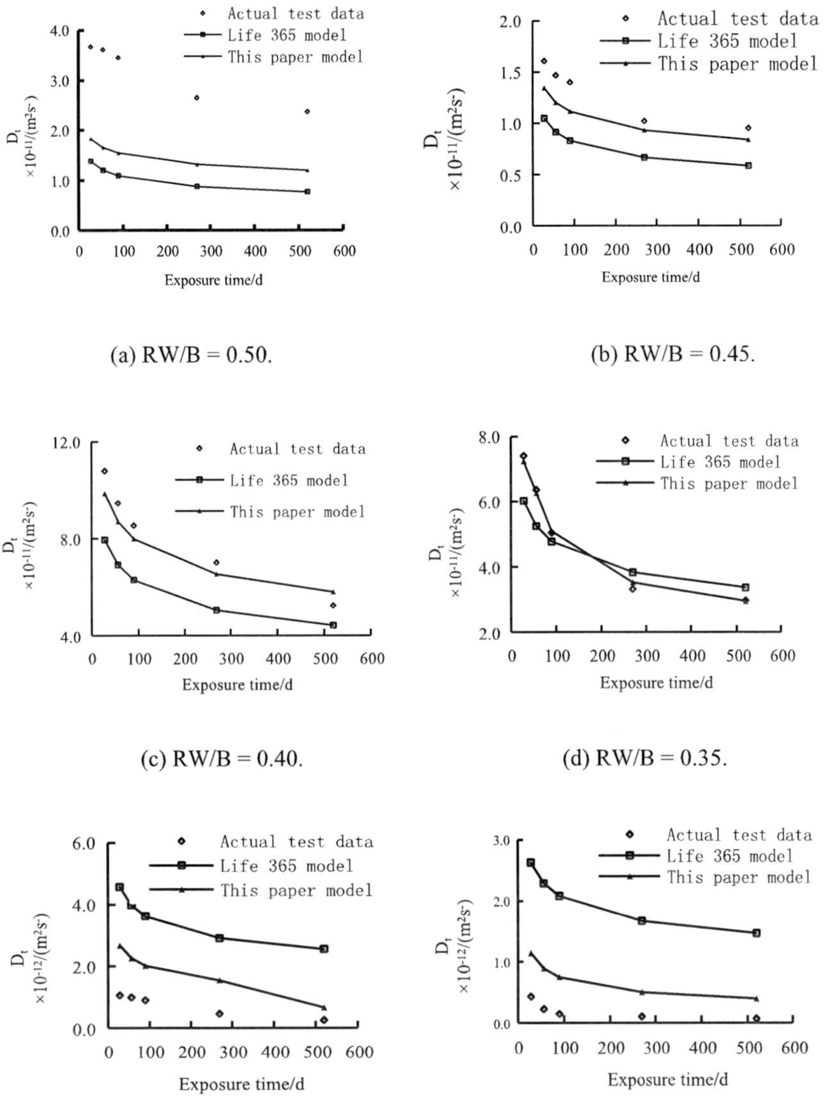

(a) RW/B = 0.50.

(b) RW/B = 0.45.

(c) RW/B = 0.40.

(d) RW/B = 0.35.

(e) RW/B = 0.30.

(f) RW/B = 0.25.

Fig. 4. Comparison between different prediction models and measured values of concrete chloride-ion diffusion coefficient. (a) RW/B = 0.50. (b) RW/B = 0.45. (c) RW/B = 0.40. (d) RW/B = 0.35. (e) RW/B = 0.30. (f) RW/B = 0.25.

model increases rapidly with decreasing water-to-binder ratio, while the increase of the value predicted by the proposed model is relatively flat. It can also be seen that when the water-to-binder ratio is greater than 0.35, the values predicted by both the Life365 model and the proposed model are smaller than the measured values. Moreover, the values predicted by the model proposed in this paper are closer to the measured values. When the water-to-binder ratio is 0.35, the values predicted by the Life365 model and the proposed model are consistent with the measured values; when the water-to-binder ratio is less than 0.35, the values predicted by the Life365 model and the proposed model are larger than the measured values, but the values predicted by the model proposed in this paper are closer to the measured values. The reason for these results is that the Life365 model does not consider the effect of solution chloride-ion concentration on the chloride-ion diffusion coefficient of concrete. It can be seen that, when compared with the Life365 model, the proposed model is better adapted to include the effects of the chloride-ion diffusion coefficient of concrete in salt-lake environments.

4 Conclusion

This paper presented an experimental study on concrete in chinese salt-lake solution. Models of CCDC in chinede salt lake were investigated. The following conclusions can be drawn from this study.

(1) When concrete is immersed in salt-lake solution, the chloride-ion diffusion coefficient of the concrete increases with increasing chloride-ion concentration and water-to-cement ratio in the salt-lake solution, and decreases with the delay in the corrosion time. The chloride-ion diffusion coefficient of concrete has a power–function relationship with solution chloride-ion concentration, water-to-binder ratio, and exposure time. The time-dependence coefficient m decreases with increasing chloride-ion solution concentration and water-to-binder ratio.

(2) Based on the measured data reported in this paper, a multi-factor coupling calculation model for the chloride-ion diffusion coefficient of concrete is proposed by considering factors like water-to-binder ratio, chloride-ion content, and time-varying characteristics in the solution.

(3) Under different water-to-binder ratios and corrosion time conditions, the proposed multi-factor coupling calculation model of the concrete chloride-ion diffusion coefficient is more suitable for salt-lake environments than the Life365 model, which is critical to informing the durability design and service life of concrete structures under salt-lake environments. The results of the analysis also provide important boundary condition.

Acknowledgements. The author thanks Hongfa Yu for valuable discussions on the experimental results.

References

1. Zhang, L.M., Yu, H.F., He, Z.M.: Chloride ion diffusivity of salt lake concrete. J. Central South Univ. (Sci. Technol.) **42**(6), 1752–1755 (2011)
2. Zhang, S.P., Xiao, Y.D.: Emphasis on material corrosion in west environment. Mater. Prot. **35**(7), 58–60 (2002)

3. Liu, L.X.: Brief introduction on the study of erosion and prevention of concrete in salt lake and saline area of Chaerhan. Chaidamu J. Build. Mater. **4**(4), 395–400 (2001)
4. Zheng, H.B., Dai, J.G., Li, W.H., et al.: Influence of chloride ion on depassivation of passive film on galvanized steel bars in concrete pore solution. Constr. Build. Mater. **166**(3), 572–580 (2018)
5. Lee, H.S., Yang, Y.M., Jitendra, K.S., et al.: Corrosion mitigation of steel rebars in chloride contaminated concrete pore solution using inhibitor: An electrochemical investigation. Constr. Build. Mater. **173**(6), 443–451 (2018)
6. Abd, E.M.: Four-years carbonation and chloride induced steel corrosion of sulfate-contaminated aggregates concrete. Constr. Build. Mater. **163**(2), 539–556 (2018)
7. Mejlbro, L.: The complete solution of Fick's second law of diffusion with time-dependent diffusion coefficient and surface concentration. Durability of concrete in saline environment, Cement AB, Danderyd, pp. 127–158 (1996)
8. Pack, S.W., Jung, M., Song, H.W., et al.: Prediction of time dependent chloride transport in concrete structures exposed to a marine environment. Cem. Concr. Res. **40**(2), 302–312 (2010)
9. Wang, Y.Z., Wu, L.J., Wan, Y.C., et al.: Prediction model of long-term chloride diffusion into plain concrete considering the effect of the heterogeneity of materials exposed to marine tidal zone. Constr. Build. Mater. **159**(1), 297–315 (2018)
10. Zhang, J.Z., Zhao, J., Zhang, Y.R., et al.: Instantaneous chloride diffusion coefficient and its time dependency of concrete exposed to a marine tidal environment. Constr. Build. Mater. **167**(4), 225–234 (2018)
11. Yang, L.F., Ma, Q., Yu, B.: Analytical solution and experimental validation for dual time-dependent chloride diffusion in concrete. Constr. Build. Mater. **161**(2), 676–686 (2018)
12. Yu, H.F., Sun, W., Yan, L.H., et al.: Study on prediction of concrete service life I: theoretical model. J Chin. Ceramic Soc. **30**(6), 686–690 (2002)
13. Yu, H.F., Sun, W., Ma, H.Y., et al.: Study on prediction of concrete service life II: application and test of model. Chin. Ceramic Soc. **30**(6), 691–695 (2002)
14. Tang, L.P., Joost, G.: On the mathematics of time-dependent apparent chloride diffusion coefficient in concrete. Cem. Concr. Res. **37**(4), 589–595 (2007)
15. Song, H.A., Shim, H.Y.B.: Service life prediction of repaired concrete structures under chloride environment using finite difference method. Cem. Concr. Compos. **31**, 120–127 (2009)
16. Life-365 Service Life Prediction Modle. http://www.life-365.org/download.html
17. Mangat, P.S., Molloy, B.T.: Prediction of long-term chloride concentration in concrete. Mater. Struct. **170**(24), 338–346 (1994)

Urban Planning and Traffic Management

Analysis of Streets and Lanes of Traditional Villages in Nanyuan of Zhuozhang River Basin Based on Space Syntax

Shuai Sun(✉), Ziying Li, and Debin Gao

North China University of Technology, Beijing 100144, China

Abstract. As one of the important birthplaces of the Central Plains civilization, the Zhuozhang River Basin in Shanxi has a long history of human habitation. These villages have rich street patterns and different development trajectories. The streets and lanes of the traditional villages in Nanyuan of the Zhuozhang River Basin can be divided into network, branch, belt, cross, semi-circular, and comb-shaped. This article analyzes Zhuozhang Henan based on the axis analysis method in the space syntax. The street and lane pattern of the original traditional villages are classified and sorted, including the degree of selection, integration, and intelligibility analysis of the accessibility of traditional streets and lanes and the characteristics of spatial structure, providing feasible guidance for the renewal and protection of historical villages.

Keywords: Space syntax · Traditional village · Street and lane pattern

1 Space Syntax

Since the theory of space syntax has been proposed by Professor Bill Hillier, it has used the mathematical model establishment and quantitative analysis of the architectural space form [1]. After years of practical testing, a large number of urban design cases have been used to revise the method, and the space syntax theory has developed into research. A new empirical analysis tool for urban space [2]. There are three main spatial syntax methods: axis method, convex space method, and horizon analysis method [3]. This research is to obtain the local CAD topographic map through the investigation of the research area, establish the spatial syntax axis diagram model based on the "most and least" standard, and then import the Depthmap software for data analysis to obtain the degree of selection, integration, and intelligibility. The space syntax data of, the data interpretation of the space [4].

By using the axis analysis method in the space syntax, the typical cases of the Zhuozhang River's southern source watershed are selected to analyze the street space. The degree of selectivity refers to the number of times a certain axis is selected as the shortest path from all spaces to all other spaces [5]. The higher the degree, the higher the ability to attract people; the degree of integration refers to the standardized distance from any space to all other spaces. The higher the degree of integration, the stronger the accessibility of the space; the degree of intelligibility refers to the degree of people's understanding of the space. The higher the space, the easier it is to perceive.

P. Mendonça and N. D. Cortiços (Eds.): ICAMC 2021, LNCE 226, pp. 275–285, 2022.
https://doi.org/10.1007/978-3-030-94514-5_29

2 Street and Lane Pattern Type

2.1 Mesh Type

The streets and lanes of mass-shaped traditional villages are often networked. Such relatively regular villages are less affected by the geographical environment. Most of them are located on relatively flat and wide ground, formed in plain areas or mountain basins, and have a networked street and lane pattern. There are two situations. One is that the village space is evenly divided by the main alleys in the village. The secondary streets and alleys are in random and natural shapes. The roads extend in all directions. The alleys that seem to have no access can actually be connected to each other. The road system is often responsible. For the functions of villagers' transportation, social interaction, and holding of activities, the handover methods of streets and lanes are mostly in the shape of "cross", "well", and "D". This type of traditional village structure is characterized by rich layers of village space, a progressive pattern from the main alley to the secondary alley to the branch alley and then to the courtyard space. The streets and lanes are wide and narrow, and there is more space for turning. Another situation is that there is no obvious main street in the village, the road form is more random, the street space has strong accessibility, and the street and lane handover methods are mostly "cross" and "D" shapes. This form of handover is convenient. The people stayed and talked at the intersection.

Take Zhangjiagou Village as an example. According to the field survey and related data, the road axis diagram of Zhangjiagou Village is drawn in cad and imported into Depthmap software. Zhangjiagou is analyzed through 3 points of selection, integration, and intelligibility. Research and analyze the street space inside the village (Fig. 1).

Selectivity. According to the quantitative analysis map of the streets and lanes of Zhangjiagou Village, it can be seen that there are two red axes with the highest selectivity. The main roads are the roads that villagers pass most frequently, which can attract a large number of people and have a high degree of choice (Fig. 1a and 1b).

Integration Degree. The higher the integration degree, the higher the accessibility of the axis is, and it is the most core space. According to the quantitative analysis map of the streets and lanes of Zhangjiagou Village, it can be found that in the global integration degree, there are two red axes with high integration degree. One is located on the northern side of the village and the other is located in the center of the village. They are easy to attract people. High accessibility, is the core of the integration of the village. Comparing with the local integration degree, it is found that the axis with the highest integration degree changes slightly, and the connection between the whole and the local core area is not enough (Fig. 1c and 1d).

Comprehensibility. Generally, when the fit is higher than 0.5, the comprehensibility is high. According to the quantitative analysis diagram of street space, it can be seen that the fit of Zhangjiagou Village is 0.47, indicating that its spatial comprehensibility is good, and people have a good perception of the spatial form of the village in it, and the possibility of getting lost is not high (Fig. 1e).

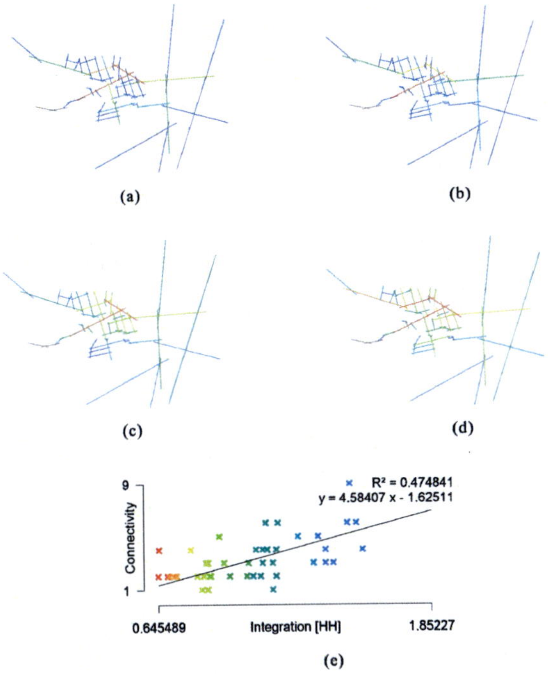

(a)Total Choice (b)Partial Choice r3

(c)Total Integration (d)Partial Integration r3

(e)Intelligibility

Fig. 1. Quantitative analysis of streets and lanes in Zhangjiagou Village (Picture Source: Research Team)

2.2 Branch Type

In the spatial structure of the village, the pattern of tree-shaped streets and lanes is mainly represented by one or two main streets connecting the entire village. The branch lanes extend from the main lane to both sides and are directly connected with the courtyard space. The external spatial system of the village is in form. The appearance is dendritic. The main characteristics of the village spatial structure of the branch-shaped streets and lanes pattern of Nanyuan in the Zhuozhang River Basin are shown in the following points: First, the overall street and lane structure has a distinct primary and secondary structure. The main body is the main traffic road of the village. The production and life of the villagers are directly related to the main street. The branch lanes extend from the main lane to both sides and are basically perpendicular to the main lane, connecting the courtyard space. The branch lanes are mostly parallel to each other, and there is less correlation. The road crossing forms are mostly in the shape of "T" and "cross". This kind of intersection makes the village traffic smooth and pedestrians see through. Second, the village has rich spatial levels, and the main lane is

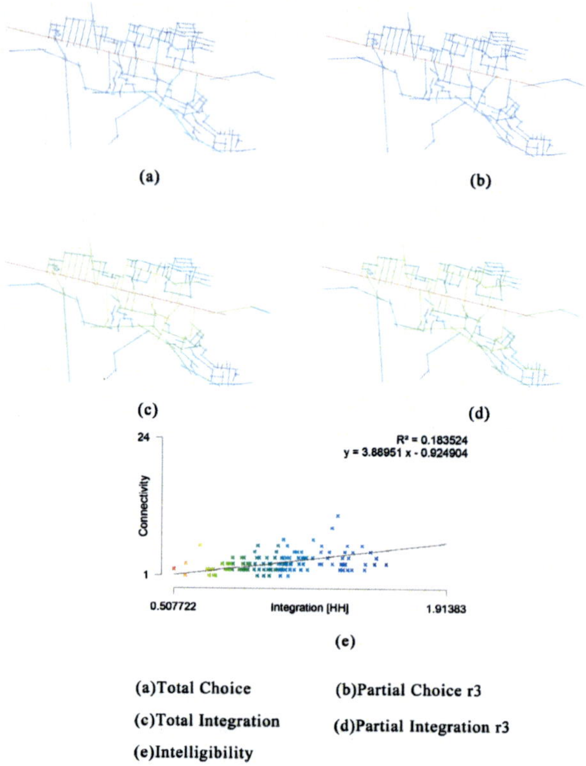

Fig. 2. Quantitative analysis of street space in Donghuo Village (Image source: research team)

connected with the branch lanes, forming a spatial system of "main lane-branch lane-courtyard space".

Taking Donghuo Village as an example, based on field research and related data, the road axis diagram of Donghuo Village is drawn in cad and imported into Depthmap software. The internal streets of Donghuo Village are analyzed through 3 points of selectivity, integration, and intelligibility. Lane space for further research and analysis (Fig. 2).

Selectivity. According to the quantitative analysis map of the street and lane space of Donghuo Village, it can be seen that there is only one red axis with the highest selectivity. Superimposed with the current situation map of the village, it can be found that this axis is located on the main street of the village and has a good attraction for people flow. Therefore, it has a high degree of selectivity. Reachability is high, connectivity is good, and the flow of people is relatively large; the remaining axes are cyan and blue with low connection values. These axes are the roads leading to the courtyard space in the village, or are far away from the center of the village. Remote and low accessibility (Fig. 2a and 2b).

Integration Degree. In the global integration diagram, the warmer the axis color, the higher the integration degree, the easier it is to attract people. There is one red axis, which is located on the main road and adjacent to an activity space in the center of the village. The village can be seen the central location is the core space of Donghuo Village. Comparing the local integration map, it is found that the core part of the integration has not changed, indicating that the main street has a strong function of gathering people (Fig. 2c and 2d).

Comprehensibility. Generally, when the fit is higher than 0.5, the comprehensibility is high. According to the quantitative analysis diagram of street space, it can be seen that the fit of Donghuo Village is 0.18, indicating that its spatial comprehensibility is low, and people have a vague perception of the spatial form of the village in it, and the possibility of getting lost is greater (Fig. 2e).

2.3 Belt Type

In a village with a strip-shaped street pattern, the main spatial feature presented is that a main street runs through the entire village. This main street bears the functions of village traffic, villagers' communication, and people's activities. The houses are along the main street. Layout on one side or on both sides, this street lane pattern village is mainly characterized by the following spatial morphology: First, the main lane in the village occupies a dominant position, and the role of the branch lane cannot be compared with that of the main lane. Lanes are formed under specific historical factors or topographical conditions; secondly, people's homes are distributed along the main street, and will be constantly changed with the development of the village to adapt to environmental changes. Generally, there are also very conspicuous important distributions on the main lane. Nodes. Generally, there will be obvious structures or open spaces at the village head, tail, or some key points in the village, such as Buddhist halls, temples, etc.; third, most of the village roads are a single space directly from the main lane to the courtyard. It is expressed by the specific characteristics of the striped street pattern in Fangdai Village.

According to the field survey and related data sorting, draw the road axis diagram of Fangdai Village in cad, and import it into Depthmap software to study and analyze the street space of Fangdai Village through three points of selectivity, integration and intelligibility (Fig. 3).

Selectivity. According to the quantitative analysis diagram of the streets and lanes of Fangdai Village, it can be seen that there is 1 red axis and 2 green axes with the highest selectivity. These three are located on the main roads in the village, and this road runs through the whole area. The village is the only way for the village to connect with the outside world. It has a high ability to attract people to gather, so it has the highest degree of choice (Fig. 3a and 3b).

Integration Degree. In the global integration degree map, the warmer the axis color, the higher the integration degree, and the easier it is to attract people. According to the quantitative analysis map of the street space of Fangdai Village, it can be found that the orange-red axis with high integration degree is exactly The east-west main streets

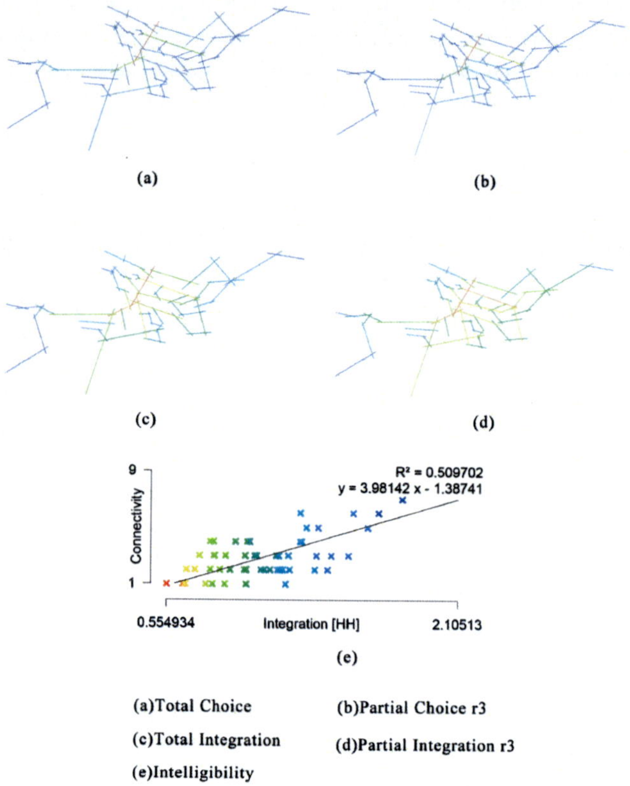

Fig. 3. Quantitative analysis of street space in Fangdai Village (Image Source: Research Team)

running through the village; the yellow axis with higher integration is the secondary streets extending from the main street to north and south, with higher accessibility; the blue axis with the least integration is farther from the main street and can be reached Low sex and low traffic. Comparing the global integration degree with the local integration degree, it can be found that the axis with the highest integration degree remains unchanged, and the connection between the whole and the local core area is very strong (Fig. 3c and 3d).

Comprehensibility. Usually when the fit is higher than 0.5, the comprehensibility is high. According to the quantitative analysis of street space, Fangdai Village has a fitting degree of 0.51, indicating that its spatial comprehensibility is very good. People in it have a clear perception of the spatial form of the village and are unlikely to get lost (Fig. 3e).

2.4 Cross Type

In a village with a cross-shaped street and lane pattern, the main spatial feature presented is that the village has two vertical main streets, the east and west main streets

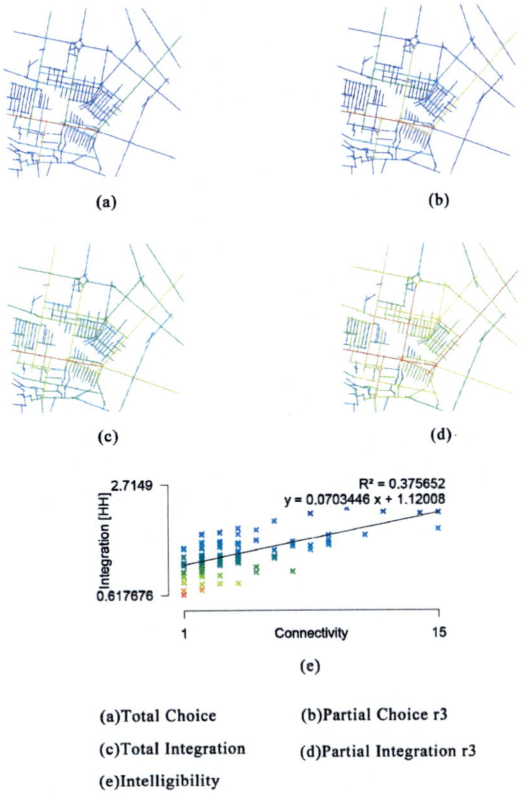

Fig. 4. Quantitative analysis of street space in Yincheng Village (Image source: Research Team)

and the north and south main streets, forming the main spatial system skeleton, dividing the village into several large areas. In the area, branch alleys are used for internal separation and connection. The cross-shaped spatial structure of the traditional villages in Nanyuan of the Zhuozhang River Basin has very clear village spatial characteristics. This article takes Yincheng Village as an example to illustrate the specific characteristics of the cross-shaped street pattern.

An east-west horizontal axis runs through the whole village in Yincheng Village. In order to understand the street space of Sangzi more intuitively, according to field investigation and related data, the road axis diagram of Sangzi Village is drawn in cad and imported into Depthmap software. The six points of degree, connection degree, control degree, integration degree, depth value and intelligibility degree are used to research and analyze the street space of Cuijiazhuang Village (Fig. 4).

Selectivity. According to the quantitative analysis map of the streets and lanes of Yincheng Village, there is 1 red axis in the global selection degree, which has a high potential for attracting people to pass through; there are 5 axes with a higher degree of selection, which are yellow, indicating these streets. Lane space has a good potential to attract people to pass through (Fig. 4a and 4b).

Integration Degree. In the integration map, the warmer the axis color, the higher the integration, and the easier it is to attract people. In the global integration map of Yincheng Village, there are a total of 5 orange-red axes. These axes represent the highly integrated street space in the village. They are located at the cross street of the village and are the main streets with high accessibility and easy to attract people. The yellow-green axis with a higher degree of integration is the secondary streets. These streets extend from the main road to both sides and are connected to the courtyard space and have high accessibility. The other blue and blue axes are located at the periphery of the village. The main street and the village center are far away, with high privacy and low integration. Comparing the local integration degree with the global integration degree, it is found that the core part of the integration degree has basically not changed. The axis with the highest degree of global integration has the highest degree of local integration. It is the Da Shizi Street in Yincheng Village, which is very attractive. The ability to flow (Fig. 4c and 4d).

Comprehensibility. According to the quantitative analysis graph, the fit degree of Yincheng Village is 0.37. Generally, when the fit degree is higher than 0.5, the comprehensibility is high. It can be seen that the overall spatial comprehensibility of Yincheng Village is poor, and people have an unclear sense of the space in the village and are more likely to get lost (Fig. 4e).

2.5 Semi-ring Type

Hedong Village uses two east-west roads as the main street, and a north-south road connects the main road to form a semi-circular street pattern. According to field investigation and related data sorting, draw the road axis map of Hedong Village in cad and import it into Depthmap software to study and analyze the street space of Hedong Village through three points of selectivity, integration and intelligibility (Fig. 5).

Selectivity. According to the quantitative analysis map of the streets and lanes of Hedong Village, it can be seen that the red axes with the highest degree of selection are located on the two main roads. These two roads are the only roads for villagers to enter and exit the village, so they have a higher degree of selection (Fig. 5a and 5b).

Integration Degree. In the global integration map, the warmer the color of the axis, the higher the integration, and the easier it is to attract the flow of people. Corresponding to the plan of the village, it can be found that the orange-red axis is the two north-south main streets of the village, which is consistent with the actual situation; The yellow axes with higher degrees are secondary streets and lanes, which are distributed by main roads and have higher accessibility; the other blue-green axes are less integrated, and these axes are all branch lanes extended from secondary streets. It is far away from the main street and public space, has low accessibility, and is not easy to attract people. By comparing the degree of global integration and the degree of local integration, it is found that the two main streets are still the most integrated places, indicating that the two main roads in the village have a high flow of people (Fig. 5c and 5d).

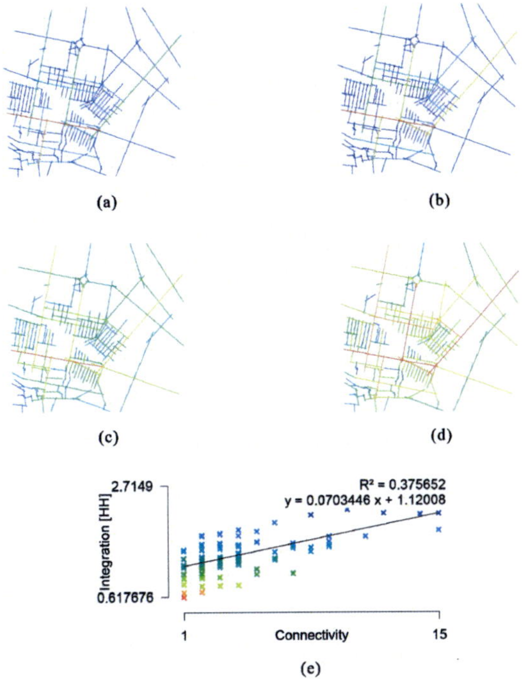

(a)Total Choice (b)Partial Choice r3

(c)Total Integration (d)Partial Integration r3

(e)Intelligibility

Fig. 5. Quantitative analysis of street space in Yincheng Village (Image source: Research Team)

Comprehensibility. Generally, when the degree of fit is higher than 0.5, the intelligibility is high. According to the quantitative analysis diagram of street space, it can be seen that the fit degree of Hedong Village is 0.44, indicating that its spatial comprehensibility is average. People in it have a vague perception of the spatial form of the village and may get lost (Fig. 5e).

2.6 Comb Type

In order to understand and evaluate the street space of Pingjiazhuang Village more intuitively, according to the field investigation and related data, the road axis diagram of Pingjiazhuang Village is drawn in CAD and imported into the Depthmap software. Research and analysis on the street space of Pingjiazhuang Village (Fig. 6).

Selectivity. According to the quantitative analysis map of the streets and lanes of Pingjiazhuang Village, the warmer the color of the axis, the higher the degree of selection. There is a total of 1 red axis. This road is the only way to enter and exit the village, and it is also the main road in the daily life of the villagers. Therefore, they It has a high degree of choice and has a good ability to attract people (Fig. 6a and 6b).

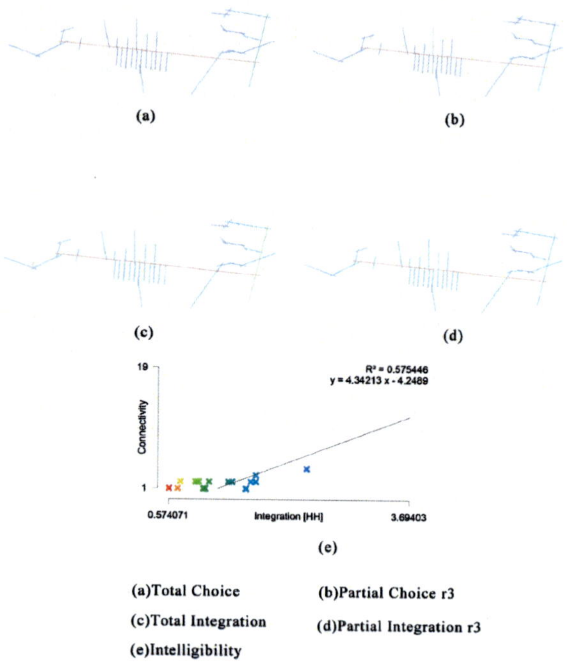

Fig. 6. Quantitative analysis of street space in Pingjiazhuang Village (Image source: Research Team)

Integration Degree. In the global integration map, the warmer the axis color, the higher the integration, and the easier it is to attract people. There is a total of 1 orange-red axis, which represents the streets connecting the roads inside and outside the village, which can attract a large number of people and has high accessibility; the green axis with higher integration is the secondary streets. These streets are divided by the main. These roads are distributed, connected with the courtyard space, and have high accessibility; the rest of the blue axis is located close to the mountain, far away from the residential buildings and the main street, and it is not easy to attract people. By comparing the global integration degree and the local integration degree, we can find that the core part of the integration degree has basically not changed. The axis with the highest degree of integration in the global degree is also the highest in the degree of local integration. It is an important channel for Pingjiazhuang Village to connect with the outside world. High ability to attract traffic (Fig. 6c and 6d).

Comprehensibility. According to the quantitative analysis graph, it can be seen that the degree of fit of Pingjiazhuang Village is 0.57, and when the degree of fit is generally higher than 0.5, Highly understandable. It can be seen that the overall space of Pingjiazhuang Village is very understandable, and people have a clear sense of the space in the village. Less likely to get lost (Fig. 6e).

3 Summary

The south source of the turbidity domain is located in the southeast of the Loess Plateau. There are many and extensive mountainous areas in the southeast of the Loess Plateau. The traditional mountain-backed mountains and hilly areas within the resource flow form mountain valleys, mid-valleys, riversides, and four corners of the ground. And the influence form of historical factors is divided into group type, belt type and fan shape.

Under the combined influence of a few natural and social environmental factors in various villages with different forms, various forms of lanes have been formed, mainly including mesh, street, belt, cross-shaped, semi-circular, and clustered changes., And have different distribution characteristics and street and lane handover forms. Generally speaking, the community is conducive to the scope of the community. The traditional village alleys are organically combined with the surrounding environment of the community, and the furniture is flexible, which connects the courtyard space in the village to form a street texture with regional characteristics.

Acknowledgment. Sponsored by the National Natural Science Foundation of China (51808004).

References

1. Guo, K., Cao, Y.: Application of planning evaluation of historical block based on space syntax—case study: third streets and two alleys historical block, Nanning. IOP Conf. Ser. Earth Environ. Sci. **768**(1), 012139 (2021)
2. Zou, H., Liu, Y.: A comparative study on the spatial characteristics of traditional residential houses in Zhangguying and Laodong villages based on space syntax. IOP Conf. Ser. Earth Environ. Sci. **769**(3), 032078 (2021)
3. Fabiana, B.: Space syntax and buried cities: the case of the Roman town of Falerii Novi (Italy). J. Archaeol. Sci. Rep. **35**, 102712 (2021)
4. Hong, C., Ya, Y., Hu, T., Hu, X., Mai, X.: Research of small valley towns based on spatial syntax taking Xuecheng as an example. Curr. Urban Stud. **7**(3), 449 (2019)
5. Li, X., Lv, Z., Zheng, Z., Zhong, C., Hijazi, I.H., Cheng, S.: Assessment of lively street network based on geographic information system and space syntax. Multimed. Tools Appl. **76** (17), 17801–17819 (2015). https://doi.org/10.1007/s11042-015-3095-2

Urban Rehabilitation Based on Uses and Social Diversity: A Case Study in Porto, Portugal

Andreia Sousa and Paulo Mendonça(✉)

School of Architecture, Art and Design, Lab2PT, University of Minho, Azurém Campus, 4800-058 Guimarães, Portugal
mendonca@eaad.uminho.pt

Abstract. Many historical cities present the difficult challenge to maintain at the same time its built heritage and its living values, two valences that, together, represent their essence. The intense tourist demand of accommodation is the main cause to the imbalance verified between these valences, which manifests in numerous urban rehabilitation interventions currently carried out in downtown Porto, the second biggest city in Portugal. This problem leads to results such as gentrification, social exclusion, and also the transformation of buildings into housing that does not fit the "inhabitants" way of living or their needs because they are not thought of in this way. An exponential investment in tourist accommodation and revitalization of the image of the historic center, without a parallel investment in housing and equipment dedicated to serve the local population is the general thought observed. In this paper is proposed a strategy for urban rehabilitation based on occupancy diversity, both in use and social aspects. With the integration of a flexible and adapted strategy, the objective is to prevent social exclusion and gentrification, at the same time requalifying the historical center and providing access to it for all, creating a balance between the built and the intangible heritage.

Keywords: Social diversity · Diversity index · Sustainability · Rehabilitation

1 Introduction

Porto, the second biggest city in Portugal, whose historical centre was classified by UNESCO as heritage site in 1996 [1], presents nowadays significant tourist activity [2], with its positive aspects but also problems.

The positive aspects are linked with cultural diversity and revenues that tourists bring, that directly support commerce and services, but also allow indirect improvements. The rehabilitation of housing has increased over the last years, largely due to the demand for tourism in the city [3]. As there is no space to build new buildings, the rehabilitation was inevitable, not only to lodge tourists, but to attract the return of inhabitants to buildings that were previously abandoned.

However, most interventions present problems associated with the loss of identity of the city downtown, that affect not only the built heritage but also the intangible

P. Mendonça and N. D. Cortiços (Eds.): ICAMC 2021, LNCE 226, pp. 286–296, 2022.
https://doi.org/10.1007/978-3-030-94514-5_30

heritage, being the still existent residents and their traditional links that more significantly suffer from this, as mentioned in several studies reporting the case of Porto and other cities subject to the same phenomena [2, 4]. A consequence of the applied housing policy strategy [3], or the lack of it, is the gentrification and decentralization of the residential function [2]. From 1981 onwards, there was a decrease in the resident population in Porto, mainly in the historic centre. At the same time, there is an increase in non-resident occupants, especially in the historical centre. The term "gentrification" was coined by Ruth Glass, an émigré sociologist who came to Britain from Germany in the 1930s that was intrigued by the social and housing changes which she saw taking place in Inner London in the 1950s [5]. It has been since then identified in several cities and profusely researched by several authors [6–8].

These values led to the logical conclusion of the need to change priorities, with regard to the rehabilitation, based on tourist accommodation promoted by private investments, which, due to the financial crisis that affected Portugal in the last decades, was presented as the only possible option. A study carried out in the mass tourism resorts of the Spanish islands of Mallorca and Tenerife indicate that an economic recession causes a significant growth of the residents' support towards tourism [9]. However, in a growing prosperity period, the unsustainable transformation of neighbourhoods into extensive touristic accommodations, began to be detrimental to the city's social cohesion [10]. Thus, it is nowadays identified the need for a housing policy for the rehabilitation of the historic centre of Porto that considers social integration between local inhabitants and tourists and, at the same time, responds to the problems of housing cost increase [3]. The maintenance of the residential function of the historical centres and of the associated social and economic environments is pivotal for the maintenance of their cultural values, on which their economic values are based [2, 11].

2 Proposed Strategy

This paper proposes an intervention strategy for Porto historical center, presenting 4 case study blocks. In order to be able to develop this strategy, a survey was conducted in 2018 [12] in order to understand the current situation of their use and occupation, based on the opinion and concerns of its residents.

As a result of this survey, the vacancy of the Porto historical center blocks could be verified in regard to permanent residents, given that out of 171 buildings, with an average of 4 dwellings per building, only 32 surveys were collected, mostly by reasons of vacancy or occupation of use for tourist accommodation. There were also cases of non-adherence and momentary absence. The issue of non-adherence also indicated distrust or disinterest from some residents.

With regard to occupation, it was noticed that, in general, the blocks are quite unoccupied, as it can be seen on Fig. 1. The occupation considered only refers to permanent housing, with the exclusion of tourist accommodation.

This vacancy does not come only from the development of tourist accommodation in the center, as downtown Porto had been in need of repopulation for some time, mainly due to the degradation of buildings, however, with the unbalanced development of investment in the tourism, the situation has not improved and has even got worse for local population.

Despite the situation in which Porto's real estate still finds, in which the buildings need in-depth rehabilitation actions, the study revealed that the people who participated are mostly satisfied with their homes, with 88% completely satisfied a 0% not satisfied. It is assumed that this satisfaction went beyond the physical condition of the house, and came, even more strongly, from the appreciation of the residents for the quality of life, functional and social, that they manage to have living in downtown Porto.

This fact is verified in most of the resident population that is in the age group of more than 65 years, mostly retired who have been in the same house for more than 40 years, which naturally justifies this appreciation for the place where they grew up. Younger residents, in a minority, are already in rehabilitated buildings and, therefore, their satisfaction is more reflected in the obvious quality of the dwelling.

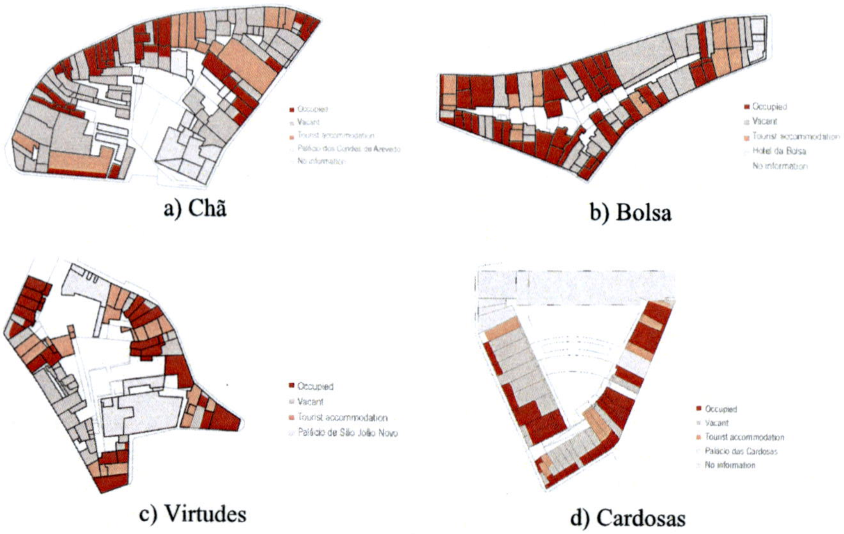

Fig. 1. Occupancy in the blocks analyzed (Adapted from [12]).

In a more concrete trajectory of the positive and negative aspects of housing, it was answered, although with a higher percentage of positive aspects, clearly what is needed to maintain and change, both in terms of the dwelling and the block.

Thus, with regard to dwelling, the appreciation for its location in the block was mentioned more often. It is followed by the visual quality of the exterior of the building and the exterior space.

And finally, the parameters: "own housing", with the almost inexistence of residents in this situation; and "low income".

The negative aspects, although minor, still reflect some concerns of the residents, such as the existence of pathological manifestations, mostly infiltration and condensation, and accessibility to the dwelling, mostly without lift and, therefore, not being adapted to the needs of elderly inhabitants.

Despite the limited financial capacity of most residents, 78% of the participants are willing to resolve the existing problems in the dwellings, as they take care of their homes and, as shown by the surveys, they are often carrying out works in the home at their own expense. Of course, the obvious answer to the financial "unavailability" on the landlords is clearly caused by the low rents, that do not allow to raise funds to solve the problems that occur in the building, which is one of the problems to be taken into account in the elaboration of the proposed intervention strategy.

Another parameter studied was the need of improvement of the block, thus, the negative aspects mentioned were mostly referring to the issue of local depopulation, the issue of isolation, given that they depended on the company of their neighbours and now feel increasingly isolated from this type of interaction, because, no matter how friendly and helpful the tourist may be, he will never be a present neighbour, as their stays are short.

Still in favour of the greater influx of tourists living in the center, the lack of peace and lack of cleanliness in the streets was also reported. The lack of services and commerce was also one of the points frequently mentioned, with people commenting more on the lack of services, such as bakery, grocery store, among others.

The lack of parking is always recurrent due to the situation of the historic center, but despite being a negative aspect, residents have shown that they adapted to this condition and make more use of public transport, which is somewhat accessible, not so much within the blocks, but on the outskirts close to public spaces and main streets.

According to the issue of uses diversity, which was also studied (Fig. 2), we were able to understand, very clearly, the high proportion of tourist accommodation. It should be mentioned that all the dwellings that were not possible to confirm another occupation were considered as rented dwellings (private sector), so the real proportion of tourist accommodation may be even higher. In the Quarteirão das Cardosas the verified percentage of occupied dwellings is so low that it was not possible to obtain accurate data on a large percentage of dwellings.

The existence of tourist accommodation is naturally predominant, due to the profit margin, which is much higher compared to the conventional house renting, which often has very low values due to the Portuguese tenancy law, which protects elderly people who have lived in the house for several years, keeping the initial contract value practically unchanged, but that, due to real estate speculation, would now be significantly higher if new contracts were made.

Fig. 2. Diversity of uses in the blocks analyzed [12].

Thus, a comparison was made between the profit margin of conventional renting in the private sector (Table 1) and tourist accommodation (Table 2), in which the highest net profitability for is clearly perceived. Despite the tourist accommodation have a higher tax rate (which can range from 35% to around 45% for, depending on the cases, in relation to the estimated 25% in conventional rent) it is still more profitable, at least when the demand for tourism is growing. This study was carried out in 2018, Covid-19 pandemic crises caused some problems of decrease in tourist visits in the year of 2020, so the analysed scenario doesn't reflect the current situation, that however is now in 2021 again improving.

Table 1. Profitability of property in conventional rent

Typology (average areas from survey)	Monthly rent (€)	Acquisition cost (€)	Net profitability (%)	Average profitability (%)
T0 (35 m^2)	700	89 600	6,75	**4,45**
T1 (42 m^2)	650	96 000	5,85	
T2 (111 m^2)	950	195 000	4.21	
T3 (146 m^2)	1 900	275 000	5,97	
T4 (204 m^2)	2 600	575 000	3,91	

Table 2. Profitability of property in tourist accommodation

Typology (average areas from survey)	Monthly rent (€)	Net rent (−30% for property management services)	Acquisition cost (€)	Net profitability (%)	Average profitability (%)
T0 (35 m^2)	1797	1258	89 600	10,95	**14,74**
T1 (42 m^2)	1605	1124	96 000	9,13	
T2 (111 m^2)	6038	4227	195 000	24,15	

The calculation of the net profitability performed consisted firstly in the collection of databases of examples available in the housing sale and rental market (T0, T1, T2, T3 and T4 typologies), provided by [13], and examples of tourist accommodation (T0, T1, T2), provided by [14].

Next, the local accommodation values were converted, from which a percentage of 30% of the value per month was withdrawn, since most of them use the services of an outside company for maintenance and customer service purposes. and cleaning the fraction. This value can vary between 25 to 30%, but it was decided to deduct the maximum value to understand the most unfavourable situation, and the purpose will be to understand the best way so that there are no losses for the owner.

Finally, an online simulator [15] was used to calculate the net profitability, changing the income tax rate values (28% for renting and 35% for tourist accommodation).

In relation to the used area, the tourist accommodation also allows the insertion of more users in less area, of approximately 50%. To obtain the same, the average area of housing in the Quarteirão da Bolsa in rental situation was calculated. For this purpose, the floor area of each housing multiplied by the number of floors was used, based on data available in [14].

A tourist tax of 2 euros/day is also applied, which is normally dissolved in the value of the accommodation paid by the tourist.

There is, however, an aggravating factor for owners who practice the local accommodation service with regard to capital gains, and their taxation increases from 50% to 95% in case of sale, but they apply equally even when there is just a change from the use of local accommodation to another (example: housing) without selling the property. Therefore, the reformulation of uses regarding existing local accommodation is significantly burdened. In terms of uses, the Bolsa Block is more balanced in terms of the percentage of "rented housing", "tourist accommodation", and "inhabited buildings", which is not the case of the Chã Block, where a more balanced distribution of uses is not anymore possible to consider. The conversion of the tourist accommodation already implemented in this block will be very costly due to the investment made and the already mentioned tax penalization.

The intervention proposal relies on the concepts of: participative architecture; increased flexibility, both spatial, as well as uses and users; self-construction, reintroduced with a new approach; as well as an economic approach providing access to all social classes.

The implementation of participative architecture was initiated with the carrying out of surveys, and the initial decision-making process that try to respect the wishes and needs felt by the inhabitants, of course, complemented with technical opinion on the best way to solve it. It does not end there, it is proposed to accompany the residents interested in the design and implementation of the project, in order to have an opinion that is close to their wishes, which may enrich the design with the adjustment to each household needs.

According to the previous premise, the concept of self-construction is also encouraged in two ways: on one hand, to the new owners when occupying unoccupied dwellings, to those who come from countries where it is usual practice, so as to serve as demonstration practice.

Another aspect is volunteer work in conjunction with initiatives such as "Porto Amigo" [16], which attracts volunteers to carry out self-construction actions when residents cannot, for example, elderly people.

Then, but always interconnected with the previous ones, it is intended to explore flexibility in three aspects:

The first is related with the uses, in order to create a balance between the residents, keeping their house, and touristic exploitation, so that owners receive the necessary financial return to allow the rehabilitation and maintenance of the building, but still preserving their tenants at affordable prices, and the amenities necessary for the proper functioning of the block, since the circulation of cars and transport is hampered. This balance prevents gentrification.

The second strand comes as a consequence of the first, and falls on the flexibility of the users of the block, which varies between new resident families, elderly people who are already integrated, students and young people starting their professional life and tourists.

The third aspect refers to the flexibility of the building's interior and exterior, allowing the articulation of all these uses, and the optimized use of the useful space, since it is scarce.

Finally, but implied in the other proposals, it is necessary to think about the entire development of the project according to the economic capacity of each household, so that all social classes can have the opportunity to integrate the blocks, preventing social exclusion.

With the application to the exemplary case, it is clear that the flexibility of the proposal is necessary, as it has to adapt to each specific case, for example, in the general context of the block, with regard to its uses, it is needed to understand which uses are already being implemented in each building, in order to maintain them, whether rented, own property or tourist accommodation, and so the rest is where the proposal can be implemented, and it is intended to be a guide to follow and not a review of the existing one.

Thus, in buildings that were previously only intended for housing, the integration of new families is maintained, on rent basis, stipulating a base income so that there is a balance between the upper, middle and lower classes.

In buildings where part of the development was previously intended for services, it is intended to resume this use on the ground floor (as already observed), given the lack of services indicated by the residents, in the surveys carried out, and by the analysis of local amenities in fault.

In cases where the integration of accommodation for the upper class is not possible, as it is already fully occupied, the improvement would depend on the state, which would benefit the owners in this situation, for example with the exemption from property taxes throughout the period in which the building served the use of rented housing to lower income households, in this way the owners would continue to have sufficient profit margin to maintain the existing uses and would not feel penalized.

In a closer analysis of the building, it is possible to define a base strategy for the distribution of the uses, and whenever there are services in the building, they will be located on the lower floor, then from the dwelling to the lower, middle and lower levels. higher for the high stratum, due to the possibility of benefiting from better views over the city, better sun exposure conditions and privacy, this to justify the increase in the income value.

There is another condition of uses distribution, only applied to the parcels that have a patio, which allow the temporary expansion of the building's living space. This space, which can be developed in 20% of the street area (according to the Municipal Regulating Plan), is intended for shared use by all building users, when the ground floor area is sufficient to include at least one dwelling or services. And it is for private use when the ground floor area is smaller than that necessary for the integration of a dwelling or service. This expansion is carried out through a retractable system that closes the space when necessary (mainly winter) and allows its opening to enjoy the public exterior space. This strategic application allows an optimized use of the existing space, complemented by a flexible interior design and promotes the integration between the building occupants. The disposition of uses allows a fair balance, both for residents and owners, which allows a net income that supports the integration of lower social classes, as well as the permanence of existing uses so that, on the one hand, it does not impose the exile of more people to the suburbs, as well as not harming the owners by trying to keep them in their house.

3 Analysis of Results

From this study, through the surveys carried out, calculation of net profitability and in situ analysis, conclusions are reached on what is necessary to intervene, and in what way.

Finally, we tried to verify whether the strategy would work, in percentage terms, by calculating the index of population density, diversity of uses and social diversity, comparing the current with the proposed for each block under study.

For this, in the case of the population density index, the average number of inhabitants per accommodation was taken from the PORDATA website [17], obtaining the value of inhabitants per m^2, which divided by the total area of the block results in the population density index.

In the case of the diversity index, was carried out a search for the index that better fits the context approached, in which the following indices were considered: Margalef's Diversity, Gleason's Diversity, Menhinick's Diversity, McIntosh's Diversity, Simpson's Diversity and Shannon-Wienner Diversity.

Of these, those that were considered most appropriate to the case study were the Simpson and Shannon-Wienner indexes, which were found to be quite similar. However, the difference between the two is that the Shannon-Wienner index has a broader use, measuring the order and disorder observed in a system [18], and that is why it was chosen to calculate the diversity index.

To calculate the Shannon-Wienner index, the various use or social categories are accounted for, and the formula is applied, which consists of subtracting the amount of fractions from category A by the total fractions of all categories, multiplying by the logarithm of the proportion of categories in relation to the total number of categories and subtracting the previous one again. After applying the formula to the various categories, they are added together, obtaining the diversity index. This index was obtained through a calculation tool available in [19].

Thus, it proceeds to the enunciation of a strategy, for an exemplary block, which relies on a participative architecture, proposing a concept of flexibility, both spatial, as well as of uses and users, which tries to integrate lost concepts of self-construction, reintroduced with a new approach, as well as an economic thinking of the proposal providing accessibility to all.

In this way, what at this moment is the distrust and disinterest in the other, felt and practiced by the local inhabitants, is reverted to an increasingly solidary and united community, given that everyone's interests will be receiving due attention.

Bolsa, Virtudes and Chã that at the moment were blocks with a slight unbalance of social classes, become blocks with the integration of various types of households, from various age groups, social classes and nationalities, as shown in Fig. 3a, with the calculation of the social diversity index (going from 0.850 to 1.110) always taking into account the preservation of residents who are already residents and the uses already inserted. In Cardosas Block the improvement was more radical, as the initial unbalance was absolute, all the occupied dwellings were high standard (Initial Social Diversity index of 0).

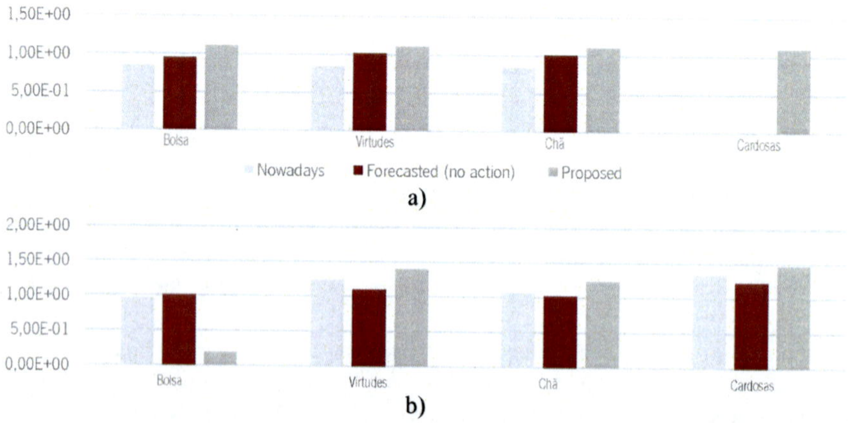

Fig. 3. Diversity indexes evaluated for the analysed blocks: a) social, and b) uses (Adapted from [12]).

Cardosas is at this moment a block with a lack of services and commerce dedicated to basic necessity goods and an excess of services and commerce of luxury goods. The first intervention proposed for this block is to implement the amenities in fault, which are required by the residents. In this specific case of lack of amenities, there are other services, which were mentioned by elderly residents that would be important to consider, such as: seamstress, shoemaker, drugstore, grocery store, among others, as within the blocks there is no place for implementing large supermarkets. To verify the uses diversity, the Shannon-Wienner diversity index was used, from which the following results were obtained for the Chã Block: 1.050 for the results of what currently exists and 1.190 for what is proposed (Fig. 3b). However, for reasons of adjustment of the index to the reality verified in the block, it was decided to place the categories of rented housing and own housing separately, since this index takes as a principle that the categories should be equivalent to each other. In addition to the comparison between what is currently verified and what is proposed, a forecast was also made for a future in which the strategy is not applied, which means a decrease in the density of uses, in the case of this block, however there is always an improvement by implementing the strategy.

For what is currently seen as very degraded or abandoned houses, its rehabilitation is proposed with a sense of flexibility of spaces that allow the integration of various uses and the optimization of existing areas in order to provide greater profitability to the owner, so that the latter, in turn, makes available and is willing to integrate and maintain the lower social classes, without being penalized for this.

4 Conclusions

The concept of community is important when the objective is to implement strategies that try to benefit everyone, both residents, owners and tourists, so that the economic and social development of the city can be possible. This paper presents a proposal of urban rehabilitation based on uses and social diversity, using 4 case study blocks located in the City centre of Porto, Portugal. The study of the blocks allowed to have a perception of the state of the buildings, uses, occupations, social relations, in order to understand the most appropriate strategy to be implemented. The final objective of this proposal is to: allow tourists to enjoy the city centre, with the benefit of sufficient accommodation for their stay; promote the integration of local commerce and services of prime necessity goods, allow owners to keep their investments yielding; while enabling access to housing for residents of all classes. Keeping the genuine identity of the city is only possible if the conservation of its built heritage is also accompanied by the enhancement of its intangible heritage.

References

1. UNESCO: Historic Centre of Oporto, Luiz I Bridge and Monastery of Serra do Pilar. https://whc.unesco.org/en/list/755/. Accessed 26 Aug 2021
2. Gusman, I., Chamusca, P., Fernandes, J., Pinto, J.: Culture and tourism in Porto city centre: conflicts and (im)possible solutions. Sustainability **11**, 5701 (2019). https://doi.org/10.3390/su11205701

3. Branco, R., Alves, S.: Urban rehabilitation, governance, and housing affordability: lessons from Portugal. Urban Res. Pract. **13**(2), 157–179 (2020). https://doi.org/10.1080/17535069.2018.1510540
4. Koens, K., Postma, A., Papp, B.: Is Overtourism overused? Understanding the impact of tourism in a city context. Sustainability **10**, 4384 (2018). https://doi.org/10.3390/su10124384
5. Hamnett, C.: Gentrification and the middle-class remaking of inner London, 1961–2001. Urban Stud. **40**(12), 2401–2426 (2003)
6. Zukin, S.: Gentrification: culture and capital at the urban core. Ann. Rev. Sociol. **13**, 129–147 (1987)
7. Helms, A.C.: Understanding gentrification: an empirical analysis of the determinants of urban housing renovation. J. Urban Econ. **54**(3), 474–498 (2003)
8. Land, M.V.D., Curley, A., Eijk, G.V.: Gentrification and neighbourhood change. Int. Encycl. Hous. Home **2**, 275–279 (2012)
9. Garau-Vadell, J.B., Gutierrez-Taño, D., Diaz-Armas, R.: Economic crisis and residents' perception of the impacts of tourism in mass tourism destinations. J. Destin. Mark. Manag. **7**, 68–75 (2018)
10. Harrill, R.: Residents' attitudes toward tourism development: a literature review with implications for tourism planning. J. Plan. Lit. **18**(3), 251–266 (2004)
11. Ko, D.-W., Stewart, W.P.: A structural equation model of residents' attitudes for tourism development. Tour. Manag. **23**(5), 521–530 (2002)
12. Soares, A.: Participated architecture in the historical center of Porto. Integrated Master Thesis in Architecture, University of Minho, Guimarães, Portugal (2018)
13. Imovirtual. https://www.imovirtual.com/. Accessed 26 Aug 2021
14. Booking. https://www.booking.com/. Accessed 26 Aug 2021
15. COMO FAZER: Simulador para cálculo de Rendimento de Arrendamento de Imóveis. http://comofazer.pt/simulador-para-calculo-de-rendimento-de-arrendamento-de-imoveis/. Accessed 20 June 2018
16. Porto Amigo. https://coesaosocial.cm-porto.pt/pessoas-idosas/porto-amigo. Accessed 20 June 2018
17. PORDATA. https://www.pordata.pt. Accessed 02 Oct 2018
18. Kajtazi, B.: Measuring Multifunctionality of Urban Area, 1st edn. ITC, Enschede (2007)
19. Easycalculation. https://www.easycalculation.com/statistics/shannon-wiener-diversity.php. Accessed 20 June 2018

Check for updates

Spatial Influencing Factors of Traditional Villages in Southern Zhuozhang River Basin

Shuai Sun(✉), Debin Gao, and Ziying Li

North China University of Technology, Beijing 100144, China

Abstract. Through historical ancient books and on-site exploration, this paper collected the folk custom remains, Buddhist buildings, and ancient business roads, ancient smelting and mining sites and ancient battlefields of 27 traditional villages in the Zhuozhang River's Nanyuan River Basin in Shanxi Province, forming an impact. A historical dataset of the spatial influence of traditional villages in Zhuozhang Henan. Collect the elevation, aspect and slope of the village location through GIS, and analyze the correlation degree that affects the village space together with the data set. Provide new influencing factors basis for the formation and distribution of traditional villages.

Keywords: GIS analysis of traditional villages · Village space · Relativity

The Zhuozhang River Basin in Shanxi Province has a long history of human habitation and is one of the important origins of the Central Plains civilization. Various important cultural relics are scattered in the ancient villages and towns along the Zhuozhang River Basin [1]. They have a long history, are numerous, and are well preserved. They are one of the treasure houses of Chinese civilization.

Although these ancient villages and towns are scattered in space, different in individual forms, and different in development trajectory, they are all affected by the unified geographical, economic, religious, and cultural background of the Zhuozhang River Basin. Starting from the perspective of watersheds, studying the common influencing factors of the formation and evolution of ancient villages and settlements in this area, through the comparison of individuals and groups, and the comparison of horizontal space and vertical time, it is more conducive to reveal the ancient times in this important traditional cultural area [2]. The spatial morphological characteristics of villages and towns [3].

1 Valley Geomorphic Factors

1.1 Terrain

The southern source of the Zhuozhang River Basin is located on the southeast side of the Loess Plateau, surrounded by mountains. The topography is generally high in the north, low in the south, high in the east and west, and low in the central valley, forming the Shangdang Basin. The altitude is between 1500 m and 1800 m. The vegetation coverage in the basin is relatively good, and the forest distribution is mostly

P. Mendonça and N. D. Cortiços (Eds.): ICAMC 2021, LNCE 226, pp. 297–311, 2022.
https://doi.org/10.1007/978-3-030-94514-5_31

concentrated in the west and southeast [4]. The traditional villages in the southern source of the Zhuozhang River Basin have the characteristics of linear distribution relying on the Zhuozhang River Valley and are affected by the natural factors of the characteristic river valley landform environment. The basic functions of traditional village settlements, such as residential life, agricultural production, material trade, etc., are also restricted by the topography of the valley, and will form settlement types such as river valley platform settlements, river valley cliff settlements, and river valley hilly settlements [5].

The elevation distribution map of the traditional villages in Nanyuan of the Zhuozhang River Basin was drawn by Arc GIS software (Fig. 1). According to different altitudes, the topography and landforms should be less than 500 m (extremely low correlation value is 0) 500–900 m (low correlation value is 1), 900–1200 m (medium correlation value is 2), and greater than 1200 The meter (high correlation value is 3) is divided into plain areas, low mountain areas, middle mountain areas, and high mountain plateau areas, and the number of traditional villages in each range is counted (Table 1).

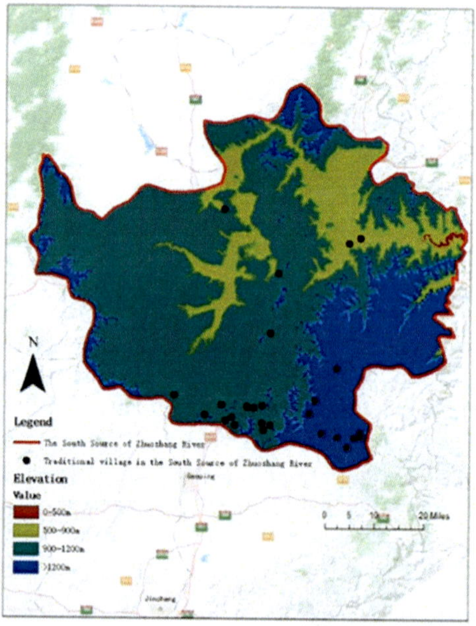

Fig. 1. Elevation distribution map of traditional villages

The results show that the 27 traditional villages in the southern source of the Zhuozhang River Basin are all located in mountainous areas, with an average elevation value of 1084 m, which is slightly lower than the average elevation of the southern source of the Zhuozhang River Basin. Most of the traditional villages are located in the Zhongshan area between 900–1200 m. Among them, 2 villages are located in low mountain areas with an altitude of 500–900 m, accounting for about 7.5% of the total number of traditional villages in the basin, and 8 villages are located in high mountain areas with an altitude of more than 1,200 m, accounting for approximately 30% of the total number of traditional villages in the basin. The complexity of the topography of the mountain area makes the outside influence less on the village, and provides a natural protective barrier for the village, so that it can form a unique customs and culture and continue to develop.

1.2 Aspect Gradient

The amount of sunlight and precipitation that traditional villages can receive is affected by the aspect of the slope. According to the aspect value, the aspect can be divided into eight directions, namely the north slope (0°–22.5°, 337.5°–360°) and the northeast slope (22.5°–67.5°), east slope (67.5°–112.5°), southeast slope (112.5°–157.5°), south slope (157.5°–202.5°), southwest slope (202.5°–247.5°), west slope (247.5°–292.5°) and northwest slope (292.5°–337.5°). According to the aspect value, the aspect can be roughly divided into shade and sun. The aspect values between 0°–90° and 270°–360° are shade slopes, and the aspect values between 90°–270°. The value is a sunny slope (Fig. 2).

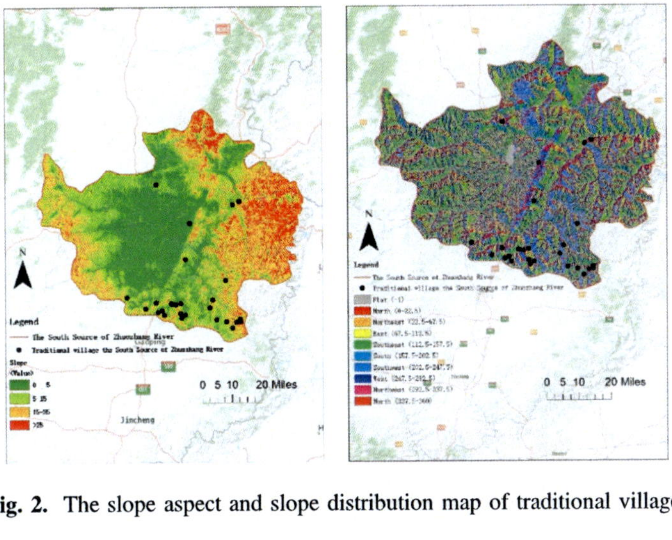

Fig. 2. The slope aspect and slope distribution map of traditional villages

The slope also has an important influence on the site selection of traditional villages. Generally speaking, the larger the slope value, the more difficult it is to build the village. The smaller the village is, the village is easier to defend against aggression and is more secure. The smaller the slope value, the convenient transportation in the village and the larger the village scale. This chapter divides the slope into flat slopes less than 5° (low correlation value is 1), gentle slope 5–15° (medium correlation value is 2), and medium slope 15–25° (high correlation value is 3). The grade is analyzed and summarized (Table 1).

In the Zhuozhang River Basin, 12 traditional villages in Nanyuan are on sunny slopes and 15 are on shady slopes. Among them, the number of traditional villages in the northwest slope is the largest. It is different from the traditional Chinese village site selection concept, which is backed by mountains and waters and shaded to the sun. There may be two reasons: First, because the southern source of the Zhuozhang River Basin was the southern gate of the ancient Shangdang area, it occupies an important strategic position. Due to military defense factors, it is necessary to build villages in the border area on a shady slope. Second, the main crops in the southern source of the Zhuozhang River Basin, prickly ash and corn, are crops that are light-loving, and the villagers are more willing to reclaim the precious sunny slope land into farmland.

1.3 River Water Source

The Zhuozhang River Basin has a semi-humid climate, with hot, rainy and concentrated summers. Villages are usually formed on alluvial fans of rivers and riparian terraces, showing a "near-water" distribution. On the one hand, the reason is that the terrain is flat and the villagers' production and life are convenient to get water. On the other hand, water transportation is a convenient mode of transportation in mountainous areas (Fig. 3).

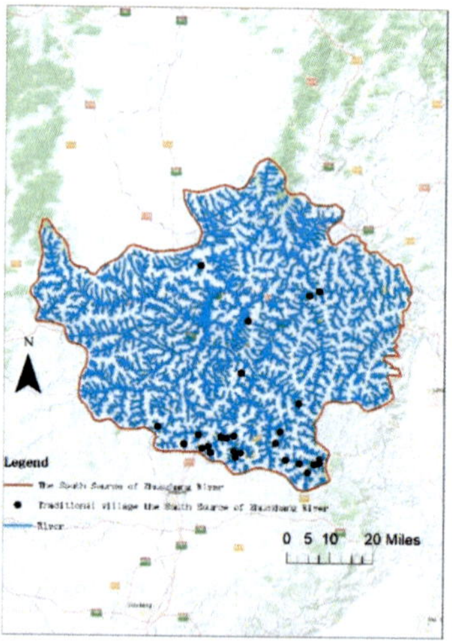

Fig. 3. Relationship between traditional villages and rivers

Use Arc GIS software to get the distance between each traditional village in the basin and the river. The correlation between each village and the river is scored according to the distance. The distance to the river is 0–150 m (high correlation value is 3), 150–650 m (medium correlation value is 2), and greater than 650 m (low correlation value is 1) (Table 1).

2 Economic Development Factors

2.1 Mineral Resources

The Zhuozhang River Basin is rich in coal and iron mineral resources, and there are a large number of iron smelting sites, copper smelting sites, coal mines, and iron ore factories. According to the "Jincheng County Chronicles", the use of coal was as extensive as salt, coal was an essential fuel for people's daily life, and ironware was an important commodity in the market (Fig. 4).

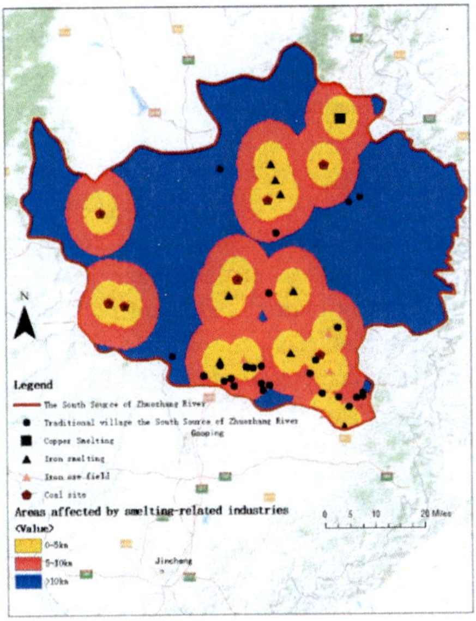

Fig. 4. Distribution map of mineral resources

Calculate the distance between each village and mineral resources, and score the correlation between each village and mineral resources. Among them, the distance to the mineral resource point or smelting point is 0–5 km as high correlation (the value is 3), 5–15 km is medium correlation (value 2), and more than 15 km is low correlation (value 1) (Table 1).

22% of the villages are within 5 km of the iron ore, 56% of the villages are within 5–15 km of the iron ore, and 22% of the villages are 15 km away. The distance to the mines is the closest. The distance between Sangzi No. 1 and Sangzi No. 2 villages is only about 2 km from the iron ore, and the distance between Juzhai Village and Yincheng Village and the iron ore is about 5 km. They are important distribution centers for iron smelting and iron sales in Luzhou center.

2.2 Lu Shang Business

Lushang business refers to the merchants who operated coal, iron, silk, salt and other resources in the Lushang Mansion (now Changzhi City, Shanxi Province) in ancient times. The Zhuozhang River Basin is an important trade channel connecting Shanxi, Henan and Hebei provinces. As a pioneer of Shanxi merchants, Lu merchants in the Ming and Qing Dynasties played an important role in regional trade activities. The location of the village and the layout of the buildings will revolve around ancient commercial activities. One is that iron smelting and silk production are the main industries in the village, and handicraft buildings such as blacksmith shops and handicraft workshops will be formed in the village. The second is that the large commercial households in the village have strong economic strength, and the houses built are generally large in number, complex in layout, and conspicuous in location. The correlation between each village and Lushang business is scored according to the Lushang business-related building remains. Among them, more remains are considered as high correlation (value 3), and less remains are considered as medium correlation (value 2) and no remains are considered as low correlation (The value is 1) (Table 1).

2.3 Ancient Commercial Road

Convenient transportation conditions can promote the prosperity and development of the regional economy, and the location of the village near the road is conducive to commercial trade and goods distribution. After sorting out ancient books such as "Manuscripts of Qing History" and "Tongzhi of Shanxi", the ancient business road in the Zhuozhang River Basin was centered on Luzhou, and the road network was more abundant than that in the Ming Dynasty. It was used by merchants in the Qing Dynasty to transport tea, Hedong salt, and iron goods. Important passage for other goods (Fig. 5).

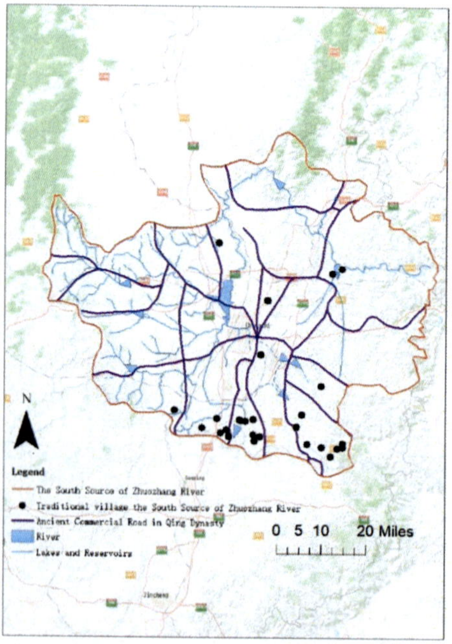

Fig. 5. Traffic map of main roads in the Qing Dynasty

The distance between the villages and the road in the Qing Dynasty was reclassified with a gradient of 3 km, and the distance distribution map between the traditional village and the main road in the Qing Dynasty was obtained. The distance between the traditional village and the ancient road was within 10 km, and about 74% of the villages. The distance to the ancient road is within 5 km, and the closest distance is only 0.5 km. Correlation analysis based on distance: 0–3 km is high correlation (value 3), 3–6 km is medium correlation (value 2), 6–9 km is low correlation (value 1) (Table 1).

3 Social and Human Factors

3.1 Folklore Remains

Folklore is a cultural phenomenon that a social group has lived together for a long time and gradually formed. Folk customs include a wide variety of material life customs (such as folk production customs, folk life customs) and intangible cultural customs (such as folk beliefs, festivals and customs, etiquette systems). Traditional villages, as people's living places, their layout, residential buildings, and village culture will all be affected by the regional folk ritual system. Through data sorting and on-site investigation, it is found that many architectural relics under the influence of the regional folk ritual system remain in the village. And cultural heritage. According to the number of folk custom remains, each village and its correlation are scored. Among them, many remains are high correlation (value 3), more remains are medium correlation (value 2), and few remains are low correlation (value 1) (Table 1).

3.2 Markets and Temple Fairs

A bazaar is a form of activity where people gather regularly for commodity transactions, and is a traditional form of rural trade organization in China. The bazaar usually also contains religious activities, festivals, commemorative activities, and folk entertainment. The local residents use the market to solve the actual needs of daily production and life. Most of the 27 traditional villages in the southern source of the Zhuozhang River Basin are located within 15 km of the market. Among them, there are 8 villages within a range of 0–5 km, 10 villages within a range of 5–20 km, and 4 villages within a range of 10–15 km.

Temple fair customs are closely related to the religious activities of Buddhism and Taoism. At the same time, it also has folk belief activities. There is a record in ancient books that "no temple cannot make a village, there must be a meeting if there is a temple". The temple fair relies on Buddhist ancestral temples, Taoist temples, archways, and guild halls as village nodes. The temple fairs are laid out linearly along ancient streets and ancient commercial roads as a whole, and pay attention to traditional handicrafts and folk activities. Temple fairs can not only promote the development of the village economy, but also strengthen the cohesion and consciousness of rural residents. There are 2 villages within a range of 0–5 km from the temple fair, 6 villages within a range of 5–15 km from the temple fair, 4 villages within a range of 15–20 km from the temple fair, and 4 villages within a range of 15–20 km from the temple fair. 15 villages above 20 km (Fig. 6).

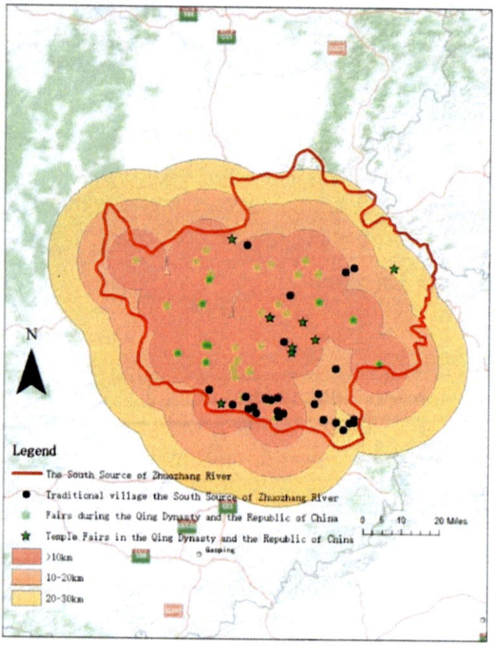

Fig. 6. Distribution map of fairs and temple fairs in the Qing Dynasty and the Republic of China

Calculate the distance between each village and the nearest market and temple fair through ArcGIS. 0–10 km away from the market and temple fair is high correlation (value 3), and 10–20 km is medium correlation (value 2). 20–30 km is low correlation (the value is 1). It can be seen from the graph data that most villages are not far away from the market, and there are regional market activities in Yincheng Town, Xihuo Town, Jiazhang Town, and Southern Song Township in Shangdang District. The distance between the village and the temple fair is farther (Table 1).

3.3 Buddhist Architectural Influence

The Shangdang area in the southeastern part of Shan xi Province has been a hub for the spread of Chinese Buddhist culture from northwest to southeast since ancient times. Located in Pingshun, Lucheng and other counties of Changzhi City, the Zhuozhang River Valley, known as Luchuan in ancient times, has preserved numerous Buddhist buildings that have existed for more than a thousand years since the Tang Dynasty. The common people have good expectations for life, such as harvesting crops, prospering family business, and living in peace. Their spiritual life depends on the blessing of the Buddhist gods, so important Buddhist buildings are usually the core space of the spiritual life of the residents of ancient villages and towns (Fig. 7).

The number of Buddhist remains is scored for the correlation between each village and Buddhism, where there are more remains as high correlation (value is 3), less remains are middle correlation (value 2), and no remains are low correlation (value is 1) (Table 1).

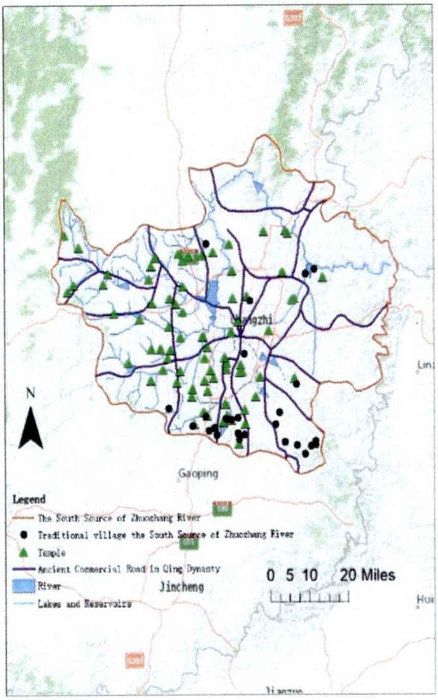

Fig. 7. Distribution map of Buddhist buildings

4 Factors Affecting Military Activity Sites

Remains of ancient battlefields, ancient passes, ancient marching routes, and Tibetan soldiers caves are also one of the historical factors that have influenced the spatial form of ancient villages and towns in the Zhuozhang River Basin. The southeast of Shanxi Province, as a critical and dangerous place, has had many large-scale battles in history. It has always been a place of military disputes. The ancient villages and towns in the area have always had defensive functions, forming defenses relying on topography and geomorphology, relying on street pattern defense, and relying on the ancient village defense space system of building courtyard defense (Fig. 8).

Luzhou (now Changzhi, Shan xi) is a battleground for military strategists. There are many ancient battlefields in the Zhuozhang River Basin. Therefore, the surrounding villages pay more attention to the defense function in terms of village morphology and site selection. The correlation analysis is carried out by counting the distance between the village and the ancient battlefield. The distance 0–10 km is high correlation (value 3), the distance 10–20 km is medium correlation (value 2), and the distance is 20 thousand. Above meters is low correlation (value 1) (Table 1).

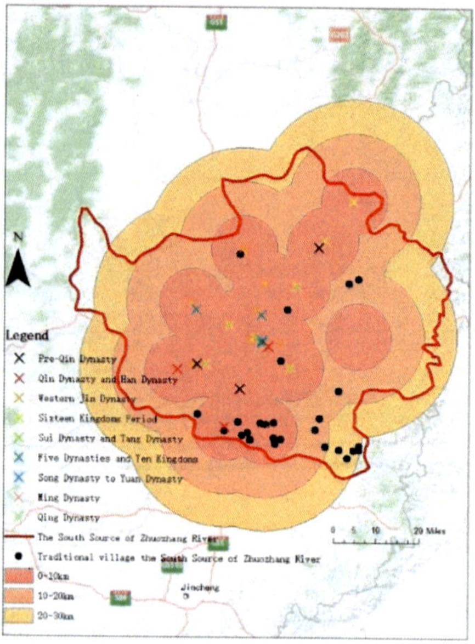

Fig. 8. Spatial distribution analysis of battlefields in the southern source of Zhuozhang River Basin

5 Comprehensive Analysis

Through the analysis of the spatial morphological characteristics of the ancient villages and towns in Nanyuan of the Zhuozhang River Basin under the influence of four major factors including river valley geomorphic factors, economic development factors, social and human factors, and military activity sites, the correlation scores of various factors can be obtained (Table 1).

According to 0–1, it is divided into low correlation, 1–2.2 is divided into medium correlation, and 2.2–3 is divided into high correlation. According to the average correlation of each influencing factor, it can be seen from the analysis of secondary factors: terrain slope, Buddhist remains are low Relevant factors; topographical elevation, river system, s melting industry, ancient business roads, fairs and temples are medium-related factors; Lushang, folk activities, and ancient battlefield sites are high-related factors. Analyzed from the first-level factor indicators: valley geomorphic factors, economic development factors, and social and human factors are all moderately correlated, and valley geomorphic factors have a relatively low impact on the village space, while economic and human factors have a relatively high impact on the village space. Military factors are highly correlated, which is related to the high-density distribution of ancient battlefield sites in the region.

According to the analysis of the total relevance scores of the villages, the average relevance score of the 27 sample villages is 21.5. Therefore, less than 20 are divided into low-relevant villages (6), 20–22 are divided into middle-relevant villages (1), and 22 points or more. Highly relevant villages (10). According to statistical analysis, the overall spatial formation of traditional villages in Nanyuan of the Zhuozhang River Basin is related to the selected four factors. This study demonstrates the effective influence of related factors on traditional villages (Fig. 9) (Table 2).

Through the regression analysis of SPSS, Correlation of elevation values, Slope correlation, River distance correlation, Relevance of smelting industry, Relevance of Lushang Relics, Correlation of Ancient Commercial Road, Relevance of Folklore Relics, Relevance of fairs and temple fairs, Buddhism relevance, Ancient Battlefield correlation is used as the independent variable, and Total correlation is used as the dependent variable for linear regression analysis. As can be seen from the above table, the model formula is: Total correlation = 0.005 + 1.004 * Correlation of elevation values + 0.997 * Slope correlation + 1.000 * River distance correlation + 0.997 * Relevance of smelting industry + 1.003 * Relevance of Lushang Relics + 0.995 * Correlation of Ancient Commercial Road + 0.993 * Relevance of Folklore Relics + 1.002 * Relevance of fairs and temple fairs + 1.000 * Buddhism relevance + 1.003 * Ancient battlefield correlation, The R-square value of the model is 1.000, which means that the independent variable can explain in 100.0% of the change in the dependent variable. When the F test was performed on the model, it was found that the model passed the Ftest ($F = 7759.065$, $p = 0.000 < 0.05$), which means that the model construction is meaningful. The final specific analysis shows that: The regression coefficient value of Correlation of elevation values is 1.004 ($t = 38.081$, $p = 0.000 < 0.01$), which means that Correlation of elevation values will have a significant positive influence on the total correlation.

The regression coefficient value of Slope correlation is 0.997 ($t = 58.589$, $p = 0.000 < 0.01$), which means that Slope correlation will have a significant positive influence on Total correlation. The regression coefficient value of River distance correlation is 1.000 ($t = 59.601$, $p = 0.000 < 0.01$), which means that River distance correlation will have a significant positive influence on total correlation. The regression coefficient value of Relevance of smelting industry is 0.997 ($t = 55.609$, $p = 0.000 < 0.01$), which means that Relevance of smelting industry will have a significant positive influence on Total correlation.

The regression coefficient value of Relevance of Lushang Relics is 1.003 ($t = 45.605$, $p = 0.000 < 0.01$), which means that Relevance of Lushang Relics will have a significant positive influence on Total correlation.

The regression coefficient value of Correlation of Ancient Commercial Road is 0.995 ($t = 36.570$, $p = 0.000 < 0.01$), which means that Correlation of Ancient Commercial Road will have a significant positive influence on Total correlation.

The regression coefficient value of Relevance of Folklore Relics is 0.993 ($t = 43.070$, $p = 0.000 < 0.01$), which means that Relevance of Folklore Relics will have a significant positive influence on the total correlation.

The regression coefficient value of Relevance of fairs and temple fairs is 1.002 ($t = 25.173$, $p = 0.000 < 0.01$), which means that Relevance of fairs and temple fairs will have a significant positive influence on the total correlation.

The regression coefficient value of Buddhism relevance is 1.000 ($t = 52.621$, $p = 0.000 < 0.01$), which means that Buddhism relevance will have a significant positive influence on total correlation.

Table 1. Relative scores of various factors in the village

Village name	Valley geomorphic factors			Economic development factors			Social and human factors			Military factors	Total correlation
	Correlation of elevation values	Slope correlation	River distance correlation	Relevance of smelting industry	Relevance of Lushang Relics	Correlation of ancient commercial road	Relevance of Folklore Relics	Relevance of fairs and temple fairs	Buddhism relevance	Ancient battlefield correlation	
Sangzi village	2	1	3	3	3	2	3	2.5	3	3	25.5
Sangzi second village	2	1	2	3	3	2	3	2.5	3	3	24.5
Xidui village	2	1	3	2	3	2	3	2	3	3	24
Donghuo village	2	2	2	2	3	3	2	2	1	3	22
Pingjiazhuang village	2	1	2	2	2	2	3	2	2	3	21
Zhangjiagou village	2	2	3	2	3	1	3	3	3	3	25
Taiyizhang village	2	3	2	2	3	2	3	2.5	3	3	25.5
Zhao village	2	2	3	2	3	3	3	2.5	2	3	25.5
Dongqili village	3	2	2	3	2	2	2	2	2	2	22
Hedong village	3	1	2	2	2	2	2	1	1	1	17
Shuzhang village	3	2	3	2	2	1	3	1	2	1	20
Dahui village	3	2	2	2	3	2	2	1	2	1	20
Zhaishang village	2	1	1	2	2	3	3	2.5	3	3	22.5
Juzhai village	2	1	2	3	2	3	2	2.5	1	3	21.5
Nansong village	2	1	2	2	2	3	2	2.5	1	3	20.5

(continued)

Table 1. *(continued)*

Village name	Valley geomorphic factors			Economic development factors			Social and human factors			Military factors	Total correlation
	Correlation of elevation values	Slope correlation	River distance correlation	Relevance of smelting industry	Relevance of Lushang Relics	Correlation of ancient commercial road	Relevance of Folklore Relics	Relevance of fairs and temple fairs	Buddhism relevance	Ancient battlefield correlation	
Xilingdi village	3	2	1	2	2	3	2	1.5	1	2	19.5
Guazhang village	3	2	2	3	3	2	3	1.5	1	2	22.5
Shenbei village	3	3	3	3	2	2	1	1	1	1	20
Nanzhangdian village	2	2	2	1	2	2	3	2.5	2	3	21.5
Xinan village	1	3	3	1	1	1	2	2	2	2	18
Tujiao village	1	1	1	1	2	2	2	2	2	2	16
Bayi village	2	2	1	3	2	2	3	2.5	2	3	22.5
Xiling village	2	2	1	2	2	3	1	3	1	3	20
Zhong village	2	2	1	1	3	3	2	3	2	3	22
Yincheng village	2	1	3	3	2	2	2	2	3	3	23
Fangdai village	2	2	3	2	2	2	2	1	1	1	18
Cuijaizhaung village	3	2	2	2	2	2	2	1.5	1	2	19.5
Mean	2.2	1.8	2.1	2.2	2.3	2.2	2.5	2	1.9	2.4	21.4
Itemized average	2.03			2.23			2.13			2.4	

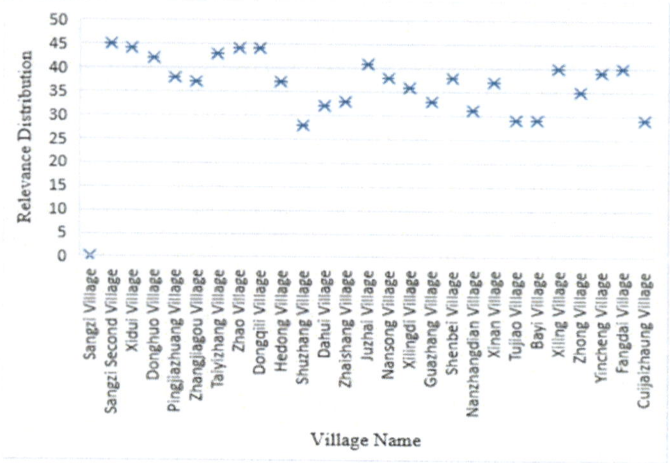

Fig. 9. Overall correlation of spatial influencing factors of traditional villages in Nanyuan, Zhuozhang River Basin

Table 2. Regression test analysis

Parameter estimates (n = 28)									
			Unstandardized coefficients						Standardized coefficients
B	t	p	VIF	R^2	Adj R^2	F		Std. error	Beta
Constant	0.005	0.117	-	0.044	0.965	-	1	1	F = (10, 17) = 7759 P = 0.000
Correlation of elevation values	1.004	0.026	0.224	38.081	0.00	2.679			
Slope correlation	0.997	0.017	0.252	58.589	0.00	1.438			
River distance correlation	1	0.017	0.29	59.601	0.00	1.836			
Relevance of smelting industry	0.997	0.018	0.255	55.609	0.00	1.629			
Relevance of Lushang Relics	1.003	0.022	0.215	45.605	0.00	1.719			
Correlation of Ancient Commercial Road	0.995	0.027	0.239	36.57	0.00	3.313			
Relevance of Folklore Relics	0.993	0.023	0.241	43.07	0.00	2.437			
Relevance of fairs and temple fairs	1.002	0.04	0.251	25.173	0.00	7.716			
Buddhism relevance	1	0.019	0.309	52.621	0.00	2.673			
Ancient battlefield correlation	1.003	0.036	0.308	27.76	0.00	9.578			
Dependent variable: total correlation									
D-W: 1.012									
*p < 0.05 **p < 0.01									

The regression coefficient value of Ancient battlefield correlation is 1.003 (t = 27.760, p = 0.000 < 0.01), which means that Ancient battlefield correlation will have a significant positive influence on Total correlation.

The summary analysis shows: Correlation of elevation values, Slope correlation, River distance correlation, Relevance of s melting industry, Relevance of Lushang Relics, Correlation of Ancient Commercial Road, Relevance of Folklore Relics, Relevance of fairs and temple fairs, Buddhism relevance, Ancient battlefield correlation All will have a significant positive influence on Total correlation. The multicollinearity of the model is tested and found that there are VIF values in the model greater than 5 but less than 10, namely Relevance of fairs and temple fairs and Ancient battlefield correlation are highly correlated.

Acknowledgment. Sponsored by the National Natural Science Foundation of China (51808004).

References

1. Qi, L., Liang, W., Xi, M.: The public space pattern research of Guangfu traditional villages based on spatial syntax: a case study of Huangpu Village in Guangzhou City, China. IOP Conf. Ser. Earth Environ. Sci. **267**(6), 062033 (2019)
2. Trisnawarman, D., Rusdi, Z.: Participatory GIS for collaborative decision making development of village resource potential. IOP Conf. Ser. Mater. Sci. Eng. **508**(1), 012117 (2019)
3. Kozak, I., Balaniuk, I., Szelenko, D., Balaniuk, S., Kozak, H.: Traditional village system – case study from the Krempna Commune (Poland). Ekológia (Bratislava) **38**(1), 87–100 (2019)
4. Priyanka, M.: Morphological analysis of village Bilawal, Badhra block, Charkhi Dadri, Haryana (India). Int. J. Res. Soc. Sci. **8**(5), 352–362 (2018)
5. "The archaeological picture went blank": historical archaeology and GIS analysis of the landscape of the Palestinian Village of Tell eṣ-Ṣâfi. Near Eastern Archaeology (2018)

Ecotown: A Case Study in Tropical Savanna Climate

Ben Cevallos and Javier Urquizo(✉)

Escuela Superior Politécnica del Litoral, ESPOL, FIEC,
Campus Gustavo Galindo, Km. 30.5 vía Perimetral, Guayaquil, Ecuador
{benceval, jurquizo}@espol.edu.ec

Abstract. An ecotown is designed and built under the principles of living with the environment and reducing greenhouse gas emissions through energy production from renewable sources. This research shows the electrical and hydraulic parameters and characteristics of an ecotown located in tropical savanna climate. The design and modelling of electricity systems and the energy analysis of a typical house was carried out considering factors such as lighting, ventilation, geographic orientation, and energy envelopes. On the other hand, the design, dimensioning, and calculation of the underground grids' electrical system in the ecotown that includes centralized photovoltaic generation and the distribution grids connected to public grid were performed in compliance with national and international codes. The design of the potable water and wastewater disposal system must include the grey water treatment and rainwater collection system as to an alternative to reduce the water consumption in the ecotown. Finally, we performed an economic analysis of the distributed generation and grey water treatment processes to determine their profitability for investors' benefit. Implementing an ecotown type urban project was proven to be both technically and economically possible in tropical savanna climates.

Keywords: Ecotown · Grey water · Distributed generation · Wastewater · Underground electrical distribution grid · Microgrid

1 Introduction

The United Nation estimated that by 2020 more than half of the human population lived in cities. This percentage will increase to wo-thirds i.e., 6.5 billion people will live in cities. Understanding that sustainable development not only focus in improving the quality of living but at the same time preserve the physical and environmental assets of the urban spaces. The rapid growth of cities is the result of population growth and migration from rural causing a increase of services, especially in emerging word [1]. Buildings and cities design with a sense of innovation and green economy is known as ecotown, this type of new system has many benefits, among others: efficient land use, habitat preservation and restoration, efficient transport management, efficient use of water, energy efficiency and improvement of the people's wellbeing [2].

Throughout the 1980s and early 1990s, the term ecotown remained as a concept, collection of ideas, and proposals on sustainable urban planning, with few practical

P. Mendonça and N. D. Cortiços (Eds.): ICAMC 2021, LNCE 226, pp. 312–324, 2022.
https://doi.org/10.1007/978-3-030-94514-5_32

examples. Since the mid-2000s, the concept seems to have become increasingly global and trending in recognizing the severity of climate change and energy improvement [3]. Consequently, green ecotown cities are developing in countries where the prevailing climate is cold.

In order to verify that the home is energy efficient, i.e., in the United Kingdom, there are two known standards: the PassivHaus UK standard [4] and the Energy Saving Trust (EST) standards, which can satisfy the Code for Sustainable Homes [5]. Examples are: EST energy efficiency in loft extensions [6], EST improving airtightness in dwellings [7], EST enhanced construction in thermal bridging and airtightness [8] and EST in-situ monitoring of efficiencies of condensing boilers [9]. This research is set about reflecting on and answering a number of preliminary questions that have not been explored in detail before and do not have a definitive answer in relation to the ecotown in tropical savanna climates.

Buildings have a long-life especially if they are built with concrete walls and brick walls as in Guayas-Ecuador. It is, therefore, important to analyse how buildings will response to climate change in the future and assess the possible changes in energy use and the corresponding carbon footprint. Local Authorities (LAs) require a spatial understanding of carbon and end-use energy demands in sub-city areas to provide evidence for future needs in planning local end-use energy services and distributed generation infrastructure. LAs have identified the domestic sector as a priority and developed a growing interest in modelling the energy use of their housing stock at sub-city areas. A priori hypothesis is that climate change would lead I northern countries to less energy use for winter heating and more for cooling during the summer months. The extent of reduction in heating and increase in cooling varies from one region/climate to another and depends in countries like Ecuador on local climate conditions and existing energy efficiency measures [10].

Tropical savannah climate presents high air temperatures throughout the year. Guayas has a geographic location that the design considers architectural bioclimatic strategies to adapt the building to its climate zone, in many periods of the year. It is necessary to use artificial conditioning systems to bring back internal air temperature to the habitability criteria. Which makes it necessary to provide buildings with greater resilience to climate [11].

This research is set about reflecting on and answering several preliminary questions that have not been explored in detail before and do not have a definitive answer in relation to the ecotown in tropical savanna climates i.e., house monitoring, type of materials, renewable energy, water, greywater, blackwater systems and the location.

1.1 Housing

The house ought to consider a real-time energy monitoring system which be accessible to low-income households and have operational facilities as well as services to provide thermal comfort, ensure protection from natural or human-caused environmental hazards.

1.2 Type of Housing Materials

For the selection of building construction materials, the embodied energy, its building's fabric construction, the size of ventilating and air conditioning equipment, and their final disposal or reuse of used-water must be considered [12].

1.3 Renewable Energy

The energy provided to ecotown must not have a significant impact on systems, human health, and climate change. For more than fifty years, nearly everything in our built environment uses chemicals derived from fossil fuels. However, using those materials has a side effect on local air pollution that may cause respiratory problems, the reduction of resources and increase of greenhouse gases to the atmosphere.

1.4 Water

The analysis of the ecotown will be performed according to the Ecuadorian regulations due to the particularity of its geographical location and ensuring that the drinking water is adequate for human consumption and that the disposal of wastewater is carried out in an eco-friendly and efficient manner and without impacts on the ecosystem.

1.5 The Place

Being on the equatorial line Ecuador's location produces little seasonality throughout the year; therefore, there are only two defined seasons: wet or winter and dry or summer. In the Coastal region, the rainy season begins in December and lasts until May, and the dry season takes place between June and November. Its topography also influences the climate. The lower elevation, the higher average temperature. On the other hand, altitude also affects the amount of rainfall because cold air has little ability to retain moisture. Therefore, the highest areas have low rainfall while the low ones up to 2,000 m presents wide variation in the amount of rain throughout the year [13]. The case study is in the coastal region of Ecuador, shown in Fig. 1.

Fig. 1. Satellite image of the location of the ecotown project from Google EarthTM

The ecotown is initially made up of 70% residential, 15% roads, 10% commercial, and 5% recreational areas, as shown in Fig. 2, which gives us a total of 258 lots and 88 commercial parking spaces.

This project develops and applies reliable electrical design and water management methodologies to an ecotown located in arid and dry climates that guarantees the safety, quality, and reliability of sustainable services. This document is written as follows. Section 2 describes the proposed methodology for sustainable design of electrical, drinking water and wastewater disposal systems. Section 3 presents the results. Section 4 discusses the results. Finally, Sect. 5 concludes this document.

Fig. 2. Ecotown architectural scheme

2 Methodology

The methodology implemented for the analysis of this document was based on the following steps:

i. The electrical design of a typical house was carried out to obtain its monthly demand. Also, using the energy modelling tool DesingBuilder (DB) [14], energy consumption measurements were performed.
ii. The ecotown electricity demand [15, 16] was obtained from the individual calculation of the demand of the different stages of the project (residences, public lighting, social club, shopping centres, recreational areas, and potable and wastewater treatment plants)
iii. Applying the techniques for designing underground electrical energy distribution systems, the transformers and feeders of the underground medium and low voltage networks and public lighting were dimensioned. The microgrid scheme was configured with a ring system [17], considering a centralized distributed photovoltaic generation calculated according to the regulations and demand capacity.

iv. The analysis and design of the drinking water and wastewater disposal systems were carried out to size the anaerobic systems used to treat black and grey water [18].

In step i. the monthly demand for standard housing was calculated using standard formulae. Using DB, energy consumption was calculated considering the lighting, air conditioning, geographical location, and comfort of the house. In step ii. the ecotown total electricity demand was calculated using traditional formulae for multiple users. In step iii. National and International standards were used for the design of underground electric power distribution systems. (reference) In step iv. like step iii. for the design of potable water and wastewater disposal systems. This research uses international standards.

3 Results

This section shows the results of the proposed methodology on the case study.

3.1 Monthly Demand Per House

Table 1 shows the results of the monthly demand for a standard 81.75 m^2 standard house. The total daily demand is 23.33 kWh, and the Monthly demand is 759.78 kWh.

Table 1. End-use demand for a standard house

Load type	Watts	Coincidence factor (CF)	kW	Hour/day	kWh/day
Lightning	640	0.5	0.32	4	1.28
Current socket	1,470	0.4	0.59	2	1.18
Refrigerator	400	1	0.40	8	3.20
Washing machine	200	1	0.20	2	0.40
Electric stove	2,880	1	2.88	3	8.64
Electric shower	1,100	1	1.10	0.5	0.55
Air conditioner	3,600	0.75	2.52	4	10.08

3.2 Standard House Orientation, Ventilation, and Cooling Design

The house's modelling in DB showed that the best orientation must be North to South [19]. With this, it avoids direct exposure to the sun in the morning and the afternoon. This is shown in Fig. 3a. The monthly analysis of the typical house, showing the air temperature, radiant temperature, operating temperature, outside dry bulb temperature, relative humidity, and hours of discomfort with natural ventilation and mechanical ventilation, hours of discomfort presented by the simulator are the hours during periods when the humidity level and/or operating temperature are not within some of the comfort zones established in the ASHRAE 55 Standard [20] are shown in Fig. 3b and c. and the data of the cooling design for the standard house [21] is shown in Fig. 3d.

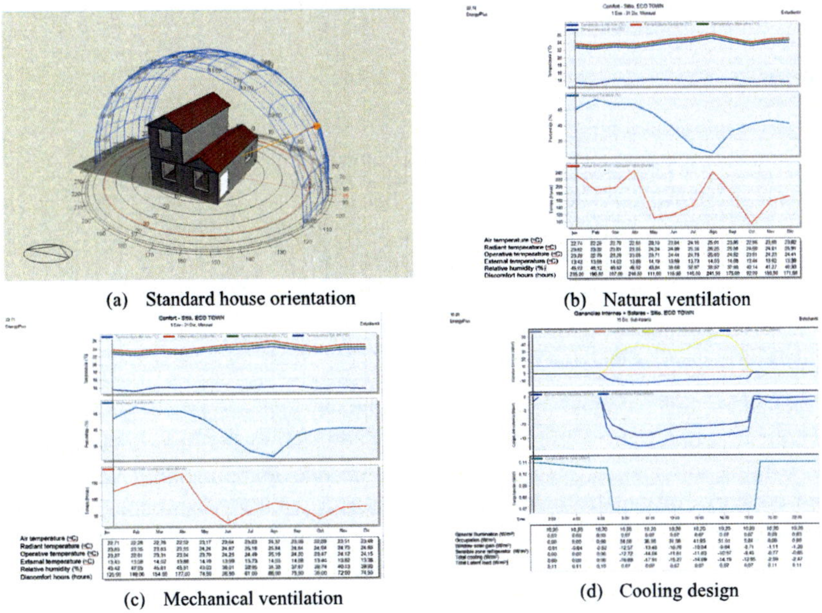

(a) Standard house orientation

(b) Natural ventilation

(c) Mechanical ventilation

(d) Cooling design

Fig. 3. Ecotown comfort data

3.3 Transformer Selection for Underground Distribution

The transformers for underground distribution are shown in Table 2. Household demand is 5.5 kW.

Table 2. Transformers by design

Bock number	Lot number	Partial demand (kW)	CF	Demand kVA	kVA (PF = 0.85)	Transformer capacity (kVA)	Transf. number
B	19	104.5	0.41	43.12	50.73	75	1
C	22	121.0	0.41	49.39	58.11	75	2
D	12	66.0	0.43	28.49	33.52	50	3
E	15	82.5	0.42	34.76	40.89	50	4
F	10	55.0	0.44	24.31	28.60	37.5	5
G	34	187.0	0.40	74.47	87.61	100	6
H	14	77.0	0.42	32.67	38.44	50	7
I	8	44.0	0.46	20.13	23.68	37.5	8
J	26	143.0	0.40	57.75	67.94	100	9
K	20	110.0	0.41	45.21	53.19	75	10
L	18	99.0	0.41	41.03	48.27	75	11
M	12	66.0	0.46	28.49	33.52	50	12
N	9	49.5	0.45	22.22	26.14	37.5	13
O	7	38.5	0.47	18.04	21.22	37.5	14

3.4 Monthly Electricity Demand

The monthly electricity demand is shown in Table 3.

Table 3. Monthly electricity demand

Residential load	Public lights	Security gates	Social club	Retail	Big commerce	Public Parks	Water/waste
520.8	22.0	3.21	7.76	42.0	6.12	30.1	12.0

3.5 Underground Electricity Distribution

The secondary low voltage network is 120/240 V and is shown in Fig. 4a. The underground public light is Ling Emitting Diode (LED) technology with controls for roads, parking lots and recreational areas. The design of the medium voltage underground networks of the Ecotown that corresponds to 13,800 V and consider the centralized generation through photovoltaic panels is shown in Fig. 4b.

(a) Low voltage underground distribution

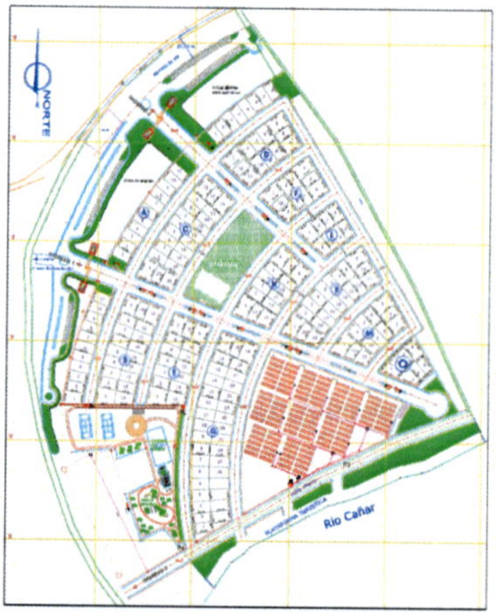

(b) Medium voltage underground distribution

Fig. 4. Undergound electricity distribution

3.6 Single Line Diagram

In this project, the use of 2,592 crystalline photovoltaic panels of 340 Wp was considered, which will be in the urbanization as a photovoltaic generator park, in the blocks called J, L, and N of the urbanization, which occupies an area of approximately 11,632.25 m^2, with the aim of not affecting the homes of the aforementioned blocks, the generating park could be located outside the urbanization on adjacent land, but this would be an additional analysis that is not within the scope of this study. The distributed centralized photovoltaic generation of the micro-grid constituted by bidirectional commercial metering, transformers, switches, and distribution switch, is shown in Fig. 5.

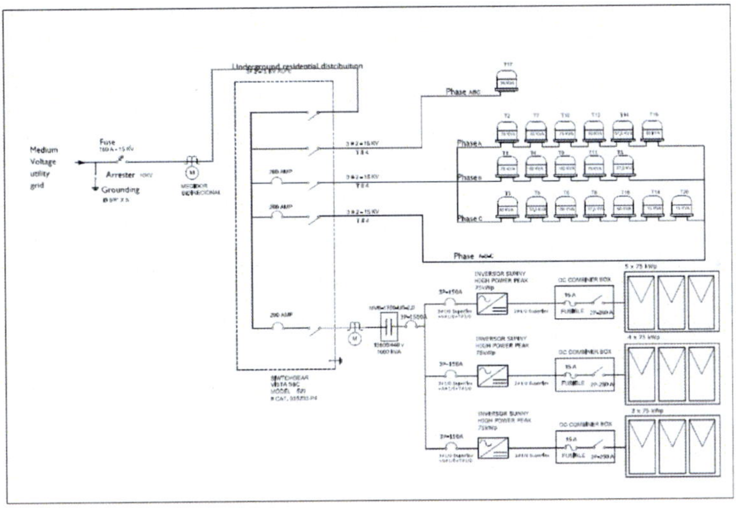

Fig. 5. Single line diagram

3.7 Potable, Black, and Grey Water Systems

The design for the ecotown drinking water system is shown in Fig. 6a and it is composed of the connections, cistern, pumping equipment, distribution network, meters, and household connections. Through a grey water system, the treated water is deposited in a cistern suitable for the consumption of toilets, urinals, cleaning floors, and irrigation of green areas. The water distribution network in the urbanization is shown in Fig. 6b. Through a grey water system, the treated water is deposited in a cistern suitable for the consumption of toilets, urinals, cleaning floors, and irrigation of green areas. The water distribution network in the urbanization is shown in Fig. 7.

3.8 Anaerobic Treatment System

The Anaerobic Treatment System for Blackwater system is made up of the following components: two-chamber septic tanks where a biological sedimentation treatment takes place, where the organic matter decomposes, creating sludge which settles to the bottom and the foams are suspended, then the waters They pass through communicating vessels and gravity towards the upward flow anaerobic filter where the gravel retains microorganisms on its faces and goes up to a clarifier, then another filter of varied gravel passes. Between the septic tank and the anaerobic filter, there is a dry chamber for maintenance. Also, there is a measuring and sampling box according to the designs designed for this system type. The Anaerobic Treatment System for Grey water system is made up of the following components as shown in Fig. 8, one-chamber septic tank where a biological sedimentation treatment takes place, where organic matter decomposes, creating sludge which sediments to the bottom and the foams are suspended, then the waters pass through communicating vessels and gravity towards the upward flow anaerobic filter, where the gravel retains microorganisms on its faces and goes up to a clarifier, then passes to the varied gravel filter. Between the septic tank and the anaerobic filter, there is a dry chamber for maintenance. Also, there is a measuring and sampling box according to the designs showed for this system type.

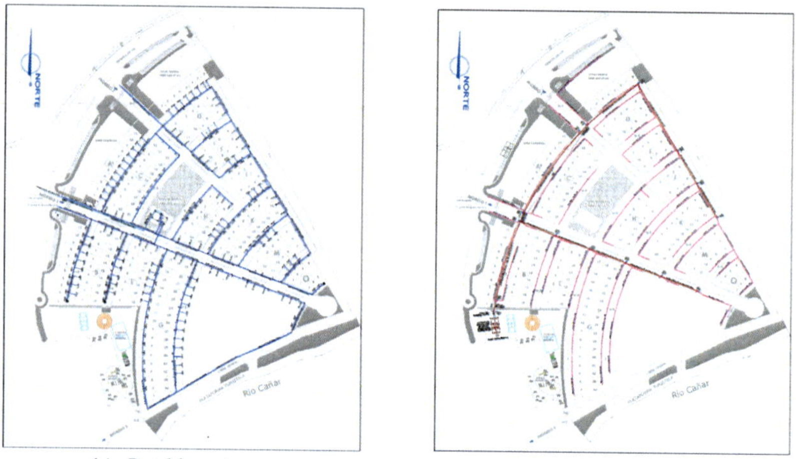

(a) Potable water system (b) Black water system

Fig. 6. Potable and black water systems

Fig. 7. Treated water distribution

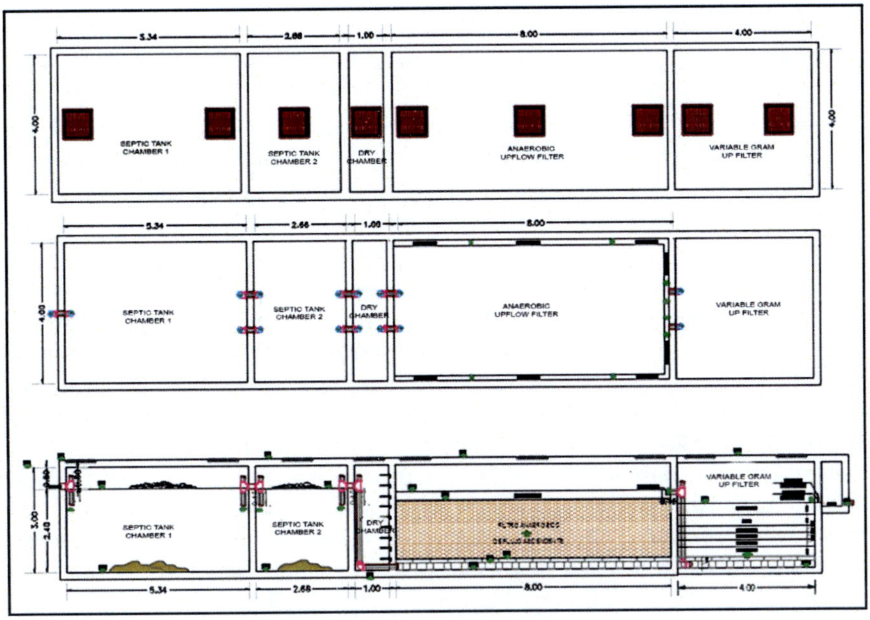

Fig. 8. Grey water anaerobic treatment system

4 Discussion

The study for the development of an ecotown project in tropical savanna climates presents the following elements for the discussion.

The three main influences on the average electricity consumption were found to be: increase in average size of dwelling, the energy intensity of electrical appliances i.e., the increase use of appliances and the social and cultural household activities known as behaviour [22], in Ecuador mainly because of the use of air conditioning systems. Increasing comfort-related preferences result in increased electricity consumption [23].

Decentralised renewable electricity in ecotowns could be supplied not only by utilities or independent local renewable generation sources but also by electricity regulated consumers who are allowed to generate their own electricity supply or to the grid using a feed-in tariff mechanism. Policy and regulatory interventions such as feed-in tariff and net metering schemes would also be required to provide incentives to consumers to invest in renewable energy technologies and be allowed to interconnect to the grid. The National Regulation No. ARCERNNR-002/21 An electric power generation plant that is developed by an Enabled Distributed Generation Company (EDGC) is considered as a distributed generation plant if it meets the following conditions: a) It has a nominal capacity equal to or greater than 100 kW and less than 10 MW; b) It connects close to consumption; c) It is connected to medium voltage or high voltage networks below 138 kV of a distribution system; d) Uses a non-conventional renewable energy source. The plants that are included in the Electricity Master Plan may also use fossil fuels as a source primary energy; e) It is built, operated, maintained, and administered by EDGCs, in accordance with the terms established in this Regulation; and f) Does not include power plants owned by self-generators or groups emergency generators.

A critical area of interest is the effect of voltage regulation in the distribution network with high penetration of renewables. One possible solution is to provide a local storage. For the forecasted demand an appropriate tap setting of the on-load tap changer (OLTC), flexible capacitors or other active and passive reactive compensators must be dispatched.

An adequate solution for human behaviour is to consider controls on lights or even in washing machines especially in the peak hours of a day. If both indoors and outdoors in public places are non-efficient lights, lots of energy is wasted. One possible solution is that the actual time for these streetlights to be switched on should be calibrated to save energy [24], and if additionally, those lights are energy efficient, this could prevent excessive energy wastage, which will also be better for the environment.

Greywater and rainwater reuse is becoming an important factor for potable water saving in emerging countries [25, 26]. In this research rainfall also is used for water irrigation reuse or for firefighting purpose. Considering greywater and rainwater combination shows an increase of potable water efficiency, and even for firefighting purposes. Water reuse on a household level could be realized as service water in case of toilet flushing, water for washing machines, dish washers or irrigation as well. Therefore, it is economically profitable to develop a grey water collection and distribution system.

Toilet waste (blackwater) or urine can be separately collected. Co-treatment of blackwater and organic household waste can be used both using the high calorific property of methane as a fuel for cooking or fertilizer in farms. Source separation (blackwater/greywater) opens for exiting urban sustainability applications. In emerging countries households blackwaters are sent to the ecosystem after to receive anaerobic treatment.

5 Conclusions

The centralized photovoltaic generation project in the eco-town is profitable and guarantees a return on investment in the short term, unfortunately not having local market prices limits us from being able to carry out a more complete and real analysis. The geographical location of Ecuador take advantage of for the development of projects based on non-conventional renewable energies such as photovoltaic energy. The ecotown model can be put into practice in countries with a tropical savanna climate. The Ecuadorian regulations must be updated based on the experiences that can be obtained in the development of similar projects in the region. It is required to have a detailed study on the potential of increasing demand for flexibility in the energy system. Also, technologies that produce a balance between demand and supply provide system flexibility. There is a potential to convert a large amount of excess power from renewables to hydrogen batteries and store it in the system. The study and design of potable water systems and wastewater disposal guarantees that the ecotown will have drinking water necessary for human consumption, the sewage systems will have a treatment process that avoids impacts on the environment, the rainwater will be collected and properly conducted to the discharge points and the grey water will be collected and treated so that it can be used in the irrigation systems of green areas and firefighting systems, in addition to being an economically profitable system and with time we will recover your investment, this research argues that over time national regulations should incorporate it as part of the design of the systems of an urbanization, public or private buildings.

References

1. Simon, D., et al.: Developing and testing the urban sustainable development goal's targets and indicators – a five-city study. Environ. Urban. **28**(1), 49–63 (2015)
2. Trier, C., Maiboroda, O.: The green village project: a rural community's journey towards sustainability. Local Environ. **14**(9), 819–831 (2009)
3. Warwick, E.: Policy to reality: evaluating the evidence trajectory for English eco-towns. Build. Res. Inf. **43**(4), 486–498 (2015)
4. Liang, X., Wang, Y., Royapoor, M., Wu, Q., Roskilly, T.: Comparison of building performance between conventional house and passive house in the UK. Energy Procedia **142**, 1823–1828 (2017)
5. MHCLG: Code for sustainable homes technical guide - code addendum. The Ministry of Housing, Communities and Local Government's (2014)
6. Energy Saving Trust: Energy Efficiency Best Practice in Housing: Energy Efficient Loft Extensions, London (2005). Contract No. CE120
7. Energy Saving Trust: Improving Airtightness in Dwellings, London (2005). Contract No. GPG224
8. Energy Saving Trust: Enhanced Construction Details: Thermal Bridging and Airtightness, 2009 edn, London (2009). Contract No. CE302
9. Energy Saving Trust: The Rise of the Machines - A Review of Energy Using Products in the Home from the 1970s to Today (2011)
10. Wan, K.K., Li, D.H., Pan, W., Lam, J.C.: Impact of climate change on building energy use in different climate zones and mitigation and adaptation implications. Appl. Energy **97**, 274–282 (2012)
11. Invidiata, A., Ghisi, E.: Impact of climate change on heating and cooling energy demand in houses in Brazil. Energy Build. **130**, 20–32 (2016)
12. Desogus, G., Mura, S., Ricciu, R.: Comparing different approaches to in situ measurement of building components thermal resistance. Energy Build. **43**(10), 2613–2620 (2011)
13. Adger, W.N., Arnell, N.W., Tompkins, E.L.: Adapting to climate change: perspectives across scales. Glob. Environ. Chang. **15**(2), 75–76 (2005)
14. Fathalian, A., Kargarsharifabad, H.: Investigating the effect of different energy saving strategies on energy rating of building by design builder software (case study: office building). J. Environ. Sci. Technol. **22**(7(98) #b001410), 199–214 (2020)
15. Bayulken, B., Huisingh, D.: Are lessons from eco-towns helping planners make more effective progress in transforming cities into sustainable urban systems: a literature review (part 2 of 2). J. Clean. Prod. **109**, 152–165 (2015)
16. Keirstead, J., Shah, N.: Urban Energy Systems: An Integrated Approach. Routledge, Oxfordshire (2013)
17. Schmitt, L., Kumar, J., Sun, D., Kayal, S., Venkata, S.M.: Ecocity upon a hill: microgrids and the future of the European City. IEEE Power Energ. Mag. **11**(4), 59–70 (2013)

18. Zuo, J., Yan, Q.: Recycling inter-feed relationship between wetlands' protective using and eco-town construction. Adv. Mater. Res. **361–363**, 1085–1090 (2012)
19. Fernandez-Antolin, M.-M., del Río, J.M., Costanzo, V., Nocera, F., Gonzalez-Lezcano, R.-A.: Passive design strategies for residential buildings in different Spanish climate zones. Sustainability **11**(18) (2019)
20. de Dear, R.J., Brager, G.S.: Thermal comfort in naturally ventilated buildings: revisions to ASHRAE Standard 55. Energy Build. **34**(6), 549–561 (2002)
21. Weber, C., Shah, N.: Optimisation based design of a district energy system for an eco-town in the United Kingdom. Energy **36**(2), 1292–1308 (2011)
22. Urquizo, J., Calderón, C., James, P.: Using a local framework combining principal component regression and Monte Carlo simulation for uncertainty and sensitivity analysis of a domestic energy model in sub-city areas. Energies **10**(12) (2017)
23. Urquizo, J., Calderón, C., James, P.: Modelling household spatial energy intensity consumption patterns for building envelopes, heating systems and temperature controls in cities. Appl. Energy **226**, 670–681 (2018)
24. Singh, M., Aggarwal, A. (eds.): Design and implementation of energy efficient illumination control device. In: 2017 International Smart Cities Conference (ISC2), 14–17 September 2017
25. Fittschen, I., Niemczynowicz, J.: Experiences with dry sanitation and greywater treatment in the ecovillage Toarp. Sweden. Water Sci. Technol. **35**(9), 161–170 (1997)
26. Mourad, K.A., Berndtsson, J.C., Berndtsson, R.: Potential fresh water saving using greywater in toilet flushing in Syria. J. Environ. Manag. **92**(10), 2447–2453 (2011)

Machine Learning to Assess Urbanistic Development in the South Pole of Lima City

Huber Nieto-Chaupis(✉) and Anthony Alfaro-Acuña

Universidad Autónoma del Perú, Panamericana Sur, Km. 16.3, Villa el Salvador, Lima, Perú
anthony.alfaro@autonoma.pe

Abstract. We employ Machine Learning through the Mitchell's criteria to carry out an assessment on the potential spatial configurations at the south pole of Lima city, at Perú. Based at both qualitative and quantitative facts, an model has been proposed that targets to measure the success of spatial expansion of districts based at distances and number of habitants. In this manner Machine Learning appears as a robust tool with capabilities to anticipate the possible achievements as well as issues along the time the city is under spatial growth. The efficiency of sustained growth is measured in terms of success probability. Therefore, we can claim that the ongoing growth of Villa el Salvador engages to some extent the philosophy of Mitchell's criteria.

Keywords: Machine learning · Urban cities · Latin American cities

1 Introduction

At the decade of 70s, Lima city had a noteworthy increase on its perimeter particularly at the north and south poles, as noted also nowadays in [1]. This social phenomenon was done as consequence of the internal immigration of country. In this way the city had a well-defined geometrical morphology as perceived as an inclined rectangle. However, the expansion of south pole has exhibited a edge-like morphology. A district that calls our attention is the one of Villa el Salvador (VES in short), in the which has had an interesting evolution in the following points:

- Number of habitants;
- Urbanistic;
- Transportation;
- Commerce.

In fact, VES has shown to be a vigorous district from all social and economic angles [2]. Nowadays, one finds well sustained arguments to anticipate a further expansion in both housing and habitants from the fact that new technologies, advances in communications as well as economy opportunities have arrived to VES [3]. Under his view we can also expect a solid and sustained growth of habitants and housing as well [4].

P. Mendonça and N. D. Cortiços (Eds.): ICAMC 2021, LNCE 226, pp. 325–337, 2022.
https://doi.org/10.1007/978-3-030-94514-5_33

In this paper, attention is paid to the usage of Machine Learning algorithms [5–8] as a tool to carry out analysis of spatial expansion of VES. Here, some coherent assumptions have been used. Essentially, we expect that such expansion might be gradual in time and satisfying social and economic laws and that are totally supported by urbanistic principles. Precisely, the usage of Machine Learning demands to apply abstracts conceptions of spatial expansion. Thus, more than a predictive tool, it plays the role as a statistical tool that allows to arrive to accurate urban topologies. From this we might to acquire a quantitative knowledge for future expansion and to make a direct relationship with similar cases that have appeared in Lima city but in different districts of opposite pole.

Therefore, emerges the question: **Can Machine Learning be used as a reliable tool to assess urbanistic and people displacement in big cities?** [9–12] In this manner the aim of this paper is twofold: (i) In one hand, to propose a concrete algorithm that enables us to predict elongation spatial, and (ii) to provide an approximated final perimeter of south pole of Lima city as consequence of the expected spatial expansion. In second section the theory of Mitchell's criteria are presented.

In third section, the district of VES is studied through the Mitchell's criteria. Here topologies are presented. Finally, the conclusion of paper is drawn.

2 The Mitchell's Criteria in a Nutshell

It is possible to resume the philosophy of Machine Learning through the view of Tom Mitchell [13] whose theory is a compactified manner of the essence of the action of learning by any system. In fact, this kind of compactification can be engineered to different areas, in all those where it is expected that the system can efficiently learn of well-defined strategies to face subsequent ones. Thus, the well-known Mitchell's criteria are as follow:

1. **Task:** Any system in universe has one or various tasks that allow to system to develop in a coherent manner.
2. **Performance:** Once the task is deeply understood, the system requires to set a solid and robust strategy to solve the task in an optimal manner and denoting an acceptable performance in terms of success of task solution.
3. **Experience:** If system was able to drawn a strategy to solve a task, then it is said that the system has achieved to learn a path whose end is the solution of task. Moreover, the route of solution can also be applied to other different tasks with different purposes. The experience can be measured in terms of efficiency and mathematically can be written as:

$$\mathcal{E} = \frac{x}{x+y} \tag{1}$$

With "x" the positive actions, and "y" the wrong ones. On the other side, the number of wrong actions can be dependent on the positive ones. Thus, we can rewrite Eq. 1 as:

$$\mathcal{E} = \frac{x}{x + g(x)}. \tag{2}$$

Here "$g(x)$" means that the wrong ones depend on the correct ones. A reasonable relation might be done through the negative exponential $Exp[-\beta x]$ whose meaning is a follow: while the number of positive actions does increase then the number of wrong ones decreases as well. Thus, while the number of positive actions go up then for large values of "x" one would expect that the efficiency of system acquires a stable value and therefore it is free of stochastic oscillations. One can claim that if $x \approx 1$ then the system has learned to solve that task, so that we imminently expect the replication of process for next task. In Fig. 1 left panel, Eq. 2 and its derivative are plotted. It tells us that the change on $\mathcal{E}$ put down the learning, so that the derivative cannot be useful to assess systems. Instead of this, the derivative can be seen **as a kind of error of the action of learning**.

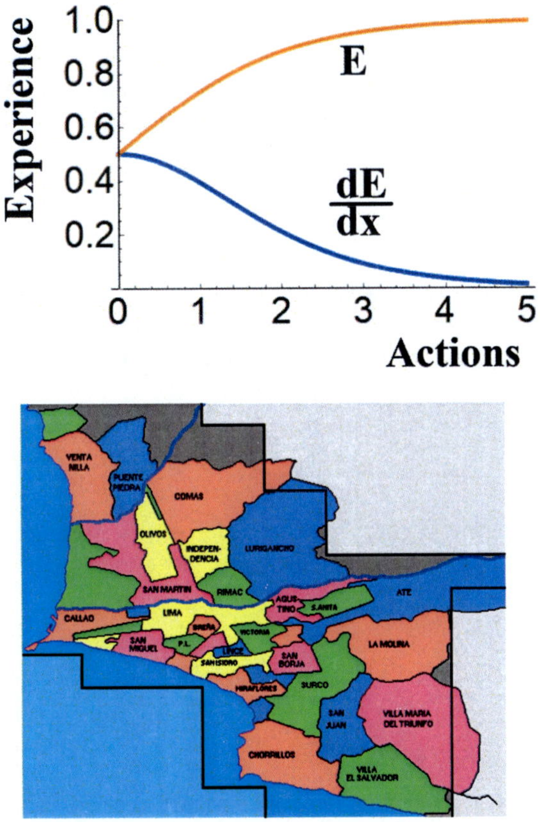

Fig. 1. (Left) Plot of experience (as done in Eq. 2) and its derivative indicating that it is below one half the efficiency of system to accomplish its task. (Right) Current map of Lima city, showing its districts. VES is located at the down as seen at the perimetric sector of south side.

3 Modeling Spatial Expansion of Villa El Salvador

It is an area of approximate 35.5 km^2 located at the south pole of Lima city and it has a population of order of 400K habitants [14]. VES was founded at the beginning of 80s. It is estimated that it is still under a demographic growth. VES has not been alien to a constant development as perceived along the last years. Internal immigration and the arrival of families, have supposed interesting projections in the next 10 years. It is noteworthy that VES has nowadays a strong participation at the economy of Lima city as seen at the industrialization of zones, intense presence of malls, several markets, as well as VES form part of the electric train network, fact that makes more dynamical the economy. At the other hand, it should be noted the presence of state and private universities (totalizing 4). In this manner one can anticipate that VES is probably running through the same path of economic axis as seen also at the case of district Los Olivos at the north pole of Lima city.

In Fig. 1 (Right) one can see the map of Lima city. One can see the location of VES at the south pole. The perimetrical districts appear to be biggest than those located at the center. In fact, it seems to be that Lima city have been experiencing a kind of isotropic expansion, in the sense that there is **not any preferred pole**. Indeed, one can see that exists there a kind of inverse proportionality in the sense that the number of habitants H is proportionally inverse to their age of district: "T". This can be mathematically expressed as:

$$T = \frac{\chi}{H}, \tag{3}$$

with χ a constant with units expressed in habitants per unit of time. Aside we can see that there is a kind of inverse proportionality between the location of districts and the center of city [15]. Thus, while the district is far from the center, its age is short. In this manner one has that:

$$T = \frac{\eta}{D}, \tag{4}$$

with η is a constant whose units are expressed in distance per unit of time. Inserting (4) in (3) one arrives to a demographic definition given by:

$$H = \frac{D\chi}{\eta}. \tag{5}$$

3.1 The Machine Learning Prediction

Equation 5 constitutes an interesting window for ends of usage the Mitchell's criteria and therefore to estimate the spatial growth of VES in time as done in [16] for example, it is mandatory then to fix below a list of qualitative facts that are evidences from experience:

1. People requires fast route in order to achieve optimal displacements.
2. People exhibits strong interest to live near to schools and markets.
3. People seek places of strong security,
4. Families seeks housing with neighbors exhibiting common interests.
5. Evidence that projects prefer to build vertical housing in new areas.
6. Perception after the buy of housing is crucial for next projects.

Consider an initial cluster for instance a big block of houses namely **A** then it is a first source to carry out lateral expansion as seen at Fig. 2 (Up). In this manner one can define the Mitchell's criteria as follows: (i) **Task**: self-sustainable expansion, (ii) **Performance**: methods and actions that optimize harmony against any risk in families, (iii) **Experience**: Has improved the economy, after spatial expansion? With all these criteria then one can carry out predictions. Consider the evolution of total area **At** given by the rectangle perimeter and its expansion at the time Δt and Δt + Δt:

$$\begin{aligned} \mathbf{At}(t = 0) &= \mathbf{A} \\ \mathbf{At}(t + \Delta t) &= \mathbf{A} + \mathrm{a}_1 + \mathrm{a}_2 \\ \mathbf{At}(t + \Delta t + \Delta t) &= \mathbf{A} + \mathrm{a}_1 + \mathrm{a}_2 + \mathrm{a}_{11} + \mathrm{a}_{22} \end{aligned} \tag{6}$$

Clearly, we can see from Eq. 6 when time is running, then one would expect the increasing of **at** as seen in Fig. 2 middle. Thus, from the computational point of Machine Learning Task: this encompasses the following algorithm:

ALGORITHM **TASK**

```
1   DO j = 1, J
2    T = T + Δt T
3     IF p > P THEN
4       DO m = I, M
5         a(j, m) = a(j, m-1) + a(j, m)
6           IF m=M THEN
7             A_T = a(j, m)
8           ENDIF
9       ENDDO
10    ENDIF
11  ENDDO
```

Above is given the pseudo code that estimates the total area $\mathbf{A_T}$ (line-7) under the requirement that p > P (line-3) with P the probability of success during the process of expansion: it is the success as to security, health and food services, as well as economy evolution and new infrastructure of new blocks. The fact that the algorithm requires of a probability, then below a methodology to derive the "urbanistic" probabilities are presented.

3.2 Derivation of Probability

From Eq. 6 we can infer that spatial expansion would have to be analogue to habitants as well. With this we have from Eq. 5 that $H = H_0 - \frac{D\chi}{\eta}$ that implies that $H = H_0\left(1 - \frac{D\chi}{\eta H_0}\right)$. With the assumption that $\eta H_0 \gg D\chi$ then one can apply the exponential approximation of number of habitants that yields

$$H = H_0 e^{-\frac{D\chi}{\eta H_0}}. \tag{7}$$

In the same manner from Eq. 5 follows $D = \frac{H\eta}{\chi}$ so that the exponential method leads to write.

$D = D_0\left(1 - \frac{H\eta}{\chi D_0}\right)$ that as previously done, one arrives to:

$$D = D_0\left(1 - \frac{H\eta}{\chi D_0}\right)$$

$$D \Rightarrow D_0 e^{-\frac{H\eta}{\chi D_0}}. \tag{8}$$

One can see that the mathematical structure of Eq. 8 and Eq. 7 can be also seen as probabilities. On the other hand, the Mitchell's criteria demand to impose a well-defined performance. Below the pseudocode corresponding to the quantitative estimation of performance is written:

ALGORITHM **PERFORMANCE**

```
1   DO k = 1, K
2    T = T + Δt T
3     IF H > H0  THEN
4     IF D > D0 THEN
5       DO n = I, N
6         a(f, n) = a(k n-1) + a(k, n)
7           IF m=N THEN
8             AT = a(k, n)
9           ENDIF
10       ENDDO
11     ENDIF
12  ENDDO
```

As seen above, the lines 3 and 4 makes the question about whether H and D from Eq. 7 and Eq. 8 are greater than their previous values H_0 and D_0.

For instance, consider a block at a time T_0 a number of 1K habitants and subsequently we carry out a measurement that yields 1K + δ, then it triggers the scenario of a possible expansion.

It should be noted that δ is a small fraction for instance for this exercise a number might be of 100 or around.

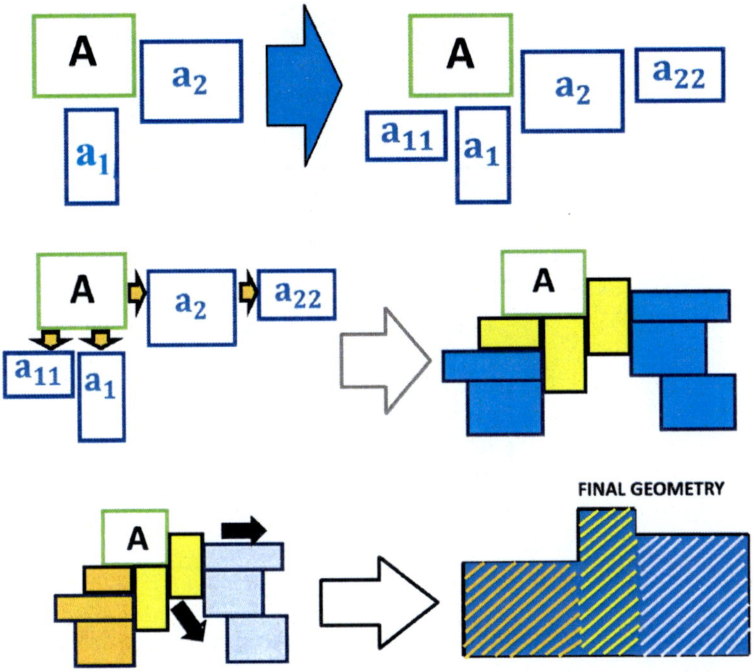

Fig. 2. Possible topologies of blocks aggregation dynamics at the district of VES. Top: the square "A" denoting the very beginning and its first forms of expansion a1 and a2. Middle: reordering of block in according to preferences, down: final perimeter showing approximately a compact geometry.

3.3 The Machine Learning Experience

One can find that for values of distances "*D*" the values of $H = H_0 e^{-\frac{D\chi}{\eta H_0}}$ might be acceptable to increase from its initial values. It is also viewed as:

$$H = H_0 e^{-\frac{(D+\delta D)\chi}{\eta H_0}}. \tag{9}$$

It is expected that while the number of habitants go up, then there is a probability to increase the housing, either rent or buy. A criterion to decide the implementation of housing projects is the verification of the existence of young families that are in the phase of renting (logically one expects that young families exhibit a minimal income to embark for housing purchase).

Particularly in VES, one expects a kind of **isotropic expansion**. The topological representation is approximately schematized in Fig. 2 (down) where not any preferred direction is opted for spatial expansion. (For example, construction of condos is done in allowed areas that are prepared by urbanistic projects). The concerns that can be considered to decide a condo project can be for instance: (i) Transportation, (ii) Markets, and (iii) Security. As seen in the scheme Fig. 3 (down), the new areas can be the spatial continuity of previous blocks. Below, the pseudocode corresponding to the Machine Learning experience.

ALGORITHM **EXPERIENCE**

```
DO q = 1, Q
  T = T + (qΔt) T
    Ps = Ps (η, χ)
      IF q = Q THEN
        IF P > Ps  THEN
          g = G(P)
        ENDIF
      ENDIF
  ENDDO n = I, N
END
```

One can define a probability $\mathbf{P_S}$ is now depending on the values of η and χ in the which one has information on the habitants and distances both per unit of time. With this one can compare with a level of standards that must satisfy to future habitants. Therefore, Machine Learning opts by the best scenario. By "best" scenario one can understand: (a) Large areas for condos, (b) Near to roadways, and (c) Secure zones. According to Google-Map (Fig. 3 left), one can see that the new areas and the creation of block have followed a logic that to some extent and within the errors, would be interpreted as a three-steps action: task, performance and experience, as commonly behaves an algorithm of Machine Learning. It is interesting to see that the creation of new blocks can be following the accomplish of a "Urbanistic Task", having employed a well-defined performance, and that can be repeated again depending on the experience of new habitants as well as the social impact of infrastructure. Arrows are indicating the spatial expansion that might be done in the following periods. To note that it would also imply the growth of Villa el Salvador as district. In this way, one would expect that South Lima is also growing due to these processes of expansion that does not only goes to the urban sector, but also is running as a housing-habitant pairs. In (Fig. 3 right) we can see that the spatial expansion of South Lima implies imminently the change of boundaries between neighbor districts as seen VES and Lurín.

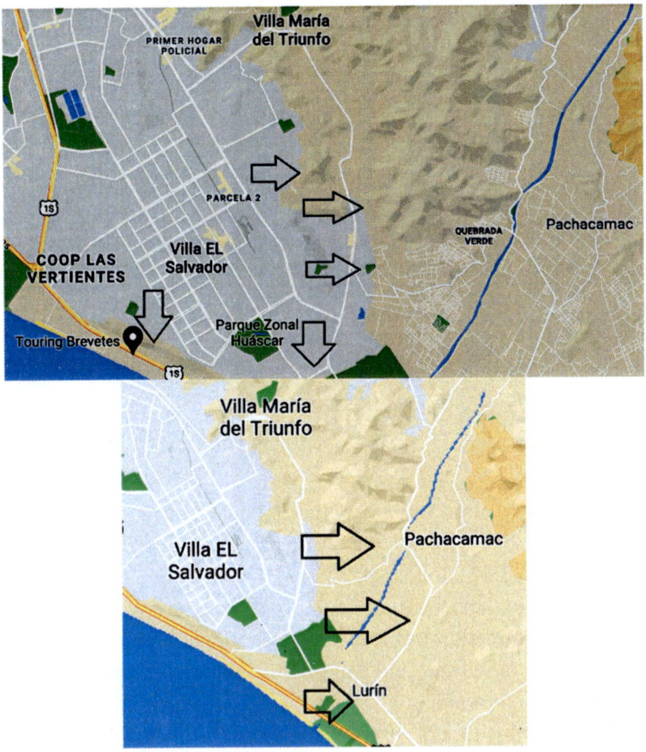

Fig. 3. Google Maps [17], indicating the direction of urbanistic growth: (Left): expansion might be done along the east sector of VES. (Right): A predictive expansion of VES that reaches to extend its limits with neighbors districts.

Mathematically speaking, from Eq. 7 and Eq. 8 we can arrive to the Machine Learning experience of distance in terms of efficiency as defined in Eq. 2:

$$\mathcal{E}_D = \frac{D_0 e^{-\frac{H\eta}{\chi D_0}}}{D_0 e^{-\frac{H\eta}{\chi D_0}} + d}. \tag{10}$$

This can also be extended for the efficiency of growth of number of habitants. Logically, the increasing of number of habitants leads to the expansion of area for housing. Therefore, the efficiency here means how the increasing of people can be efficient to implement and realize housing projects. From this one arrives to (with δD an inherent systematic error):

$$\mathcal{E}_H = \frac{H_0 e^{-\frac{(D+\delta \mathrm{D})\chi}{\eta H_0}}}{H_0 e^{-\frac{(D+\delta \mathrm{D})\chi}{\eta H_0}} + h} \tag{11}$$

3.4 The Success Probability of Spatial Expansion

Finally, one can wonder: How to measure the efficiency of urbanistic projects once there is a firm conjunction of causes? Although this work does not target to this, one can answer this in terms of product of probability from Eq. 10 and Eq. 11:

$$\boldsymbol{P} = \mathcal{E}_H \otimes \mathcal{E}_D = \frac{D_0 e^{-\frac{H\eta}{\chi D_0}} \quad H_0 e^{-\frac{(D+\delta D)\chi}{\eta H_0}}}{D_0 e^{-\frac{H\eta}{\chi D_0}} + d \, H_0 e^{-\frac{(D+\delta D)\chi}{\eta H_0}} + h} \tag{12}$$

That can be written in a more compact manner as:

$$\boldsymbol{P} = \mathcal{E}_H \otimes \mathcal{E}_D = \frac{H_0 D_0 e^{-\frac{H\eta}{\chi D_0} - \frac{(D+\delta D)\chi}{\eta h}}}{\left(D_0 e^{-\frac{H\eta}{\chi D_0}} + d\right)\left(H_0 e^{-\frac{(D+\delta D)\chi}{\eta H_0}} + h\right)} \tag{13}$$

Denoting the efficiency of complete spatial expansion is a success probability $\boldsymbol{P}$. It should be noted the presence of δD that is incorporated at Eq. 11 and Eq. 12 as a systematic error at the spatial expansion.

Although it is not clear the independence of this error, we can argue that it is a consequence of all those uncertainties that might to appear along the implementation of housing projects. On the other side we can also to adjudicate also to the apparition of this uncertainty the social noise that would be around of housing expansion. As experienced at 2020 (because the Covid-19 pandemic), the frozen of housing projects can also be seen as a firm noise that has stopped growth urban in a partial manner. Thus, one can wonder: How δD can be measured accurately? For this, it might be necessary to measure the rate of annual growth, with this one can predict the next values, so that we can carry out quantitative comparisons between them. In Appendix, the ensembled Machine Learning is proposed.

Clearly, not exact estimates can be extracted. In Fig. 4 a Google-Maps is depicted. It is interesting that there is a kind of isotropy along the east-south side of VES as seen in the apparition of small blocks at the sector denominated "BARRIO 2". Finally, the Machine Learning probability of success for a spatial extension of VES is a multi-variable function that can be written in a closed-form and inspired from Eq. 1, as:

$$\boldsymbol{P}(H,D) = \frac{\dfrac{H_0 D_0 e^{-\frac{H\eta}{\chi D_0} - \frac{(D+\delta D)\chi}{\eta H_0}}}{\left(D_0 e^{-\frac{H\eta}{\chi D_0}} + d(D)\right)\left(H_0 e^{-\frac{(D+\delta D)\chi}{\eta H_0}} + h(H)\right)}}{\dfrac{H_0 D_0 e^{-\frac{H\eta}{\chi D_0} - \frac{(D+\delta D)\chi}{\eta H_0}}}{\left(D_0 e^{-\frac{H\eta}{\chi D_0}} + d(D)\right)\left(H_0 e^{-\frac{(D+\delta D)\chi}{\eta H_0}} + h(H)\right)} + \boldsymbol{G}(H,D)} \tag{14}$$

In Fig. 5 the simulation of Eq. 13 under the assumption of the simplest model where $\frac{\eta}{\chi D_0} \approx \frac{\delta D \chi}{\eta H_0} \approx 1$. As well as $\boldsymbol{G}(H,D) \approx Sin[H * D]$, and $\frac{H_0}{D_0} \approx 1$, $\frac{d(D)}{D_0} \approx \frac{h(H)}{H_0} \approx 0$ has yielded a certain similarity (qualitative) with Fig. 4 where the east-south sector of VES exhibits a sustainable growth.

Fig. 4. Block's dynamics and the isotropy of spatial growth in VES. From Google Maps [17].

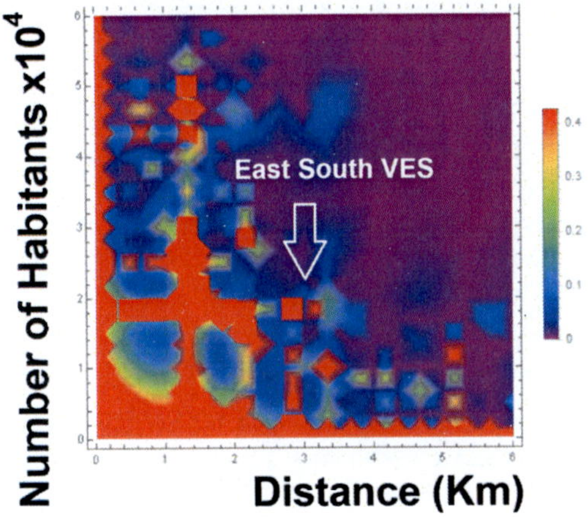

Fig. 5. Simulation of Eq. 13 showing the level of coincidence with east-south of map as shown in Fig. 4. It was done with the package Wolfram Mathematica.

4 Conclusion

In this paper, the social phenomenon of urbanistic growth has been studied through the Tom Mitchell criteria inside the framework of Machine Learning. Thus, a reduced mathematical model was presented. The corresponding algorithms for task, performance and experience, have been developed. The model of blocks appears that engages the recent blocks dynamic in VES. Clearly one arrives to the fact that independent dynamics of new urbanistic might be following rules that matches well to the criteria of Mitchell. In future work, the problem of learning will be faced. How dynamics of spatial expansion can exhibit a kind of learning to opt for future decision shall be seen in a subsequent work.

Apendix: Ensambled Machine Learning Algorithm

Below the unified algorithm that manages the decision to enlarge a perimetric existent area. This algorithm is in essence based on probabilities. The setpoint probabilities have turned out to be of order of 0.85, and 0.75. This last one that is equivalent to express 3/4 can be interpreted as from 4 constructions (building, house or condo) 3 of 4 can be replicated in adjacent areas.

```
TASK
SEARCHING HOUSING
DO k = 1, K   FOR ALL SCENARIOS
   DEFINE A SCENARIO

   PERFORMANCE
   SEARCHING THE BEST PERFORMANCE FOR DECISION
    DO q = 1, Q
      SERVICES
       IF (PROBABILITY OF SUCCES > 0.85) THEN
         DEFINE CONSTRUCTION
       ENDIF
      DO l = 1, L
        DISTANCES
         IF (PROBABILITY OF SUCCES > 0.75) THEN
         CONFIRM CONSTRUCTION
        ENDIF
       DO n = 1, N
        SOCIAL ADVANTAGES
          IF (PROBABILITY OF SUCCES > 0.75) THEN
          CONFIRM CONSTRUCTION
          ENDIF
        DO m = 1, M
         AVERAGE ALL PROBABILITIES
    EXPERIENCE
           IF (PROBABILITY OF EXPERIENCE > 0.75) THEN
           TASK IS SOLVED
            ENLARGIN GEOGRAPHICAL PERIMETER
          ENDIF
        ENDDO
      ENDDO
     ENDDO
    ENDDO
   ENDDO
  ENDDO
```

References

1. Alfaro-Acuña, A., Parodi-Chávez, J.: Evaluation of the commercial and financial viability of a housing real estate project in the Puente Piedra district, Peruvian University of Applied Sciences (2013). https://repositorioacademico.upc.edu.pe/handle/10757/273560
2. Rondinel-Oviedo, D., Sarmiento-Pastor, J.: Water: consumption, usage patterns, and residential infrastructure. A comparative analysis of three regions in the Lima metropolitan area Water: consumption, usage patterns, and residential infrastructure. A comparative analysis of three regions in the Lima metropolitan area. Water Int. **45**(4) (2020). https://doi.org/10.1080/02508060.2020.1830360
3. Rondinel-Oviedo, D.R., Schreier-Barreto, C.: Methodology for selection of sustainability criteria: a case of social housing in Peru. In: Brinkmann, R., Garren, S. (eds.) The Palgrave Handbook of Sustainability, pp. 385–409. Palgrave Macmillan, Cham (2018). https://doi.org/10.1007/978-3-319-71389-2_21
4. Guerrero, L.A., Maas, G., Hogland, W.: Waste Management, vol. 33 no. 1, pp. 220–232 (2013). https://doi.org/10.1016/j.wasman.2012.09.008
5. Mitchell, T.M., Schwenzer, G.M.: Applications of artificial intelligence for chemical inference XXV. A computer program for automated empirical 13C-NMR rule formation. Org. Magn. Reson. **11**(8), 378–384 (1978)
6. Mitchell, T., Steinberg, L., Shulman, J.: A knowledge-based approach to design. IEEE Trans. Pattern Anal. Mach. Intell. (1985)
7. Mitchell, T., Keller, R., Kedar-Cabelli, S.: Explanation-based generalization: a unifying view. Mach. Learn. **1**(1) (1986)
8. Mitchell, T.: Does machine learning really work? AI Mag. **18**(3), 11–20 (1997). Invited paper, AAAI Press, Fall
9. Chirkin, A.M.: Concept of interactive machine learning, in urban design problems. https://core.ac.uk/download/pdf/228470726.pdf
10. Avital, M., Te'Eni, D.: From generative fit to generative capacity: exploring an emerging dimension of information systems design and task performance. Inf. Syst. J. **19**, 345–367 (2009)
11. Batty, M.: Exploring isovist fields: space and shape in architectural and urban morphology. Environ. Plann. B Plann. Des. **28**(1), 123–150 (2001)
12. Peponis, J., Wineman, J., Mahbub Rashid, S., Hong, K., Bafna, S.: On the description of shape and spatial configuration inside buildings: convex partitions and their local properties. Environ. Plann. B **24**(1997), 761–782 (1997)
13. Michalski, R.S., Carbonell, J.G., Mitchell, T.M. (eds.): Machine learning an artificial intelligence approach
14. INEI: Compendio Estadístico. https://www.inei.gob.pe/media/MenuRecursivo/publicaciones_digitales/Est/Lib1714/Libro.pdf
15. Liu, L., Wei, W., Zhao, D., Ma, H.: Urban resolution: new metric for measuring the quality of urban sensing. IEEE Trans. Mob. Comput. **14**(12), 2560–2575 (2015). https://doi.org/10.1109/TMC.2015.2404786
16. Mohammadjafari, S., Roginsky, S., Kavurmacioglu, E., Cevik, M., Ethier, J., Bener, A.B.: Machine learning-based radio coverage prediction in urban environments. IEEE Trans. Netw. Serv. Manag. **17**(4), 2117–2130 (2020). https://doi.org/10.1109/TNSM.2020.3035442
17. https://www.google.com/maps/@-12.2193218,-76.9377212,14z

Intersection Safety Improvement Using Crash Modification Factors

Azad Salim Abdulhafedh(✉), Mudhafer H. Selman, and Heleen Faris Sadeeq

Highway and Bridges Engineering Department, Duhok Polytechnic University, Duhok, Iraq
mudhfar.hamid@dpu.edu.krd

Abstract. Road crashes are among the main causes of death in the world. In order to minimize the number of crashes, suitable safety countermeasures must be installed at road networks. In the City of Duhok located in the Iraqi Kurdistan Region, the number of registered vehicles has increased dramatically due to the economic growth and stability after 2005, which resulted in significant increase in vehicle crashes. This paper aims at introducing the concept of the Crash Modification Factors (CMFs) as an important analytical tool on measuring the road safety and suggesting suitable countermeasures that might help in improving the safety through the reduction of crashes. A case study of Zari land intersection in Duhok City is presented in the paper. Crash data used in the analysis are obtained from the Duhok Directorate of Traffic Police for the years 2017, 2018, and 2019. The expected number of crashes at the intersection is calculated using the predictive method of the Highway Safety Manual, and three safety countermeasures are suggested to enhance the intersection safety. Results show that applying the suggested safety countermeasures would decrease the number of crashes by 42% from 12 crashes/year to 7.0 crashes/year. This implies that the CMFs can effectively assess the safety situation at any intersection or road segment, which can be addressed by the installation of proper safety countermeasures to improve the safety at the site.

Keywords: Crash modification factors · Road safety · Duhok city · Iraqi Kurdistan region

1 Introduction

A crash modification factor (CMF) indicates a quantitative change in safety when a particular safety countermeasure is applied to the road segment or intersection with respect to the base condition [1–17]. The crash modification factors are available at the CMF clearing house website (http://www.cmfclearinghouse.org), which is maintained by the US Federal Highway Administration (FHWA). This website compiles documented CMFs and a searchable database that can be easily used to identify the required CMFs to meet the user's needs. In addition, each CMF has a "star rating" from 1 to 5 indicating the quality or confidence in the results of the study producing the CMF. In the Iraqi Kurdistan Region, the Ministry of Health has reported that on a daily average,

P. Mendonça and N. D. Cortiços (Eds.): ICAMC 2021, LNCE 226, pp. 338–346, 2022.
https://doi.org/10.1007/978-3-030-94514-5_34

3 people were killed and other 28 people were injured by road crashes. The lack and poor quality of the road safety in Kurdistan Region road networks maximize the loss of lives and injuries in road accidents. The existing road geometric design errors such as the width of the lanes, shoulders, the median barrier, the highway curves, and the lack of traffic signs, had inversely affected the road safety [18–20].

2 Methodology

The predictive method of the Highway Safety Manual (HSM) is used in this paper to determine the expected number of crashes at the intersection from the following steps [1]:

2.1 Crashes Due to Multiple - Vehicle Collisions

$$N1 = EXP[a + b(\ln(AADTmajor) + c(\ln(AADTminor)] \quad (1)$$

Where:

$N1$ = predicted number of crashes from multiple-vehicle collisions (crashes/year);

$AADT_{major}$ = annual average daily traffic volume (vehicles/day) for major road (both directions of travel combined);

$AADT_{minor}$ = annual average daily traffic volume (vehicles/day) for minor road (both directions of travel combined); and a, b, c = constants from Table 12 - 10 in the HSM.

2.2 Crashes Due to Single Vehicle Collisions

$$N2 = EXP[a + b(\ln(AADTmajor) + c(\ln(AADTminor)] \quad (2)$$

Where:

N_2 = predicted number of crashes from single-vehicle collisions (crashes/year);

The sum of predicted crashes from the above cases is determined as in 'Eq. (3)':

$$N\ total = N1 + N2 \quad (3)$$

The intersection related CMFs in the HSM predictive method are obtained from the appropriate formulae and tables in the HSM. A combined CMF is calculated by multiplying all CMFs together, then this should be multiplied by the total predicted crashes to obtain the expected crashes. After which, the expected number of crashes should be compared to the observed (actual) crashes at the site. If the number of observed crashes is lower than the expected crashes (by at least 10% as a threshold of the HSM), then we would conclude that the safety condition at the intersection is good and no safety improvement is necessary. However, if the number of observed crashes is higher than the expected crashes (by at least 10%), then we would conclude that the safety condition at the intersection is critical and there is a need of safety improvement. The final stage would be the suggestion of safety countermeasures which are provide in CMFs clearinghouse website.

3 Zari Land Intersection as a Case Study in Duhok City

Zari land intersection was selected as a case study in Duhok City. The layout is shown in Fig. 1. The major approach of the intersection is located along Qazi Mohamad road, which is a 6- lanes divided highway. The minor approach is located along Amad street, which is also a 6-lanes divided highway. The annual average daily traffic for the major approach was measured in the field and found to be 28,842 veh/day. The annual average daily traffic for the minor approach was found to be 25,833 veh/day. All four approaches are equipped with left-turn and right-turn lanes. Protected signal phasing is provided for all left-turn movements. Lighting and red light cameras are provided at the intersection site.

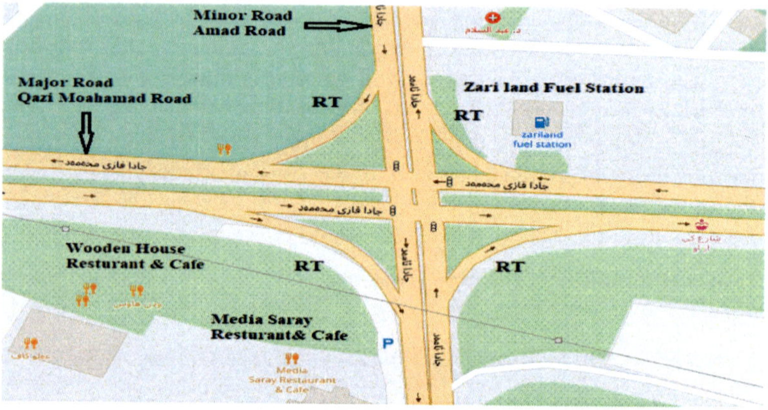

Fig. 1. The layout of Zari Land Intersection in Duhok City.

The observed vehicle crashes at the intersection for the study period (the years 2017, 2018, 2019) are obtained from the Directorate of Traffic Police in Duhok City, as shown in Table 1.

Table 2 shows the total number of predicted crashes from both Multiple and Single-Vehicle collisions, calculated by applying Eqs. 1, 2, and 3.

The individuals CMFs, and the combined CMF are calculated using the appropriate formulae and tables in the HSM, as shown in Table 3 below.

Table 1. Observed vehicle crashes at the intersection for the years 2017, 2018, and 2019

Year	Observed (actual) vehicle crashes	Crash severity level	
		FI (Fatality, Injury)	PDO (Property Damage Only)
2017	10	3	7
2018	12	5	7
2019	13	6	7
Average	11.7 use 12.0	4.7 use 5.0	7

Table 2. Predicted crashes at the intersection

Vehicle Collision type	Value of Predicted Crashes (crash/year)
Multiple (N_1)	10.331
Single (N_2)	0.617
Total (N_{total})	10.948

Table 3. The individual and combined CMFs

CMF Number	CMF Name	CMF Value
CMF1	Left-Turn Lane	0.66
CMF2	Left-Turn Signal Phasing	0.781
CMF3	Right-Turn Lanes	0.85
CMF4	Right-Turn on Red	1
CMF5	Lighting	0.911
CMF6	Red-Light Cameras	1.012
Combined CMF	CMF1 * CMF2* ……* CMF6	0.404

The expected number of crashes at the intersection is then determined as follows: N expected = N predicted * Combined CMF = 10.948 * 0.404 = 4.422 crashes/year or nearly 5.0 crashes/year, which should be compared to the observed crashes at the site as shown in Table 4 below:

Table 4. Expected number of crashes and the observed crashes at intersection.

Observed crashes	Expected crashes	Conclusion
12	5	Since observed crashes > expected crashes, then this intersection needs safety improvement

4 Results and Discussion

Single vehicle accidents involve only one car. These types of accidents often involve instances in which the driver is unable to control, because of weather conditions, distracted drivers, or mechanical defects. Multi-Vehicle accidents involve two or more vehicles, which could result in more severe fatal-injury victims, and vehicle damages. Therefore, we can see that the expected number of crashes from multi-vehicle collisions (10.331) is much higher than the expected number of crashes from single-vehicle collisions (0.617) at the intersection. The total predicted number of crashes at the intersection (from both multi-vehicle and single-vehicle) is found to be (10.948 crashes/year). The combined CMF is found to be 0.404. Applying this combined CMF to the total predicted number of crashes would determine the expected number of crashes at the intersection, which is (4.422 crashes/year or nearly 5.0 crashes/year). Each geometric or operation feature has its own CMF in the HSM tables and formulae. For example, the CMF of the left turn movement is (0.66), the CMF of the Red-Light camera is (1.012), and the CMF of the presence of the lighting system is (0.911). We can realize that the existence of left turn lanes at all approaches has a CMF of less than 1.0, which can positively decrease potential crashes at the intersection by 34% (i.e., 1.0–0.66 = 0.34). The existence of lighting system has a CMF of less than 1.0, which can positively decrease potential crashes by 8.9%. But the existence of red-light cameras has a CMF of greater than 1.0, which could increase potential crashes because of the concerns that red light cameras usually scare drivers (who want to avoid a ticket) into more sudden stops, which may increase the risk of rear-end collisions at the intersection. The total expected number of crashes is found to be (5.0 crashes/year) after considering all geometric and operation features based on the HSM predictive method. This number must be compared to the observed (actual) crashes, which is (12 crashes/year) that were obtained from the Traffic Police records. Since the number of the observed crashes is greater than the number of expected crashes, hence, we conclude that the safety condition of the intersection is critical and needs to be improved. Therefore, we must find suitable safety countermeasures that are available at the cmf clearinghouse website. This implies that the intersection has some safety problems that need to be fixed to increase the safety at this intersection, which would reduce the numbers of crashes, therefore some countermeasures are suggested for safety improvement, using the cmf clearinghouse website.

5 The Proposed Safety Countermeasures

Safety countermeasures are the best solutions for improving the safety conditions at intersections and streets. Each countermeasure has an accompanying crash modification factor (CMF) that represents the effectiveness of implementing the countermeasure at the site. Safety countermeasures are selected from the cmf clearinghouse website based on the cost-effective approaches, the applicability of using the countermeasure, the number of quality stars of the accompanying CMFs, and engineering judgment and expertise. There are many safety countermeasures available at the website for improving the safety of intersections and streets, but the cost of implementing the countermeasure is a major factor that should be considered in selecting a specified countermeasure. For example, converting a signalized intersection to an innovative modern intersection might needs more right of way in the field, which might not be available. In addition, the engineering judgement is very important in the selection process. Experienced professionals with updated knowledge can make more suitable selection than non-experienced personnel.

Selecting and implementing several countermeasures might be more effective than just one, however, using more than one and multiplying all CMFs together would produce overlapping in the results. In order to avoid this undesirable consequences, it is recommended not to use more than three CMFs concurrently at the same time [1]. Therefore, we could use up to three CMFs as maximum selection criteria. After a careful search at the cmf clearinghouse website, and considering all the above selection factors, the following countermeasures are selected as shown in Table 5:

1. **Coordinating the Traffic Signals of the Intersections along the Major Road (CMF = 0.79)**
   This safety countermeasure is referenced at the CMF clearinghouse website [21]. The CMF rating is 3 stars, which is very good ranking. The CMF of this countermeasure is (0.79) in urban areas.
2. **Improve signal visibility (CMF = 0.867)**
   This safety countermeasure is referenced at the CMF clearinghouse website [22]. The CMF rating is 5 stars, which is an excellent ranking. The CMF of this countermeasure is (0.867) in urban areas. Increasing traffic light visibility will assist in reducing accidents at intersections controlled by traffic lights. The most efficient way to achieve this objective is to install back plates on all traffic lights and frame these back plates with a highly visible retro-reflective edge.
3. **Install Automated Speed Camera at Signalized Intersection (CMF = 0.76)**
   This safety countermeasure is referenced at the CMF clearinghouse website [23]. The CMF rating is 3 stars. The CMF of this countermeasure is (0.76) in urban areas. Unlike the Red Light Cameras, intersection speed cameras have not commonly been used. Automated speed cameras are similar in concept to automated red light cameras. However, speed cameras monitor the speed of traffic to enforce the speed limit instead of monitoring the red light violations. Both types can be installed at intersections to improve the safety.

Combined CMF = 0.79 * 0.867 * 0.76 = 0.521 crash/year

$$\begin{aligned} N\,expected &= (Observed\,Crashes) * (Combined\,CMF) \\ &= 12.0 * 0.521 = 6.25\,\text{crashes/year, use } 7.0\,\text{crashes/year} \end{aligned} \quad (4)$$

The reduction of crashes will be 42% after applying all the above mentioned CMFs at the intersection, which is an extremely encouraging outcome.

Table 5. Proposed safety countermeasures

Safety countermeasure	Value of CMF	HSM # of Stars
Coordinating the Traffic Signals of the Intersections along the Major Road	0.79	3
Improve signal visibility	0.867	5
Install Automated Speed Camera	0.76	3
Combined CMF	0.521	

6 Conclusion

Crash Modification Factors (CMFs) are important tools to assess the safety of a site by estimating the change in the expected number of crashes when a specific safety countermeasure is implemented. The HSM predictive method was used in this paper to evaluate the safety condition of 4-leg signalized intersection in Duhok City, and it was found to be critical, because the observed (actual) crashes at the site exceeded the expected number of crashes obtained from the predictive method. As a result, suitable safety countermeasures were selected from the cmf clearinghouse website, maintained by the US Federal Highway Administration. The combined CMF was found to be 0.404. Applying this combined CMF to the total predicted number of crashes would determine the expected number of crashes at the intersection, which was (4.422 crashes/year or nearly 5.0 crashes/year). The observed (actual) number of crashes at the intersection was found to be 12 crashes/year. Since the observed number of crashes (12 crashes/year) was greater than the expected number of crashes (5 crashes/year), we concluded that the safety condition at the intersection was critical and needs to be improved by applying effective safety countermeasures, which can be selected from the cmf clearinghouse website. Three countermeasures were selected:

1) Coordinating the Traffic Signals of the Intersections along the Major Road
2) Improving signal visibility.
3) Install Automated Speed Camera at Signalized Intersection.

The combined CMF was determined to be (0.521). The expected number of crashes was determined by multiplying the observed number of crashes by the combined CMF, and it was found to be (6.25 crashes/year, or nearly 7.0 crashes/year). Hence, the reduction of crashes was 42% after applying all the above mentioned CMFs at the intersection, which is a promising outcome.

References

1. The Highway Safety Manual (HSM) by AASHTO. Washington, DC (2014)
2. Al-Marafi, M.N., Somasundaraswaran, K., Ayers, R.: Developing crash modification factors for roundabouts using a cross-sectional method. J. Traffic Transp. Eng. **7** (3), 362–374. https://doi.org/10.1016/j.jtte.2018.10.012

3. Shahdah, U., Saccomanno, F., Persaud, B.: Integrated traffic conflict model for estimating crash modification factors. Accid. Anal. Prev. **71**, 228–235 (2014)
4. Hoogendoorn, S., Daamen, W., Hoogendoorn, R., Goemans, J.: Assessment of dynamic speed limits on freeway a20 near Rotterdam, Netherlands. Transp. Res. Rec. J. Transp. Res. Board **2380**, 61–71 (2013)
5. Essa, M., Sayed, T.: Traffic conflict models to evaluate the safety of signalized intersections at the cycle level. Transp. Res. Part C Emerg. Technol. **89**, 289–302 (2018). https://doi.org/10.1016/j.trc.2018.02.014
6. Elvik, R., Vaa, T.: The Handbook of Road Safety Measures. Elsevier Science, Burlington, MA (2004)
7. Karmacharya, R., Dissanayake, S.: Calibration of highway safety manual predictive methods for 4-leg signalized intersections at urban and suburban areas in Kansas. In: Dept. of Civil Engineering, Kansas State Univ., 2118 Fiedler Hall, Manhattan, KS 66506. International Conference on Transportation and Development (2018)
8. Montella, A., Persuad, B., D'Apuzzo, M., Imbriani, L.: Safety evaluation of an automated section speed enforcement system. Presented at the 91st Annual Meeting of the Transportation Research Board, Paper No. 12-0226, Washington, D.C. (2012)
9. Elvik, R., Vaa, T.: Handbook of Road Safety Measures. United Kingdom, Elsevier, Oxford (2004)
10. Bonneson, J.A., Zimmerman, K., Fitzpatrick, K.: Roadway safety design synthesis. Report No. FHWA/TX-05/0-4703-P1. Texas Department of Transportation, Austin, TX, November 2005
11. Williamson, M.R., Fries, R.N., Qi, Y., Mandava, P.: Identifying the safety impact of signal coordination projects along urban arterials using a meta-analysis method. J. Traffic Transp. Eng. **6**, 61–72 (2018). Department of Civil Engineering
12. Institute of Transportation Engineers: Toolbox of Countermeasures and Their Potential Effectiveness to Make Intersections Safer. Sheet no. 8, ITE, FHWA (2004)
13. Zegeer, C.V., Bushell, M.: Pedestrian crash trends and potential countermeasures from around the world. Accid. Anal. Prev. **44** (1) (2012). https://doi.org/10.1016/j.aap.2010.12.007
14. Lord, D., Bonneson, J.: Development of accident modification factors for rural frontage road segments in Texas. Transp. Res. Rec. J. Transp. Res. Board **2023**, 20–27. Transportation Research Board, Washington, D.C. (2007)
15. Chen, X., Qi, Y., Lu, Y.: Safety impacts of using short left-turn lanes at unsignalized median openings. In: TRB 93rd Annual Meeting (2014)
16. Mohammadi, M., Samaranayake, V., Bham, G.: Safety effect of missouri's strategic highway safety plan-missouri's blueprint for safer roadways. In: TRB 93rd Annual Meeting (2014)
17. Karmacharya, R., Dissanayake, S.: Calibration of highway safety manual predictive methods for 4-leg signalized intersections at urban and suburban areas in Kansas. In: International Conference on Transportation and Development: Connected and Autonomous Vehicles and Transportation Safety (2018)
18. Mohammed, H., Jaff, D., Schrock, S.: The challenges impeding traffic safety improvements in the Kurdistan Region of Iraq. Transp. Res. Interdiscip. Perspect. (2019).https://doi.org/10.1016/j.trip.2019.100029
19. Jaff, D.: A public health initiative to address road traffic accidents in the Kurdistan Region of Iraq. Middle East J. Family Med. **7** (2018)
20. Jrew, B., Msallam, M., Khaled, S., Abojaradeh, M.: Analysis and evaluation of traffic accidents for principle urban streets in Arbil City in Iraq. Accid. Anal. Prev. **10**, 118–131 (2017)

21. Williamson, M.R., Fries, R.N., Qi, Y., Mandava, P.: Identifying the safety impact of signal coordination projects along urban arterials using a meta- analysis method. J. Traffic Transp. Eng. **6**, 61–72 (2018)
22. El-Basyouny, K., Sayed, T., El Esawey, M., Pump, J.E.: Investigating the effect of collision aggregation on safety evaluations using multivariate linear intervention models: case study of signal head upgrade program. Presented at the Transportation Research Board 91st Annual Meeting, Paper No. 12-2093, Washington, DC, 22–26 January 2012
23. Vanlaar, W., Robertson, R., Marcoux, K.: An evaluation of Winnipeg's photo enforcement safety program: results of time series analyses and an intersection camera experiment. Accid. Anal. Prev. **62**, 238–247 (2014)

The Urban Underpass Design Based on AASHTO Specifications

Heng Mi[1(✉)], Chenchen Lu[2], Xuetao Zhao[1], Bing Zhou[3], and Tun Li[1]

[1] CCCC First Highway Consultants Co., Ltd., Xi'an, China
mih@ccroad.com.cn
[2] Chang'an University, Xi'an, China
[3] China Road and Bridge Corporation, Beijing, China

Abstract. With the development of economics and growth of population, the city development tends to be more three-dimensional. Underpasses can reduce land occupation and minimize interference with surrounding buildings, therefore, so underpasses are increasingly used in urban transportation infrastructure construction. According to the AASHTO design specifications, this paper summarizes procedures and key points of the urban underpass design issues. Based on an actual project, the design load, loading mode, flexural resistance analysis, shear resistance analysis are carried out, and the relevant analysis results are compared between American AASHTO specifications and Chinese JTG specifications. The result of this paper can be applicable to similar project construction.

Keywords: Urban underpass · Design · AASHTO specifications

1 Introduction

Three-dimensional transportation is a new urban planning mode and a common way to improve urban transportation. With the development of economics and growth of population, the city development tends to be more three-dimensional. The core concept of three-dimensional transportation is to allocate traffic flows in different plane, so as to achieve the maximum use of urban space. Underpass is a manifestation of three-dimensional traffic, which can be set up in parallel with existing urban roads, thereby can reduce land occupation and minimize interference with surrounding buildings. Therefore, it is increasingly used in urban transportation infrastructure construction [1–6].

According to the AASHTO LRFD Bridge design specifications [7], this paper summarizes procedures and key points of the underpass design. Based on an actual project, the design load, loading mode, flexural resistance analysis, shear resistance analysis are carried out, and the relevant analysis results are compared between American AASHTO specifications and Chinese JTG specifications [8]. The result of this paper can be applicable to similar project construction.

P. Mendonça and N. D. Cortiços (Eds.): ICAMC 2021, LNCE 226, pp. 347–358, 2022.
https://doi.org/10.1007/978-3-030-94514-5_35

2 Case Study and Design Comparison

The Pushkin project is located in the Addis Ababa, capital of Ethiopia. The total route length is 2.3 km, and the typical cross section is 30 m wide with 6 lanes in both directions, The project can greatly alleviate traffic congestion in the central and western regions of Addis Ababa city, The key engineering node of the project is the underpass between TA RD and MS RD, which is located directly under the BRT lane, the total length of the underpass is 316 m. The layout of the underpass is shown in Fig. 1.

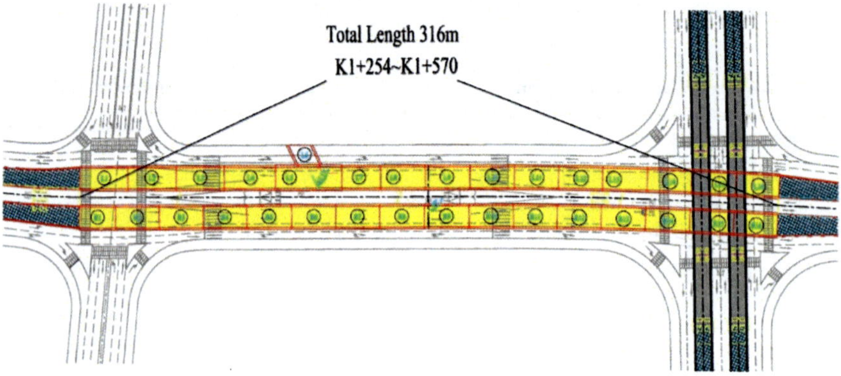

Fig. 1. General layout of the underpass

The cross section of underpass is 9 m (Width) × 8 m (Height) as shown in Fig. 2, the structure type is reinforced concrete, fill height above the underpass is 2–4 m.

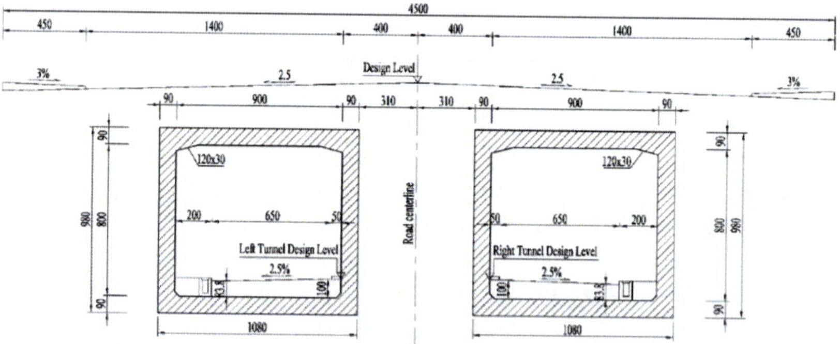

Fig. 2. Typical cross section of the underpass

2.1 Design Loads

Vehicular Load. According to the AASHTO codes, for underpass buried in the soil, the design truck load needs to be considered. The design truck load has a front axle load of 35 KN, a middle and rear axle load of 145 KN, the distance between front and middle axle is 4.3 m, and the distance between front and middle axle is 4.3–9.0 m.

The wheel width of the truck load is W = 510 mm, and the wheel length is L = 250 mm. For underpass with a fill height greater than 600 mm, the distribution width of the wheel load in the soil layer increases with the thickness of the fill height. The distribution width of the wheels is $L_w = W + 1.15\ h$, and the distribution length of the wheels is $L_L = L + 1.15\ h$, h is the fill height. If the distribution widths between the wheels overlap, the superimposed influence needs to be considered Fig. 3.

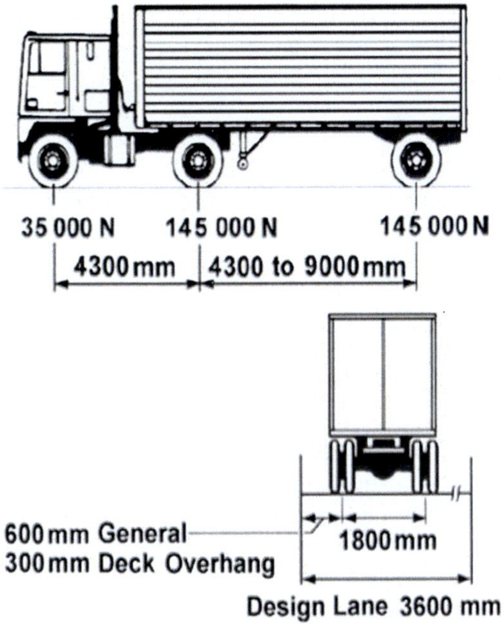

Fig. 3. Design truck load

For the underground structure, the dynamic load allowance IM can be calculated according to Eq. 1:

$$IM = 33\left(1.0 - 4.1x10^{-4}D\right) > \ = 0 \tag{1}$$

Where:

IM = Dynamic load allowance

D = Fill height of the slab (mm)

According to Eq. 1, the dynamic load allowance does not need to be considered when the fill height is greater than 2440mm, which is a little different from the Chinese JTG code, in Chinese JTG code, of which when fill height is greater than 500mm, the dynamic load allowance can be neglected.

For this project, the maximum filling height of the underpass is h = 4.5 m, and two lanes are arranged horizontally. It can be seen from Fig. 4 that the distribution width of a wheel in the horizontal direction $L_w = 510\ mm + 1.15 \times 4500\ mm = 5685\ mm >$

wheelbase = 1200 mm or 1800 mm, so in the horizontal, the two lanes are overlapped, and the total distribution width after lateral superposition is L_w = (1800 + 1200 + 1800) + 510 + 1.15*4500 = 10485 mm.

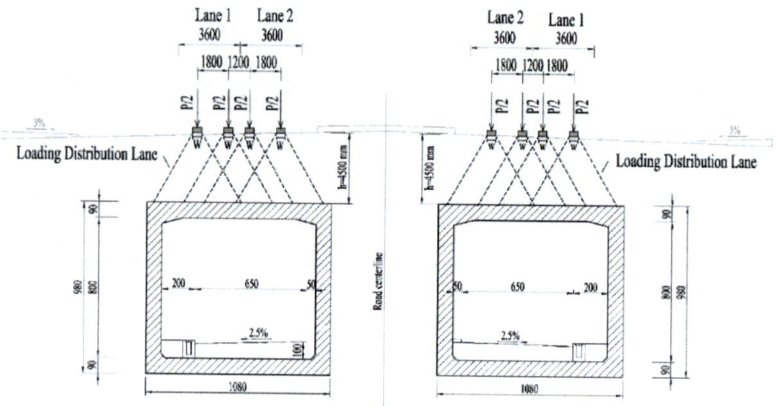

Fig. 4. Lateral loading layout

In the same way, the longitudinal distribution width of a wheel = 250 mm + 1.15 ×4500 mm = 5425 mm > longitudinal wheelbase = 4300 mm, there is overlap in the longitudinal distribution, the total longitudinal distribution width L_L = (4300 mm + 4300 mm) + 250 mm + 1.15*4500 mm = 14025 mm.

According to Eq. 1, the dynamic load allowance of the vehicle is equal to 0 under the filling of 4500 mm. In summary, there are two vehicles arranged horizontally in the distribution width boundary. The front axle, middle axle, and rear axle in the longitudinal length are all involved in the verification. The value of the vehicle load acting on the top of the underpass is $p_L = \sum P/(L_w \times L_L) = 2\ x(35000 + 145000 + 145000)/(10485\,\text{mm x}\,14025) = 0.00442\,\text{Mpa} = 4.42\,\text{KN/m}^2$Vertical Earth Pressure.

According to the AASHTO LRFD, the vertical earth pressure can be calculated by Eq. 2:

$$p_Z = \gamma_s g z \left(10^{-9}\text{Mpa}\right) \tag{2}$$

Where:

p_Z = Vertical earth pressure (Mpa)

γ_s = Density of soil (kg /m^3)

z = Depth from soil surface to top slab (mm)

G = Acceleration of gravity (mm^2)

For this project, z = 4500 mm, $\gamma_s = 1800\,\text{kg/m}^3$, according to Eq. 2, p_Z = 0.081 Mpa = 81 KN/m^2.

Horizontal Earth Pressure. According to the AASHTO LRFD, the horizontal earth pressure can be calculated by Eq. 3:

$$p_H = k\gamma_s gz\left(10^{-9}\text{Mpa}\right) \tag{3}$$

Where:
p_H = Horizontal earth pressure (Mpa)
K = Horizontal earth pressure factor
γ_s = Density of soil (kg /m^3)
z = Depth from soil surface to top slab (mm)
G = Acceleration of gravity (mm^2)
For the buried structure like underpass, the K can be calculated by Eq. 4 and 5:

$$k_a = \frac{sin^2\left(\theta + \varnothing_f'\right)}{\xi[sin^2\theta \sin(\theta - \delta)]} \tag{4}$$

$$\xi = \left[1 + \sqrt{\frac{\sin\left(\varnothing_f' + \delta\right)\sin\left(\varnothing_f' - \beta\right)}{\sin(\theta - \delta)\sin(\theta + \beta)}}\right]^2 \tag{5}$$

Where:
δ = Friction angle between abutment and soil (°)
β = Angle between the surface of filled soil and horizontal surface (°)
θ = Angle between the surface of abutment and horizontal surface (°)
$\varnothing_f'$ = Effective angle of internal friction (°)
For this project, $\delta = 17°, \beta = 0°, \theta = 90°, \varnothing_f' = 30°$, so the K = 0.3, the horizontal earth of abutment top = 0.3 × 81 KN/m2 = 24.3 KN/m^2, the horizontal earth of abutment bottom = 0.3×81 KN/m^2 = 24.3 KN/m^2 = 0.3×257.4 KN/m^2 = 77.2 KN/m^2.

Live Load Surcharge. For this project, the traffic direction is perpendicular to the span direction and parallel to the abutment. According to the AASHTO specification, the equivalent height of soil for vehicular load heq can be taken based on Table 1, the LS = 0.3×18 KN/m^3*0.6m = 3.24 KN/m^2.

Table 1. Equivalent height of soil for vehicular load Heq

Abutment Height(mm)	h_{eq}(mm) Distance from wall backface to edge of traffic	
	0.0 mm	≥ *300* mm
1500	1500	600
3000	1050	600
≥ 6000	600	600

2.2 Load Model

According to the above calculation, take the 1m slab in longitudinal direction to analysis, the loading mode of the underpass is shown in the Fig. 5.

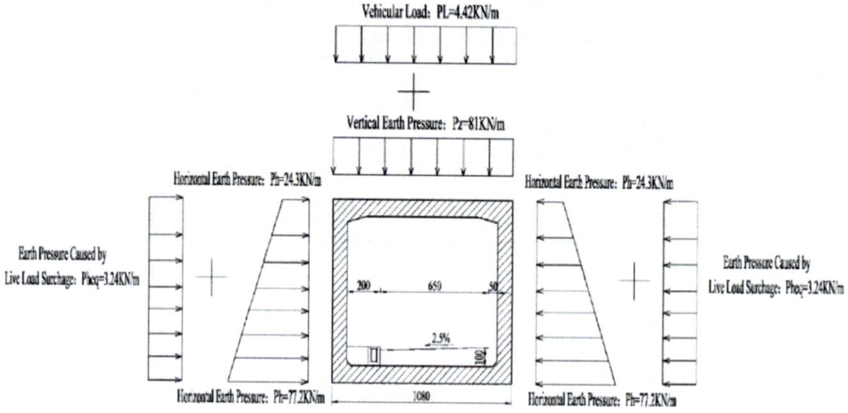

Fig. 5. Loading model

2.3 Boundary Stimulation

The underpass traverses the black cotton soil layer which is an unique expansive soil in east Africa. According to field experiments, the bed coefficient is 8000 KN/m^3, the boundary base can be simulated by elastic support, and the elastic support stiffness of each node can be taken as 8000 KN/m.

2.4 Calculation Result

The Bending Moment of Top Slab. The bending moment of top slab is shown in Table2 and Table 3.

Table 2. The bending moment at span middle

Loading	Self weight (KN.m)		Vertical earth pressure (KN.m)		Horizontal earth pressure (KN.m)		Live load surcharge (KN.m)		Vehicular load (KN.m)	
Limit state	Load factor	Value	Load factor	Value	Load factor	Value	Load factor	Value	Load factor	Value
ULS	1.25	187	1.35	582.9	1.5/0.9	−129.4	1.5	−8.7	1.75	31.8
SLS	1.0	187	1.0	582.9	1.0	−129.4	1.0	−8.7	0.8	31.8

Table 3. The bending moment at corner point

Loading	Self weight (KN.m)		Vertical earth pressure (KN. m)		Horizontal earth pressure (KN.m)		Live load surcharge (KN.m)		Vehicular load (KN.m)	
Limit state	Load factor	Value	Load factor	Value	Load factor	Value	Load factor	Value	Load factor	Value
ULS	1.25	−87.9	1.35	−396.6	1.5/0.9	−104.8	1.5	−6.8	1.75	−21.6
SLS	1.0	−87.9	1.0	−396.6	1.0	−104.8	1.0	−6.8	0.8	−21.6

The Bending Moment of Bottom Slab. The bending moment of bottom slab is shown in Table 4 and Table 5.

Table 4. The bending moment at span middle

Loading	Self weight (KN.m)		Vertical earth pressure (KN. m)		Horizontal earth pressure (KN.m)		Live Load surcharge (KN.m)		Vehicular load (KN.m)	
Limit state	Load factor	Value	Load factor	Value	Load Factor	Value	Load Factor	Value	Load Factor	Value
ULS	1.25	−273.1	1.35	−450	1.5/0.9	116.5	1.5	7.1	1.75	−81.6
SLS	1.0	−273.1	1.0	−450	1.0	116.5	1.0	7.1	0.8	−81.6

Table 5. The bending moment at corner point

Loading	Self weight (KN.m)		Vertical earth pressure (KN. m)		Horizontal earth pressure (KN.m)		Live load surcharge (KN. m)		Vehicular load (KN.m)	
Limit state	Load factor	Value	Load factor	Value	Load factor	Value	Load factor	Value	Load factor	Value
ULS	1.25	200	1.35	248.6	1.5/0.9	115	1.5	7.2	1.75	84. −6
SLS	1.0	200	1.0	248.6	1.0	115	1.0	7.2	0.8	84.6

The Shear Force Calculation Result. The shear force slab is shown in Table 6 and Table 7.

Table 6. The Shear force of top slab at support

Loading	Self weight (KN.m)		Vertical earth pressure (KN. m)		Horizontal earth pressure (KN.m)		Live load surcharge (KN. m)		Vehicular load (KN.m)	
Limit state	Load factor	Value	Load factor	Value	Load factor	Value	Load factor	Value	Load factor	Value
ULS	1.25	116	1.35	399	1.5/0.9	0	1.5	0	1.75	21.8
SLS	1.0	116	1.0	399	1.0	0	1.0	0	0.8	21.8

Table 7. The Shear force of bottom slab at support

Loading	Self weight (KN.m)		Vertical earth pressure (KN. m)		Horizontal earth pressure (KN.m)		Live load surcharge (KN. m)		Vehicular load (KN.m)	
Limit state	Load factor	Value	Load factor	Value	Load factor	Value	Load factor	Value	Load factor	Value
ULS	1.25	244	1.35	335	1.5/0.9	0	1.5	0	1.75	63.3 –
SLS	1.0	244	1.0	335	1.0	0	1.0	0	0.8	63.3

The Load Combination Result. The load combination result is shown in Table 8.

Table 8. The load combination result

Location	Top slab			Bottom slab		
Limit state	Bending moment at span middle (KN.m)	Bending moment at support (KN.m)	Shear force at support (KN)	Bending moment at span middle (KN.m)	Bending moment at support (KN.m)	Shear force at support (KN)
ULS	960	−851	722	−987	917	868
SLS	666	614	532	−672	638	629

2.5 Flexural Resistance

When calculating the flexural resistance in AASHTO, it is first necessary to judge whether the tensile strain of the reinforcement reaches the value specified by the code, and then the component resistance coefficient and the reinforcement stress can be determined. The calculation formula of bending resistance is given based on the AASHTO when the tensile strain meets the requirements.

Flexural Resistance and Check. According to the AASHTO LRFD Bridge Design Specifications, the flexural resistance of rectangle section can be obtained by Eq. 6 to 9:

$$M_r = \varnothing M_n \tag{6}$$

$$M_n = A_s f_s \left(d_s - \frac{a}{2}\right) \tag{7}$$

$$a = \beta c \tag{8}$$

$$c = A_s f_s / 0.85 f'_c \beta b \tag{9}$$

Where:
M_n = Nominal resistance (N-mm)
$\varnothing$ = Resistance factor

M_r = Factored nominal resistance (N-mm)
A_s = Area of tensile reinforcement (mm^2)
f_s = Stress of tensile reinforcement (Mpa)
a = Depth of the equivalent stress block (mm)
c = distance from extreme compression fiber to the neutral axis(mm)
β = Stress block factor, when $f_c' < = 28Mpa$, $\beta = 0.85$,when $f_c' < = 28Mpa$, $\beta = 0.85 - 0.35f_c' - 28 \geq 0.65$.
B = width of the compression face of the member (mm)
d_s = distance from extreme compression fiber to the centroid of the tensile Reinforcement (mm)
f_c' = specified compressive strength of concrete for use in design (Mpa), it is cylinder strength, the relationship between the cylinder strength and cube strength can be taken as $f_c' = 0.8f_{cu}$.

For this project, 1m length underpass was taken to analysis, Resistance factor $\varnothing = 0.9$, width of the section b = 1000 mm, the depth of the section h = 900 mm, $d_s = 860$ mm, C30 concrete is adopted, $f_c' = 0.8f_{cu} = 0.8 \times 30 = 24$Mpa, 9-D28 G420 bar is adopted of tensile reinforcement, $f_s = 420$Mpa, the factored nominal resistance is shown in Table 9, the nominal resistance calculated based on Chinese JTG code is also given:

Table 9. Flexural resistance

Location	Bending moment and flexural resistance for top slab				Bending moment and flexural resistance for bottom slab			
Code name	Bending moment at span middle (KN.m)	Flexural resistance Mn (KN. m)	Bending moment at support (KN.m)	Flexural resistance Mn (KN. m)	Bending moment at span middle (KN.m)	Flexural resistance Mn (KN. m)	Bending moment at support (KN.m)	Flexural resistance Mn (KN. m)
AASHTO LRFD	960	1651	−851	1651	−987	1651	917	1651
Chinese JTG		1426		1426		1426		1426

It can be seen from the above tables that the flexural resistance of the underpass both meets the requirements of AASHTO LRFD and Chinese JTG standards, but for the same material and reinforcement, the flexural resistance calculated by the AASHTO code is greater than the Chinese JTG code, the main reason is that, in the AASHTO LRFD code, the specified value or yield strength of concrete and reinforcement in ULS is directly used when calculating the flexural resistance, for Chinese JTG code, the stress of concrete and reinforcement in ULS should be divided by the corresponding material safety factor.

2.6 Shear Resistance and Check

According to the AASHTO LRFD, the shear resistance can be obtained by Eq. 10 to 15:

$$v_u = \frac{V_u}{\varnothing b_v d_v} \tag{10}$$

Where:
v_u = Shear stress (Mpa).
$\varnothing$ = Resistance factor.
b_v = effective web width taken as the minimum web width.
d_v = effective shear depth (mm), it need not be taken to be less than the greater of 0.9 de or 0.72 h

$$dv = \frac{M_n}{A_s f_s} \tag{11}$$

When $v_u \leq 0.18 f_c'$, the shear resistance can be taken as the lesser of:

$$V_n = V_C + V_s \tag{12}$$

$$V_n = 0.25 f_c' b_v d_v \tag{13}$$

Where:
V_n = Nominal shear resistance (N).
V_r = Factored shear resistance (N).
V_c = Shear strength provided by the concrete (N).
V_s = Shear strength provided by the shear reinforcement(N)

$$V_c = 0.083 \beta_c \sqrt{f_c'} b_v d_v \tag{14}$$

$$V_s = \frac{A_v f_y d_v (\cot\theta + \cot\alpha)\sin\alpha}{s} \tag{15}$$

Where:
f_c' = specified compressive strength of concrete for use in design
A_v = Area of shear reinforcement (mm^2)
f_y = The stress of shear reinforcement (Mpa)
S = Spacing of transverse reinforcement (mm)
α = Angle of inclination of transverse reinforcement to longitudinal axis (°)
β and θ is corresponding section factor, for normal calculation, $\beta = 2$, $\theta = 45°$.

For this project, 1m length underpass was taken to analysis, Resistance factor $\varnothing = 0.9$, width of the section b = 1000 mm, the depth of the section h = 900 mm, bv = 1000 mm, $f_c' = 0.8 f_{cu} = 0.8 \times 30 = 24$Mpa, 4-D12 bar is adopted for shear reinforcement, s = 300 mm, $f_y = 420Mpa$, the factored nominal shear resistance is shown in Table 10, the nominal resistance calculated based on Chinese JTG code is also given:

Table 10. Shear reinforcement

Location	Shear force and shear resistance for top slab		Shear force and shear resistance for bottom slab	
Code name	Shear force (KN)	Shear resistance (KN.m)	Shear force (KN)	Shear resistance (KN.m)
AASHTO LRFD	722	1006	868	1006
Chinese JTG		875		875

It can be seen from the above tables that the shear resistance of the underpass both meets the requirements of AASHTO LRFD and Chinese JTG standards, but for the same material and reinforcement, the shear resistance calculated by the AASHTO code is greater than the Chinese JTG code, the main reason is that, in the AASHTO LRFD code, the specified value or yield strength of concrete and reinforcement in ULS is directly used when calculating the shear resistance, for Chinese JTG code, the stress of concrete and reinforcement in ULS should be divided by the corresponding material safety factor.

3 Conclusion

This paper summarizes procedures and key points of the urban underpass design. Based on an actual project, the design load, loading mode, flexural resistance analysis, shear resistance analysis are carried out, and the relevant analysis results are compared between American AASHTO specifications and Chinese JTG specifications. The following conclusions can be drawn:

A. When designing the underpass, according to the AASHTO code, the impact load does not need to be considered when the fill is greater than 2440 mm
B. According to the AASHTO LRFD, the equivalent height of soil for vehicular load should be determined based on the direction of traffic and the height of underpass.
C. For the same material and reinforcement, the flexural and shear resistance calculated by the AASHTO code is greater than the Chinese JTG code, the main reason is that, in the AASHTO LRFD code, the specified value or yield strength of concrete and reinforcement in ULS is directly used when calculating the flexural and shear resistance, but for Chinese JTG code, the stress of concrete and reinforcement in ULS should be divided by the corresponding material safety factor.

References

1. Heng, C., Sun, S., Zhou, Z., Zhang, J.: Prediction of surface settlement with ultra-shallow-burial and large rectangular cross-section urban underpass. KSCE J. Civil Eng. **23**(11), 4641–4650 (2019). https://doi.org/10.1007/s12205-019-2406-y
2. Liao, S.-M., Cheng, C.-H., Chen, L.-S.: The planning and construction of a large underpass crossing urban expressway in Shanghai: an exemplary solution to the traffic congestions at dead end roads. Tunn. Undergr. Space Technol. **81**, 367–381 (2018)
3. Correia, J.R., Arruda, M.R.T., Branco, F.A.: Structural assessment of reinforced-concrete arch underpasses subjected to vehicular. Overloads. J. Perform. Constructed Facil. **28**(2), 321–329 (2014)
4. Tan, S.B., Yang, K.S.: Laterally loaded piles for a vehicular underpass. J. Geotech. Eng. **110** (2), 318–330 (1984)
5. Mori, H.: Some case records of stability and settlement of embankment of the soft. Ground. In: Proceedings of 6th Asian Regional Conference on Soil Mechanics and Foundation Engineering, vol. 2 (1979)
6. Ballesteros-Tajadura, R., Santolaria-Morros, C., Blanco-Marigorta, E.: Influence of the slope in the ventilation semi-transversal system of an urban tunnel. Tunn. Undergr. Space Technol. **21**(1), 21–28 (2006)
7. AASHTO: AASHTO LRFD Bridge Design Specifications ,8th ed, Wasgington, DC (2017)
8. Ministry of Transport of People's Republic of China, 2018, Specifications for Design of Highway Reinforced Concrete and Prestressed Concrete Bridges and Culverts, Beijing (2018)

Building Energy Conservation and Construction Management

Effect of Roof Types on Energy Use in Residential Buildings in Cold Climates

Rashid Gaitov[1], Serik Tokbolat[2(✉)], Mohamad Mustafa[3], and Rajnish K. Calay[3]

[1] Nazarbayev University, 53 Kabanbay Batyr Ave, Nur-Sultan 010000, Kazakhstan
[2] Nottingham Trent University, 50 Shakespeare Street, Nottingham NG1 4FQ, UK
serik.tokbolat@ntu.ac.uk
[3] UiT The Arctic University of Norway, 8514 Narvik, Norway

Abstract. The role of residential building envelope components is significant in terms of supporting a structure and transferring associated loads, providing aesthetic appearance, and controlling the flows of matters and energy. The control function is particularly important in terms of energy use as poorly designed and constructed envelopes can negatively affect the overall performance of a building. Among others, roofs play a critical role due to their area of coverage, direct interaction with precipitation, and a significant share of total heat transfer. This is especially true in case buildings located in countries with cold climate conditions. This study aimed to investigate the impacts of different roof types on the energy use of residential buildings in Nur-Sultan city, Kazakhstan. Moreover, it aimed at performing a cost analysis to compare the four roof types. The building models with roof types such as flat roof, green roof and gable roof with finished and unfinished attics were simulated and compared in terms of their energy use. The findings indicate that the most energy and cost efficient two-storey building has gable roof with finished attic. The green roof is the most energy efficient choice for a one-storey building. It consumes 4.5% less energy and will pay off in 9 years.

Keywords: Roof types · Energy consumption · Cold climate · Building energy

1 Introduction

Roofs have traditionally played a critical role in the lives of humans from the early ages. The major function of any roof has always been to protect the inhabitants from precipitation and to provide shelter from the natural environment [1]. Ancient people built basic shelters and primitive huts. Central Asian nomads, for example, lived in so called "yurts", which are common in Kazakhstan even at present and popularized as a part of the national heritage. The yurts have a round roof structure and 2 m height [2]. Turkey and Mexico are the countries known for their elongated houses with flat roofs. The roofs played a role of not only protection, but also as a suitable space to spend time with family [1].

P. Mendonça and N. D. Cortiços (Eds.): ICAMC 2021, LNCE 226, pp. 361–368, 2022.
https://doi.org/10.1007/978-3-030-94514-5_36

A roof is commonly known as a building component located on the top of a building that serves to protect it from the precipitation and to keep the comfortable indoor temperature [1]. There are several types of roofs which vary depending on several parameters such as shape, slope and materials used in the construction [1]. This study considers the commonly used two roof types such as “flat” and “gable” roofs, or so called “pitched roofs”. At present, flat roofs are rising in popularity due to their modern designs. The flat roofs tend to have a slope not higher than 5% [1]. Such roofs are also recognized for their higher levels of insulation that help to maintain the building temperature at desired levels and makes the heating of a building easier [3]. The additional advantage is the space on the roof which can be used for leisure and gardening activities, the latter leading to creation of a green roof “coating”.

The flat green roofs are currently gaining popularity around the world, particularly in densely populated cities such as, for example, Singapore, San Francisco, Beijing, Osaka, etc. [4, 5]. The green roof is a type of flat roofs that has a vegetation on top. To be called a green roof, it should be covered by vegetation partially or entirely over the insulation membrane [5]. The insulation and soil layers lead to lower levels of energy use and efficient harnessing of rainwater [5]. A recent study conducted in Singapore has shown that the green roofs could save up to 14.5% of the consumed energy [3]. Moreover, green roofs are ecologically friendly and can serve as gardens spaces, which are quite useful in densely populated cities where there is not much agricultural land. According to recent studies, green roofs are very sustainable in terms of usage of rainwater with the most vegetated green roofs re-using up to 87% of the rainwater, whereas the average green roofs re-use about 83% of water [4, 5]. In addition, such roofs are advantageous from the financial perspective. The rainwater re-use, energy savings and unwanted emissions reduction allowed saving about US $1.04 million (M), $0.87 M and $0.09–0.41 M pear year respectively in Washington DC, USA [6].

The gable roofs, also recognized as pitched roofs, are one of the most common type of roofs in Kazakhstan [7]. Due to the complexity of the construction process, the construction cost of the pitched roof tends to be higher than of the other roof types. However, the main benefit of the gable roof is that the structure of the roof creates a space under the roof, attic, which could be used as a living or storage area, thus, increasing the overall useful area of the building [8]. However, there are cases when the space under the pitched roof is left unfinished and unused. This creates a source of inefficiency in terms of energy use needed for heating and cooling.

There is a lack of research and literature around the topic of roof types and their impact on energy use and overall sustainability in Kazakhstan. Considering the specific climate conditions of Kazakhstan as well as the gap in the field, this article aimed to conduct a comparative analysis in which the assessment has been done for different roof types and their respective energy consumption levels. Two cases, specifically, with finished and unfinished attics were compared using numerical energy simulation tools to analyze the amount of energy used. Moreover, the study analyzed two types of flat roofs (flat/green) and compared them with the gable roof. The findings are deemed to be useful for construction industries of developing countries in terms of selection of roof types. Also, the study compared the expenses such as the construction cost and annual energy consumption expenses of all four roof types.

2 Methodology

2.1 Location

The model analysis may vary due to the choice of a location. In this study, the location was chosen as Nur-Sultan city (formerly, Astana), the capital of Kazakhstan. The city is located in the northern Kazakhstan, on the banks of the Ishim river [9]. The city is ranked as the second coldest capital in the world [9]. Nur-Sultan has a very dry climate with high temperatures during summers and very low temperatures in winter. The lowest monthly-average temperature in Nur-Sultan is –24 °C in January and the highest is 33 °C in June and July. However, the temperature extremes may go down to −34.7 °C in winters and up to 36.6 °C in summers (Table 1).

Table 1. Location details.

Page	Mean	Min	Max
Dry Temp. (°C)	4.3	−34.7	36.6
Wet Temp. (°C)	−2.4	−38.4	20.2
Relative Humidity (%)	64.9	10	100
Snow Depth (cm)	33.3	0	85

The city is surrounded by the hills from the East and West and located in a shallow basin. The capital city's altitude varies from 300 up to 500 m and averages at 347 m. The city has lines of trees on the North, West and East that protect it from the strong winds pertinent to the locality [9]. Nur-Sultan's cold weather during winters is affected by the Asiatic high, the pressure system that covers the region mostly in winter. It dominates starting from September, peaks in December and ends by May [9]. During this time, cold and dry winds move into the area from the East. In addition, the inversions caused by the Asiatic high save moisture and pollution and with the contribution of the shallow basin in which Nur-Sultan is located lower the visibility and result in cold air [9].

2.2 Software and Analysis

In this study, the site, source, and heating energy consumptions were calculated using the EnergyPlus software. Also, the Building Energy Optimization (BEopt) energy simulation tool was used for the comparison of roof types and obtaining more accurate results. BEopt is a software used to find the most energy efficient building design [10]. The software allows the user to select the options to be included in the optimization. The design process allows to simply create a model and changing parameters such as, for example, wall and ceiling materials, HVAC system, windows, etc. The analysis results show the energy use, construction cost and allows choosing the optimal design in terms of energy use and cost efficiency according to the specific location [10].

BEopt applies the DOE2 and TRNSYS engines for simulation purposes and automates the process of finding the most efficient design by means of a sequential search strategy [10]. The discrete building methods allow identifying the most energy and cost-efficient design options. BEopt runs various scenarios defined by the user and identifies the most energy efficient one. The software calculates the energy use and savings according to a reference model, the user's own design or designed according to the integrated in the software Building America Benchmark [10].

BEopt 2.8 version uses a linear search method and finds positive and negative interactions between different categories, providing an optimal combination of options [10]. The approximate costs of the building constructions were calculated via BEopt 2.8.0 function. All of the materials' prices were taken as standard values inserted in the software. The authors of the research have previously used various simulation tools to perform energy performance simulations [11–15].

2.3 Building Model

This study aims to analyze the average annual energy use and cost of the average-sized two-story house in Kazakhstan. The typical house for Kazakhstan was assumed to have a gable roof and unfinished attic (Fig. 1.b). It was compared to building models with a flat roof, flat green roof, and gable roof with finished attic (Fig. 1).

a) Flat roof/green roof b) Gable roof with unfinished attic c) Gable roof with finished attic

Fig. 1. Simulated building types.

The gable roof with unfinished attic is considered as a building with two floors of living space and a not usable attic. This is a commonly used type of roof and is included in this analysis to show the excess amount of energy that is dissipated comparing to different house types. All the building models have the same living area equal to 251 m^2 (Table 2). The building plan was generated automatically by the software and includes 3 bedrooms and 2 baths. All the building models were assumed to have the same parameters and components such as wall thickness, window openings, ceiling, floors, etc. The only structural difference between the models is the type of materials used in roof coating as the flat roofs tend to have better insulation.

Table 2. Simulation details.

Parameter	Value
Building Area	251 m^2
Duration of simulation	8760 h (1 year)

The two models with gable roofs (Figs. 1.b, 1.c) both have standard R-30C Fiberglass Batt insulation with R-value of 30.6 (Tables 3, 4). Flat roofs were insulated with R-38C Fiberglass insulation, which has twice-higher R-value (61.2). The gable roofs were coated with dark-colored asphalt shingles, which have high absorptivity and emissivity (0.92 and 0.91 respectively). The flat roof has a PVC coating with low absorptivity (0.1) and emissivity (0.3), whereas the green roof has both absorptivity and emissivity values equal to 0.8. The rest of the materials are the same for all the models and were chosen based on the B10 Benchmark.

Table 3. Differences in roofing materials

Parameter	Flat green roof	Flat roof	Unfinished attic	Finished attic
Roof insulation	R-38C Fiberglass Batt	R-38C Fiberglass Batt	R-30C Fiberglass Batt	R-30C Fiberglass Batt
Roof coating	Green roof coating	PVC	Asphalt Shingles, Dark	Asphalt Shingles, Dark

Table 4. Properties of the materials used in roof insulations (obtained from the BEopt 2.8)

Material type	R-Value	Absorptivity	Emissivity
R-38C Fiberglass Batt	61.3	–	–
R-30C Fiberglass Batt	30.6	–	–
Asphalt Shingles, Dark	–	0.92	0.91
PVC	–	0.1	0.3
Green roof coating	–	0.8	0.8

3 Results and Discussion

3.1 Energy Consumption

Results show that one of the most common types of roofs in Kazakhstan, the gable roof with the unfinished attic, is the least financially preferable type of roofs and has the highest construction cost. In addition, it is the least ecologically friendly and has the highest annual energy consumption values. Table 5 shows the annual energy consumption of four different roof types. The flat roof has 2.9% higher annual site energy and 4% higher heating energy amounts than the flat green roof. EnergyPlus simulation showed that the flat roof has 286.9 GJ of source energy, 2.04% higher amount than the flat green roof model. However, the BEopt simulation provided only 0.28% difference.

The most energy-efficient type of roofs was found to be the gable roof with the finished attic. In case if the attic is used as a living space, the roof structure has a significant effect on the energy and cost savings. The flat green roof has a small difference from the flat roof in terms of the annual energy consumption. From the economic perspective, it will take about 9 years to return the installation cost of the green roof with the money saved from energy.

Table 5. Results obtained from 1-year simulation

	Flat green roof	Flat roof	Unfinished attic	Finished attic
Site Energy [GJ]	201.62	207.61	215.59	170.91
Source Energy [GJ]	280.95	286.81	295.86	245.35
Energy Use for Heating [GJ]	146.96	153.12	160.93	117.39
Source Energy (BEopt) [GJ]	276.3	277.1	291.2	244.7

The results show that the unfinished attic model has the highest site, source and heating energy use equal to 215.6 GJ, 295.9 GJ and 160.9 GJ respectively. The space under the roof left as a non-residential leading to 20.7% increase in terms of site energy, 17.1% increase in source energy and 27.1% increase in heating energy consumptions comparing to the model with finished attic.

However, the gable roof is the most preferable only as a two-floor house. For the one-floor house, more than US $11,000 could be reduced from the construction price by building the 130 m^2 house with flat or the green roof instead of the gable roof with the unfinished attic.

Table 6. Differences in expenses depending on the roof type

Type of expense	Flat green roof	Flat roof	Unfinished attic	Finished attic
Installation cost [USD]	51951	50935	62825	46083
Annual Energy costs [USD]	2465	2576	2682	2323

Table 6 provides the construction and the annual energy consumption costs for four different roof types. The unfinished attic model has the highest installation and annual energy costs, equal to US $62,825 and $2,682 respectively. It has 26.7% higher construction cost and equivalent of US $360 higher annual energy consumption cost than the finished attic model. The green and flat roofs have medium construction costs totaling to US $51,951 and $50,935 respectively. As for the annual energy consumption, flat green roof has $111 lower costs than the flat roof.

Since the gable roof is the most difficult in terms of construction, there is no obvious economical explanation for its popularity amongst the residential house owners. One of the reasons for the popularity of the gable roofs in Kazakhstan could be the lack of knowledge in the construction sphere. Most of the people that build houses stick to the time-tested gable roof construction. As the results indicate, the selection of this roof type leads to extra expenses. This study should be followed by conducting the physical analysis of the similar-sized houses with different roof types. The software may have errors and the physical analysis will validate the results. The source energy values show that depending on the software, the results may have significant changes. Therefore, to obtain the more accurate results, further research using different software is preferable.

4 Conclusions

In this study, the flat roof, green roof, and gable roof with finished and unfinished attics were compared. The results show that the gable roof with an unfinished attic has the highest installation and energy consumption costs. Conversely, the most energy efficient and low-cost type of roofs is the gable roof that has a finished attic. A one-storey house with a finished attic is the best choice for a medium-sized family. However, the gable roof is the most preferable only as a two-storey house. For the one-storey house, more than $11000 could be reduced from the construction price by building the 130 m^2 house with a flat or a green roof instead of the gable roof with an unfinished attic. The comparison of the green and flat roofs shows that green roof has a slight benefit in terms of energy efficiency. From the financial perspective, the green roof savings will pay back the installation cost in about 9 years.

References

1. Ulubeylia, S., Kazazb, A., Erb, B., Birgonulc, M.T.: Comparison of different roof types in housing projects in turkey: cost analysis. Procedia. Soc. Behav. Sci. **119**, 20–29 (2014)
2. National Geographic. https://www.nationalgeographic.org/encyclopedia/yurt/. Accessed 24 May 2021
3. Assessment of the contributions of different flat roof types to achieving sustainable development. Build. Environ. (141), 182–192 (2018)
4. Wonga, N.H., Cheonga, D.K.W., Yana, H., Soha, J., Ongb, C.L., Siab, A.: The effects of rooftop garden on energy consumption of a commercial building in Singapore. Energy Build. **35**, 353–364 (2003)
5. VanWoert, N.D., Rowe, D.B., Andresen, J.A., Rugh, C.L., Fernandez, R.T., Xiao, L.: Green roof stormwater retention: effects of roof surface, slope, and media depth. J. Environ. Qual. **34**(3), 1036–1044 (2005)
6. Niu, H., Clark, C., Zhou, J., Adriaens, P.: Scaling of economic benefits from green roof implementation in Washington. DC. Environ. Sci. Technol. **44**(11), 4302–4308 (2010)
7. Owens Corning. https://www.owenscorning.com/en-us/roofing/tools/roof-types. Accessed 24 May 2021
8. Wang, S., Shen, Z.: Effects of roof pitch on air flow and heating load of sealed and vented attics for gable-roof residential buildings. Sustainability **4**(9), 1999–2021 (2012)
9. Higdon, M.: Astana, Kazakhstan: A Full-Year Study (2006)
10. Christensen, C., Anderson, R., Horowitz, S., Courtney, A., Spencer, J.: BEopt(TM) Software for Building Energy Optimization: Features and Capabilities. US Department of Energy (2006)
11. Tokbolat, S., Al-Zubaidy, S., Badr A.: Low energy design for future housing developments in Kazakhstan: case study. Energy Effic. **9**, 211–222 (2016)
12. Tokbolat, S., Tokpatayeva, R., Al-Zubaidy, S.: The effects of orientation on energy consumption in buildings in Kazakhstan. J. Sol. Energy Eng. Trans. ASME (Am. Soc. Mech. Eng.) **135**(4), 0199–6231 (2013)
13. Gaitov, R., Tokbolat, S., Zhong, H., Mustafa, M.: Investigation of the impacts of design parameters on thermal performance of buildings in cold climate. In: E3S Web of Conferences. EDP Sciences, vol. 246, p. 07002 (2021)

14. Tokbolat, S., Naizabekov, Y., Mariani, S.: Energy performance of different facade types in residential buildings. In: E3S Web of Conferences. EDP Sciences, vol. 172, p. 24010 (2020)
15. Nadeem, A., Abzhanov, Y., Tokbolat, S., Mustafa, M., Sørensen, B.R.: The impacts of climate zone, wall insulation, and window types on building energy performance. In: El Dimeery, I., et al. (eds.) Design and Construction of Smart Cities. JIC Smart Cities 2019. Sustainable Civil Infrastructures, pp. 270–277. Springer, Cham (2021). https://doi.org/10.1007/978-3-030-64217-4_31

Assessment of the Effect of Architecture Morphology on Building Energy Demands Using BIM, Example of Kabul City Residential Buildings

Ahmad Walid Ayoobi[1,2], Mehmet Inceoğlu[3], and Oğuz Çolak[4(✉)]

[1] Department of Architecture, Graduate School of Sciences, Eskisehir Technical University, Eskisehir, Turkey
a_w_a@eskisehir.edu.tr
[2] Department of Architecture, Faculty of Construction, Kabul Polytechnic University, Kabul, Afghanistan
[3] Department of Architecture, Faculty of Architecture and Design, Eskisehir Technical University, Eskisehir, Turkey
[4] Department of Mechanical Engineering, Engineering Faculty, Eskisehir Technical University, Eskisehir, Turkey

Abstract. Architecture morphology as a wide concept encompasses all aspect of building design like form, shape, windows, etc., and all these constituents have a vital role in building energy demands. Considering the rapid growth of urbanization and human activities, ecosystem is faced with destruction. Buildings as the most resource consumer take the first position in ecosystem demolition.

Buildings morphological analysis is one of the methods to assess buildings energy demands and to provide the opportunity to select the most favorable morphology. This study aims to assess architecture morphology on residential building energy efficiency using BEM in Kabul city.

In this study, we made an effort to review most of the energy efficiency analysis associated with the literature. Consequently Revit Architecture 2018 for 3D modeling and Green Building Studio GBS as the whole building energy simulation engine that estimates the building energy performance based on the effects and interactions of building form, materials, systems, usage, and climate were utilized. Kabul city was selected as the case study of this work. The findings of this study show that old or courtyard houses are characterized by the most energy-efficient morphology compared to the three main types of building morphology located in Kabul city. The buildings morphology has a significant impact on building energy demand and, the comparison of three buildings morphology in Kabul city confirmed that the building morphology could decrease energy demand up to 44%.

Keywords: Morphology · Building · Energy demand · Energy efficiency · Kabul city · BEM tools

P. Mendonça and N. D. Cortiços (Eds.): ICAMC 2021, LNCE 226, pp. 369–380, 2022.
https://doi.org/10.1007/978-3-030-94514-5_37

1 Introduction

Morphology in architecture is the study of evolution in the forms and environment where it is built. According to [1], morphology has a vital role in the architecture designing process that can make a linkage between buildings and their environment. In the contemporary age, the ecosystem is being destroyed due to the rapid growth of urbanization and non-standard building designs resulting in the reduction of human life wellbeing. More than 40% of the total world's energy consumption is originated from the buildings sector [2–8]. Also, energy consumption in buildings, including ventilation, heating, and cooling systems accounts for more than 60% of total buildings energy demands [9–11].

According to [12, 13], the morphology of buildings has a significant impact on building cost and energy efficiency. In architecture design, the morphological analysis gives a chance to architects to create freely and systematically many solutions for designing project in preliminary stages according to predefined needs, creative imagination, knowledge and skill [14]. Several studies have been carried out on the impact of buildings morphology on energy demands and application of Building Energy Modeling BEM for buildings morphology analysis according to energy efficiency; There is a strong relationship between building morphology and energy consumption that paying more attention to morphology can help to elevate the energy efficiency [15]. To control the energy requirements and efficiency in building design, various factors e.g. glazing, elements and components, orientation, shape, and size of the building have to be considered [16]. The optimized morphology of the buildings environment can have a significant impact of above 20% on energy consumption reduction [17]. An appropriate building morphology can provide more energy-efficient forms than conventional ones [18].

Kabul city as a the capital of Afghanistan has an arid and semiarid climate with hot summers and cold winters [19]. The wide weather fluctuations during the year have resulted in higher energy consumption for heating and cooling of housing [20]. From the housing type aspects, the most famous type of housing in Afghanistan is a building with a private surrounded yard where more than 95% of the population of the country is living [21]. In Kabul city, three types of residential buildings morphology with the private yard are found e.g. old or courtyard houses, modern, and contemporary types that each of them has its morphology and provides specific energy efficiency features. According to [22], currently, residential buildings in Kabul city during the warm seasons consume electricity and natural gas and in cold seasons, wood, coal, charcoal along natural gas are the main energy sources, cause to produce smoke resulting in outdoor pollution. In addition, the rapid growth of urbanization and non-standard buildings design raised environmental pollution and caused the degradation of natural recourses [23].

A diverse of studies have been conducted on building sustainability and energy demands in Kabul city as [16, 20, 21, 24–27]. All the aforementioned literature focused on urban morphology and energy efficiency in buildings. However, there is not any study carried out to investigate the morphology of the buildings as a vital parameter of passive design and energy efficiency, specifically in Kabul city. Furthermore, the impact of typical building morphologies on energy efficiency in Kabul city has not been assessed comparatively. In addition, there is lack of study on the favorable selection of a typical building form to be adaptive with Kabul environmental condition from energy perspective.

This study aims to highlight the impact of building morphology on energy demand and efficiency, and to characterize the weakness and strengthen points of Kabul city residential buildings morphology considering energy consumption. Lastly, the residential buildings morphology in Kabul city is compared in terms of energy efficiency to target the most favorable morphology.

2 Material and Methods

Kabul city as a merge point of diverse cultures and ethnic groups in the country is selected as an input in this study. The city includes three types of buildings morphology with private yard; The first type is known as the old type which can be found over the old parts of the city (Serai Lahori, Deh Afghana, etc.) and is followed by courtyard house form where the building encloses a courtyard with rooms which are built around a courtyard in two or three floors and have been regulated according to diurnal rotation, climate and culture [28]. Each room has a window through the courtyard and has independent access to the courtyard. The second type is considered as a modern or new type of houses which is distributed all around the city and is currently perceived as the most famous types of houses type in the city. In this type, the building is located in the center of the yard and surrounded by walls. Each room has an independent connection with outside by windows but due to the limited orientation of rooms, some of them do not have a view through the yard and cause to decrease in the quality of rooms according to passive energy and daylighting. Therefore, based on the importance of the rooms and environmental conditions, the main rooms such as living room, guest room, bed room etc. are placed to the north side or yard side, while the sub-spaces e.g. toilets, stairs, kitchens, etc. are oriented to other sides.

The third type is contemporary houses which is following the second type morphology, however, based on some evolutions, and has been emerged after civil wars in 2001 characterized by the impacts of regional countries. This issue has caused many living problems to Afghanistan families from environment, economy, and comfortability perspectives. Building is generally design based on regional country culture and climate which are facing with lack of enough courtyard space; represent massif construction, fitted with colored glasses facades, deep awning elements, and following hot climate architecture style. The architecture diversity in Kabul city between new and contemporary houses morphology reflects the integration of different identities seeking to catch up with modernity, at the same time preserving their traditional way of life [29]. In this study, the effect of buildings morphology on energy consumption was analyzed on three types of typical buildings morphology in Kabul city. The pre-existed buildings with the most famous type of housing morphology in Kabul city (modern type) have been selected, subsequently, the pre-existed buildings morphology was justified to another two morphology, (old and contemporary). Three buildings morphology has been analyzed in the same location with the same condition for the comparison purposes (Table 1).

Table 1. Description of Pre-existed and proposed models.

Profile	Pre-existed Building	Proposed Morphologies for Pre-existed Building	
	Modern Type	Contemporary Type	Old/Courtyard Type
	Model 1	Model 2	Model 3
3D Illustration			
Ground Floor Plan	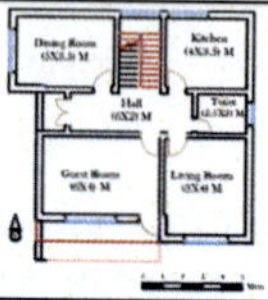	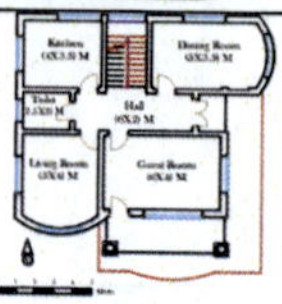	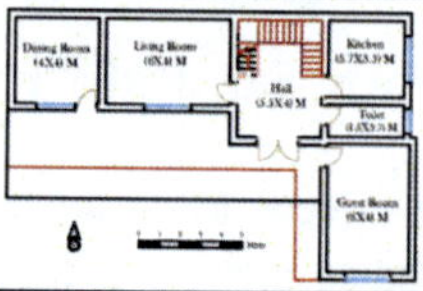
First Floor Plan	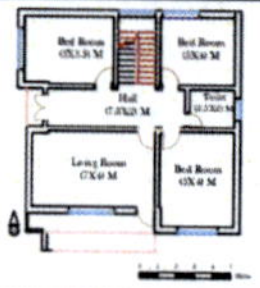	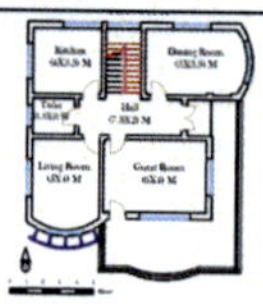	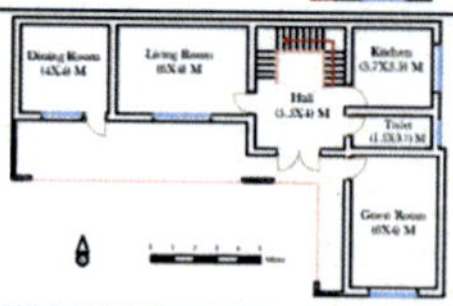
Window relative to room size	25%	35%	25%
Orientation	One side window room	Two side window room	One side window room
Area		189 m2	
Volume		532.63 m3	

Rivet Architecture 2018 was used for modeling the projects and preparing the inputs for Green Building Studio (GBS) using the DOE 2.2 dynamic thermal whole building energy simulation engine to estimate building energy use and operating costs. The estimations are carried out based on the effects and interactions of building form, materials, systems, usage, and climate (Fig. 1).

2.1 Energy Analysis/BEM Tools

BEM tools are powerful to study the energy performance and thermal comfort during the building's life-cycle [30]. It is found that 24% and 33% of energy loads can be reduced in buildings by using a simulation- based optimization method [31]. BEM in architectural design focuses on how design elements and building morphology can lead to energy savings, and help you reduce the energy footprint of your buildings. Many BEM tools have been developed that a more comprehensive list of programs can be found at the U.S DOE's Energy Efficiency and Renewable Energy Building Energy Software Tools Directory [32].

Among BEM tools, Rivet Architecture is considered as a top-rated software provided by National Building Specification (NBS) [33], and Green building studio (GBS) plaguing have been added to Rivet for buildings energy simulation [34]. Autodesk Green Building Studio is web-based energy analysis software that can help architects and designers perform whole-building analysis, optimize energy efficiency, and work toward carbon neutrality earlier in the design process. In this study, GBS add-on to Revit was used for the analysis of all three types of building morphology (Model 1–3) being contributed to simulate the annual and monthly energy consumption of each model and also we compared them to find the most energy-efficient morphology of houses at different climate conditions.

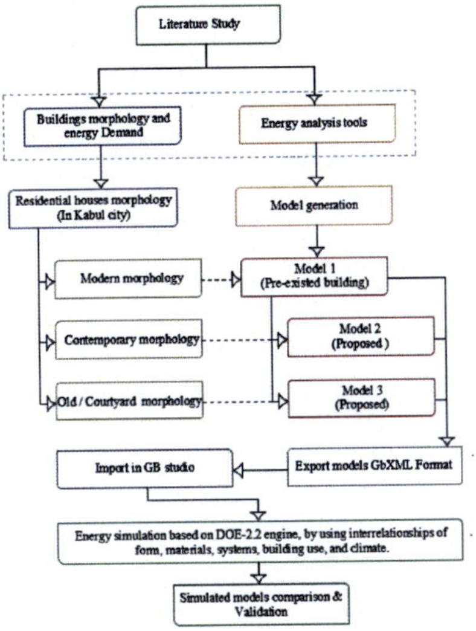

Fig. 1. Study flow chart

2.2 Model Generation and Energy Simulation

A typical pre-existed residential building (Model 1) located in Kabul city between (34° 31′23.00″N) latitude and (69° 6′14.20″E) longitude with two other proposed models (2 & 3) were selected as an input model for analysis. The pre-existed model has been designed by Kabul city municipality with the modern type of morphology, and the rest of the two other models (Model 2&3) have the same specification of Model 1 with old and contemporary morphology as described in (Table 1). Normally each model was designed on two floors to contain the ground floor with a guest room, living room, dining room, kitchen, toilet, and stairway to access the first floor. The first floor consists of a living room, three bedrooms, and a toilet. 3D modeling, selection of geographical location, assigning the construction material, and selection of HVAC system of all proposed and pre-existed models was carried out in the Rivet 2018 environment as depicted in (Table 2).

In addition, the climatic condition of the study area was analyzed in Rivet architecture, the average annual wind direction in Kabul city is most often from the north for five months, from May 3 to October 6, with a peak percentage of 73% on July 13 and from October 6 to May 3 for seven months the wind is most often from the west with a peak percentage of 45% on January [35]. Based on the analysis, the weather within the study area reaches +38 °C during the summer season and to −14 °C during winter as the result is also confirmed by [22].

Table 2. Description of the selected energy analysis parameters

NO	Parameters	Description
1	Location	District-5, Kabul city (34°31′23.00″N) lat, (69° 6′14.20″E) lon
4	HVAC	Central VAV, HW Heat, Chiller 5.96 COP, Boilers 84.5 eff (Default)
5	Wall	Brick masonry + Plaster, 35 cm thickness
6	Window	Single glazed windows - domestic
7	Door	Wood panel door
8	Roof	RCC + Low insulation
9	Floor	RCC + Un isolated

The outputs of gxml files from Rivet were sent to GBS for energy simulation, respectively. Finally, GBS tool automatically analyzed the models separately, and the models associated energy consumptions percentages for electricity is described in (Fig. 2a), while the Fuel gas is stated in (Fig. 2b).

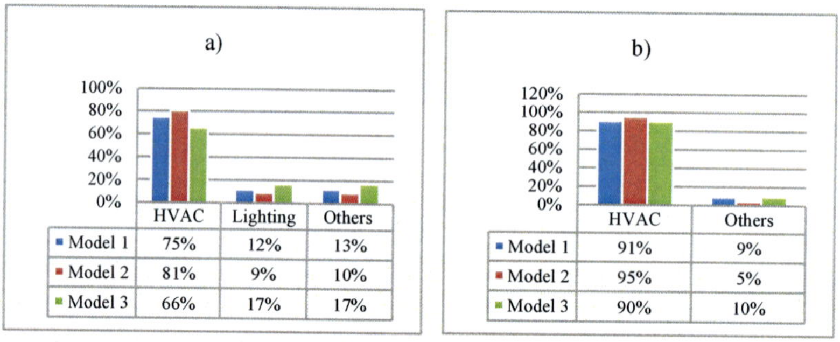

Fig. 2. Energy consumptions percentages a: Electricity consumption for each model, b: Fuel gas consumption for each model.

Based on this analysis, and selected (HVAC) system for each model, electricity is basically used for space cooling, lighting and some other sub-issues such as pump & Aux, misc, equipment, fans, rarely space heating, whereas, the fuel gas is used for heating and hot water purposes. In this study, we are considering on energy demands of buildings from heating and cooling perspectives.

The building design strategy and morphology have a significant link with buildings energy demand and sustainability which can be achieved by appropriate architecture design and energy analysis programs at the preliminary steps of design. In addition, building morphology without space cooling and heating have a critical effect on other issues of building such as lighting, ventilation, etc. which is shown in (Fig. 2 a and b), in this study, more focused was done on space heating and cooling as the highest energy consumer in buildings.

3 Results and Discussion

Considering the geographical location of Kabul city in both the northern and eastern hemispheres; the building's opening should be faced to the south to get the most benefit from solar energy [16]. The amount of energy consumption in buildings at different period of the year depends on building windows orientation and morphology. Architecture forms and patterns must expose a good sun light and wind orientation for better ventilation in winter and nominal ventilation in summer seasons to achieve the buildings energy efficiency or passive energy demands.

In this study, a residential building in Kabul city with two proposed morphology as described in (Table 1 and 2) was analyzed to achieve the most energy-efficient morphology. The obtained results of the annual energy intensity for one pre-existed building (Model 1) and two other proposed models (Model 2,&3) at the same location with different morphology for estimated life cycle (30 years span) is shown in (Fig. 3).

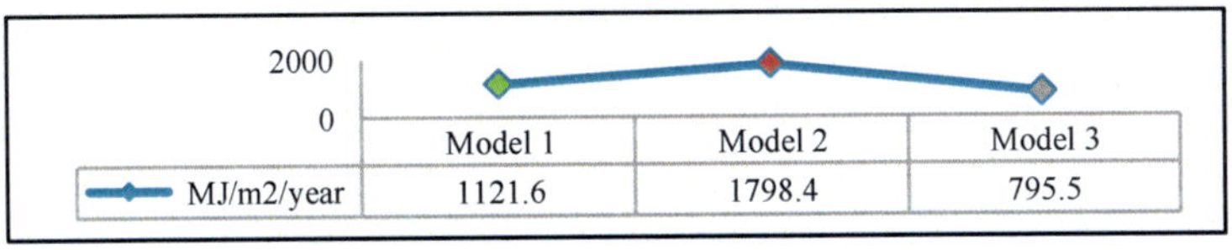

Fig. 3. Diagram showing the annual energy intensity of each model

[13, 36, 37] stated that optimization and energy efficiency strategies can be achieved by defining optimum orientation, shape and forms, compactness and window percentages. In this study, buildings orientation, windows dimensions and architecture pattern or morphology for sunlight control were mostly focused during the energy efficiency analysis by GBS. According to comparative diagram (Fig. 3), old type courtyard house designed with Afghanistan traditional architecture style (Model 3) with 25% window size relative to room size, and orientation of all room to south shows the lowest energy demand (795.5 $MJ/m^2/yr$) and achieved the first position of energy efficiency. In addition to the old type of courtyard houses in Kabul city, another two types of morphology such as modern and contemporary types can be found which has been established according to cultural influences. The architecture diversity in Kabul city between new and old-style reflects the mixture of different Afghan identities seeking to catch up with modernity while at the same time preserving their traditional way of life [29]. In this study, (Model 1) as a typical modern type of residential building for Kabul city designed by Kabul municipality with 25% window size relative to room size, obtained the second position from an energy efficiency perspective with (1121.6 $MJ/m^2/yr$) energy as shown in (Fig. 3).

In contemporary days, the decorative massive facade, a balcony and colored glazing and wider windows represent contemporary hoses morphology in Kabul city [29]. This type of morphology has been identified by regional country influences after 2001 and has become the famous type of houses in Kabul city. According to [21], the influenced architecture and lifestyle by unsuitable regional and international architecture style in Kabul city clash with local culture and climate. In this study, Model 2 with 35% window size relative to room size, representing the contemporary building morphology which is proposed for the same location to compare with old and new types of residential building morphology. Based on the analysis, comparative results between the proposed models are shown in (Fig. 3). Models 2 achieved the final position with (1798.4 MJ/m^2/yr) energy demand, respectively. The results show that this type of morphology is not suitable with Kabul city local climate and culture, therefore, must be changed or emerged with old and new types of building morphology.

Based on comparative analysis of distinct types of residential buildings morphology in terms of energy efficiency in Kabul city using GBS tools, most parts energy consumption in buildings are dedicated to space cooling and heating demand which can be controlled by appropriate design strategy and selecting a suitable morphology.

According to obtained result the buildings energy demand significantly depends on climate condition and building morphology. Each model with its morphology shows the wide range of energy consumption for heating, cooling, lighting and ventilation demands during the summer and winter seasons.

4 Validation

In this study, the obtained results from BIM tools were validated using the actual electricity consumption bills. The actual electricity consumption of pre-existed model (Model 1) was compared with simulated electricity consumption of the same model.

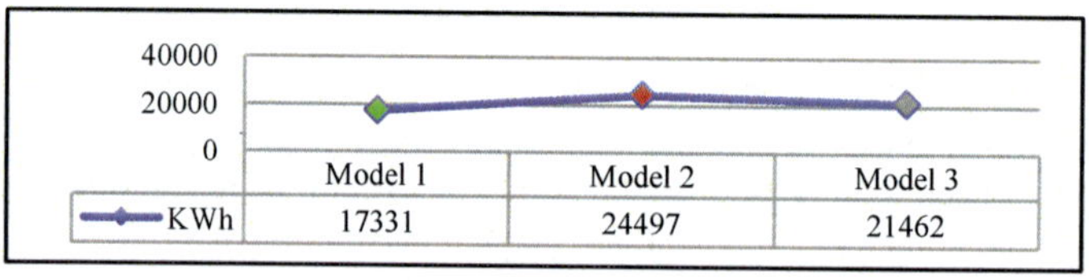

Fig. 4. Annual electricity consumption of each model according to GBS simulation

Electricity consumption of each simulated model by GBS is shown in (Fig. 4) that simulated annual electricity consumption of a pre-existed building (Model 1) is 17331 KWh, while the actual annual electricity consumption of building from previous records based on Afghanistan Electricity board/electricity bells was 4362 KWh, between (2019–2020). The actual electricity bills for every 60 days are shown in (Table 3) of an existed building (Model 1).

Table 3. Actual electricity consumption of pre-existed (Model 1) between (2019–2020) based on electricity bills.

Periods	Electricity consumption for 60 days (KW/h)
1	851
2	499
3	538
4	374
5	1001
6	1099
Total	4362

The simulated values of electricity using GBS is 4 times greater than the actual bill values. This is because of electricity power cut up to 15 h per day on peak consumption time e.g., in summers daytime and winters in night times [38]. This made the people to use alternative energy sources as private electricity (generator), wood and domestic techniques. Therefore, a coefficient factor was added to validate the result. The simulated electricity consumption for (Model 1) estimate for 24 h is 17331 KWh. Considering the above 12 h electricity usage, the obtained value is multiplied by a coefficient factor (0.5) and based on power cut effect on peak time is multiplied by a coefficient factor (0.5). If data assumes consumption for only 12 h, then the simulated values are for 24 h is 17331 KWh. For 12 h of supply, 0.5 is subtracted from multiplier 1, also another 0.5 is deducted to the new multiplier due to the power cut at peak times, so the final multiplier of 0.25 is used to validate the actual situation of the building. The revised simulated value of electricity 4333 KWh (17331*0.25) seems very close to the actual values. The actual values are 0.7% greater than the revised simulated values. The allowable percentage error between the simulated and the actual data must be ±15% only then the software tool used for the simulation is reliable [39, 40].

There are many other errors which are out of focus; the actual orientation of the building, the number of people living in the building varies from month to month as also their financial stature, and there are many assumed factors in (GBS) such as the real building elements, heat transfer factors (U values) which might not be compatible with Kabul city buildings facility.

5 Conclusion

There are diverse factors affecting buildings sustainability and energy efficiency such as climate conditions, passive design behavior, buildings envelop, heating and cooling systems, building services, occupants' energy demand etc. Moreover, architectural morphology factors including form, glazing, shape, patterns, facade elements etc., have essential impacts on buildings energy demands.

Kabul city includes three types of houses morphology; The first type is known as old type or courtyard houses style, the second type is considered as a modern or new type of houses which represent the integration of old styles, and the third type is the contemporary type which has been established over the last years and following the neighboring countries morphology. Basically each of them introduces a specific morphology and represents distinct energy efficiency in different period of the year.

Based on BEM analysis, old type or courtyard houses forms according to the orientation of all room to the south exposed to good sunlight and having appropriate passive ventilation systems and façade elements were found as the most energy-efficient building morphology in Kabul city. The new type of houses morphology based on limited rooms orientation to the south side and similar passive strategy with courtyard morphology secured the moderate morphology in terms of energy demands. The contemporary houses according to Kabul city environmental condition suitability and following neighbor countries morphology with decorative façade and more glazing percentage features show the most energy consumer morphology in Kabul city. Based on a comparison of the annual energy intensity of each model in (Fig. 3) it is confirmed that energy consumption of the most energy consumer model (model 2) is 44% greater than most energy-efficient model (model 3). Therefore building morphology as a most important element of passive design strategy has a significant impact on building energy efficiency and sustainability. It is proposed to consider the building morphology following the environment condition at the essential stages of architecture design to find the most suitable energy-efficient morphology.

Acknowledgment. We would like to thank Professor Dr. Bashir Ahmad KAZIMEE for his valuable comments and helping us to improve the paper.

References

1. Hanson, J.: Morphology and design reconciling intellect, intuition, and ethics in the reflective practice of architecture (2001)
2. Saadatian, O., Haw, L.C., Sopian, K., Sulaiman, M.Y.: Review of windcatcher technologies. Renew. Sustain. Energy Rev. **16**, 1477–1495 (2012). https://doi.org/10.1016/j.rser.2011.11.037
3. Harish, V.S.K.V., Kumar, A.: A review on modeling and simulation of building energy systems (2016). https://www.sciencedirect.com/science/article/pii/S1364032115014239?casa_token=IzqJP6gu1wsAAAAA:3hjcRzR4rEFSaCdQFKyknhhb2LaaAiq_dMbHW3tFo_vikz_QG0HFO0GV1rSG-Paacw7Zv0RD6M_M, https://doi.org/10.1016/j.rser.2015.12.040
4. Khamma, T.R., Boubekri, M.: Statistical analysis of impact of building morphology and orientation on its energy performance. J. Eng. Archit. **5**, 15–25 (2017). https://doi.org/10.15640/jea.v5n1a2
5. Kaza, N.: Understanding the spectrum of residential energy consumption: a quantile regression approach. Energy Policy (2010).https://doi.org/10.1016/j.enpol.2010.06.028
6. Ewing, R., Rong, F.: The impact of urban form on U.S. residential energy use. Hous. Policy Debate. **19**, 1–30 (2008). https://doi.org/10.1080/10511482.2008.9521624
7. Ratti, C., Baker, N., Steemers, K.: 2005 Energy consumption and urban texture. Energy Build.https://doi.org/10.1016/j.enbuild.2004.10.010
8. Steemers, K.: Energy and the city: density, buildings and transport. Energy Build. 3–14 (2003). https://doi.org/10.1016/S0378-7788(02)00075-0
9. Adamczyk, J., Dylewski, R.: The impact of thermal insulation investments on sustainability in the construction sector (2017). https://doi.org/10.1016/j.rser.2017.05.173

10. Hoseini, G., et al.: Sustainable energy performances of green buildings a review of current theories, implementations and challenges. Renew. Sustain. Energy Rev. **25**, 1–17 (2013)
11. Tunali, S.: Using building energy simulation methods as a decision support system in the building design process (2012). https://tez.yok.gov.tr/UlusalTezMerkezi/tezSorguSonucYeni.jsp
12. Catalina, T., Virgone, J., Iordache, V.: Study on the impact of the building form on the energy consumption. In: Proceedings of the Building Simulation 2011 12th Conference International Building Performance Simulation Association, pp. 1726–1729 (2011)
13. AlAnzi, A., Seo, D., Krarti, M.: Impact of building shape on thermal performance of office buildings in Kuwait. Energy Convers. Manag. (2009). https://doi.org/10.1016/j.enconman.2008.09.033
14. Prokopska, A.: Application of morphological analysis methodology in architectural design. Acta Polytech. **41** (2001). https://doi.org/10.14311/194
15. Alhassan, M.K.A.: A study on the effect of building morphology on energy consumption of detached houses in Sudan (2020). https://www.researchgate.net/publication/342159732, https://doi.org/10.13140/RG.2.2.34567.01442
16. Yarramsetty, S., Rohullah, M.S., Sivakumar, M.V.N., Anand Raj, P.: An investigation on energy consumption in residential building with different orientation: a BIM approach. Asian J. Civ. Eng. **21**(2), 253–266 (2019). https://doi.org/10.1007/s42107-019-00189-z
17. Calcerano, F., Martinelli, L.: Numerical optimisation through dynamic simulation of the position of trees around a stand-alone building to reduce cooling energy consumption. Energy Build. (2016). https://doi.org/10.1016/j.enbuild.2015.12.023
18. Okeil, A.: A holistic approach to energy efficient building forms. Energy Build. (2010). https://doi.org/10.1016/j.enbuild.2010.03.013
19. Qutbudin, I., et al.: Seasonal drought pattern changes due to climate variability: case study in Afghanistan. Water (Switzerland). **11**, 1096 (2019). https://doi.org/10.3390/w11051096
20. Fazeel Hosham, A., Kubota, T.: Field study of passive techniques and adaptive behaviour in the traditional courtyard houses of Kabul. IOP Conf. Ser. Earth Environ. Sci. **294**, 1–10 (2019). https://doi.org/10.1088/1755-1315/294/1/012039
21. Sharifzai, M.S., Kitagawa, K., Halimee, M.K., Habib, J., Sakaguchi, D.: A Comparative study of afghan traditional and contemporary courtyard housing regarding affordable planning and sustainability. World Acad. Sci. Eng. Technol. **10**, 412–417 (2016). https://zenodo.org/badge/, https://doi.org/10.5281/zenodo.1123721,. svg
22. Mehrad, A.T.: Causes of air pollution in Kabul and its effects on health. Indian J. Ecol. **47**, 997–1002 (2020)
23. Wafa, W., Hairan, M.H., Waizy, H.: The impacts of urbanization on Kabul City's groundwater quality. Int. J. Adv. Sci. Technol. **29**, 10796–10809 (2020)
24. Rostami, R., Khoshnava, S.M., Lamit, H., Streimikiene, D., Mardani, A.: An overview of Afghanistan's trends toward renewable and sustainable energies. Renew. Sustain. Energy Rev. **76**, 1440–1464 (2017). https://doi.org/10.1016/j.rser.2016.11.172
25. Temori, M.O.: Energy efficiency guidebook for buildings. Deutsche Gesellschaft fur Internationale Zusammenarbeit (GIZ) GmbH (2017)
26. Hidayat, O., Kajita, Y.: Application of sustainable development principles in developing countries; focus on Kabul City urban planning. J. Eng. Res. Appl. **9**, 30–41 (2019). https://doi.org/10.9790/9622
27. Kazimee, B.A.: Sustainable reconstruction and planning strategies for Afghan cities: conservation in cultural and environmental heritage. WIT Trans. Ecol. Environ. **93**, 49–59 (2006). https://doi.org/10.2495/SC060051
28. Kazimee, B., Mcquillan, J.: Living traditions of the afghan courtyard and Aiwan. Tradit. Dwell. Settl. Rev. 23–34 (2002)

29. Issa, C., Kohistani, S.: Kabul's urban identity: an overview of the sociopolitical aspects of development. Asien **104**, 51–64 (2007)
30. Maile, T., Fischer, M., Bazjanac, V.: Building energy performance simulation tools - a life-cycle and interoperable perspective. Cent. Integr. Facil. Eng. 1–49 (2007)
31. Foroughi, R.: New approach in designing a kinetic window shading using optimization methods (2018)
32. BEST Directory: BEST Directory: Building Energy Software Tools. https://www.buildingenergysoftwaretools.com/. Accessed 15 Nov 2020
33. National Building Specification 2018. The National BIM Report 2018. Natl. Build. Specif. (2018)https://doi.org/10.1017/CBO9781107415324.004
34. Autodesk: Revit | BIM Software | Autodesk. https://www.autodesk.com/?_ga=2.41620645.607518862.1613829219-489894097.1613829219
35. Weather-Spark: Average Weather at Kabul International Airport Afghanistan. https://weatherspark.com/y/148954/Average-Weather-at-Kabul-International-Airport-Afghanistan-Year-Round. Accessed 14 Mar 2021
36. Menezo, C., Lepers, S., Depecker, P., Virgone, J.: Design of buildings shape and energetic consumption. Build. Environ. **36**, 627–635 (2001)
37. Ling, C.S., Ahmad, M.H., Ossen, D.R.: The effect of geometric shape and building orientation on minimising solar insolation on high-rise buildings in hot humid climate. J. Constr. Dev. Ctries. **12**, 27–38 (2007)
38. DABS: Da Afghanistan Breshna Sherkat (DABS). https://main.dabs.af/News/NewsDetail/2064
39. Maamari, F., Andersen, M., de Boer, J., Carroll, W.L., Dumortier, D., Greenup, P.: Experimental validation of simulation methods for bi-directional transmission properties at the daylighting performance level. Energy Build. **38**, 878–889 (2006). https://doi.org/10.1016/j.enbuild.2006.03.008
40. Reeves, T., Olbina, S., Issa, R.: Validation of building energy modeling tools: Ecotect^{TM}, Green Building Studio^{TM} and IES^{TM}. In: Proceedings of the Winter Simulation Conference (WSC) (2012). https://doi.org/10.1109/WSC.2012.6465223

Thermal Comfort and Energy Consumption in Healthcare Buildings – A Review

Siyu Ma[1,2], Yuanli Ma[1(✉)], Qianxi Zhang[1,3], Wu Deng[1], Jun Lu[1], and Tongyu Zhou[1]

[1] Department of Architecture and Engineering, University of Nottingham, Ningbo, China
{saxsm4, zyl5469}@nottingham.edu.cn
[2] School of Architecture and Transportation Engineering, Institute of Architecture, Ningbo University of Technology, Ningbo, China
[3] School of Design, Institute of Architecture and Urban Planning, Ningbo Tech University, Ningbo, China

Abstract. As special public buildings, healthcare building have high requirements for thermal comfort and are also large energy consumption. This paper reviews the thermal comfort and energy saving of healthcare buildings. It discusses the importance of thermal comfort in healthcare buildings and the current research in different regions. In order to solve the problem of thermal discomfort, air conditioning energy consumption is the most important part of healthcare buildings' total energy consumption. Finally, this paper puts forward the prospect: analysing the quantitative relationship between thermal comfort and energy consumption is meaningful to energy saving of healthcare buildings.

Keywords: Healthcare building · Thermal comfort · Energy saving

1 Introduction

With the continuous and rapid development of China's economy and the government's attention to people's livelihood issues, healthcare buildings are in a period of rapid expansion. As a special public building, healthcare building has complex space [1] and diverse occupants [2]. Healthcare buildings have strict thermal environment requirements. American Society of Heating, Refrigerating and Air-Conditioning Engineers (ASHRAE) explicitly states that healthcare buildings should pay attention to indoor temperature and humidity to avoid the rapid growth of bacteria and viruses, especially for spaces with a high degree of air cleanliness ASHRAE [3].

On the other hand, healthcare buildings are big energy consumers. According to China Association of Building Energy Efficiency's latest report, the energy consumption of public buildings was 238 million tons of carbon emission, taking up 38% of the total operational energy consumption with only 17% of the construction area [4]. Moreover, the energy consumption of healthcare buildings in China is 1.6 to 3 times that of general public buildings, far exceeding the average level of developed countries [5]. Optimising the energy consumption of healthcare buildings can contribute to China's carbon neutral target.

P. Mendonça and N. D. Cortiços (Eds.): ICAMC 2021, LNCE 226, pp. 381–390, 2022.
https://doi.org/10.1007/978-3-030-94514-5_38

Therefore, healthcare buildings have two contrasting goals: to keep the indoor thermal environment comfortable and, in the meantime, to reduce energy consumption. On the one hand, a good and stable indoor thermal environment requires energy consumption; on the other hand, optimised energy strategies should not affect indoor thermal comfort [6]. Therefore, research on thermal comfort and energy consumption of healthcare buildings has great significance in order to achieve indoor thermal comfort and energy reduction in tandem.

2 Thermal Comfort

Thermal comfort refers to the psychological state that people are satisfied with the thermal environment [3]. Since the factors and conditions affecting human thermal comfort are very complex, many experimental studies have been conducted since the 1920s. Based on the interaction of different factors, people have put forward many evaluation indexes of thermal comfort [7]. As a result, various countries and institutions formulated the thermal comfort standard to provide reference and basis for specific thermal state values for indoor thermal environment design [8]. These standards are based on Fanger's model described in detail in Sect. 2.1. ASHRAE standard and ISO standard are currently widely used in the world as thermal comfort indicators.

Thermal comfort model is a mathematical model for human evaluation of thermal environment. After more than 40 years of development, the thermal comfort model has become more and more elaborate. At present, the commonly used thermal comfort models include PMV-PPD model, multi-node multi-segment model, and adaptive model.

2.1 PMV-PPD Model

Generally, Fanger's predicted mean vote (PMV) model put forward in 1970 is the most widely used thermal comfort model [9]. Fanger's model is based on data from 1396 healthy young subjects aged between 20 and 60 and is used in an invariant air-conditioned environment in moderate thermal climate space [10]. Fanger applied logic to derive three conditions affecting thermal comfort: clothing, activities, and environment and carried out a curve regression analysis of thermal comfort with six main physical quantities: air temperature, air velocity, radiant temperature, humidity, metabolic rate, and level of clothing [11].

2.2 Multi-node Multi-segment Model

Limited by requiring a constant environment, the accuracy of PMV-PPD varies widely among ventilation strategies, building types, and climate groups [10]. However, the multi-node multi-segment model focuses more on the human body's response to the instantaneous temperature change, therefore, is more advantageous to the transient state [6].

As early as 30 years before Fanger proposed the PMV-PPD model, Gagge had already proposed the two-node model from another perspective. The two-node model abstracts the human body into two concentric zones: the layout (skin), which is considered to account for 10% of the body's mass, and the core, metabolic heat is produced in the core thermal compartment [12]. However, Gagge's two-node model was too crude. Until 1971, when Stolwijk proposed a more detailed multi-segment model, the multi-point multi-segment model had just taken shape and began to show its advantages in the evaluation of transient thermal comfort [13]. Stolwijk's model consists of five cylinders and a ball representing the torso, limbs, and head. Furthermore, each segment is abstracted into concentric cylinders made up of core, muscle, fat, and skin [14]. After Stolwijk, more and more accurate multi-point and multi-segment models were proposed, among which the Berkeley Comfort Model and AUD Model were relatively famous.

2.3 Adaptive Model

Compared with the PMV-PPD model, the multi-point and multi-segment model can simulate the transient thermal comfort of instability. However, humans have the ability to adapt to the environment and are not passive recipients of the thermal environment [3]. Therefore, an adaptive empirical model based on the observation of human behaviour is proposed. Compared with the multi-point and multi-segment model, it also considers the influence of culture, climate, society, psychological and behavioural adaptation factors on thermal comfort [15].

As early as in the 1970s, Humphreys proposed this thermal adaptive theory from the phenomenon that the actual thermal comfort of users in naturally ventilated buildings is higher than predicted by PMV [16]. The basic assumption of this theory is that if the thermal environment does not satisfy people's thermal comfort, people's behaviour will instinctively adjust to the direction of comfort [17]. Due to the complexity and uncontrollability of human behaviour, the adaptive model is difficult to have a clear formula and model like the PMV-PPD model or the multi-point and multi-segment model [15]. The adaptive model is based on a large number of experimental data and is drawn according to the Black Box theory [18].

3 Thermal Comfort in Healthcare Buildings

Medical research indicates that good care environments can reduce the 30-day mortality and improve healthcare staff's work quality. Aiken et al. conducted a study on the correlation between work environment, nurse' working outcomes, and patients' treatment outcomes of 10,184 nurses and 232,342 surgical patients. Studies showed that hospitals with suitable care environments had a 14% lower 30-day mortality rate than those with poor care environments Aiken, Clarke [19]. Friese et al. studied the influence of average reported patients per nurse, proportion of nurses per hospital with bachelor's degree or higher, and nurse practice environment on working efficiency based on the datasets of Pennsylvania cancer centres. They found that all the indicators that can measure the quality of healthcare staff's work in favourable environments are far lower than those in unfavourable or mixed working environments Friese, Lake [20]. Aiken et al. analysed and studied the workforce stability of nurses in 1,150 hospitals in the United States and Europe. It was concluded that the most critical factors affecting nurses' work efficiency were nurse staffing and work environments. Among them, improving working environments was a relatively low-cost strategy for improving healthcare. In addition, different from nurse staffing which had better ratios depending on the country and the type of hospital, improving the working environment was universal in improving staff's working efficiency Aiken, Sermeus [21]. Blum and Wang investigated the working efficiency of healthcare staff in a Parrish medical centre in the USA before and after improving the working environment. They found that after improving the natural lighting, noise interference, thermal comfort, and air quality, the working efficiency of healthcare staff increased by 10%, and turnover intention decreased by 14% Blum and Wang [22]. In addition, Australian authoritative medical guidelines, the New South Wales Guidelines (2019), also pointed out the importance of thermal comfort in healthcare buildings. The improper indoor thermal environment can lead to medical staff's mistakes in prescribing or injecting [23]. To sum up, a good care environment directly affects the work efficiency of healthcare staff and the health of patients. As one of the three main elements of indoor environment, thermal environment is vital in healthcare buildings.

There are quite a few researches on thermal comfort of healthcare buildings abroad. For example, Khalid et al. studied the comfortable temperature and airflow acceptance rate of patients, visitors, and nurses in the inpatient ward of a private hospital in Malaysia. The comfortable temperature for patients, visitors, and nurses is 25.9 °C, 25.5 °C, and 23.5 °C, respectively; the acceptance of airflow is higher than 91% for all occupants Khalid, Zaki [24]. Lomas and Giridharan measured and simulated the thermal environment of free-running wards in a British hospital and predicted the improvement of the resilience of the ward to climate change by electric fans in scorching years. The experimental results showed that in the case with fans, the period satisfying BSEN15251 Thermal Comfort Categories I was 12% more on average than that of the case without the fan Lomas and Giridharan [25]. Balaras et al. investigated and studied the thermal comfort and energy consumption of HVAC in 20 operating rooms of 10 hospitals in Greece. The results show that the indoor temperature range is 13.8–28.9 °C, averaging 22 °C; relative humidity range is 13–80%, averaging 37%; the number of air changes per hour range is 3.2–58 ACH, existing the problem of insufficient indoor air exchange and poor control of thermal conditionsBalaras, Dascalaki [26]. Van Gaever et al. compared the thermal comfort of healthcare staff and thermal comfort standards in the operating room of four hospitals in Belgium. He used the standard ISO 7730 to predict the general thermal comfort of all surgical staff in the operating room. The survey results showed that only the surgical assistance thought the thermal environment comfortable; the anesthetist and nurse thought the environment was too cold, while the surgeon and surgeon with lead apron thought the environment was slightly hot. Van Gaever et al. concluded that the thermal comfort of different types of healthcare staff varies greatly, and the current technical means cannot make the indoor thermal environment meet all occupants' requirements Van Gaever, Jacobs [27]. These studies cover temperature, humidity, air-flow rate and pollutant concentration. Studies conducted comprehensive field investigation within the healthcare facilities, and then evaluate whether the thermal environment met the guidelines.

In China, the research team in Chongqing University has done a series of in-depth research in this field. Feng examined the applicability of the PMV-PPD model of patients in summertime and wintertime in Chongqing. He predicted a preferred temperature of 22.2 °C for the winter and 25.1 °C for the summer. However, in the analysis of humidity, only a humidity sensation vote was made. Feng concluded that in summer, humidity is moderate, but in winter, it is slightly too low [28]. Liu identified the thermal comfort range for patients with different health and activity level in summer and winter in various departments. The thermal comfort range of patients is 20.1–23.8 °C in winter and 25–26.7 °C in summer [1]. Yao studied occupants' thermal comfort in hospital wards in asymptomatic high altitude areas in Xining. It was found that with the increase of altitude, occupants' thermal neutral operative temperature decreases in summer and winter [29]. In addition, Chen conducted simulation analysis and research on applying cold radiation air conditioning in healthcare buildings in Guangzhou in the summer [30]. Jin studied the influence of the outer envelope of healthcare buildings on energy-saving and thermal comfort in winter in the severe cold area of Northeast China [31]. Peng investigated the thermal environment of maternity waiting rooms in Guangzhou in summer, spring and winter, and found that the thermal environment basically met pregnant women's requirements [32]. Table 1 shows the domestic research on the thermal comfort of healthcare buildings. In summary, China is a vast country, and its climate varies significantly between different regions. Therefore, the research on thermal comfort varies greatly in different regions. The research on cold areas is more focused on winter, while the research on hot summer and warm winter areas is more focused on summer. There are relatively few studies on transition season and humidity.

Table 1. Summary of domestic research on thermal comfort of healthcare buildings.

Author Name (year)	Research Area	Research Contents		Research Seasons		
		Temperature	Humidity	Summer	Winter	Transition Season
Liu (2014)	Chongqing	√	√	√	√	×
Feng (2015)	Chongqing	√	×	√	√	×
Jin (2016)	Severe cold area	√	×	×	√	×
Yao (2017)	Xining	√	×	√	√	×
Chen (2017)	Guangzhou	√	√	√	×	×
Peng (2020)	Guangzhou	√	×	√	√	√

4 Energy Saving in Healthcare Buildings

In terms of energy consumption composition of healthcare buildings, HVAC system plays a dominant role in the energy consumption of healthcare buildings. Li and Lv conducted an extensive survey on the energy consumption of general hospitals in China. Air-conditioning systems consumption accounted for 65% of the total energy consumption, medical equipment for 10%, lighting for 9%, medical disinfection for 8%, elevators for 2%, and others for 6% Li and Lv [33]. According to the energy consumption analysis of more than 30 healthcare buildings in Shanghai, the consumption of air-conditioning systems accounted for about 56.7% of the total energy consumption, while in the United States, the proportion was only 36%–48% [34]. Hu and Yang believed this is one of the main reasons why China's medical buildings consume so much more energy than those in the United States. In conclusion, the energy consumption of air-conditioning system in healthcare buildings can be optimised to a large extent. Therefore, it is significant to study the relationship between thermal comfort and energy consumption for building energy conservation.

Energy saving of healthcare buildings can be divided into active energy saving and passive energy saving. The active energy saving of healthcare buildings can be realized by two ways: reducing the average energy consumption of air conditioning equipment and optimizing the energy structure [35]. Xie et al. Conducted energy-saving optimisation on the temperature and humidity control of air conditioning in clean operating rooms in summer and winter. They proposed energy saving schemes for the actual operation to reduce the energy consumption of the air conditioning system as low as possible while ensuring the thermal comfort and the air cleanliness [36]. Moreno et al. installed a sensible heat recovery system to the air conditioning in the operating room of a Spanish hospital and optimised the air flow rate. This system recycled heat while maintaining air quality. And the results showed that circulating 25% of the air saved 24.1% of energy consumption [37]. It can be seen that the most effective way of active energy saving at present is to optimise the running status of HVAC quantitatively, such as setting reasonable temperature, humidity and air return volume, and managing the running time scientifically.

Passive energy saving usually starts from three aspects: roof, exterior wall and exterior window. Table redesigned an energy-efficient hospital in Abu Dhabi. He optimised the shading, exterior wall, exterior window and roof of the hospital based on IES software analysis. And it was concluded that cool roof was the most potential optimisation strategy - after using cool roof, energy consumption was reduced from 29 kW to 21 kW [38]. Zhang and Gu used eQUEST to simulate and analyse the renovation of roof, exterior wall and exterior window of a hospital in Shanghai. They concluded that in the case of retrofitting an existing building, adding maintenance structural insulation was the most economical approach. When the insulation layer is 60 mm thick, the energy saving rate is the highest, reaching 4.39% [39]. Cesari et al. studied the impact of exterior window design on energy consumption in hospitals in four Italian cities. They analysed the influences of the window-wall ratio (WWR), exterior window orientation, and the U value and g value of glass on the energy consumption of heating, cooling and lighting. It is concluded that the thermal insulation performance of exterior window is more important for the energy saving of Italian hospitals because the heating energy consumption accounts for 82% of the total energy consumption [40]. The three parts of passive energy saving, although the impact of the roof on the thermal environment is greater than the exterior wall than the exterior window [5], different projects tend to optimise the most economical part according to their own characteristics.

In conclusion, energy saving strategies for healthcare buildings are summarized as Table 2. Active energy saving involves the field of HVAC, including air-conditioning system Settings; Passive energy saving can be divided into three categories: roof, exterior wall and exterior window. In addition to the important influencing factor of roof, the orientation, U value, shading coefficient, and WWR are all important factors affecting energy consumption.

Table 2. Summary of energy saving in healthcare buildings.

	Categories	Specific strategies
Active energy saving	HVAC system	Operative temperature
		Operative humidity
		Air return volume
		Operation mode
Passive energy saving	Roof	Cool roof
	Exterior wall	U value
		Orientation
		Shading coefficient
	Exterior window	Window to wall ratios (WWR)
		U value
		Orientation

5 Discussion and Future Research Works

From all the reviews so far, there has been considerable research in the area of thermal comfort and energy efficiency in healthcare buildings. However, few studies have developed a quantitative approach to optimise the design strategies considering thermal comfort and energy consumption in tandem. Based on this, our future work is threefold:

- Thermal comfort model of healthcare building

Different from the multi-node multi-segment model and adaptive model, the PMV model does not take into account the self-regulation of human body to the thermal environment. APMV model is an adaptive model based on PMV model, which is suitable for steady thermal environment and takes into account people's adaptation to the environment, and can more accurately reflect the real thermal sensation of occupants. Therefore, it is important to establish an aPMV model suitable for healthcare buildings to improve the later optimisation accuracy.

- Quantitative relationship between thermal comfort and energy consumption

At present, few studies have been conducted to research the quantitative relationship between thermal comfort and energy consumption. However, the establishment of a quantitative relationship between uncomfortable time and energy consumption is the basis for subsequent optimisation.

- Multi-objective optimisation of thermal comfort and energy consumption

Multi-objective optimisation based on the quantitative relationship between thermal comfort and energy consumption can provide data support for the formulation of green building guidelines for healthcare buildings.

6 Conclusions

This article has reviewed the literature related to thermal comfort and energy saving in healthcare buildings. The conclusion are:

- Thermal comfort of healthcare buildings is vital to improve the working efficiency of healthcare staff and reduce the medical negligence.
- The research on thermal comfort of healthcare buildings in different climate zones in China has different emphases and results.
- There is a lot of energy saving space for healthcare buildings to optimise the energy consumption of air conditioning system while ensuring thermal comfort.
- Energy saving of healthcare buildings can be started from four aspects: HVAC, roof, exterior wall and exterior window.

References

1. Liu, X.: The study on indoor thermal environment and comfort In: Hospital Wards. Chongqing University (2014)
2. Del Ferraro, S., et al.: A field study on thermal comfort in an Italian hospital considering differences in gender and age. Appl. Ergon. **50**, 177–184 (2015)
3. ASHRAE: IEEE/ASHRAE Guide for the Ventilation and Thermal Management of Batteries for Stationary Applications. IEEE, pp. 1–116 (2018)
4. CABEE: China Building Energy Consumption Research Report (2020)
5. Shi, Y.: Research on energy consumption distribution and energy-saving design of building layout in the 3a hospitals in cold regions. Xi'an University of Architecture & Technology (2020)
6. Cowie, A.R.: Numerical optimisation of building thermal and energy performance in hospitals. University of Leeds (United Kingdom), Ann Arbor (2017)
7. Taleghani, M., et al.: A review into thermal comfort in buildings. Renew. Sustain. Energy Rev. **26**, 201–215 (2013)
8. De Dear, R.J., Brager, G.S.: Thermal comfort in naturally ventilated buildings: revisions to ASHRAE Standard 55. Energy Build. **34**(6), 549–561 (2002)
9. van Hoof, J.: Forty years of Fanger's model of thermal comfort: comfort for all? Indoor Air **18**(3), 182–201 (2008)
10. Cheung, T., et al.: Analysis of the accuracy on PMV – PPD model using the ASHRAE global thermal comfort database II. Build. Environ. **153**, 205–217 (2019)
11. Fanger, P.O.: Thermal comfort. Analysis and applications in environmental engineering. Thermal comfort. Analysis and applications in environmental engineering (1970)

12. Du, H., Yang, C.: Re-visitation of the thermal environment evaluation index standard effective temperature (SET) based on the two-node model. Sustain. Cities Soc. **53**, 101899 (2020)
13. Roelofsen, P.: A comparison of the dynamic thermal sensation between the modified Stolwijk model and the Fiala thermal physiology and comfort (FPC) model. Intell. Build. Int. **12**(4), 284–294 (2020). London
14. Robert, M.M.: Thermal Comfort: Analysis and Applications in Environmental Engineering, pp. 116–117. Wayne State University Press, Detroit (1973)
15. Yao, R., Li, B., Liu, J.: A theoretical adaptive model of thermal comfort – adaptive predicted mean vote (aPMV). Build. Environ. **44**(10), 2089–2096 (2009)
16. Brager, G.S., De Dear, R.J.: Thermal adaptation in the built environment: a literature review. Energy Build. **27**(1), 83–96 (1998)
17. Nicol, J.F., Humphreys, M.A.: Adaptive thermal comfort and sustainable thermal standards for buildings. Energy Build. **34**(6), 563–572 (2002)
18. Zhang, S., Lin, Z.: Adaptive-rational thermal comfort model: adaptive predicted mean vote with variable adaptive coefficient. Indoor Air **30**(5), 1052–1062 (2020)
19. Aiken, L.H., et al.: Effects of hospital care environment on patient mortality and nurse outcomes. J. Nurs. Adm. **38**(5), 223 (2008)
20. Friese, C.R., et al.: Hospital nurse practice environments and outcomes for surgical oncology patients. Health Serv. Res. **43**(4), 1145–1163 (2008)
21. Aiken, L.H., et al.: Patient safety, satisfaction, and quality of hospital care: cross sectional surveys of nurses and patients in 12 countries in Europe and the United States. Bmj (2012)
22. Blum, A., Wang, X.: How hospital design saves lives. Design changes can cut infection rates, lower pyhsician errors, improve staff performance. Urban Environ. Des. **Z3**, 116–117 (2011)
23. NSW: NSW Guidelines (2019)
24. Khalid, W., et al.: Thermal comfort requirements for different occupants in Malaysian hospital in-patient wards. J. Adv. Res. Fluid Mech. Therm. Sci. **43**(1), 128–140 (2018)
25. Lomas, K.J., Giridharan, R.: Thermal comfort standards, measured internal temperatures and thermal resilience to climate change of free-running buildings: a case-study of hospital wards. Build. Environ. **55**, 57–72 (2012)
26. Balaras, C.A., Dascalaki, E., Gaglia, A.: HVAC and indoor thermal conditions in hospital operating rooms. Energy Build. **39**(4), 454–470 (2007)
27. Van Gaever, R., et al.: Thermal comfort of the surgical staff in the operating room. Build. Environ. **81**, 37–41 (2014)
28. Feng, H.: Study on the thermal comfort. In: Hospital Wards. Chongqing University (2015)
29. Yao, D.: The study on indoor thermal environment and human thermal comfort in asymptomatic high altitude wards. Chongqing University (2017)
30. Chen, Y.: The application research of ceiling radiant cooling panels system in medical building in Guangzhou. Guangzhou University (2017)
31. Jin, W.: Research on energy efficiency design of medical building envelope in the severe cold areas. Harbin Institute of Technology (2016)
32. Peng, T.: Study on human thermal comfort in obstetric waiting area of hospitals in Guangzhou. Guangzhou University (2020)
33. Li, Z., Lv, F.: Discussion on air conditioning design and energy saving for hospital buildings. Heat. Vent. Air Cond. **39**(04), 21–23 (2009)
34. Hu, Y., Yang, G.: Energy usage and saving in hospitals. Heat. Vent. Air Cond. **39**(04), 1–4 (2009)
35. Zhaoliang, Y.: Potential analysis and method research of hospital building energy-saving. Chang'an University (2012)

36. Yong, X., Hailong, S., Xiaoming, Y.: Energy efficiency design of air cleaning system in hospital clean operating room based on humidity priority control. Heat. Vent. Air Cond. **47** (09), 99–104 (2017)
37. Moreno, G.S.-B., et al.: Sustainable solutions for thermal energy saving in hospital operating theatres. In: E3S Web of Conferences. EDP Sciences (2019)
38. Taleb, H.M.: Enhancing the skin performance of hospital buildings in the UAE. J. Build. Eng. **7**, 300–311 (2016)
39. Yuhan, Z., Pingdao, G.: Building energy consumption of a hospital in hot-summer and cold-winter region based on the characteristics of enclosure structure. Build. Energy Environ. **37** (12), 98–101+6 (2018)
40. Cesari, S., et al.: The energy saving potential of wide windows in hospital patient rooms, optimizing the type of glazing and lighting control strategy under different climatic conditions. Energies **13**(8), 2116 (2020)

Passive Energy-Saving Technologies for Low-Rise Residential Buildings Based on Building Prototyping: Cixi City as a Case

Yuanli Ma[1], Dongjie Ye[2(✉)], Siyu Ma[1], Qianxi Zhang[1,3], Yuanda Hong[1,4], Jing Xie[1], and Wu Deng[1]

[1] Department of Architecture and Built Environment, University of Nottingham Ningbo China, Ningbo, China
zyl5469@nottingham.edu.cn

[2] Department of Architecture and Art, Zhejiang Business Technology Institute, Ningbo, China

[3] Institute of Architecture and Urban Planning, NingboTech University, Ningbo, China

[4] Shanghai Daren Construction Technology Group Co. Ltd, Shanghai, China

Abstract. Under the background of energy saving and emission reduction, the application of passive energy saving technology and the development of green building have reached a necessary stage. While, different design technologies perform differently in different building types. This research using a hybrid clustering method and generated three prototypical low-rise residential building models, whose information were collected from 13 small residential districts in Cixi City. In addition, on this basis, it made a sensitivity analysis of four passive energy-saving technologies based on the passive house concept's principles. As a result, the impact factor of each technology on three building models were proposed. The results indicate that the thermal insulation of the external wall has the largest effects and the building airtightness should also be focused for energy saving.

Keywords: Energy saving · Prototyping · Residential buildings · Sensitivity analysis · Hot summer cold winter

1 Introduction

With economic growth and the improvement of people's living standards, building energy consumption increases substantially. The severe reality of high energy consumption, low energy efficiency and energy shortage has become a significant factor that restricting social sustainable development. Worldwide, the construction industry already consumes more energy than other major industries, such as industry and transportation [1, 2]. In China, the primary energy consumption of building energy accounts for about 28% of the country's total energy consumption. According to statistics, by 2017, buildings consumed in total 947 million tons of standard coal

P. Mendonça and N. D. Cortiços (Eds.): ICAMC 2021, LNCE 226, pp. 391–398, 2022.
https://doi.org/10.1007/978-3-030-94514-5_39

equivalent (tce). In addition, the buildings carbon emissions were 2.044 billion tons CO2, accounting for 19.5% of the total energy-related carbon emissions in China [3, 4].

In the past years, the governments of developed countries have successively formulated the development goal of building towards zero energy consumption. The application of "nearly zero energy consumption buildings" is an important stage on the path to achieving the goal. In China, the government also decided to achieve a carbon peak by 2030 and achieve carbon neutrality by 2060 On September 2020. Therefore, buildings are the field with the greatest potential for saving energy, it is urgent to put forward a series of energy-saving technologies through scientific innovation.

The development of low-energy buildings develops better in cold region and has already built several successful pilot projects. However, as one of the five major climatic zones in China, the development of low-energy buildings in the hot summer and cold winter (HSCW) zone was still in the beginning stage. The HSCW zone is a transition zone of cold regions and hot regions, characterized by high temperature and humidity in summer and cold in winter. This specific condition leads to the requirements of both heating and cooling for air condition systems. In addition, there were many disputes on the design strategies due to the differences in climate characteristics and living habits between the HSCW zone and Europe or the cold zone in China. For now, there are no low energy buildings in Ningbo. Although Ningbo is generally part of the region of the HSCW zone, it also has its unique climatic condition, architectural traditions and energy usage habits. Lang et al. [5] and Theodoridou et al. [6] believed that architectural typology is the prerequisite for defining a comprehensive architectural classification scheme, and there is a dialectical relationship with technology, history, politics and social process. In addition, some research has found that different buildings have different performance even using the same design strategies [7]. Therefore, it is necessary to generate representative building prototypes and analyse each passive design strategy by sensitivity calculation.

Cixi City, under the jurisdiction of Ningbo, is an important industrial and commercial city in the Yangtze River Delta region (see Fig. 1). It is under rapid development, therefore, contains buildings with a wide span of years from 1993 to 2021. On the climate, it has the coldest month is January with an average daily temperature of 3.3 °C and the hottest month is July with an average daily temperature of 32 °C. The relative humidity in summer can range from 85% to 100%. This research aims to provide an applicable design package for low-rise residential buildings using Cixi City as an example. It can be divided into two parts: generating building prototypes and making sensitivity analysis of passive design strategies on the energy consumption performance.

Fig. 1. The location of Cixi City.

2 Low-Rise Building Prototyping of Cixi City

By the definition of the Uniform standard for the design of civil buildings (GB50352-2019), a low-rise residential building has 1 to 3 the number of floors. In total, there are 499 small residential districts (SRDs) in Cixi city, which is too large to investigate. Therefore, the hybrid method proposed by Ma et al. [8] was used to narrow down the samples. The least number of samples required can be determined by Eqs. (1), (2), and (3).

$$\bar{x} = \sum\nolimits_{i=1993}^{2021} x_i \cdot \frac{n_i}{499} = 2011.784 \tag{1}$$

$$\sigma = \sqrt{\frac{1}{n-1} \cdot \sum\nolimits_{i=1993}^{2021} (x_i - \bar{x})^2} = 7.178 \tag{2}$$

$$n = \frac{\sigma^2}{\frac{E^2}{z^2} + \frac{\sigma^2}{N}} = \frac{7.2^2}{\frac{4^2}{1.96^2} + \frac{7.2^2}{499}} \approx 12.14 \tag{3}$$

Where, σ is the standard deviation of the construction year of SRDs;

E is the sampling error range;

z is the z-value of confidence coefficient, in this research, the z-value is 1.96 based on Ornstein [9];

n is the number of SRDs.

Table 1. The timeframe information of selected SRDs.

Timeframe	Total number of SRDs	Proportion	Number of SRDs selected
Before 1993	1	0.2%	0
1994–2005	109	21.8%	3
2006–2015	184	36.9%	5
After 2016	205	41.1%	5

As a result, totally 13 SRDs were selected as samples (see Table 1). Then, the information of all individual low-rise residential buildings was measured using the following performance indicators:

The height, the shape coefficient (Cf), the window-to-wall ratio (WWR) of each facade, and the area of each floor.

- Height
  The height of buildings can be used as one of the significant standards for civil building classification. The difference in the number of floors leads to the distinction in the level of sunlight received by the building, and also affects the natural ventilation conditions to a certain extent.
- Window-to-wall ratio
  WWR as an index of building geometry represents the ratio of the glazed area to the external wall area of the buildings. For residential buildings located in the HSCW climate zone, the buildings were designed to prevent solar heat in summer but try to gain solar heat in winter by a large WWR on the south facade and minimize WWR on the west and east facade [10].
- Shape coefficient and the area of each floor
  The shape coefficient is one of the critical factors of evaluating building energy efficiency [11]. The definition of this index is the ratio of the external surface area of a building in contact with the outdoor atmosphere and the volume enclosed by it. In residential buildings, the layout of each floor is similar, mainly determined by the length and width of the building. Although most residential buildings can be summarized simply in geometric shapes, the simple use of length, width and height will lead to too compact or scattered clustering results, resulting in inaccurate results. According to ASHRAE standard 90.1, Goel et al. [12] suggest using the combination of single-storey area and shape coefficient of buildings to express the size of buildings in prototype development. In this study, the building shape coefficient and the area of each floor are also used as a factor for building prototyping.

Clustering the data of low-rise buildings by the k-means method [13] and elbow method [14], the prototypical building models were put forward as Table 2 shown. It can be concluded in three types [15]:

- Single-family houses: a one or two-story detached or semi-detached house.
- Terraced houses: a one or two-story single house connected to other houses.
- Multi-family houses: small buildings with a limited number of apartments (up to 15 apartments and 2–5 floors, or 16–20 apartments and 2–4 floors).

Table 2. Generated model information by k-means method.

Clusters	Building types	Height (m)	WWR E	WWR S	WWR W	WWR N	Cf	Area of each floor (m^2)	Sketches of model
Model 1	Single-family houses	9.1	0.12	0.31	0.18	0.28	0.47	132.88	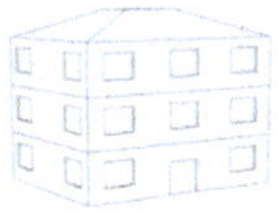
Model 2	Terraced houses	8.9	0.24	0.26	0.15	0.23	0.40	216.06	
Model 3	Multi-family houses	9.2	0.16	0.36	0.18	0.19	0.36	482.35	

3 Sensitivity Analysis of Passive Design Strategies on Energy Consumption

Based on the design principles of "passive house" [16], the thermal insulation of the external wall (Uwall), roof (Uroof), window (Uwindow), and the building airtightness level were conducted as variables for sensitivity analysis of each model proposed.

By Importing the building model into IES-VE energy simulation software, the physical model can be therefore transferred into energy models. Then, each variable was tested in four levels based on the regulation of the Design standard for energy efficiency of residential buildings in hot summer and cold winter zone (JGJ134) and the Design standard for energy efficiency of residential buildings (DB 33/1015-2015). The performance of each level increased from level 1 (the worst) to level 4 (the best), which was shown in Table 3. In order to eliminate the effect caused by the magnitude and the unit, each parameter was dealt with dimensionless data standardization by the z-score method (Using Eq. 4). The results for multivariate regression were listed in Table 4.

$$x^* = \frac{x - \bar{x}}{\sigma} \tag{4}$$

Where, $\bar{x}$ is the mean of the original data, σ is the standard deviation of the original data.

Table 3. The level of design strategies used for simulation.

Design strategies	Uwall (W/m °C)	Uroof (W/m °C)	Uwindow (W/m °C)	Airtightness (h-1 @N50)
Level 1	2	1.1	2.9	1
Level 2	1.6	0.9	2.6	0.75
Level 3	1.2	0.7	2.3	0.5
Level 4	0.8	0.5	2.0	0.25

Table 4. The data processed by z-score method

Design strategies	Uwall	Uroof	Uwindow	Airtightness
Average value	0.90	1.00	2.25	0.63
Standard deviation	0.52	0.52	0.65	0.32
Standardized level 1	1.16	1.16	1.16	1.16
Standardized level 2	0.39	0.39	0.39	0.39
Standardized level 3	−0.39	−0.39	−0.39	−0.39
Standardized level 4	−1.16	−1.16	−1.16	−1.16

These parameters and their related performance levels were then permutated and combined in the following $L_{16}(4^5)$ orthogonal table. The effects on building energy consumption of each scenario can be further simulated by IES-VE software. The regression process was done separately for each kind of model by STATA software using multivariant regression, and the results were put forward as Fig. 2 shown.

Fig. 2. Impact factor of each design strategy on building energy consumption

Figure 2 demonstrates the impact factors of each design strategy among 3 types of building models. In the Rader chart, a higher than 0 value means the design strategy can decrease energy, or adverse. The larger the value is represented this strategy can save more energy as the level increase.

4 Discussion

Combined the effects of heating and cooling together, total building energy consumption is a commonly used indicator for assessing the building performance of interior systems. The above results are all the influences on building energy consumption within the limited range, and the influences may change out of the bounds. Generally, in the HSCW zone, the U-value of external wall and airtightness level are two design strategies that have larger effects than others. While as the number of households increases, the proportion of external walls decreases, so does the impact on building energy consumption. In addition, the airtightness of the building has a greater impact on energy consumption than the thermal performance of the roof and windows and remains highly sensitive. Compared with external wall insulation and airtightness, the influence of window heat transfer coefficient on building sensitivity is relatively low, but its influence on each model is stable. Specific analysis of each type of architectural model is as follows:

For model 1: Uwall > airtightness > Uwindow > Uroof
For model 2: Uwall > airtightness > Uroof > Uwindow
For model 3: Uwall > airtightness > Uwindow > Uroof.

To conclude, for low-rise buildings, the thermal insulation of the external wall has the largest effects, and the influence of building airtightness level on building energy saving cannot be ignored. In addition, although the roof occupies a relatively small ratio. The thermal insulation for the roof is as important as external wall thermal insulation.

5 Conclusion

This study collected the information of low-rise buildings from 13 SRDs in Cixi City based on an architectural prototyping method and established three types of low-rise residential building models. On this basis, it also carried out the sensitivity analysis of four passive energy-saving measures for energy saving. On the prototypical residential building model generating, three physical building models were generated by the k-means clustering method, using the number of building floors, floor area, shape coefficient and window-wall ratio as parameters. These three architectural models contain basic physical architectural form information, which was in accordance with the evaluation results provided by the current "*Energy Saving Design Standard for Residential Buildings in Hot Summer and Cold Winter areas (JGJ134-2010)*", thus verifying the correctness and feasibility of these models. On the sensitivity analysis, gradient change was used to evaluate the energy performance of each energy-saving measure, and a regression model was established to obtain the sensitivity of each factor. Finally, suggestions were given for its application in low-rise residential building types in Cixi City, which is helpful to provide a reference for other kinds of buildings and further urban sustainable development.

Acknowledgments. The research is acknowledged by Cixi Science and Technology Bureau's research project (Project code: CN2019006).

References

1. Zuo, J., Zhao, Z.: Green building research–current status and future agenda: a review. Renew. Sustain. Energy Rev. **30**, 271–281 (2014)
2. Pérez-Lombard, L., Ortiz, J., Pout, C.: A review on buildings energy consumption information. Energy Build. **40**, 394–398 (2008)
3. Harkouss, F., Fardoun, F., Biwole, P.H.: Passive design optimization of low energy buildings in different climates. Energy **165**, 591–613 (2018)
4. Yu, S., Eom, J., Zhou, Y., et al.: Scenarios of building energy demand for China with a detailed regional representation. Energy **67**, 284–297 (2014)
5. Lang, D.H., Kumar, A., Sulaymanov, S., et al.: Building typology classification and earthquake vulnerability scale of central and South Asian building stock. J. Build. Eng. **15**, 261–277 (2018)
6. Theodoridou, I., Papadopoulos, A.M., Hegger, M.: A typological classification of the Greek residential building stock. Energy Build. **43**, 2779–2787 (2011)
7. Filogamo, L., Peri, G., Rizzo, G., et al.: On the Classification of large residential buildings stocks by sample typologies for energy planning purposes. Appl. Energy **135**, 825–835 (2014)
8. Ma, Y., Deng, W., Xie, J., Heath, T., Xiang, Y., Hong, Y.: Generating prototypical residential building geometry models using a new hybrid approach. Build. Simul. **15**(1), 17–28 (2021). https://doi.org/10.1007/s12273-021-0779-6
9. Ornstein, M.: A Companion to Survey Research. Sage, Los Angeles (2013)
10. Hong, Y., Ezch, C.I., Deng, W., et al.: Correlation between building characteristics and associated energy consumption: prototyping low-rise office buildings in Shanghai. Energy Build.**217**, 109959 (2020)
11. Qi, F., Wang, Y.X.: A new calculation method for shape coefficient of residential building using Google Earth. Energy Build. **76**, 72–80 (2014)
12. Goel, S., et al.: Enhancements to ASHERE Standard 90.1 Prototype Building Models. U.S Department of Energy PNNL-23269 (2014)
13. Huang, T., Liu, S., Tan, Y.: Research based on k-means clustering algorithm. Comput. Technol. Dev. **021**(007), 54–57, 62 (2011)
14. Liu, F., Deng, Y.: Determine the number of unknown targets in open world based on elbow method. IEEE Trans. Fuzzy Syst. **1** (2020)
15. Ballarini, I., Corgnati, S.P., Corrado, V.: Use of reference buildings to assess the energy saving potentials of the residential building stock: the experience of TABULA project. Energy Policy **68**, 273–284 (2014)
16. Ma, Y., et al.: Passive house in China's hot summer cold winter zone. IOP Conf. Ser. Earth Environ. Sci. **798**, 012002 (2021)

Thermal Comfort and Adaptive Behaviors in Medical Buildings: A Literature Review

Chaoju Wang[1], Tongyu Zhou[1](✉), Yinxue Lyu[2], Wu Deng[1], and Sung-Hugh Hong[1]

[1] The Department of Architecture and Built Environment, The University of Nottingham Ningbo China, 199 Taikang East Road, Ningbo 315100, China
Tongyu.Zhou@nottingham.edu.cn
[2] The Sino-British College, University of Shanghai for Science and Technology, 1195 Middle Fuxing Road, Shanghai 200031, China

Abstract. Thermal comfort in medical buildings is more complex than in ordinary buildings. The purpose of this article is to review the literature on human thermal comfort in the medical building environment and related adaptability studies. It also discusses some typical research cases worldwide. From the review, the paper concludes that identifying different thermal comfort conditions among different occupants in medical buildings is essential. Protocols for spatial design for healthcare-related institutional facilities deserve a sophisticated understanding of the patients' needs in every way so that architects and designers can help to design a better thermal environment for rehabilitation. In addition, it is found that there is a lack of a framework enabling multidisciplinary investigations, which is considered essential for the study of thermal comfort in medical buildings.

Keywords: Thermal comfort · Adaptive behaviors · Medical buildings · Built environment

1 Introduction

Thermal comfort in medical buildings has received more and more attention in recent years [1]. Florence Nightingale, the founder of modern nursing, was the earliest proponent of the necessity of thermal comfort in a healing environment [2]. A large number of studies are committed to providing optimized thermal environments to accelerate the recovery of patients and improve the work efficiency of medical staff. Due to the particularity of medical buildings, their requirements for comfort are often different from other types of public buildings. For instance, in medical buildings, the thermal comfort of patients and staff may be possibly applied to a lower level to reduce the risk of surgery and avoid infection [1]. For another example, a reasonably designed medical working environment can improve the efficiency of work [3]. In recent decades, there have been many well-known green building certification systems that have standards specifically for medical buildings and facilities, such as the LEED (Leadership in Energy & Environmental Design) for Healthcare [4], BREEAM (Building Research Establishment Environmental Assessment Method) Healthcare [5], and

P. Mendonça and N. D. Cortiços (Eds.): ICAMC 2021, LNCE 226, pp. 399–407, 2022.
https://doi.org/10.1007/978-3-030-94514-5_40

Evaluation Standard for Green Hospital Building (GB/T 51153-2015) in China [6]. Table 1 shows the details of key thermal factors in medical buildings from the applications handbook of healthcare facilities published by ASHRAE (American Society of Heating, Refrigerating and Air-Conditioning Engineers) [7]. The regulation of temperature and humidity limits the reproduction and activity of bacteria so as to reduce the risk of infection and the spread of disease in medical buildings.

Table 1. Key thermal factors in medical buildings by ASHRAE [7]

Function space	Patient room	New-born nursery suite	Labour/delivery/recovery	Patient corridor	General office area
Design temperature, °C	24 ± 1	22–26	24 ± 1	–	≤24
Relative humidity %	30 (winter), 50 (summer)	30–60	30 (winter), 50 (summer)	–	30–60

Thermal comfort has an important impact on human behavior and psychology. At the same time, people also make a series of adaptive behaviors in response to the current thermal comfort level. There are a lot of studies focusing on this aspect. According to the research from Skoog et al., the adaptability of people in medical buildings is different from that in residential or office environments [8]. The survey pointed out that although the activity level of staff may vary through the day, they rarely have the opportunity to adjust their clothing all of the time. Patients can easily control their clothing levels, but they lack the authority to adjust to environmental facilities such as air conditioning and ventilation systems.

The purpose of this study is to comprehensively review and summarize the research results related to the thermal environment in medical buildings in recent 20 years. It has been noted that, due to the complexity of medical buildings, most of the research could only cover a part of the spaces. This paper begins with a review of the different indices used to measure thermal comfort in medical spaces. Then it summarizes the most studied spaces and their focus of the selected research. In addition, it is worth noting that patients and medical staff tend to have different perceptions, needs and adaptive behaviors to the thermal environment, which are also discussed in this work.

2 Thermal Comfort Index Models

Thermal comfort research integrates several sciences, such as physiology, building physics, mechanical engineering, and psychology. According to Nicol's handbook in 1993, understanding the importance of thermal comfort has three reasons [9]:

- Provide people with satisfactory conditions,
- Control energy consumption,
- Make suggestions and develop standards.

In recent decades, researchers have developed nearly 60 thermal comfort index models, each of these indices have their advantages and drawbacks [10]. Thermal indices are widely used in terms of analysis on human heat exchange, designing building environments, and forecast purposes [11]. These indicators include atmospheric parameters, such as wind speed, temperature, humidity, solar radiation, and physiological parameters such as metabolic rate and etc. For example, the Universal Thermal Climate Index (UTCI) introduced in 1994 is based on the multi-node dynamic thermal physiology Fiala model. It takes dry bulb temperature, relative humidity, solar radiation, and wind speed [12]. Standard effective temperature (SET) is a reasonable and most comprehensive comfort temperature index [13]. It is calculated by two physiological parameters (skin temperature and skin moisturization) [14]. Physiological equivalent temperature (PET) is a thermal index derived from the heat-balance model of a human body [15]. Comparing to PMV, PET is more focusing on personal states and geographical conditions. Fanger [16] proposed the predictive mean vote (PMV) to predict the average of individuals by using four environmental parameters (including dry blub temperature, mean radiant temperature, wind speed and relative humidity) and two human parameters (clothing insulation and metabolism). PPD (predicted percent of dissatisfied) is extended from PMV, whose scope ranges from 0 to 100 [17].

Table 2. Comparing thermal perceptions in various bioclimatic indices.

	PMV	PET	SET	UTCI
Adopted in references	[8, 15, 19–22]	[23]	[24]	[25]
Thermal sensation	Indices range			
Very cold (extreme cold stress)	−3	<4	–	<−40
(very strong cold stress)		–		−40 to −27
Cold (strong cold stress)	−2.5	4–8		−27 to −13
Cool (moderate cold stress/moderate hazard²)	−1.5	8–13	<17	−13 to 0
Slightly cool (slight cold stress)	−0.5	13–18		0 to +9
Comfortable (no thermal stress/no danger)	0	18–23	17–30	+9 to +26
Slightly warm (slight heat stress)	0.5	23–29		–
Warm (moderate heat stress/caution)	1.5	29–35	30–34	+26 to +32
Hot (strong heat stress/extreme cation)	2.5	35–41	34–37	+32 to +38
(very strong heat stress)		–		+38 to +46
Very hot (extreme heat stress/danger)	3	>41	>37	>+46

Source: Blazejczyk et al.[14]

Among these indices, although UTCI has been widely verified and applied worldwide in the past ten years, many studies in medical buildings still use PMV as the thermal comfort evaluation, as shown in Table 2. The possible reason is that PMV is more commonly used to evaluate buildings with mechanical heating and cooling, and its parameters are easier to collect [18].

3 Human Thermal Comfort and Adaptive Behaviors in Medical Buildings

In the field of thermal environment in medical buildings, developed countries such as the United States, Japan, and Europe started research early and have achieved many research results. Table 3 lists 10 typical case studies in medical buildings. These studies have greatly impacted the thermal comfort study of patients and medical staff and the renovation of people-oriented medical buildings.

Mora [22] studied the impact of the operating room thermal environment on patients and surgical staff. Even if the indoor environment reaches the surgeon's satisfaction with the thermal comfort conditions (e.g. air temperature about 19 °C), the patient may feel too cold. If the position of the surgical lamp is not set properly, the thermal radiation generated by it may also cause the surgeon's dissatisfaction with the thermal environment. Through Mora's experiment, it is easy to find that the thermal environment does have a certain impact on patients and staff. Such research results indicated the importance of evidence-based design (EBD) research in the medical environment [26]. Experimental methods have been used to verify the effect of beneficial elements in the space environment on patients, which can have a healing effect on patients and promote the development of a "healing environment".

Khodakarami and Nasrollahi [1] conducted a literature review of thermal comfort research in medical buildings, and concluded that thermal comfort has a great relationship with the efficiency of hospital staff. Berardi and Leoni [19] assessed the air-conditioning system and microbial contamination of a hospital in Bologna, Italy, and found that 52% of the calculated PMV values in the hospital were within a comfortable range in winter and 62% in summer. There are more microorganisms in the air than in wards and operating rooms than offices and laboratories, and it is found that the contamination of microorganisms is not caused by the air-conditioning system, but by the shift, patient turnover, quantity, and behavior of the staff in the hospital. Skoog, Fransson, and Jagema [8] pointed out that patients with low activity levels need higher operating temperatures, and patients in the same ward have different heat requirements. The authors also conducted a study at hospitals in Sweden, and noted that in winter, the indoor temperature and humidity settings of hospitals were lower than the recommended value, and the humidity was 15% lower than the recommended value. Moreover, although the humidity is within the recommended range in summer, the staff still feel dry. After comparing the temperature preference of patients and doctors in winter and summer, the authors suggested that patients and doctors cannot be considered as a whole. Hashiguchi et al. [20] introduced a humidifier in winter to increase the average relative humidity of a hospital ward and nurse station in southern Fukuoka, Japan, from 32.8% to 43.9% to meet the Hospital Engineering Association of Japan (HEAJ) recommendations. The health and comfort level of the patients were found to be significantly improved. Wagner, Byrne, and Kolcaba [27] highlighted that the patients' discomfort during the operation would affect their comfort during the entire rehabilitation process. When the patient feels cold, it will make them irritable, aggravate the pain, and reduce their satisfaction. Mazzacane et al. [28] conducted a questionnaire and actual measurement in the operating room of a hospital in Italy. They pointed out that the thermal environment parameters given by National Institute for Occupational Safety and Prevention (ISPESL) are not suitable for everyone in the operating room, because the activity level of the staff during operations is different.

In general, these studies reveal that the thermal comfort indicators for ordinary public buildings should not be directly applied in medical buildings. The needs of staff and patients in thermal comfort are also different even in the same environment. Adaptation may play an important role in this.

Humans often adapt to the environment through unconscious behaviors like sweating and shivering, and conscious behaviors like using fans and opening windows [29]. However, the thermal comfort of patients might be different from that of healthy people as the nature of physical disabilities always affects thermal physiology, thermal sensation, metabolism, blood flow, and regulatory response, such as vasomotor control of body temperature or sweating ability [30].

Parson [31] conducted a field study on physically disabled people, and the results showed that when the thermal environment does not properly meet the needs of the disabled. Their physical functions will decline, which may lead to discomfort, pain and poor quality of life. Hwang et al. [21] conducted field studies on thermal comfort in a hospital in Taiwan, involving patients' perceptions of thermal comfort. The research showed that patients' thermal sensitivity and adaptive behaviors are blunt. Hwang and Chen [32] investigated the adaptive behavior of middle-aged and elderly people in Taiwan and found that their main strategy in summer is to open windows, while in winter, they wear more clothes to keep warm. Understanding the perception of the thermal environment and related adaptive behaviors can help architects better design medical buildings while enhancing patients' comfort and mental health.

Table 3. Overview of case studies conducted in medical buildings

Reference	Object	Key findings	Function space	Occupants (staff/Patient)	Sample size	Country/region
[8]	Thermal environment	Staff and patients' perception of the indoor air temperature is higher in winter than in summer	Orthopaedic wards	Both	40 employees and 35 patients	Sweden
[19]	Thermal environment	By controlling the thermal environment, the risk of microbial contamination in hospitals can be reduced	Wards and surgeries	Both	1 department	Italy
[20]	Thermal environment	Humidifiers can improve the low-humidity environment and relieve discomfort	Sickrooms and nurse stations	Both	36 patients and 45 staff	Japan
[21]	Thermal requirement	Patients' physical strength significantly affects their thermal requirements	Medical and surgical wards	Patient	927 sets in 83 rooms	Taiwan
[22]	Thermal environment	Reducing the thermal radiation from surgical lights is favourable for the patient and all the staff	Operating room	Staff	2 rooms	Canada
[28]	Thermo-hygrometric comfort	The thermal comfort of surgical team may be affected according to their positions with respect to the scialytic lamp and their particular task	Operating room	Staff	–	Italy
[33]	Ventilation Rate	Multidisciplinary review can provide helpful changes on the ventilation recommendations	Patient rooms/labor/delivery area/air born infection isolation rooms/emergency rooms and radiology /procedure rooms/surgery and critical care/ancillary	Both	82 tested rooms	America

(*continued*)

Table 3. (*continued*)

Reference	Object	Key findings	Function space	Occupants (staff/Patient)	Sample size	Country/region
[34]	HVAC and indoor thermal conditions	Hospital operating rooms in Hellenic has common problems include air exchange, poor control on indoor thermal conditions, and insufficient ventilation	Operating rooms	Staff	20 ORs in 10 hospitals	Hellenic
[35]	Thermal comfort requirements	The thermal requirements are different between staff and patients in Iranian hospitals. The existing thermal comfort conditions are not acceptable	Wards	Both	14 rooms in 4 separate hospitals	Iran
[36]	Average heating and cooling demand	Keeping the indoor air temperature in the required range and reducing the heating and cooling demands can be achieved through simple changes to the fabric of the buildings	Whole building	Both	2 buildings	Iran

4 Discussion and Conclusion

This paper conducted a literature review of human thermal comfort studies in medical buildings over the last two decades. By reviewing the different indices used to measure thermal comfort in medical space, it is confirmed that patients and medical personnel have different needs for thermal environments. Due to the extensive application of mechanical heating, cooling and ventilating in medical buildings, the PMV index is often used for thermal comfort related research. The relevant case studies show that the thermal comfort index of ordinary public buildings should not be applied directly to medical buildings. Even in the same environment, staff and patients have different needs for thermal comfort. The optimization of medical building environment can improve the satisfaction of patients and staff, so as to reduce patients' pain and improve staffs' work efficiency.

The literature review shows that there is a lack of a framework that can provide a valid methodology for interpreting the data collected from physical measurements and their correlations with subjective thermal assessments, such as social and behavioural factors. It is of great value to combine with other professionals, such as psychologists, physiologists, physicians, and other architecture-related professionals, to develop a systematic and holistic research method. It will help to better understand thermal comfort and its impact on patients and doctors' physiological and behavioral aspects. The optimization of the thermal environment in medical buildings requires more interdisciplinary research, in combination with medical requirements of patients with different symptoms and staff in different departments.

References

1. Khodakarami, J., Nasrollahi, N.: Thermal comfort in hospitals – a literature review. Renew. Sustain. Energy Rev. **16**, 4071–4077 (2012)
2. Nightingale, F.: Notes on nursing (1947)
3. Wu, H.: Design principle about Chinese hospital with combination of medical care and pension in the twenty-first century. Int. J. Struct. Civ. Eng. Res. **10**, 7 (2021)
4. Council, US Green Building: Leadership in Energy and Environmental Design (2008). www.usgbc.org/LEED
5. Shibani, D.A., Hassan, D.: Building research establishment environmental assessment methodology on the UK residential projects. Int. J. Constr. Manag. **9**, 183–189 (2020)
6. Tang, H., Ding, J., Lin, Z.: On-site measurement of indoor environment quality in a Chinese healthcare facility with a semi-closed hospital street. Build. Environ. **173**, 106637 (2020)
7. ASHRAE: ASHRAE thermal comfort standard (2017)
8. Skoog, J., Fransson, N., Jagemar, L.: Thermal environment in Swedish hospitals: summer and winter measurements. Energy Build. **37**, 872–877 (2005)
9. Nicol, F.: Thermal Comfort a Handbook for Field Studies Towards an Adaptive Model (1993)
10. Burton, I., Ebi, K.L., McGregor, G.: Biometeorology for adaptation to climate variability and change. In: Ebi, K.L., Burton, I., McGregor, G.R. (eds.) Biometeorology for Adaptation to Climate Variability and Change, pp. 1–5. Springer, Dordrecht (2009). https://doi.org/10.1007/978-1-4020-8921-3_1
11. Pantavou, K., Santamouris, M., Asimakopoulos, D., Theoharatos, G.: Empirical calibration of thermal indices in an urban outdoor mediterranean environment. Build. Environ. **80**, 283–292 (2014)
12. Baaghideh, M., Mayvaneh, F., Shekari Badi, A., Shojaee, T.: Evaluation of human thermal comfort using UTCI index: case study Khorasan Razavi, Iran. Nat. Environ. Change **2**, 165–175 (2016)
13. Gagge, A.P., Fobelets, A.P., Berglund, L.G.: A standard predictive Index of human reponse to thermal enviroment. Trans. Am. Soc. Heat. Refrig. Air-Conditioning Eng. **92**, 709–731 (1986)
14. Blazejczyk, K., Epstein, Y., Jendritzky, G., Staiger, H., Tinz, B.: Comparison of UTCI to selected thermal indices. Int. J. Biometeorol. **56**, 515–535 (2012)
15. Esmaili, R., Montazeri, M.: The determine of the Mashad bioclimatic condition base on hourly data. Geogr. Environ. Planning **24** (2013)

16. Fanger, P.O.: Thermal comfort. Analysis and applications in environmental engineering. Thermal comfort. Analysis and applications in environmental engineering (1970)
17. Liu, S., et al.: Predicted percentage dissatisfied with vertical temperature gradient. Energy Build. **220**, 110085 (2020)
18. Alfano, F.R.A., Palella, B.I., Riccio, G.: Notes on the calculation of the PMV index by means of apps. Energy Procedia **101**, 249–256 (2016)
19. Berardi, B.M., Leoni, E.: Indoor air climate and microbiological airborne: contamination in various hospital areas. Zentralbl Hyg Umweltmed. **194**, 405–418 (1993)
20. Hashiguchi, N., Hirakawa, M., Tochihara, Y., Kaji, Y., Karaki, C.: Effects of setting up of humidifiers on thermal conditions and subjective responses of patients and staff in a hospital during winter. Appl. Ergon. **39**, 158–165 (2008)
21. Hwang, R.-L., Lin, T.-P., Cheng, M.-J., Chien, J.-H.: Patient thermal comfort requirement for hospital environments in Taiwan. Build. Environ. **42**, 2980–2987 (2007)
22. Mora, R., English, M.J.M., Athienitis, A.K.: Assessment of thermal comfort during surgical operations. ASHRAE Trans. (2001)
23. Mayer, H., Höppe, P.: Thermal comfort of man in different urban environments. Theor. Appl. Climatol. **38**, 43–49 (1987)
24. Spagnolo, J., de Dear, R.: A field study of thermal comfort in outdoor and semi-outdoor environments in subtropical Sydney Australia. Build. Environ. **38**, 721–738 (2003)
25. Jendritzky, G., de Dear, R., Havenith, G.: UTCI—why another thermal index? Int. J. Biometeorol. **56**, 421–428 (2012)
26. Zimring, C., et al.: Evidence-based design of healthcare facilities: opportunities for research and practice in infection prevention. Infect. Control Hosp. Epidemiol. **34**, 514–516 (2013)
27. Wagner, D., Byrne, M., Kolcaba, K.: Effects of comfort warming on preoperative patients. AORN J. **84**, 427–448 (2006)
28. Mazzacane, S., Giaconia, C., Costanzo, S., Cusumano, A.: A survey on the thermal conditions experienced by a surgical team. Indoor Built Environ. **16**, 99–109 (2007)
29. Udrea, I., Croitoru, C., Nastase, I., Crutescu, R., Badescu, V.: First adaptive thermal comfort equation for naturally ventilated buildings in Bucharest, Romania. Int. J. Vent. **17**, 149–165 (2018)
30. Verheyen, J., Theys, N., Allonsius, L., Descamps, F.: Thermal comfort of patients: objective and subjective measurements in patient rooms of a Belgian healthcare facility. Build. Environ. **46**, 1195–1204 (2011)
31. Parson, K.: Human Thermal Environments: The Effects of Hot, Moderate, and Cold Environments on Human Health, Comfort, and Performance, 3rd edn. CRC Press, Boca Raton (2014)
32. Hwang, R.-L., Chen, C.-P.: Field study on behaviors and adaptation of elderly people and their thermal comfort requirements in residential environments. Indoor Air **20**, 235–245 (2010)
33. Ninomura, P., Bartley, J.: New ventilation guidelines for health-care facilities. ASHRAE J. **43**, 29+30+32–33 (2001)
34. Balaras, C.A., Dascalaki, E., Gaglia, A.: HVAC and indoor thermal conditions in hospital operating rooms. Energy Build. **39**, 454–470 (2007)
35. Khodakarami, J., Knight, I.: Required and current thermal conditions for occupants in Iranian hospitals. HVAC&R Res. (2011)
36. Khodakarami, J., Knight, I., Nasrollahi, N.: Reducing the demands of heating and cooling in Iranian hospitals. Renew. Energy **34**, 1162–1168 (2009)

Thermal Performance Optimization Methods for Heat Exchangers with Accumulation in Phase Change Materials

Fernando Neto[1], Bruno Pereira[1], Rui Teixeira[2(✉)], and Rafael Rodrigues[2]

[1] Department of Mechanical Engineering, University of Aveiro, TEMA, Aveiro, Portugal
[2] Heaboo Lda, Aveiro, Portugal
rui.teixeira@heaboo.com

Abstract. The article addresses the need to improve the performance of hot water systems in buildings by proposing a novel technique based upon the use of PCM. A heat exchanger using a PCM was developed and tested. By resorting to numerical models, different assembling processes were tested in order to combine low fin-pipe thermal resistances with high output manufacturing processes. Experimental tests have shown that the system can provide instant hot water for an adequate period of time hence preventing the use of circulating loops with high installation costs and reducing the waste of energy and water.

Keywords: Phase change materials · Heat storage · Performance optimization methods · Thermal batteries · Domestic hot water · Instant hot water

1 Introduction and Objective

Buildings are responsible for 40% of the energy consumption in developed countries and 20% of this energy is wasted with domestic hot water systems. Conventional solutions to ensure instant hot water in every faucet and shower are based on recirculating loops, a solution that significantly increase the energy wasted by domestic hot water system. Few alternatives have been presented to reduce the energy wasted with recirculating loops but none of them offer the same value proposition of conventional solutions and the fact is that this inefficient solution is still a normal procedure in construction, despite it being highly inefficient. Buildings are becoming more efficient due to better insulation, optimization of equipment performance, etc. but domestic hot water distribution has been left out of this energy optimization. Heaboo proposed a disruptive approach to this known problem, an innovative technology based on Phase Change Material (PCM) that delivers hot water instantly with no use of recirculating loop, eliminating the waste of energy that represents more than 50% of the energy consumed to heat domestic water. This solution is presented on the form of a thermal battery to be installed in each compartment to ensure that when a user opens a tap the battery instantly heats the water and then, when hot water arrives, the device self-

P. Mendonça and N. D. Cortiços (Eds.): ICAMC 2021, LNCE 226, pp. 408–417, 2022.
https://doi.org/10.1007/978-3-030-94514-5_41

restore the heat for the next use. This fully passive technology has no connection to the electricity and requires no maintenance during its lifetime.

The storage of heat in the thermal battery occurs during the shower time with a small temperature differential because domestic hot water (DHW) is distributed around 50ºC and the supply temperature to the faucet must be higher than 40ºC. These specifications present a big challenge for the development of the heat exchanger regarding performance, constancy, capacity to charge the thermal battery during a shower time and also the capacity to release this heat with a power output to supply water at a temperature above 40ºC. Outlet temperature has to remain stable during the discharging phase or otherwise the thermal battery would not present a tangible value to user. The team proposed the use of a phase change material with a change set point of 45ºC and, to ensure a stable power outlet the heat exchanger was developed to guarantee that phase change occurs only in a part of the exchanger at a certain time and consequently the power output and temperature outlet remains constant. This concept was proposed and computational simulation was used to optimize geometry. When testing the first prototype created by the team it was found that the influence of the thermal resistance between fins and the pipes was significant on the operation of the heat exchanger. In this work are presented the used methods to solve this problem and a comparison of the results.

2 State-of-The Art

Although harnessing the wasted energy on the production of hot water is available commercially for non-industrial settings, their efficiencies tend to be low. The introduction of phase change materials allows the increase of the available energy storage for the recovered energy, increasing energy density and helping to bridge production and demand peaks. The biggest obstacle to the application of PCMs on the creation of thermal batteries associated with hot water is the low thermal conductivity of these materials. This low conductivity constrains the total amount of energy that the thermal battery can store, as well as the speed that the energy can be returned to the water when needed. To improve the thermal conductivity of PCM solutions, several studies have been conducted, with the addition of additives, the use of fins or the use of composite PCMs among the studied solutions. Among these, the use of fins seems to be the most used. Several geometrical types of fins can be used, with longitudinal, annular and plate fins being widely used [1]. The design parameters of the applied fins must be considered to optimize the performance of the designed systems. The heat transfer rate is directly dependent on the number of fins and their dimensions [2]. A more in-depth analysis of the mechanisms for the improvement of thermal energy transfer with the use of fins was recently reviewed by several authors [3]. When using fins to improve the thermal energy transfer between the Heat transfer fluid (HTF) and the PCM, the thermal contact resistance between the fins and the tube has a significant impact on the results along with the geometrical properties of the fins, [4]. Shobhana Singh et al. investigated this effect on the performance of a fin and tube exchanger using a 3D steady state modelling, finding an estimated 8.23% decrease in the overall heat transfer [5].

3 Experimental Method

The quality of the thermal connection between the tube and fin has a decisive impact on the performance of the heat exchanger and therefore it is crucial to find methods that can improve the thermal contact to reduce the thermal resistance in the connection and to consequently improve the thermal performance. Several options/alternatives were formulated and tested, with the requirement that the process could then be replicated in a production process. The following strategies to be tested were selected: 1) To use a fin diameter smaller than pipe diameter and proceed to assemble the heat exchangers using a press; 2) To heat up the fins and cool down the pipes so they can slide due to thermal expansion/contraction; 3) To enlarge the copper tube using a spindle to fit the copper tube to the fins with a larger diameter.

To validate the first hypothesis an assembly bench with a press was developed and the prototype was manufactured – Fig. 1 presents the bench in operation. The experimental results showed that the conceptual thermal battery can recharge during a shower time with the defined domestic hot water distribution temperature.

Fig. 1. Assembly bench with press (first manufactured prototype)

The performance results of the heat exchanger are however significantly lower than the simulated results with ideal contact conditions between fins and cooper; therefore, an alternative hypothesis was used to obtain higher performance. The second hypothesis proposed was to assemble the exchanger using heat/cooling contraction/retraction of the materials (aluminium and copper). A bench was developed to preheat the fins to a temperature of about 300ºC in a system that could then receive the copper tubes previously placed in a freezer. Figure 2 shows the bench composed by the oven and the pneumatic action system to push the pipes into the oven.

Fig. 2. Assembly bench with oven and press

Despite the initial positive results with a small heat exchanger, the bench did not work as expected since the pipes bended after the first centimetres entered the oven. Different strategies were tested but it became quite clear that the heat exchanger could not be manufactured with the full length. The third hypothesis comprises the enlargement of the copper tube after the fin assembly has been mounted using a method where the copper tube is formed. The bench includes a spinning driller attached to a pneumatic system that pushes the drill into the pipes causing a plastic deformation in the pipe, increasing the inner and outer diameters and consequently increasing the contact between fin and pipe. With this alternative it was also intended to develop a method that could be industrialized with automation. Figure 3 presents a schematic drawing of the concept developed to produce and test a prototype optimized heat exchanger.

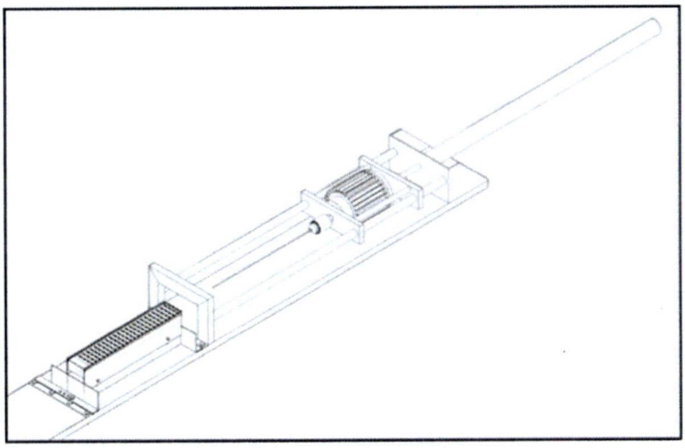

Fig. 3. Schematic drawing of the manufacturing heat exchanging bench with pipe widening

The results of this method showed important improvements on the results as presented on the experimental results chapter.

4 Optimization Strategies Based on Previous Simulations Effort

The computational model used is based on a simplified approach of the heat exchanger represented on an axisymmetric model. This model includes a set of correlations of areas and volumes to minimize the error. Figures 4 and 5 present the simplified geometric model in a two-dimensional axi-symmetric representation that significantly simplifies the computational calculation process and the three-dimensional representation of the axi-symmetric model.

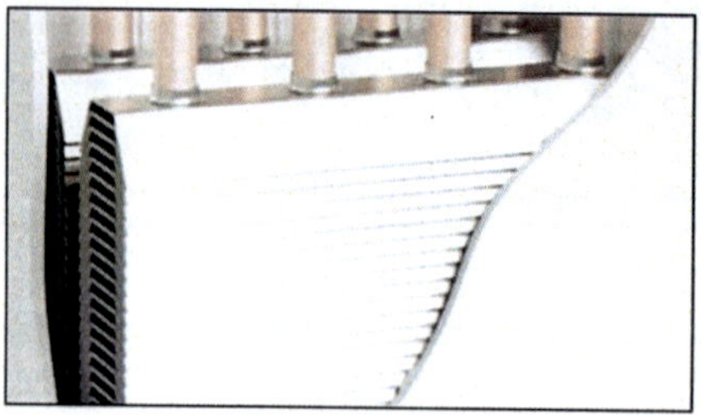

Fig. 4. Heat exchanger concept

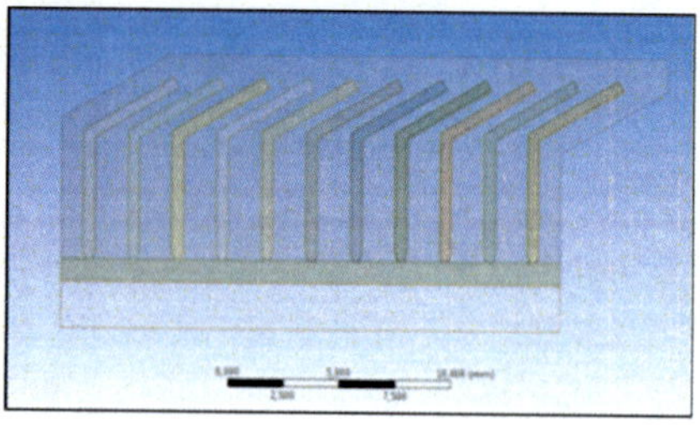

Fig. 5. Axis-symmetric model

The parametric study defined an optimized set-up that was used to create the first prototype of the thermal battery concept. This first prototype was based on a segment with geometry and specification determined by simulation but first tests demonstrated that performance was not aligned with the simulation results. After a detailed analysis, it was concluded that the thermal connection between aluminium fins and cooper pipes wasn't as good as represented by the model. To correct the model, a thermal resistance was added on the boundary aluminium/cooper to represent the inefficiency of the thermal connection. Figure 6 shows the variation profile of the water outlet temperature of the exchanger control volume considering different heat conduction coefficients for the connection tube/fin.

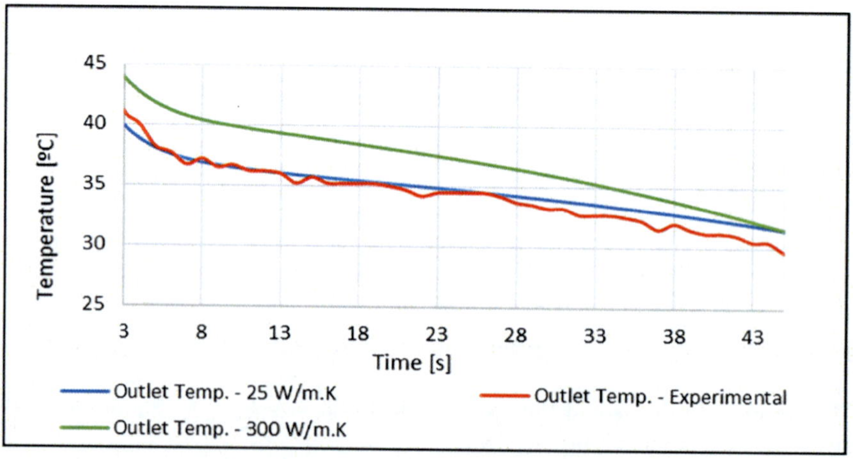

Fig. 6. Experimental results from first concept prototype and comparison with simulated case

As shown, the temperature profiles obtained experimentally and the results obtained from the computational model when considering a tube-fin equivalent heat transfer coefficient of $25 W.m^{-1}.K^{-1}$ are in close agreement, particularly in the initial part of the test, the most important for product characterization because this is the moment when most of the energy is released into the thermal battery. This experimental validation demonstrated that the computational model represents the built prototype. Also, it was noticed that thermal performance is still below what could be expected if the quality of the thermal contact fin-tube would be improved since the equivalent value of the heat transfer coefficient is about 15 times lower than the thermal conductivity of the tube material; this result highlights the low quality of the fin-tube thermal connection.

5 Experimental Results

To properly test the capacity of the thermal battery concepts, set-up conditions were used to compare the results of the three optimization strategies proposed. As referred in the experimental method section, the second optimization method proposed was impossible to implement during prototype manufacturing and therefore the validation will be focused on the two successfully strategies that results in two thermal batteries concepts: a first one produced with the optimization strategy based on the assembly press that forces the fins into the pipe with a tight connection and a second one based on a post assembly pipe widening that ameliorates the connection after assembly.

Table 1 shows the experimental plan used to characterize each of the prototype optimization strategies.

Table 1. Experimental plan with heat released results

	Optimization strategy	Flow rate	Time f/charge
	Units	*l/h*	*h*
Test 1	1st: assembly press	400	1
Test 2	1st: assembly press	400	24
Test 3	3rd: pipe widening	400	1
Test 4	3rd: pipe widening	500	1
Test 5	3rd: pipe widening	400	24
Test 6	3rd: pipe widening	500	24

For each prototype (1st: assembly press and 3rd: pipe widening), the experimental plan included the variation of the waiting time after charging (1 and 24 h). For the protype which best satisfies these requirements, a subsequent comparison was made based on flow rate variation (400 and 500 l/h). Figure 7 presents the performance response of each prototype 1 h after the thermal battery charge.

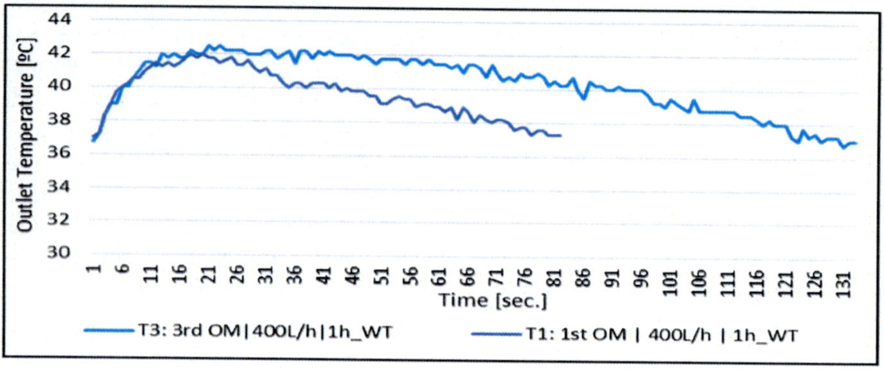

Fig. 7. Outlet temperature comparison for the two prototypes, 1h after charging

Table 2 presents the comparison of the results concerning total water volume and total heat released during the experiment.

Table 2. Experimental results sum-up of water volume and heat released

Optimization strategy	Heat released	Volume of water	Improvement (in volume)	Improvement (in heat)
Units	*kJ*	*L*	%	%
1st: assembly press	714,2	9,0		
3rd: pipe widening	1 197,7	14,1	57%	68%

The 3rd optimization strategy presents significant improvements on the stabilization and output thermal capacity of the thermal battery. The first prototype was able to heat the water flow during approximately 80 s with an outlet temperature above 37ºC. The third prototype presents a result for the same test of 130 s which represent a capacity improvement of more than 60%. Figure 8 shows a similar comparison when the time after charging was increased to 24h.

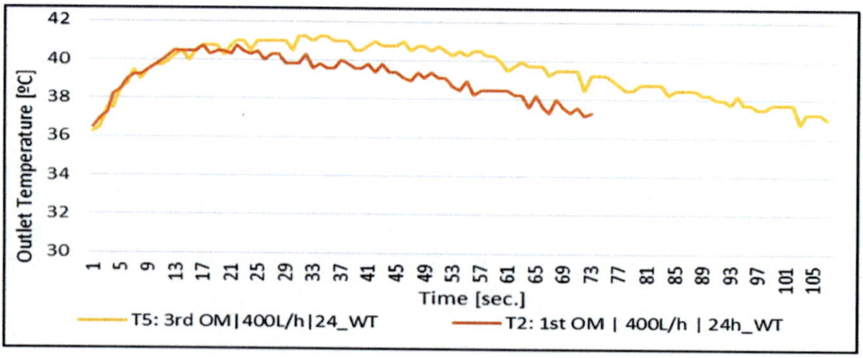

Fig. 8. Outlet temperature comparison for the two prototypes, 24h after charging

Table 3 presents the comparison of the results regarding total water volume and total heat released during the experiment.

Table 3. Experimental results sum-up of water volume and heat released

Optimization strategy	Heat released	Volume water	Volume water improvement	Heat released improvement
Units	*kJ*	*L*	%	%
1st: assembly press	617,3	7,9	–	–
3rd: pipe widening	920,3	11,4	45%	49%

If the time after charging was increased to 24 h, using the same thermal insulation pack on both prototypes, results showed that the first prototype provides instant hot water during approximately 70 s with an outlet temperature above 37°C. When testing the third prototype for the same conditions, the performance was increase to nearly 110 s, with additional 45% of water volume heating capacity and an increase of 49% of the heat released.

To understand the capacity of the prototype to deliver instant hot water with different flow rates, a new test was made using two different flow rates of 400l/h and 500l/h. The results are plotted in Fig. 9 for a waiting time from charge of 1 h.

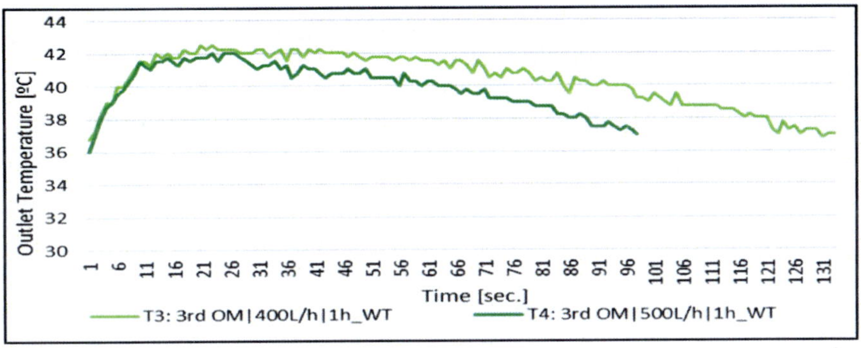

Fig. 9. Outlet temperature comparison with different flow rates, 1 h after charging

Results show that performance is constant in the first 30 s, showing that the thermal battery adjusts the power output when requested: afterwards the outlet temperature will be reduced impacting the duration of the hot water supply: for water above 37°C, the supply duration is reduced from 130 s to approximately 100 s. However, the volume of water that is heated is approximately the same. Considering the purpose of the thermal battery (to heat a volume of water stagnated in pipeline), the results show that higher flow rates have practically no impact on the capacity of heating the water in pipes. Figure 10 shows similar results when waiting time from charge is increased to 24 h.

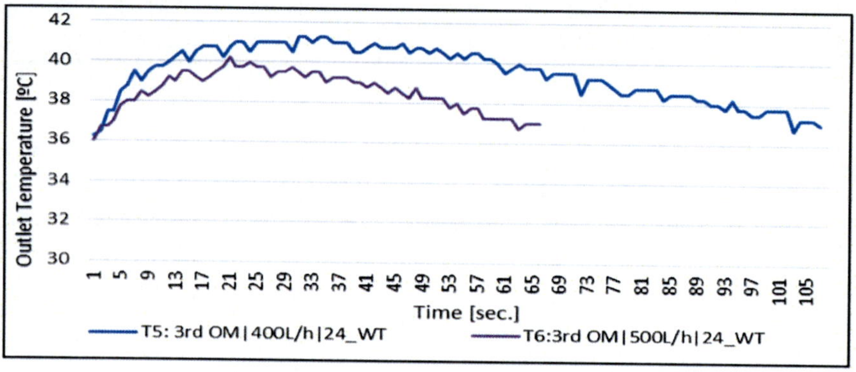

Fig. 10. Outlet temperature comparison with different flow rates, 1h after charging

In this case, the duration of water supply above 37°C is also affected by the flow rate increase. For 400l/h the system can provided water with a temperature above 37°C for nearly 110 s; when the flow rate is increased to 500l/h this capacity is reduced to nearly 70 s.

Table 4 shows the energy that was released in each prototype based on inlet and outlet temperatures and flow rate.

Table 4. Heat released by the thermal battery prototype in each test

	Optimization strategy	**Flow rate**	**Time from charge**	**Heat released** (>37°C)	**Water volume** (>37°C)
	Units	*l/h*	*h*	*kJ*	L
T.1	1st: assembly press	400	1	714,2	9,0
T.2	1st: assembly press	400	24	617,3	7,9
T.3	3rd: pipe widening	400	1	1 197,7	14,1
T.4	3rd: pipe widening	500	1	1 080,8	13,2
T.5	3rd: pipe widening	400	24	920,3	11,4
T.6	3rd: pipe widening	500	24	672,7	8,1

6 Conclusions

It was found out that low-quality contact between fins and pipes of the heat exchanger caused a substantial degradation on the performance of the concept developed. Thermal performance is a key element for the conceptual product developed because water outlet must remain above 37°C as long as possible during supply. The thermal performance influences the charging time and the consequent global efficiency of the solution and therefore it was crucial to discover an optimization strategy that could, in a compatible manner with industrialized manufacturing techniques, increase the thermal contact between the aluminium and cooper and did not interfere with the phase change material. Three optimization methods were tested but a prototype was successfully developed only for two of these strategies.

The results concerning water outlet temperature showed that the pipe widening after assembly increases the thermal performance by 55–65% when compared with the alternative technique based in press assembly. Results also show that the thermal battery concept satisfies multiple flow rates and that the thermal output is adjusted to heat request.

Acknowledgement. This research was co-financed by COMPETE 2020/Portugal 2020/EU through the project P2020 I&DT number 24215 from the promotor Heaboo, Lda.

References

1. Zayed, M.E., Zhao, J., Li, W., Elsheikh, A.H., Elbanna, A.M., Jing, L., Geweda, A.E.: Recent progress in phase change materials storage containers: geometries, design considerations and heat transfer improvement methods. J. Energy Storage **30**, 101341 (2020)
2. Abdulateef, M., Mat, S., Abdulateef, J., Sopian, K., Al-Abidi, A.A.: Geometric and design parameters of fins employed for enhancing thermal energy storage systems: a review. Renew. Sustain. Energy Rev. **82**, 1620–1635 (2018)
3. Kalapala, L., Devanuri, J.K.: Influence of operational and design parameters on the performance of a PCM based heat exchanger for thermal energy storage – a review. J. Energy Storage **20**, 497–519 (2018)
4. Taler, D., Ocłoń, P.: Thermal contact resistance in plate fin-and-tube heat exchangers, determined by experimental data and CFD simulations. Int. J. Therm. Sci. **84**, 309–322 (2014)
5. Singh, S., Sørensen, K., Condra, T.J.: Influence of the degree of thermal contact in fin and tube heat exchanger: a numerical analysis. Appl. Therm. Eng. **107**, 612–624 (2016)

Certifying Buildings in China Using Energy, Environmental, Engineering and Economic (4E) Factors: Assessing the Suitability of BREEAM, LEED and Green Star

Y. Hong[1,2], C. I. Ezeh[2], Y. Ma[1], S.-H. Hong[1], and W. Deng[1(✉)]

[1] Department of Architecture and Built Environment, University of Nottingham Ningbo China, 199 Taikang East Road, Ningbo 315100, China
wu.deng@nottingham.edu.cn

[2] Shanghai Daren Construction Technology Group Co. Ltd, 1818, Lianhang Road, Minhang District, Shanghai, China

Abstract. Insight into the suitability of building rating systems and its application is pertinent in deciding the best assessment approach that suits building evaluation standards in China. Hence, this report presents a comparative study of universally accepted building rating schemes (BREEAM, LEED and Green Star) concerning the goals, criteria and certification standards in China. The comparison was conducted by normalizing the different score ratings accredited to each assessing criteria: energy, environmental, engineering and economic (4E). Furthermore, using a highly rated hotel in China as a case study, the applicability of each rating system is discussed by the content of related documents issued by the Ministry of Housing and Urban-Rural Development of China. The obtained results establish that the 4E assessment standards in China match that of the BREEAM In-use and LEED (O + M) rating systems. Here, environmental > energy > engineering > economic benefits is observed. In the absence of a comprehensive building assessment approach in China, this report identifies interrelated factors that will guide building stakeholders towards selecting the most appropriate rating system.

Keywords: High-rise buildings · BREEAM · LEED · Green Star · Chinese 3-star · Sustainability criteria · Comparative analysis

1 Introduction

Globally, the surge in energy consumption demands appropriate measures to circumvent its continuous rise for the purpose of developing a sustainable environment [1]. Against this background, a sustainable built environment is necessary given that this sector immensely contributes to global energy consumption [2]. Also, a globally accepted assessment system and certification are required to rate buildings for effective monitoring and implementation of sustainability policies [3]. Aside from the energy implication, there are other factors to be considered to attain sustainability in the building sector. Energy, environmental, engineering and economy (4E) are the

P. Mendonça and N. D. Cortiços (Eds.): ICAMC 2021, LNCE 226, pp. 418–427, 2022.
https://doi.org/10.1007/978-3-030-94514-5_42

fundamental pillars of sustainability upon which the most common rating systems are designed despite the region/occupancy-specific definitions [3, 4].

Internationally, there are several assessment systems for sustainable buildings [2], however, only a few systems are widely acknowledged as the recognizable standards for sustainable development. These systems include the Building Research Establishment's Environmental Assessment Method (BREEAM) [2, 5], the Leadership in Energy and Environmental Design (LEED) [5, 6] and the Green Star [2, 7]. Developed in the UK in 1990, BREEAM is considered to be the most widely applied building assessment system [8]. On the other hand, the LEED rating system was developed in the US. in 1998 and serves as a sustainability assessment suite for over 30 countries [8]. Lastly, the Green Star was launched in 2003 in Australia with a broader sustainability scope, including social factors such as occupant's health and productivity [8]. In China, however, there is no standard building rating system.

Since 2003, China developed a domestic certification, 3-Star system [6]. Despite this, building owners/developers have the discretion to choose the most appropriate rating system for their building certification [6]. Therefore, it is crucial to provide insight into the goals and criteria of the different assessment rating systems. Furthermore, the suitability of each rating system and its application in China is also pertinent, as this will serve as the determining factor in deciding which rating system best suits the evaluation standards of China.

2 Brief Literature Review

During the past decades, China has promoted the construction of energy-efficient buildings, which are certified using the 3-Star rating system. In spite of this, building developers are entitled to choose the most desirable rating system to certify their building operations, especially in different climate environments. Therefore, it is important to understand the goals and criteria of the different green building rating systems in comparison to the 3-Star rating system. Internationally, the most prominent building rating systems are BREEAM, LEED and Green Star.

2.1 BREEAM In-Use

The BREEAM rating system is the first recognized building assessment approach. It was first launched in 1990 and has been updated severally, with its most recent version in 2016. At first, this rating system only covered the construction stage of new buildings but has broadened to consider the entire building's life cycle, starting from the design phase to the in-use and retrofitting period. BREEAM In-use scheme rates existing buildings under eight criteria with a maximum achievable credit score, as shown in Table 1. Also, an additional 10% credit score is allocated to the final score of the building for innovation as the innovation credit. Based on the total credit score, the following ratings are issued: Unclassified (<30 points), Pass (≥ 30 points), Good (≥ 45 points), Very good (≥ 55 points), Excellent (≥ 70 points) and Outstanding (≥ 85 points).

2.2 Leed (O + m)

Similar to BREEAM, this rating system assesses sustainability across the entire building phase, including design, construction, operation and maintenance (O + M). It is the most widely used building assessing scheme, particularly in the U.S. and Canada. The system was first developed in 1998 and has undergone several version updates. Its most recent update was in 2013 and now incorporates standardized criteria to cover different international contexts. Similar to the rating criteria for BREEAM In-use, LEED (O + M) scheme assesses eight different criteria that address key sustainability aspects of existing buildings (Table 1). The maximum total credit score is 100 with an additional 10 credit scores allotted for the "Innovation" and "Regional Priority" categories. Based on the total credit score attained, buildings are rated as LEED Certified (40–49 credit score), LEED Silver (50–59 credit score), LEED Gold (60–79 credit score) and LEED Platinum (80+ credit score).

2.3 Green Star

This rating system is one of the most adopted voluntary building environmental assessment tools developed by the Green Building Council of Australia (GBCA). The latest version was updated in 2016 and focuses on the building construction, operation and retrofitting stages. Here, nine specific criteria are assessed, each with a maximum credit score (Table 1). Based on the overall credit score (with a maximum achievable score of 100), the buildings are rated as 1–3 Stars (10–44 points) corresponding to Minimum Practice, Average Practice and Good Practice, respectively; 4 Stars (45–59 points) for Best Practice); 5 Stars (60–75) for, Excellence Level and 6 Stars (> 75 points) for World Leadership Grade.

To address which rating system is more suitable and applicable in China, this study briefly compares the BREEAM, LEED and Green Star rating systems using their goals and criteria to the certification standards in China. As a case study, a hotel in China was rated by the aforementioned rating systems for the comparisons, which can identify the key factors that will guide building owners/developers towards selecting the most appropriate rating system. In addition, by employing logistic models, this research provides the first empirical evidence regarding what factors might affect the green building certification options in the Chinese context.

3 Method

First, the sustainability criteria are classified into nine general categories: energy and atmosphere, location and transportation, sustainable site and regional priority, water efficiency, indoor environmental quality, emissions/pollutions, materials and resources, management/waste management and innovation. Following the listing of the criteria for each rating system, each criterion is then allocated (with its weighting credits) to one of the nine categories. A Pearson correlation analysis is then used to compare the general categories of each rating system to that of the 3-Star. An Excel statistical tool,

XLSTAT (version 2019.3.2) software was used for the computation of the correlation coefficients. Pearson correlation coefficient is calculated using:

$$r = \frac{\mathrm{Cov}(x, y)}{\sqrt{\mathrm{Var}(x).\mathrm{Var}(y)}} \quad (1)$$

where Cov(x, y) is the covariance of x and y variables, Var(x) and Var(y) are the variance of x variable (credit scores of BREEAM, LEED and Green Star), and y variable (credit score of 3-Star).

For further comparison simplification, the nine categories are sorted under environmental, energy, engineering and economic (4E) classification groups and their credits assigned to their respective groups. Subsequently, the allocated credits are normalized to a 1.0 (or 100%) total credit. The normalized credits serve as the indicator for comparing the rating systems. The outcome of this process is a distribution of the share of each category relative to the total number of points that could be theoretically achieved under each rating system. Furthermore, using a highly rated hotel in China as a case study, the applicability of each rating system is discussed using the comparative analysis.

4 Results and Discussions

4.1 Comparative Analysis

Table 1 represents a breakdown of the vital sustainability criteria that are assessed in each rating system. Pictorially, Fig. 1a and Fig. 1b show the average impact of each category of the sustainability criteria and the classification groups, respectively.

Table 1. Normalized weighting credits for BREEAM, LEED and Green Star in comparison to the 3-Star rating systems

Classification	Rating category	Credit Scores of common rating systems			3-Star rating system
		BREEAM In-use	LEED (O + M)	Green star	
Energy	Energy and atmosphere	0.26	0.35	0.22	0.25
Environment	Location and transportation	0.11	0.14	0.06	–
	Sustainable site & regional priority	0.10	0.09	0.05	0.10
	Water efficiency	0.08	0.11	0.11	0.15
	Indoor environmental quality	0.17	0.15	0.16	0.15
	Emissions/pollutions	0.14	–	0.05	–

(*continued*)

Table 1. (*continued*)

Classification	Rating category	Credit Scores of common rating systems			3-Star rating system
		BREEAM In-use	LEED (O + M)	Green star	
Engineering	Materials and resources	0.09	0.07	0.09	0.15
	Innovation	–	0.05	0.09	0.10
Economic	Management/Waste management	0.05	0.04	0.15	0.10
Total scores		1.00	1.00	1.00	1.00
Correlation coefficients (with 3-Star)		0.8827	0.9377	0.7565	

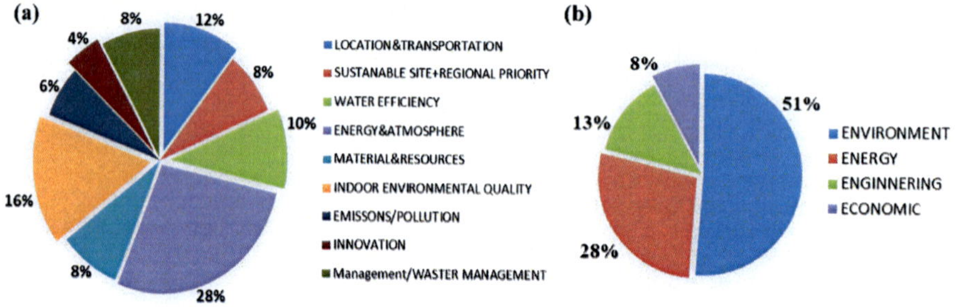

Fig. 1. Average impact of (a) the sustainability criteria and (b) the classification groups.

Depending on the rating system, the maximum credit score for each criteria varies. The variation in the credit scores highlights the impact of each criterion in the rating system. Comparing the four classification groups, the normalization results reveal that the credit score weights allocated to the groups for each rating system follow an identical trend: environmental > energy > engineering > economic factors, as observed in Fig. 2.

Comparing the rating systems within the classification groups, it is observed (see Fig. 2):

Environment perspective: BREEAM In-use > LEED (O + M) > Green Star.
Energy perspective: LEED (O + M) > BREEAM In-use > Green Star.
Engineering perspective: Green Star > LEED (O + M) > BREEAM In-use.
Economic perspective: Green Star > BREEAM In-use > LEED (O + M).

Relative to these rating systems, the 3-Star bears a close proximity with the average of the three rating systems (Fig. 1b). Overall, the environmental, energy, engineering and economic factors for the 3-Star have total credit scores of 40%, 25%, 25% and 10%, respectively. Also, the correlation analysis demonstrates that the 3-Star system is most matched with the LEED system with a coefficient of *ca.* 0.94 and least matched with the Green Star system with a 0.76 coefficient.

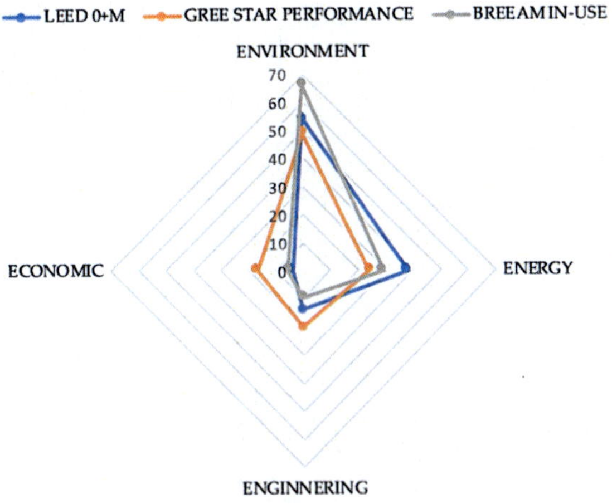

Fig. 2. Comparison of BREEAM In-use, LEED (O + M) and Green Star rating systems with regards to 4E factors.

4.2 Case Study Analysis

Description of the Building Sample. The selected building is a 3-stars rated private 28-story hotel building (Haiyue Hotel) situated in Xining City, in central China (Fig. 3). The hotel has a total construction area of 3.6 km^2. The first floor constitutes of a general service desk, a concierge department, a business center and a western restaurant. The second floor is a multi-functional hall. Above the second floor are the hotel and service rooms.

Assessment of the Hotel Building. Using the defined categories and classifications for each rating systems presented in Sect. 4.1, Haiyue Hotel was assessed to compare the different rating systems. Tables 2, 3 and 4 show the credit weight for each of the rating systems using the normalized scores. Figure 4 presents a pictorial comparison of the weighted scores concerning the classification groups.

From the obtained results, it is evident that the 4E factors of the hotel by the BREEAM In-use and LEED (O + M) rating systems are similar. Here, the observed trend is environmental > energy > engineering > economic factors, which bears a similar trend with the requirements by the 3-Star system. In the 3-Star system, buildings are required to attain certain control items and a minimum score in each category. The star level is decided by the minimum number of each category rather than by the total credits obtained [6]. Unlike in other rating systems, the rating approach in the 3-Star system promotes a better performance across all categories. Given that Haiyue hotel has a rating of 3-stars, it is logical to assume that the above trend (environmental > energy > engineering > economic) is satisfied. As such, it is it is evident that the rating of Haiyue hotel by the 3-Star system matches the overall trend by the BREEAM In-use and LEED (O + M) rating systems.

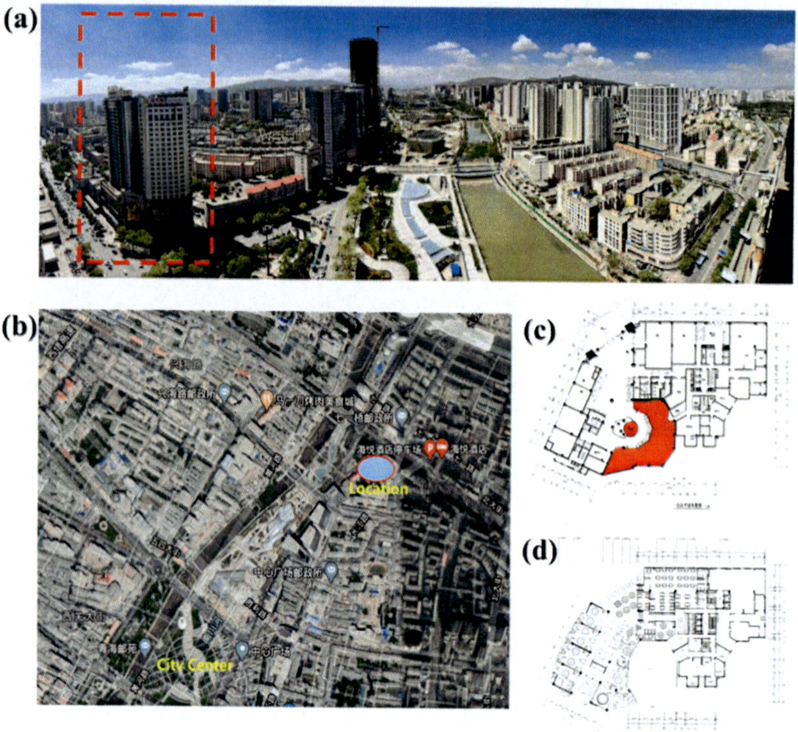

Fig. 3. Overview of the location of Haiyue hotel (a) 3D view, (b) bird-eye view, (c) 1st floor plan and (d) 2nd floor plan.

Table 2. BREEAM In-use weighed credit score for Haiyue hotel building

Criteria	Weighed Score				
	Engineering	Economic	Environmental	Energy	Total
Transport			13		13
Land use and ecology			4		4
Water			16		16
Energy				50	50
Material	21				21
Health and well-being			21		21
Pollution			6		6
Waste		1			1
Attained total score	21	1	60	50	132
Attained total score (normalized)	7.55	1.38	33.28	13.50	55.71

Table 3. LEED (O + M) weighed credit score for Haiyue Hotel

Criteria	Weighed score				
	Engineering	Economic	Environmental	Energy	Total
Location & transportation			12		12
Sustainable site			4		4
Water efficiency			6		6
Energy & atmosphere				15	15
Material & resources	5				5
Indoor environmental quality			11		11
Regional priority (management)		2			2
Innovation	3				3
Attained total score	8	2	33	15	58
Attained total score (normalized)	7.27	1.82	30	13.64	52.73

Table 4. Green Star weighed credit score for Haiyue Hotel

Criteria	Weighed score				
	Engineering	Economic	Environmental	Energy	Total
Transport			4		4
Land use and ecology			4		4
Water			7		7
Energy				10	10
Material	4				4
Indoor environmental quality			12		12
Emissions			6		6
Management		10			10
Innovation	5				5
Attained total score	9	10	33	10	58
Attained total score (normalized)	8.18	9.09	30	9.09	56.36

Moreover, it is observed that based on the BREEAM In-use and LEED (O + M) rating systems, considerations for the waste and management criteria (economic factors) by the hotel are weak when compared to other categories. However, this is not the case under the Green Star rating system, where the same criteria are more highly rated with a credit score greater than that of the engineering factor. Here, the trend, environmental > energy > economic > engineering factors, are observed. This variation within the rating systems suggests that the rating criteria under the BREEAM In-use and LEED (O + M) rating systems are more restricted than that of the Green Star. Contrarily, the energy criteria under the Green Star system is more strict than that of the other BREEAM In-use and LEED (O + M).

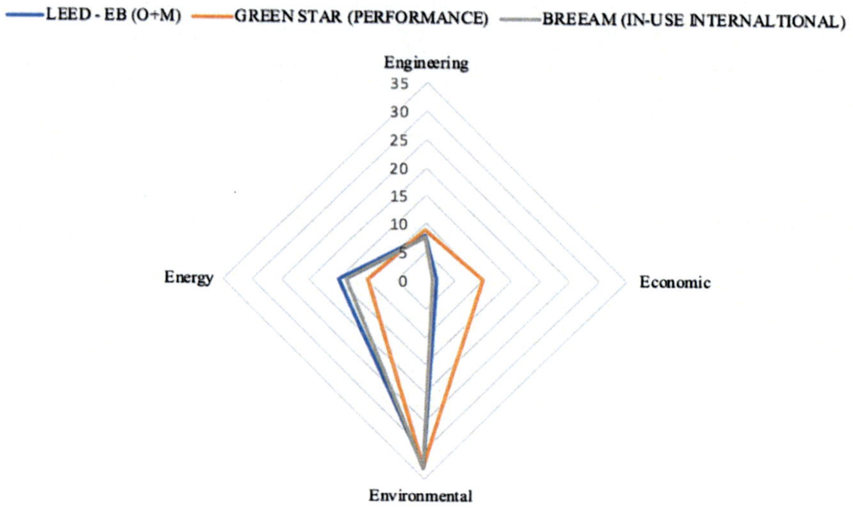

Fig. 4. Pictorial comparison of the weighted scores concerning the 4E classification groups for Haiyue Hotel using the BREEAM In-use, LEED (O + M) and Green Star rating systems.

From the environmental perspective, the hotel building displayed a similar score (~33 credit score) in all three rating systems, with the BREEAM In-use being slightly higher than the others with a normalized credit score of 33.28. For the energy factor, the scores are LEED (O + M) > BREEAM In-use > Green Star. For the economic and engineering factors, the scores are Green Star > LEED (O + M) > BREEAM In-use. Overall, based on the BREEAM In-use rating system, the total credit score of 55.71% will give the Hotel a Very Good rating. This rating corresponds to the total credit scores of 52.73% (LEED Silver) and 56.36% (5-Star (Excellent Level)) under the LEED (O + M) and Green Star rating systems, respectively. In general, understanding the variations in the different rating systems will provide building owners/developers with the best suitable rating scheme to adopt for their building certification. Also, it will offer guidance towards the necessary sustainability criteria to establish an appropriate rating standard for existing buildings.

5 Conclusion

An internationally accepted assessment and certification is required to rate existing buildings for effective monitoring and implementation of sustainability policies. Among the numerous assessment systems, the BREEAM In-use, LEED (O + M) and Green Star are, globally, the most commonly adopted. Analyzing the contents of the assessment criteria, this report compares the credit score weighting approach of these three rating systems. Based on this comparison, an assessment of the Hotel building was conducted using these three rating systems as a case study. The outcomes of the case study are as follows:

- Environmental factor accounts for the largest share with almost similar credit scores for all three systems.
- LEED (O + M) gave the highest energy factor score.
- Green Star gave the highest economic and engineering factor scores.
- The Hotel building scored relatively low in the engineering factor.

Overall, the Hotel achieved a medium level scale in this assessment A further detailed evaluation is required for a need for retrofitting.

References

1. Hong, Y., et al.: Building energy retrofit measures in hot-summer–cold-winter climates: a case study in Shanghai. Energies **12**(17), 3393 (2019)
2. Mattoni, B., et al.: Critical review and methodological approach to evaluate the differences among international green building rating tools. Renew. Sustain. Energy Rev. **82**, 950–960 (2018)
3. Lazar, N., Chithra, K.: A comprehensive literature review on development of building sustainability assessment systems. J. Build. Eng. **32**, 101450 (2020)
4. Awadh, O.: Sustainability and green building rating systems: LEED, BREEAM, GSAS and Estidama critical analysis. J. Build. Eng. **11**, 25–29 (2017)
5. Shan, M., Hwang, B.-G.: Green building rating systems: global reviews of practices and research efforts. Sustain. Cities Soc. **39**, 172–180 (2018)
6. Zou, Y.: Certifying green buildings in China: LEED vs. 3-star. J. Cleaner Prod. **208**, 880–888 (2019)
7. Illankoon, I.M.C.S., et al.: Life cycle costing for obtaining concrete credits in green star rating system in Australia. J. Clean. Prod. **172**, 4212–4219 (2018)
8. Nguyen, B.K., Altan, H.: Comparative review of five sustainable rating systems. Procedia Eng. **21**, 376–386 (2011)

The Effect of Green Building to Decrease Operational Expenditure of Property Development in Indonesia

Rezi Berliana Yasinta[1](✉), Christiono Utomo[1], and Yani Rahmawati[2]

[1] Department of Civil Engineering, Institut Teknologi Sepuluh Nopember, Surabaya, Indonesia

[2] Architecture Engineering and Planning, Universitas Gadjah Mada, Yogyakarta, Indonesia

Abstract. The green building concept is a solution that has a role in sustainable development. However, there are still obstacles in applying green building concept from the perspective of property developers, that high initial investment costs. Many theories state that implementing green buildings can reduce operational and maintenance cost. This study aims to analyze the effect of green buildings to decrease in operational expenditure on apartment properties in Surabaya, Indonesia. This research was conducted by online questionnaire with descriptive analysis and regression analysis to determine the effect of green building variables on decreased operational expenditure. The results of this study show that the variable of green building can affect to reduce operational expenditure by use of eco-friendly roof, use of eco-friendly glass, use eco-friendly light, water-saving with rainwater harvesting, and building design with renewable energy. The effect of green building decrease in operational expenditure by 58.3%. The implication of this study is strategies in planning the application of green building concepts, especially at planning and operational stages.

Keywords: Green building · Operational expenditure

1 Introduction

Construction development greatly affects environmental sustainability and quality. The construction industry in developing countries accounts for around 30% of total greenhouse gas emissions [1]. CO_2 gas emissions produced by the building sector reach 34.8% of the total energy use in the world. Buildings also use energy consumption which is quite large from the total energy use. This is proven by the highest energy use in Indonesia, namely from the household sector in Indonesia, amounting to 31.03% of the total energy use [2].

Green building is building that maximize energy savings, protect the environment, maintain health, reduce pollution, use space effectively, and harmony with nature in its life cycle. The characteristics of green buildings are environmental improvement and human health, the efficiency of natural resources and building materials, and water and energy efficiency [3]. The green building concept often has obstacles in its application.

P. Mendonça and N. D. Cortiços (Eds.): ICAMC 2021, LNCE 226, pp. 428–437, 2022.
https://doi.org/10.1007/978-3-030-94514-5_43

One of the main reasons for the obstacles to implementing the green building concept is the high initial investment cost. The high cost of capital leads to a lack of application of the green building concept [4]. Based on this, there is a debate about the high initial costs of investing in green buildings. However, there is no systematic evidence regarding the difference in the initial cost of implementing green buildings. It is only the developer's perception that the application of the green building concept requires high capital cost [5].

It can be said that implementation of the green building in Indonesia is running slowly. The slow implementation of green building in Indonesia is caused by the lack of understanding of the green building concept from stakeholders and building occupants. Other obstacles to the implementation of green building in Indonesia are the implementation of green building which burdens stakeholders, lack of awareness, high cost of implementing green building, towards changes to a healthier environment, and lack of a building management role. The lack of support from the government regarding the implementation of green building is also an obstacle in implementing green building in Indonesia. The need for support and appreciation from the government is a challenge for practitioners and building occupants to own, design, build, and implement green buildings [6].

Green building apartments use technology for energy efficient and environmentally friendly. This can be applied in green roofs, roof skylights, walls that can reduce incoming heat, photovoltaic, utility systems, and green open spaces. This is done in order to save energy by reducing the existing electrical load. Therefore, green building apartments that are environmentally friendly are an effort to overcome the problem of housing and green open space. The application of green building is increasingly widespread. This is evidenced by the results of green building certification by Green Building Council Indonesia (GBCI), it was found that 118 new buildings and 28 existing buildings have been certified [7].

2 Literature Review

2.1 Green Building

Green building is defined as building that can save resources to the maximum extent during their life cycle, including energy, land, water, materials, etc., to protect the environment, reduce pollution and serve as a comfortable living space. Until now, the development of green buildings has increased rapidly in recent years [7]. Green building is also an improvement of the environment and human health, the efficiency of natural resources and building materials, and water and energy efficiency [4]. The main objectives of implementing green buildings are minimizing environmental disturbances and waste piles, minimizing energy and other resource use, increasing the use of renewable energy, and improving human health and comfort [6].

Each country has its own rules and criteria regarding the green assessment of a building. The U.S. Green Building Council (USGBC) is the first organization formed in 1993 in the United States which is based on green buildings. USGBC establishes Leadership Energy and Environment Design (LEED) as a rating system for green building certification tools specifically for commercial, institutional, and residential projects. LEED has 6 aspects related to the implementation of green building certification assessment.

Sustainable Sites. The criteria for sustainable sites describe the prevention of pollution-related to construction, the impact of construction, alternative transportation, stormwater management, climate reduction due to construction and light pollution. Construction produces adverse effects on the environment and human health due to pollution caused during the implementation of various construction processes, such as earth moving, trucking on unpaved roads, material production and operation of construction equipment [8].

Water Efficiency. The criteria of water efficiency describe the reduction in landscape water use, reduction in indoor water use and wastewater strategy. The use of landscape water is at least 30% of the baseline and the reduction of indoor water use is at least 20% of the baseline. In this study, water efficiency criteria are divided into two categories, namely use recycled water and water-saving with rainwater harvesting [9].

a. *Use Recycled Water.* The problem of the water crisis has not been resolved properly. Various sectors such as industry, agriculture and fisheries require large amounts of water in the process. Water conservation is one alternative to overcome this problem. The problem of the water crisis has not been resolved properly. Water conservation is one alternative to overcome this problem [9].
b. *Water-Saving with Rainwater Harvesting.* Saving water is an alternative to the problem of limited water use. Efficient use of water can be done by extracting rainwater which can be used for human needs such as landscaping and watering plants [10].

Energy and Atmosphere. An explanation of energy and atmosphere, namely commissioning, optimization of energy performance throughout the building, cooling management, and use of renewable energy. The depletion of natural resources in the world is caused by the use of fossil fuels, around 80% to 95% of the world's primary energy needs. This has led to the development and use of renewable energy sources such as solar power, geothermal power, hydropower, and wind power [11].

Materials and Resources. The materials and resources criteria explain the selection of a recycling collection location, reuse of buildings, construction waste management, purchasing materials that are produced regionally, materials that can be used for recycling and materials that contain low mercury. green materials are materials that eco-friendly, healthy, and can be recycled to efficiently reduce the impact on the environment and human health during their entire life cycle [12].

a. *Use Eco-friendly Roof.* The roof serves for visual comfort, improves the health and safety of occupants. It is important to know the type of roof used in the building. Eco-friendly roof can be defined as living roofs, garden roofs, eco-roofs, vegetated rooftops. The use of the type of roof used, can have an impact on the level of sunlight, the efficiency of construction costs and the visual comfort of the occupants [13].

b. *Use Eco-friendly Glass.* The importance of building design can determine energy efficiency or reduce the regulation of energy consumption used in buildings. It aims to achieve a good thermal environment for the user, good ventilation and natural lighting [14].
c. *Use Eco-friendly Wall.* The application of a vertical greenery system is part of the use of eco-friendly walls. The vertical greenery system is formed by several basic components, namely plants, systems and media, and irrigation. The application of VGS has many benefits such as increased aesthetics and property value of buildings, improve visual quality, reduce pollution, and noise reduction [15].
d. *Use Eco-friendly Wall Paint.* Using eco-friendly paint also needs to be applied in green building materials. Wall paint has an emission of organic compounds that can have an adverse effect on building users. Volatile Organic Compounds (VOC) emitted from decorative building materials are a major source of indoor air pollution that can pose a hazard to human health [4].
e. *Use Eco-friendly Light.* Occupants' visual discomfort is sometimes caused by glare or inadequate lighting. Excessive lighting and lack of light distribution can also affect the discomfort. To reduce the impact of unfavorable visual discomfort on occupants, building operators have been entrusted with ensuring building interiors receive adequate lighting to enhance the quality of work and daily activities [7].

Indoor Environmental Quality. The explanation regarding the criteria for indoor environmental quality is minimizing factory smoke, air ventilation, indoor air quality, using low-emitting materials and the ability to control thermal and lighting systems. The quality of the indoor environment influences on the health, performance and productivity of the community around the environment [13].

Innovations. The criteria for innovations describe innovative strategies for sustainable design. Technology that is increasingly developing by applying the concept of green building affects on building design strategies. Energy-efficient green buildings require integrated designs taking into account climate, technology, operational and maintenance costs, and occupant behavior. This causes a dynamic interaction between building system integration and building users and the resulting building performance. The design of an integrated building system must review the needs of the building users because building performance objectives are aimed at building users [12].

2.2 Green Building to Decrease Operational Expenditure

Research on green buildings to decrease operational expenditure is still growing rapidly now. There is research that discusses the comparison of the Life-Cycle Cost (LCC) of green building with conventional buildings. This research method uses comparative case study analysis of two green buildings and a conventional building with the same specifications. The data needed to perform the LCC analysis was extracted through documentary analysis. The results show that the comparative analysis shows that the construction costs of green industrial buildings are 37% higher than conventional buildings. Although, the operational, maintenance and final costs of green buildings account for 28%, 22%, and 11%. This results in an overall cost savings of 21% on green buildings [16].

2.3 Conceptual of Research Model

The variables are obtained from the results of a literature study on research related to green buildings on decreasing operational expenditure. In this study, there are 3 aspects of green buildings studied, namely material aspects, water conservation, and design. Based on these 3 aspects, research variables are obtained based on literature studies of the green building concept. Each variable has an operational definition that explains how the variable can be measured. Figure 1 is an explanation of the conceptual model.

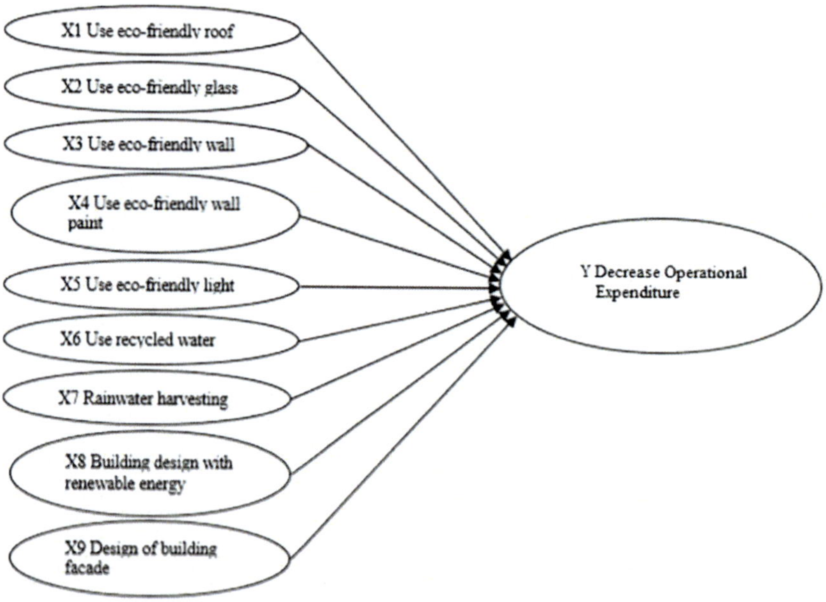

Fig. 1. Conceptual model of green building to decrease operational expenditure

3 Research Methods

3.1 Sampling and Data Collection

This study uses a survey in data collection, taking sampling from the population using an online questionnaire. This study uses a non-probability sampling technique, namely purposive sampling with the aim of determining the appropriate and competent sample with the research topic to be carried out. This data was obtained from respondents, namely Director, General Manager, Project Manager, Site Manager, Engineer, Architect, and Green Building Auditor.

The scale used in this study to measure respondents' opinions and perceptions of green building indicators for decreasing operational expenditures is using a Likert scale. The measurement scale consists of a "Agree" 1–5 scale or a "Disagree" scale. The scale on the questionnaire is used to measure the respondent's level of agreement with the green building variable on decreasing operational expenditures.

3.2 Data Analysis Method

The method used in this research is descriptive analysis and multiple regression analysis. Descriptive analysis is used to describe the relationship of green buildings to decrease in operational expenditure through the independent variable and the dependent variable, and also analyzed using a scatter diagram. This can determine whether there is a relationship between the green building variables used to decrease operational expenditure. Multiple regression analysis is used to measure the effect of the green building variable on decreasing operational expenditure. In this study there were 30 respondents with the results to be analyzed using multiple regression analysis.

4 Result

4.1 Profile of Respondents

The respondent's explanation contains profile data and answers from each respondent based on questionnaires that have been distributed online. Grouping respondents based on profession and working experience. Based on profession, there are 9 respondents who work as Director, 6 as Engineer, 5 as Project Manager, 3 as Site Manager, 3 as Green Building Auditor, 2 as Architect, and 2 as General Manager.

Based on working experience, it is divided into work experience with <1 year = 8 people, 1–3 years there are 10 people, 3–5 years there are 5 people, 5–10 years there are 6 people, and >10 years there are 4 people.

4.2 Descriptive Analysis

The descriptive analysis, it aims to determine the relationship between the green building variable and decrease in operational expenditure based on the perceptions of each respondent. This can use a scatter diagram where there are three assessment specifications, namely positive relationships, negative relationships, and unrelated. Table 1 is an explanation of the average green building variable to decrease in operational expenditure.

Table 1. Descriptive analysis

Item	Variable	X mean to Y
X1	Use eco-friendly roof	3.51
X2	Use eco-friendly glass	3.94
X3	Use eco-friendly wall	3.58
X4	Use eco-friendly wall paint	3.18
X5	Use eco-friendly lamp	4.12
X6	Use recycled water	3.76
X7	Water-saving with rainwater harvesting	3.48
X8	Building design with renewable energy	3.97
X9	Design of building facade for temperature comfort	3.58

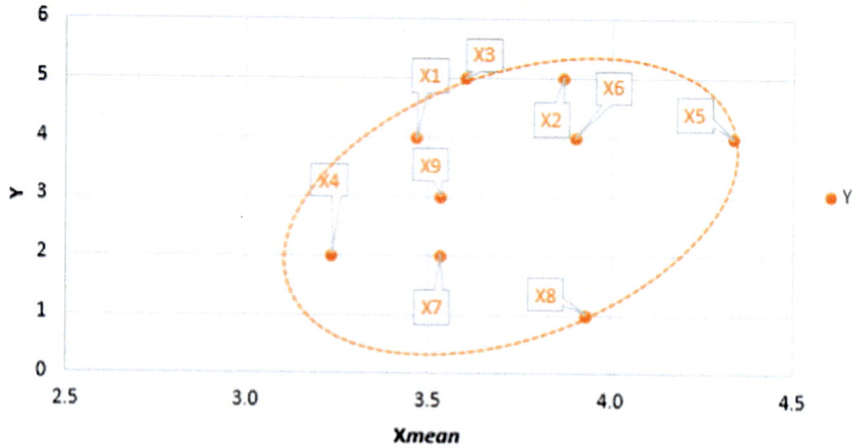

Fig. 2. Scatter diagram green building to decrease operational expenditure

Based on Table 1, scatter diagram explanation is obtained from the average green building variable on a decrease operational expenditure in Fig. 2.

Based on Fig. 2, it can be seen in the scatter diagram that there is an increase in the green building variable in X4 is 3.18 followed by an increase in the green building variable in X7 is 3.48 with an increase of 0.3. Then there was an increase in the green building variable in X9 is 3.58 with an increase of 0.1, and so on. Based on this, the relationship between the green building variable and a decrease in operational expenditure is a positive relationship. A positive relationship occurs if there is a pattern where the value of the green building variable is related to the variable value of the decrease in operational expenditure. It can be concluded that the green building variable has an effect on the decrease in operational expenditure.

4.3 Regression Analysis

Regression analysis is used to determine the effect of green building variables on a decrease in operational expenditure. Model summary of regression analysis, the variable of a decrease operational expenditure, there are model which are: R is 0.764[a]; R Square is 0.583; Adjusted R Square is 0.396 and Std. Error of The Estimate is 0.978. The value of the Y model shows that the effect of the green building variable on the decrease in operational expenditure is 58.3%, while the 41.7% decrease in operational expenditure is influenced by other variables not examined.

Meanwhile, the dependent variable coefficient is presented green building to decrease operational expenditure in Table 2.

Table 2. The coefficient of green building variable

Model	Unstd coefficients		Std. Coeff.	t	Sig	Collinearity statistics	
	B	Std. error	Beta			Tolerance	VIF
Cons.	0.097	1.011		0.096	0.092		
X1	0.207	0.221	0.220	0.938	0.035	0.380	1.631
X2	0.220	0.214	0.219	1.025	0.031	0.458	1.182
X3	−0.139	0.182	−0.144	−0.765	0.045	0.585	1.709
X4	−0.339	0.192	−0.423	−1.772	0.092	0.365	1.741
X5	0.160	0.296	0.122	0.540	0.059	0.410	1.438
X6	−0.006	0.254	−0.005	−0.023	0.098	0.427	1.344
X7	0.301	0.225	0.325	1.338	0.019	0.353	1.831
X8	0.477	0.199	0.562	2.395	0.027	0.378	1.647
X9	−0.115	0.185	−0.151	−0.619	0.054	0.350	1.857

Based on Table 2, the equation of regression equation is obtained such as:

$$Y = 0.097 + 0.207x_1 + 0.220x_2 - 0.139x_3 - 0.339x_4 + 0.160x_5 - 0.006x_6 + 0.301x_7 + 0.477x_8 - 0.115x_9 \quad (1)$$

The dependent variable to decrease in operational expenditure has a constant value of 0.097. Variable use of eco-friendly roof (X1) has an effect of 20.7% on a reduction in operational expenditure. Variable use of eco-friendly glass (X2) has a 22% effect on reducing operational expenditure. Meanwhile, variable use of eco-friendly wall (X3) and variable use of eco-friendly wall paint (X4) did not have an effect of 13.9% and 33.9% respectively on the reduction in operational expenditure. Variable use eco-friendly lamp (X5) there is an effect of 16% on a decrease in operational expenditure. Meanwhile, variable use recycled water (X6) does not have an effect of 0.6% on a decrease in operational expenditure. On variable water-saving with rainwater harvesting (X7) and variable building design with renewable energy (X8), respectively, have an effect of 30.1% and 47.7% on the reduction of operational expenditure. In variable design of building facade for temperature comfort (X9), there is no effect of 11.5% on the reduction of operational expenditure.

5 Discussion

Variable of green building that can affect to can reduce operational expenditure by use of eco-friendly roof, use of eco-friendly glass, use eco-friendly light, water-saving with rainwater harvesting, building design with renewable energy.

Selection of glass building energy performance impact due to the use of the system and the different glass coatings. The main energy requirement for using glass is the movement of glass through windows [14]. This shows that the importance of choosing the glass which has a high energy efficiency so that the impact on energy savings. Although the costs used in selecting glass during the design stage are high, in the long run, it saves energy [17].

The use of recycled water is an alternative to address water scarcity. Water use is a strategy used to increase water demand and manage wastewater properly. The use of recycled water is also used for water conservation needs [9]. Water recycling is the most effective alternative because it aims to save water resources, reduce environmental impact and reduce operational and maintenance costs [18].

Selection of building roof with thermal performance of the building. As with usage of green roofs, the coating on green roof can affect the thermal performance of building. The green roof functions to reduce solar heat and absorb rainy air. The high cost used in green roofs has a worthwhile benefit, namely that green roofs can increase the effectiveness of thermal insulation in green roof systems [13].

6 Conclusion

This study analyze the effect of green buildings on decrease in operational expenditure on apartment properties in Surabaya. Based on the research results, variable of green building that can affect to can reduce operational expenditure by use of eco-friendly roof, use of eco-friendly glass, use eco-friendly lamp, water-saving with rainwater harvesting, building design with renewable energy. The result showed that the effect of green buildings to decrease in operational expenditure was 58.3%.

It only discuss the effect of green building variables on decrease in operational expenditure. This research also only discusses the material aspects, water conservation, and green building design. Furthermore, it is necessary to conduct research using other variables related to green buildings on finance and other aspects of green buildings. This is intended to expand the field of green building which will be investigated in further research to obtain diverse and accurate results in the research that will be achieved later.

Acknowledgement. The authors express appreciation for research grant of Penelitian Dasar Unggulan Perguruan Tinggi (PDUPT) RistekBRIN 2021.

References

1. Geng, Y., Ji, W., Wang, Z., Lin, B., Zhu, Y.: A review of operating performance in green buildings: energy use, indoor environmental quality and occupant satisfaction. Energy Build. **183**, 500–514 (2019)
2. International Economy Agency. World Energy Outlook (2018)
3. Darko, A., Chan, A.: Critical analysis of green building research trend in construction journals. Habitat Int. **57**, 53–63 (2016)
4. Li, S., Lu, Y., Kua, H., Chang, R.: The economics of green buildings: a life cycle cost analysis of non-residential buildings in tropic climates. J. Cleaner Prod. **252**, 119771 (2020)
5. Chegut, A., Eichholtz, P., Kok, N.: The price of innovation: an analysis of the marginal cost of green buildings. J. Environ. Econ. Manage. **98**, 102248 (2019)
6. Wimala, M., Akmalah, E., Sururi, M.R.: Breaking through the barriers to green building movement in Indonesia: insights from building occupants. Energy Procedia **100**, 469–474 (2016)
7. Khoshnava, S.M., Rostami, R., Valipour, A., Ismail, M., Rahmat, A.R.: Rank of green building materials criteria based on the three pillars of sustainability using the hybrid multi criteria decision making method. J. Clean. Prod. **173**, 82–99 (2018)
8. Giunta, M.: Assessment of the environmental impact of road construction: Modelling and prediction of fine particulate matter emissions. Build. Environ. **176**, 106865 (2020)
9. Rossum, T.V.: Water reuse and recycling in Canada—history, current situation and future perspectives. Water Cycle **1**, 98–103 (2020)
10. Şahin, N.I., Manioğlu, G.: Water conservation through rainwater harvesting using different building forms in different climatic regions. Sustain. Cities Soc. **44**, 367–377 (2019)
11. Fan, K., Wu, Z.: Incentive mechanism design for promoting high-level green buildings. Build. Environ. **184**, 107230 (2020)
12. Sun, C., Chen, Q., Xiao, J., Liu, W.: Utilization of waste concrete recycling materials in self-compacting concrete. Resour. Conserv. Recycl. **161**, 104930 (2020)
13. Tam, V., Wang, J., Le, K.: Thermal insulation and cost effectiveness of green-roof systems: an empirical study in Hong Kong. Build. Environ. **110**, 46–54 (2016)
14. Hess, M., Bongaerts, J.C.: Differences in perception of the environmental and health impacts of decorative paints among average consumers. Sustain. **12**, 4495 (2020)
15. Kazemi, F., Rabbani, M., Jozay, M.: Investigating the plant and air-quality performances of an internal green wall system under hydroponic conditions. J. Environ. Manage. **275**, 111230 (2020)
16. Vyas, G.S., Jha, K.N.: What does it cost to convert a non-rated building into a green building? Sustain. Cities Soc. **36**, 107–115 (2018)
17. Khahro, S.H., Kumar, D., Siddiqui, F.H., Ali, T.H., Raza, M.S., Khoso, A.R.: Optimizing energy use, cost and carbon emission through building information modelling and a austainability approach : a case-study of a hospital building. Sustainability **13**, 3675 (2021)
18. Al-Kayiem, H.H., Koh, K., Riyadi, T.W.B., Effendy, M.: A comparative review on greenery ecosystems and their impacts on sustainability of building environment. Sustainability **12**, 1–26 (2020)

Challenges and Countermeasures of Project Cost Management in COVID-19 Emergency Infectious Disease Hospital

Yanan Fan(✉)

College of Urban Construction, Tianjin College, University of Science and Technology, Beijing 301830, China

Abstract. The spread of corona virus pneumonia in the world has brought serious threat to people's life and health. The models of Beijing Xiaotangshan Hospital and Wuhan Huoshenshan Hospital prove that the construction of emergency infectious disease hospital plays an important role in dealing with sudden epidemic. However, because of the differences in construction technology, design concept and construction period, the cost management experience of traditional hospital can not be simply applied to the project cost of emergency infectious disease hospital. Taking Huoshenshan Hospital in Wuhan as an example, this paper analyzes the challenges and countermeasures of project cost in emergency infectious disease hospital, which can provide reference for similar project cost management.

Keywords: Corona virus disease · Emergency infectious disease hospital · Project cost management · Challenge

1 Introduction

On December 8,2019, the first case of new pneumonia was found in Wuhan. Since then, the new corona-virus has spread around the world. As of March 11th,118,847,915 cases had been confirmed overseas, 2,630,406 had died and 90,642,972 had been cured, and 102,194 had died and 96,873 had been cured in China [1]. The rapid construction of emergency infectious disease hospitals such as Huoshenshan Hospital and Leishenshan Hospital has played an important role in the prevention and treatment of new crown disease. Different from the traditional hospital construction project, the construction period, building structure and construction technology of emergency infectious disease hospital are different, which brings new challenges to the cost management of such projects [2]. This paper takes Huoshenshan Hospital as the research object, first introduces the characteristics of emergency infectious disease hospital, then analyzes the challenges in the cost management of such projects, and finally puts forward the corresponding countermeasures.

P. Mendonça and N. D. Cortiços (Eds.): ICAMC 2021, LNCE 226, pp. 438–446, 2022.
https://doi.org/10.1007/978-3-030-94514-5_44

1.1 Characteristics of Emergency Infectious Disease Hospital

Emergency infectious disease hospitals are places that deal with all kinds of sudden public infectious disease, the most notable features of which are real time and effectiveness. The establishment of emergency infectious disease hospital is the most effective choice to control the epidemic situation when the new corona-virus wreaks havoc all over the world. The engineering characteristics of emergency infectious disease hospital in Huoshenshan Hospital of Wuhan City are analyzed in this paper.

On January 23,2020, the number of newly diagnosed cases of pneumonia in Wuhan was on the rise. Ten days later, the number was more than 20,000. The number of confirmed cases is rising, medical supplies are tight, hospital beds are tight, many suspected confirmed cases can only go home to wait, the situation is very serious. In response to the challenge of the epidemic, the city closed the next day, the construction of the Huoshenshan Hospital began, 10 days after the completion of February 3 will be put into use [3].

Huoshenshan Hospital is located in Zhiyinhu Road, Caidian District, Hubei Province (Fig. 1). It is a special emergency infectious disease hospital for treating patients with novel corona-virus pneumonia, referring to the model of "Beijing Xiaotangshan Hospital"during SARS in 2003. The emergence of Huoshenshan Hospital has greatly relieved the pressure of the traditional hospital to treat the patients with new coronary pneumonia and made it focus on rescuing the critically ill patients. Compared with traditional hospitals, Huoshenshan Hospital has the following characteristics:

Urgent Construction Period. Compared with the General Hospital, the construction period of emergency infectious disease hospital is very short. According to the normal engineering construction procedure, the construction of a hospital has to go through more than 10 stages, such as project approval, preparation of feasibility study report, Scheme design, etc., which generally takes several years, obviously can not meet the emergency infectious disease hospital rapid contracting and rapid construction requirements. According to the report, Wuhan Huoshenshan Hospital was built on the basis of Wuhan Workers sanatorium, referring to the model of Beijing Xiaotangshan Hospital during the 2003 SARS fight, and was specially used to treat patients with novel corona virus infection.

In the afternoon of January 23, 2020, CITIC Architectural Design and Research Institute received an urgent design task from Wuhan Huoshenshan Hospital, and quickly formed a project team of more than 60 people, and put into design work that night. CITIC Design completed the site leveling design drawing within 5 h of receiving the task, and gained time for the overnight construction; completed the plan design drawing within 24 h, which was approved by the Wuhan Municipal Government; after 60 h of continuous struggle, all the construction drawings. were delivered to the early morning of January 26.The total gross leasable area is 33,900 square meters with 1,000 beds. Yet a hospital of this size was built and completed in just 10 days. Therefore, how to do a good job in the project cost control at the same time has become one of the difficulties of project management and control [4].

Complex Structure Engineering. The overall layout of Huoshenshan Hospital is L-shaped, and the main consideration is to keep a certain distance from the east water system on the basis of satisfying as many beds as possible. Vulcan Mountain Hospital mainly includes the reception area, negative pressure ward building, ICU, and medical technology department. Among them, there are two negative pressure wards: No. 1 inpatient building is a single-storey, with a total of 8 buildings, and the one near the reception area is a suspected ward.; Ward No. 2 has two floors and a total of 4 buildings [5].

Fig. 1. Design renderings for Huoshenshan hospital

The main transmission route of the novel corona virus is droplet transmission and close contact, which is highly infectious. Compared with general hospitals, emergency infectious disease hospitals have higher requirements in infection control, diagnosis and treatment safety and service efficiency. With reference to the form of battlefield hospitals as a whole, the designers fully considered the construction material resources that can be purchased the fastest at that time, and adopted a modular design. It was mainly composed of reception area, negative pressure ward building, ICU, medical technology building, network computer room, central supply warehouse, and temporary waste disposal. Storage room, ambulance decontamination room, etc. Huoshenshan Hospital was strictly divided the infected area into three areas: contaminated area, potential contaminated area and clean area, and strictly required the flow of personnel. Set up personnel flow, material flow, medical waste flow routes, and set up a full set of protective measures and management mechanisms. Hospitals and departments quickly subdivided the infection wards, designed standard flow routes for people, logistics and medical waste respectively, and implemented strict management.

Shortage of Construction Resources. The construction time of emergency infectious disease hospital is often special, which is usually the special period of epidemic spreading, and it is often accompanied by the shortage of engineering materials and living materials. For example, the construction of Huoshenshan Hospital in Wuhan during the spring festival holiday period, coupled with the implementation of local traffic control in order to prevent the spread of the epidemic, From the perspective of material supply: Wuhan's closure of the city will affect the transportation, distribution and supply of medical supplies and living materials to a certain extent, and will cause a shortage of medical supplies in the short term. Huoshenshan Hospital project raw materials and labor costs have increased significantly compared to the past. At the same time, because of the epidemic prevention and control reasons, although the number of workers on site reached thousands, but the vast majority of workers still need to work more than 12 h a day.

Special Construction Technology. The special construction technology of Huoshenshan Hospital is mainly reflected in the construction technology adopted by the prefabricated building. The general engineering project uses the cast-in-place type construction, the construction period is long, the expense is big. The cast-in-place construction is the assembly of formwork at the site, the binding of steel bars, and then the pouring of concrete. Although the cast-in-place building has good overall rigidity, strong earthquake resistance and good waterproof performance, the whole project is time-consuming, labor-intensive and requires higher soil quality because of the long concrete setting and hardening time and the slow construction speed, are extremely susceptible to the external environment.

Through the standardized architectural design and modular, factory production of parts, most of the work is done in the factory to achieve the building components of the general and on-site construction of the assembly, mechanization [6]. Its remarkable characteristics include design simplification, design speed, construction speed, construction quality, construction environment improvement, labor conditions improvement, resource and energy saving, cost saving and so on. At the same time, after the external splicing, the whole lifting is carried out, and the field construction and the whole lifting are interspersed to achieve the maximum efficiency. For example, China Railway Eleventh Bureau Group Corporation was notified that Vulcan Mountain Hospital is in urgent need of a large number of welders and steel structure workers. The group organized a professional team of nearly 100 people in the shortest time, with more than 20 welding machines and rushed to the construction site of Huoshenshan immediately (Fig. 2).

Fig. 2. Hundreds of excavators are working on the construction site

Special Way of Project Contracting. In order to speed up the process of hospital contracting and construction, the construction of Huoshenshan Hospital adopts the mode of Engineering-Procurement-Construction General Contract (EPC). According to the contract, the EPCM Enterprise undertakes the design, procurement, construction and trial operation, and to be responsible for the quality, safety, cost and schedule of the contracted project. Under the model, construction units led by the Third Bureau, Wuhan Construction, Wuhan Municipal, Hanyang municipal co-organization, nearly 2,000 people on New Year's Eve team carried out 24 h uninterrupted construction. In the EPC mode, design and procurement, construction, trial operation and other business complement each other, forming a unified organism with each other. The ability of coordination between design scheme and construction technology can be cultivated through cooperative operation.

2 Challenges of Emergency Infectious Disease Hospital to Project Cost Management

2.1 Tight Construction Period Leads to Insufficient Design Work

In order to deal with the epidemic situation, compared with the traditional hospital project, the emergency infectious disease hospital has the characteristics of shorter construction period. This may lead to the early design work is not solid, the case of the project on the construction. Huoshenshan Hospital completed the blueprints within 24 h. But in the similar project, the front work is not solid may cause to the engineering quantity judgment inaccuracy, therefore should pay more attention in the site selection, the reconnaissance and the design aspect work precision (Fig. 3). If it involves hidden works, underground works, no survey and design, geographical conditions can not be judged, affect site selection, affect underground construction, so can not accurately judge the amount of work; In the architectural design of Huoshenshan Hospital, the

Fig. 3. Construction site of Huoshenshan Hospital

design experience of Beijing Xiaotangshan hospital can be fully used for reference. The Third Bureau of the China Construction Construction Bureau has urgently transferred technical cadres with rich experience in medical architecture, to Wuhan Fire Test Center, City Center for Disease Control and Prevention and all participating units throughout the design institute to jointly design drawings.

2.2 Complex Structure Produces Special Engineering Quantities and Equipment

The hospital consists of building 1 and Building 2, of which Building 1 is a single-storey building consisting of the inpatient building, the medical technology building and the ICU; Building 2 is a two-storey inpatient building; in addition, in addition, an outdoor oxygen station room, a negative pressure suction machine room, a garbage storage room, a corpse storage room and an incinerator room were also built. Huoshenshan hospital has mainly adopted the style of container temporary room, which has a 5 year service life, said prefabricated building, chief medical officer of the hospital. The underground infrastructure is built according to the amount of a floor, such as water consumption, electricity consumption, sewage treatment capacity, etc., is not enough to support multi-storey buildings, resulting in the overall project area is large, earthwork is far more than similar projects. As this project is an emergency project, some equipment and materials have also adopted special requirements, such as the installation of container buildings and the operation of construction machinery in less than 24 h due to the low temperature during the construction period, therefore, the concrete strength grade will be raised to C35. The raft of the building part of this project is 300 mm thick except for the medical technology and ICU part which is 450 mm thick. The raft foundation thickness of the equipment room is designed according to the relevant requirements provided by the factory.

2.3 The Shortage of Resources Causes the Price of Artificial Materials to Rise

The construction of the Vulcan Mountain Hospital is in the midst of an epidemic. As a result of the epidemic period of personnel and traffic control, resulting in the construction of Huoshenshan Hospital used in the construction of building materials and labor prices are higher than usual. The workers in this project, weak electricity, welders, heating, decoration and other work 24 h double shift, 1500 yuan per day, 1000 yuan per day, wages daily junction (Fig. 4). The price increase of artificial materials caused by the shortage of resources makes the project cost of Huoshen Mountain unable to simply apply the construction quota. 2.4 cost increase due to complex construction process.

2.4 Complex Construction Process Leads to Increased Cost

Compared with the traditional construction technology, the construction technology using prefabricated building leads to higher project cost. On the one hand, the construction price offered by the precast plant is often greater than the unit price of cast-in-

Fig. 4. Workers in Huoshenshan Hospital

place concrete. On the other hand, the scale of components used in Huoshenshan Hospital is relatively small, which makes the fixed cost account for a high proportion of the total cost. Finally, in view of the small number of prefabrication plants and the long distance of component transportation, the transportation cost increases correspondingly. At the same time, the large size, special shape, weight and lifting mode are different from the traditional construction mode, so there will be some vertical transport costs.

2.5 Special Contracting Methods Make Cost Control Difficult

Huoshenshan Hospital uses the EPCM EPC contract. Compared with the traditional DB contract, EPC contract is generally a total price contract. Due to the special project time limit, the contract signed is a cost-plus-fee contract, which is not conducive to the prior control of investment. On the other hand, the project bidding procedure is simplified, after the construction drawing budget, so in the implementation process the project changes frequently, the project quantity estimation is not accurate and so on.

3 Countermeasures of Project Cost Management in Emergency Infectious Disease Hospital

3.1 Adopt BIM Technology to Accelerate the Advancement of Pre-design

Due to the short construction period and heavy tasks, the Vulcan Mountain Hospital cannot go through the different stages of plan design, preliminary design, and construction drawing design in accordance with the normal design process, which may lead to a disconnect between design and construction. In the Huoshenshan Hospital project, the project decided to design, construct, modify and adjust at the same time, and the cross assault was completed. The construction period was extremely tight, and the designer's modeling and drawing were carried out together with the construction site construction of the construction party, so there was no chance of rework, and the task was very arduous. Therefore, the advantages of BIM technology should be fully utilized in the design. The design team should fully digest the design drawings, use BIM technology and other tools to complete the in-depth design of various electromechanical specialties, and integrate electromechanics, structure and decoration to form an integrated module, and conduct information simulation of the entire construction process and key processes [7]. It is no longer necessary to prepare construction drawing budget separately, and use the audited design budget as the guideline for cost management.

3.2 More Reserve Expenses to Deal with the Shortage of Resources

For an engineering project, the whole process of cost management implements the entire project, and it plays a major role in the construction of the project. The author believes that the factors that have a greater impact on the project cost are construction raw materials, construction workers' labor costs, project management costs, and tax deductions, change negotiations (change the previous design practices during the construction process or increase a part of the engineering volume). The Vulcan Mountain Hospital project has strong emergency response, large scale of construction, and tight schedule. The factories around Wuhan experienced a shortage of raw materials during prefabrication, high labor costs, and insufficient construction personnel, which had a great impact on the project. Due to the tight schedule, sufficient expenses should be reserved. According to the specific personnel and mechanical configuration of the project, material resource scheduling and transportation costs, etc., it may be considered to increase from the normal level.

3.3 Combining the Construction Technology to Reasonably Determine the Investment Estimate

The cost of prefabricated building construction is generally included in the general contracting construction cost. The price of raw materials for the entire construction is high. At the construction stage, unloading trucks-stacking-prefabricated board houses are installed one by one (control elevation pad placement-lifting, in place-temporary fixing -Decoupling, correction-anchoring reinforcement installation, combing), the capital budget should be reasonably arranged according to the special construction technology of Huoshenshan Hospital. For example, the ward area structure adopts a modular container structure, and the medical technology building and ICU adopt a light steel structure; the foundation adopts a design method combining raft and steel support, which not only protects the anti-seepage membrane of the building, but also ensures the overall Construction efficiency. Investment estimates should be based on the nature, scale, construction standards, site selection, funding sources, construction period requirements, transportation distances of components and parts, and construction site cost levels, comprehensive consideration of the increase in the cost of prefabricated buildings, technical measures and organizational measures for project implementation (Including measures taken to shorten the construction period) etc. to determine the estimated investment.

3.4 Flexible Use of Contract Pricing to Incentivize Enterprises

The contract form used in the construction of infectious disease emergency hospitals is cost plus remuneration [8]. A cost plus remuneration contract is a type of contract in which the owner pays the contractor the actual cost of the project and pays the remuneration according to a certain method agreed in advance. That is, the final contract price of the project is calculated based on the actual cost of the contractor plus a certain percentage of remuneration, and a specific contract price cannot be determined when the contract is signed, only the percentage of remuneration can be determined. The remuneration consists of management fees, profits and bonuses. This form of contract, combined with prefabricated construction, effectively shortened the construction period. Due to time constraints, if the original vouchers are temporarily missing, the construction unit can promise in advance and fill in the information within 45 days after the completion of the project.

4 Conclusions

In view of the challenges in the cost management of emergency infectious disease hospitals, this paper takes Huoshenshan Hospital as an example, first analyzes the characteristics of construction and management of Huoshenshan Hospital, then the paper makes a deep analysis of the difficulties in cost management, and puts forward some pertinent suggestions, which can provide reference for similar projects.

References

1. Devi, H.S., Devi, K.S., Devi, T.K., et al.: Knowledge of COVID-19 and death toll cases in Manipur. Arch. Current Res. Int. 36–40 (2020)
2. Cai, Y., Huang, T., Liu, X., et al.: The effects of "Fangcang, Huoshenshan, and Leishenshan" hospitals and environmental factors on the mortality of COVID-19. PeerJ **8**(2), e9578 (2020)
3. Ahmed, J.: 1.3 China's Super high Speed of Building the Huoshenshan Hospital (2020)
4. Cai, J.: Research on Project Cost Integrated Management under EPC Mode. West China University (2014)
5. Wang, Y., Zhang, L., Chan, A., Duan, W., Zhan, Y., Song, C.: Construction and management of the flow of medical care in the ward of Wuhan Huoshenshan Hospital. Chin. Nurs. Adm. **20**(11), 1675–1679 (2020)
6. Yanpeng, S., Yingjie, C.: Application of prefabricated light-steel structure movable-slab room in Wuhan Huoshenshan Hospital. Build. Econ. **41**(07), 118–120 (2020)
7. Qingrui, Z., Jiefang, T.: Application of BIM technology in prefabricated building: a case study of Wuhan Huoshenshan Hospital. Hous. Real Estate **21**, 177 (2020)
8. Li, J.: The influence of rate bidding on project cost control. Petrol. Eng. Constr. **37**(01), 77–80 +10 (2011)

Architectural History, Architectural Heritage Protection and Architectural Education

The Catholic Churches from Modern Times in the North of Shaanxi, China

Shan Huang[1(✉)] and Antonio Lopera[2]

[1] School of Mechanics, Civil Engineering and Architecture, Northwestern Polytechnical University, Xi'an, Shaanxi, China
huangshan@nwpu.edu.cn
[2] Escuela Técnica Superior de Arquitectura de Madrid, Universidad Politécnica de Madrid, Madrid, Spain

Abstract. Shaanxi province, where located the famous city Chang'an which flourished during the Tang Dynasty in China, was the first place where Christianity was introduced. The north part of Shaanxi province has a very distinctive local architectural—cave dwellings. Since 1840 when the Opium War broke out, a number of catholic churches have been built. These churches were built according to the design and under the supervision of foreign catholic missionaries, but constructed by local workers. They exemplify how western architectural culture blended in and evolved in phase with the local architectural tradition. Ever since Christian churches were introduced into China, due to political, economic, cultural and other factors, the church buildings were not as magnificent or highly decorated as they were in Europe. However, features of western church architecture were reproduced in the proportions, compositions and motifs of the buildings. In this article, the history of the evolution of the churches will be described, the surviving Catholic Churches in northern Shaanxi will be analyzed, and their development, changes in use and current reutilization will be discussed through an extensive survey of the literature and field investigation.

Keywords: Catholic Churches · Cave dwellings · North Shaanxi · China · Modern times

1 Introduction

Ever since 1840 when broke out the Opium War, China was forced to turn in the semi-colonial society under the intense pressure of western powers. The Modernization process of traditional Chinese architecture started under the impact and promote of the newly introduced western culture. The continuation of traditional Chinese architectural culture and the spreading of the western architectural culture interacted and combined which constituted the development of Chinese architecture during 1840 and 1949 (when P. R. China was founded).

The church buildings were the very first western architectural formation that was introduced in ancient China. This early introduction of western architecture has left a deep imprint. During 1840–1949, along with the spreading of Christianity all over

P. Mendonça and N. D. Cortiços (Eds.): ICAMC 2021, LNCE 226, pp. 449–463, 2022.
https://doi.org/10.1007/978-3-030-94514-5_45

China, the Christian churches were constructed in a lot of places. The western classic church formation was combined into various types of Chinese local architecture formations. The Christian Church buildings of western-Chinese combined formation built in China in 1840–1949 began the modernization process of Chinese architecture and occupied a crucial position in the Chinese architectural history. It is a significant field for research on the influence and spread of western architecture in China's inland cities. They are also the records of the architectural culture of Chinese semi-colonial society.

The north of Shaanxi province is one of the important parts of loess plateau in China. The local resident found the most suitable architectural formation for the local natural, geographic and climate condition—the cave dwelling. This architectural formation is quite common and popular in the large loess plateau of China, and also is one of the most important traditional Chinese residential forms. After the Catholic was brought into the north of Shaanxi, the foreign missionaries and the local craftsmen created the Catholic Church buildings which took advantage of both the local cave dwelling structure and many western classic church elements. The existing Catholic churches (1840–1949) built in northern Shaanxi are not only the unique western architectural formation, but also a standing evidence for the research of modern architecture research in Shaanxi province.

However, until now, there is no research on the development, the distribution and the architectural characteristic of Catholic churches built in 1840 to 1949 in northern Shaanxi. This article focus on these Catholic Church buildings not only reveals the architectural developing history of north Shaanxi in 1840–1949, but also fills the blank in the comprehensive research on how the western Christian Church buildings were blending into the various local architectural formations in different regions of China, and what the combined formations are.

2 Christianity in China

China went through 2000 years of feudal society, after the outbreak of the Opium War in 1840; it turned into a semi-colonial, semi-feudal society when the Chinese society and architecture began the modernization process. However, prior to that, there had been three appearances of Christianity already. The first appearance of a Christian mission occurred in Tang Dynasty (618–907). The Christian scripture was brought into Chang'an city (the capital city of China by then) in 635 by Alopenzz for the very first time. The Emperor encouraged the missionary work and ordered the construction of the very first Christian church in China at the northwest corner of Chang'an city. The second spread of Christianity in China occurred in the Yuan Dynasty (1271–1368), when Mongolian dominated China and the Emperor Kublai supported the missionary work. However, in China, during Tang and Yuan Dynasty, the spread of Christianity mainly relied on the deeper rooted Buddhism or the support of Mongolian noble which leading up to the end of the first two peak periods of the Christian mission when the Buddhism-Prohibit Movement was initiated in late Tang Dynasty and when the Mongolians were expelled from China at the end of Yuan Dynasty.

Christianity came to China for the third time in the 17th century, when China was on the brink of the transition from Ming Dynasty (1368–1644) from Qing Dynasty

(1644–1911). Catholicism became the main religion which was largely preached. Since 1601, the missionaries began to preach Catholicism by imparting advanced scientific knowledge and technology to people from the upper-class. With the incessant spread and development of Christianity, there formed eleven main mission regions, with about 150000 believers (1664). 40% of them were from Shaanxi Province which was more than any other regions. Since 1706, the Chinese Rites Controversy deepened the divergence between the Vatican and the Qing Government. The Qing Government began to implement a very strict Christianity Prohibition Policy, so Christianity declined in China again [1].

In 1840, China turned into the semi-colonial semi-feudal society after the First Opium War broke out. When China was defeated in 1842, the Qing Government had to sign series of unequal treaties with the western powers. As well as ceding territory and paying indemnities, there were many specific clauses which protected the freedom of Christian missions in China. The Qing Government had to rescind the Christianity Prohibition Policy. Thus Christianity began to spread in China as a strong cultural force for the very first time. The missionaries built a lot of Christian churches to expand the influence of Christian mission, which inevitably brought western architectural culture into China. Churches became the most important vector that introduced western traditional architectural culture into China, the church buildings occupied a crucial position in Chinese architecture history during Modern times (1840–1949).

3 The Preaching of Catholicism in Shaanxi Province

Catholicism was most extensively preached when Christianity came to China for the third time. It was brought into Shaanxi Province in the late time of Ming Dynasty (1368–1644). The clearest evidence was that Wang Hui, who was the first local believer of Catholicism, invited Nicolas Trigault to Shaanxi in 1625. They built up a church in Tangfang Street for missionary work. In 1627, the church was rebuilt by the German missionary Johann Adam Schall von Bell, and named Chongyi Church. This church was abandoned when the Christianity Prohibition Policy was implemented, and had been rebuilt again in 1881. It was finally destroyed in the Cultural Revolution (1966–1976). With the strong development of Catholicism and the increasing believers in Shaanxi, in 1696 when the Vatican re-divided the Chinese Mission, Shaanxi Province became an Apostolic Vicariate and one of the most important Catholic regions in China.

After the Opium War, the western powers forced ancient China to sign the unequal treaties which could protect the preaching of Christianity, large numbers of Catholic missionaries came to mainland China, and Catholicism again had a great opportunity for development. The Shaanxi apostolic vicariate was set up again (1844) and was divided into the northern and the southern apostolic vicariate (1887). Later, the northern Shaanxi apostolic vicariate had been divided into two parts (1911): North Shaanxi and Central Shaanxi. In 1946, the Northern Shaanxi apostolic vicariate became Yan'an diocese.

According to the annals of several countries, the Belgian missionaries brought Protestantism to the northwestern part of northern Shaanxi before 1900, and several protestant churches were built. However, Protestantism was never well developed there and as the Red Army entered northern Shaanxi (1935), all the churches were abandoned and prohibited, and Protestantism quickly faded away in northern Shaanxi. No more written record could be found regarding to the protestant churches built in northern Shaanxi in this period, nor could any site or old building. So only the Catholic churches in northern Shaanxi in this period are referred to [2, 3].

3.1 The Spread of Catholicism in Northern Shaanxi Province

There were three main paths of the spread of Catholicism in northern Shaanxi during late Qing Dynasty to the Republic of China.

> The first path: The northern route. In 1872, the Belgian missionary Ye Maozhi of CICM Missionaries came to Ningtiaoliang Town of Jingbian County in northern Shaanxi. It took the missionaries a very short time to build up a church at the side of a bridge. Meanwhile, they bought the land nearby and rented it to the neighbourhood in order to preach Catholicism. The arrival of Catholic missionaries in Jingbian County and the construction of the church were both the sign that Catholicism had officially landed in northern Shaanxi and the mission began. From then on, Catholicism became spread into Dingbian, Suide, Huanyuan and other counties in northern Shaanxi.
>
> The second path: The eastern route. As the outbreak of the Boxer Rebellion, the Catholic activities of northern Shaanxi didn't decline or slow down, but the missionary activity expanded. While Catholicism was developing well among the northern counties which was mentioned above, in 1900, the Italian priest Sun and friar Ma Liupin came to Tanjiapin village in Jia County from the nearby Province to avoid the pursuit by the Boxer Rebellion. Afther that, Catholicism directed by Franciscan was brought into northern Shaanxi and had a strong influence among Jia, Wubao and Shenmu counties.
>
> The third path: The southern route. Until 1905, Catholicism spread from the central part of Shaanxi to Ganquan county. Later, when Yan'an parish was set up (1911), the Catholicism of the Franciscans began to spread in Yanchang, Anding and Ansai. It continued apreading towards south to Fu, Huangling, Yijun and Hengshan counties. Figure 1 According to the Local Chronicles of Northern Shaanxi region, there were 67 churches existed before 1949 [4–7].

The North Shaanxi apostolic vicariate was redivided into Central Shaanxi apostolic vicariate and North Shaanxi apostolic vicariate (1911). The first bishop, for more than 30 years, was Spanish, Cae.lbañez Aparicio. During that period, Catholicism spread widely in northern Shaanxi. At the peak time, there were about 30 priests, 50 churches and 10000 believers.

3.2 The Distribution of Catholic Churches in Northern Shaanxi Province

When Catholicism was brought to northern Shaanxi, the missionaries built churches in many counties to hold religious activities and to serve as the center of the preaching area. The church building was not only an important symbol that Catholicism ever existed in a locality, but was also a stronghold for further spreading the religious

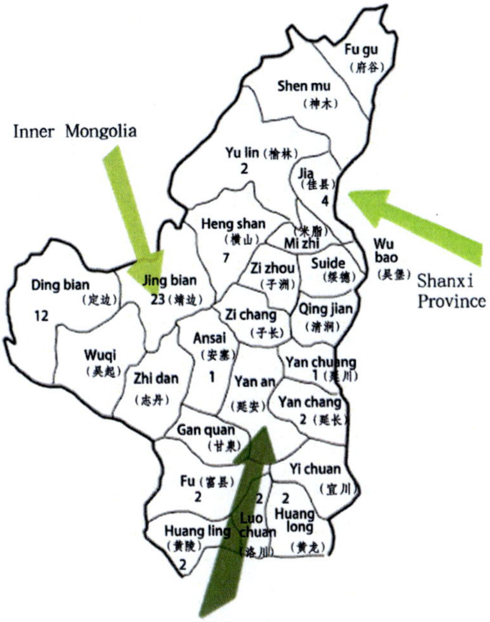

North of Shaanxi Province

Fig. 1. The sketch map of Catholicism's paths of entry and the distribution of Catholic Churchnumbers in northern Shaanxi

doctrine. So, the church was the focal point of Catholicism in a county as well as the base for spreading. In the 25th year of the Republic of China (1936), the Shaanxi Provincial government ordered a survey and registration of the Christian churches and the Mosques. The list for the investigation of Catholic churches in northern Shaanxi was recorded in Shaanxi Provincial Archives. On the basis of this data, the Catholic churches existing in northern Shaanxi before 1936 are summarized in the Table 1. Catholic churches were distributed in 60% of the counties of northern Shaanxi [8].

Table 1. List of Catholic churches existing in northern Shaanxi before 1936

Location	Name of the church	Year of construct	Actual situation
Yan'an (Fushi County)	Qiao'ergou Catholic Church	1931–1934	Occupied by the Lu Xun art academy since 1939
	The Guanguyi Catholic Church	1933	Occupied by local grain supply centre
Suide county	Baihuali Catholic Church	1922	Non-extant. No recorded data
	Xindian village Catholic Church	1928	
Hengshan County	Huanyuanbao Catholic Church	1915	Moved to Jingbian County in 1935

(continued)

Table 1. (*continued*)

Location	Name of the church	Year of construct	Actual situation
Jia County	Tanjiaping Catholic Church	1909	Still in use
	Tongzhen Catholic Church	1911	
Yulin County	Yulin Catholic Church	1914	
	North Mountain Temple Catholic Church	1932	
Jingbian County	Xiao Qiaopan Catholic Church	1874	Knocked down in 1968

4 Example Analysis of the Existing Catholic Churches in the Northern Shaanxi

As the Red Army entered northern Shaanxi in 1935 and was stationed there, religious activities were prohibited and all the Catholic churches were forced to close and were abandoned. Some of the Churches were taken over by the Communist Party of China to be reused as schools or occupied by government and local people as warehouses. Most of the churches were destroyed during the Chinese Culture Revolution (1966–1976), for various reasons. However, several Catholic churches were conserved because they were reused for other purposes by the party or the local people. Now a few of them have been returned to the church for missionary work, others have become exhibition places or are occupied for civil use. There are several examples of the existing Catholic churches in northern Shaanxi which were built up during the period 1840–1949.

4.1 The Tanjiaping Catholic Church

In Jia County, at first, Catholicism was brought in from Shanxi Province which locates in the east of Shaanxi Province. In 1900, when the Boxer Rebellion took place, the Italian Priest Friar Ma and Sun came cross the Yellow River from Shanxi Province to Tanjiaping to escape pursuit and capture during the Boxer rebellion. In this way, Catholicism was brought into Jia County. After 1900, the Franciscans began to do missionary work in the northern Shaanxi. In July 1909, the Spanish priest Nicolas Nieto came to Tanjiaping and began to build the church. However, he died of typhoid on July 5th 1910. Then construction of the church continued under the direction of Priest Wei Xiangque and was finished in early 1912 [9, 10].

This church is the very first Catholic Church in Jia County. After it was built, the management and activities were governed by 8 missionaries in succession and the missionary range covered to some of the area along the Yellow River in Shanxi Province. According to the statistics of 1937, there were more than 1000 followers belonging to the Tanjiaping Catholic Church. Since the liberation of P.R.China, the church was used by a production team and turned to a warehouse until Christianity was again granted the freedom to preach. In 1966, the rebel groups of the Cultural Revolution dug up the remains of Priest Nieto as well as a gravestone with an epitaph. The remains of priest Nieto were thrown into the Yellow River, but the gravestone was preserved by church members of Tanjiaping Church until today. There is a cross on top the gravestone, and 10 lines of Latin in the middle: (translated in Spanish).

"AQUÍ ESPERA
LA RESURRECCIÓN FUTURA
(DE) NICOLÁS NIETO
DE LA PROVINCIA DE SANTIAGO EN España
NACIDO EL AÑO DE 1881
LLEGADO A CHINA EL DIA 10 SEPT 1906
MURIÓ EN TANJIAPIN DÍA 5 JUL. 1910
DESCANSE EN PAZ"

About 190 Chinese characters were engraved in 13 lines on the bottom. The gravestone was oringinally set up in May 1912. It recorded the history from when priest Nieto arrived in Tanjiaping to how he began to built up the church. He dedicated so much that he got very sick and died of typhoid on the July 5th, 1910 when the construction of the church was not finished yet. T construction of the church was continued by Priest Wei. Priest Nieto was very kind and he helped a lot of local people, so the gravestone was set up to commemorate him.

In 1989, Priest Song was sent to Jia County from Yanan Parish to take charge of the management and activities. The reparation he did on the Tanjiaping Catholic Church was very important for the church revival and the membership. Today, the Tanjiaping Catholic Church has become a tourist site as it is the church which has the longest history in Yulin Parish. The whole building is drawn to scale in Fig. 2.

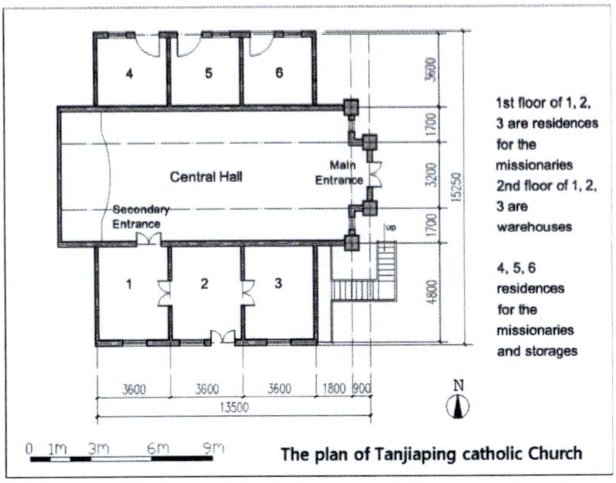

Fig. 2. Plan of Tanjiaping Catholic Church

The whole construction is of stone-arch structure, it appears a combination of typical local residence—cave dwelling and Gothic Fig. 3. The longitudinal axis is along the east-west direction with the façade facing east. The large cave in the middle is used as the main church and connected to the small caves on the south and north. The church has a rectangular plan of 15.3 m long and 6.6 m wide. The inner part is pointed-arch cave dwelling with 10 m in height. The main entrance is facing east side while the altar is located on the west. There is a second entrance on the south of the church connecting to the cave which is used as the sacerdotal residence. Two towers were built on each side of the main entrance. The gates and windows on the façade have gothic pointed arches. The cave dwellings on the south elevation contain six caves arranged in two floors. There is a L shaped stairs leads to the ground. There are three cave dwellings on the north arranged in one row and covered by a flat roof. The total area is 1360 m^2 including storehouse and the stable [11, 12].

Fig. 3. Sketch of Tanjiaping Catholic Church's cave Dwelling Structure

4.2 The Qiao'ergou Catholic Church

It is located in Qiao'ergou village, Baota district, Yan'an County which was just a small village with four families in 1911 when Yan'an parish was set up. The Spanish bishop Caesar Ibañez Aparicio went back to Europe 3 years to raise funds to build the church. In 1914, he bought about 46667 m^2 and built 52 stone caves, 17 bungalows, and 80 loess caves in succession. One large cave was used as the church, and then schools, orphanage and medical clinic were set up gradually. Construction on the Catholic Church started in 1931 and finished in 1934. During the agrarian revolution movement (1935), the priest and associates were evacuated and the schools were closed before the newly-built church was acturally used. When the Central Committee of the Communist Party of China garrisoned in Yan'an (1936), the Qiao'ergou Catholic Church was used as the auditorium of the communist party school. Then it was occupied by the Luxun art academy which belongs to the communist party of China (1939–1945). Since 1949, the Qiao'ergou Catholic Church has been conserved as a very important historical site of Shaanxi Province Fig. 4.

Fig. 4. Qiao'ergou Catholic Church, taken in 1938.

The main body of the church is of rectangular plan with 36.28 m long and 15.86 m wide. The overall area is 575.4 m^2. The bell house includes the hall in the middle and two towers on each side. The central hall and the altar are arranged from south to north. The bell tower and the central hall of the bell house are 25 m and 18 m in height respectively, while 3.83 m in deep.

The façade of the middle hall is divided into three parts. The concave semicircular arch gate at the bottom is the main entrance. Two pilasters are arranged on each side of the main entrance. There are three paratactic windows with semicircular arches in the waist part. On the top part, there is a triangle pediment with two pilasters symmetrically on each side. A traditional Chinese horizontal board inscribed with the words 'Catholic Church' in Chinese is hanging in the middle. There is a concave round window which is similar to rose window above the inscribed board.

The two bell towers are totally symmetrical, which include four parts seperately: the concave arch gate at the bottom used as the second entrance, two arch windows with an arch above midway down, and a rose shape window in the upper part, a concave arch window decorated by a European colonnade on top. The overall appearence is quite tall and straight Fig. 5, 6 and 7 [13, 14].

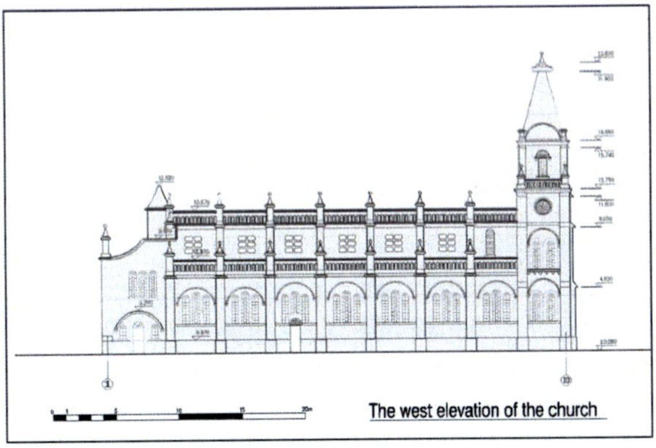

Fig. 5. West elevation of Qiao'ergou Catholic Church

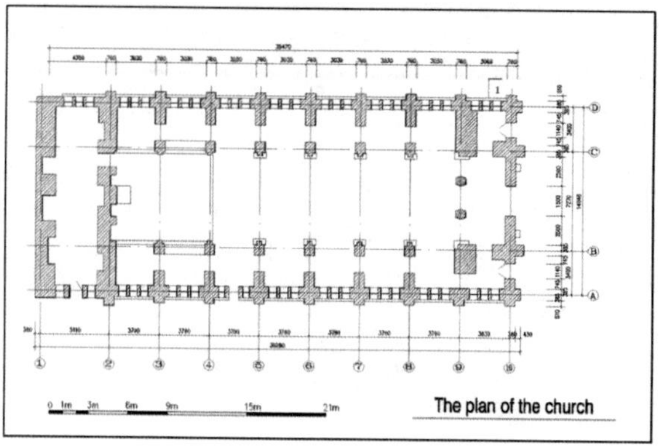

Fig. 6. Plan of Qiao'ergou Catholic Church

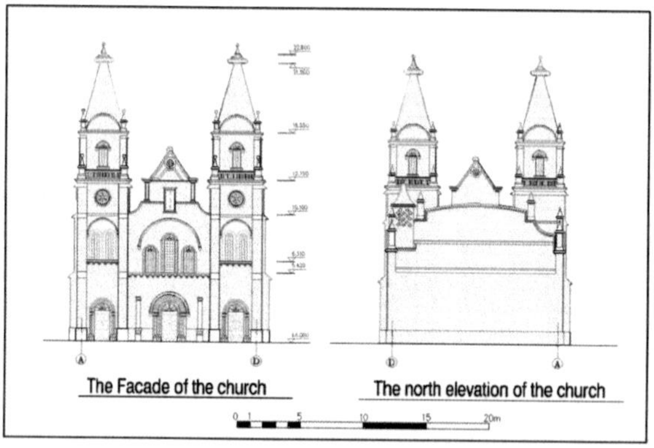

Fig. 7. Facade and east elevation of Qiao'ergou Catholic Church

The central hall of the church is about 27.35 m long and 15.8 m wide. There are twelve Corinthian columns in height of 4.5 m standing in two rows which divide the hall into three parts—the nave and two aisles. Each column has peony, plum blossom and other plant patterns engraved on head and body. The ceiling is of semicircular rib-arch structure which was constructed in stone. The central nave is 11 m × 7.2 m, while the aisle is 6.3 m × 7.2 m, which consists of successive arches with 7 sets of windows opening to east and west. Every set of window contains three small arch windows; the middle one is higher than the side ones. The altar is on the north end in length of 5.1 m. Ttraditional chinses plant and animal patterns are used in detailed stone carvings. Well-decorated pilasters and brick carved niches are standing on the north wall of the altar. Also, there is a second-entrance on the west side.

The whole bottom part of the church is constructed by stone; the other part is of grey bricks in brick-wood structure. The indoor decoration mixed both Chinese and western style, while the outdoor decoration is more similar to Romanesque-style.

4.3 The Guanguyi Catholic Church

It's locating in Baota District of Yan'an city. After Catholicism was first brought here in 1911, a Spanish priest bought an area about 33000 m^2. He built up three stone cave dwellings used as the church. A stone horizontal board inscribed the year 1926 and the sign of the Cross is hanging up. Besides, other 6 cave dwellings were built up and used as the primary school. In 1927, another 13 cave dwellings were built up, 8 of them were used for minor seminary of the parish, while the other 5 used as orphanage and medical clinic.

The existing Ganguyi Catholic Church was constructed in 1931–1934. However, after the liberation of Yan'an city (1935), the church was evacuated and the minor seminary was closed down.Then the Catholic Church was using as the First Complete Primary School in Yan'an. In 1939, the church was turned to a hospital, and then changed again into a textile factory (1944), serving the Red Army. After 1951, the church was first occupied by the government for grain storage, and then altered again into a grain-oil sale station (1953). The windows of the church were blocked in 1967 and the church was reverted to grain storage. Part of the cave dwellings were knocked down in that year.

The Ganguyi Catholic Church has been listed as one of the Municipality Protected Historic Sites since 1984, then it was restored in 1988. Now the church is kept in good condition, but the indoor facilities no longer exist. There are 22 cave dwellings and a 77 m of enclosure wall on south with the main entrance. The Chinese words 'Catholic Church' are engraved on. The church is now conserved without function, but all the other associate buildings are used as the grain storage in the town.

The church has a rectangular plan with an outward arch shaped altar at the north end. It is about 16 m wide and 31 m long. The bell tower in the middle with two aisles on the side composed the façade in vertical. The composition of the aisle elevations is completely same and symmetrical, with pointed arch underneath, triangle pediment above with rose shaped blank window engraved with traditional Chinese plant decoration. Two spires stands on both side of the pediment Fig. 8.

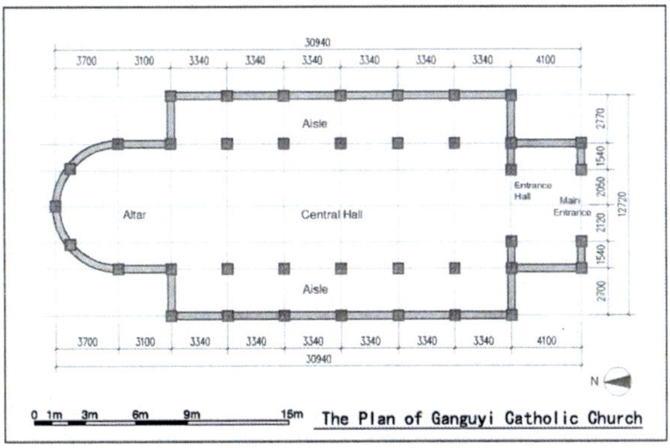

Fig. 8. Plan of Ganguyi Catholic Church

The elevation of the bell tower is the center of the composition of facade, which has been divided into three parts. The main entrance of the church is in the bottom part, which is a three layer concave pointed arch gate. Three concave pointed windows and the horizontal inscribed board is in the middle. A hexagonal bell tower standing on top, which is decorated with cornice and has pointed arch windows Fig. 9.

Fig. 9. Façade of the Ganguyi Catholic Church

The indoor space of the church is composed by the central hall, two aisles, the bell tower and the semicircular altar at the north end of the central hall. The roof truss of the altar is vaulted while the central hall and aisles is arched. Oil paintings depicting Bible stories are on the wall. There are portraits of the Virgin Mary, the Father and Jesus on the altar.

The whole church is constructed of stone-wood structure using stonesas the main constructional material. All the chapiters have are decorated the same in detail. There are spires on the roof, all the windows and gates have pointed arches and are finely engraved. They are in typical gothic style with traditional Chinese decorations.

4.4 The Xiaosuangou Catholic Church

The Xiaosuangou Catholic Church was first built up in 1932 on the hillside of Qinghuabian Town, Baota District, Yan'an City. It is a combination of Chinese and western styles, with cave dwellings body and Roman façade.

The church is composed by typical mountain cave dwelling and a Catholic Church façade constructed with bricks. It is 12 m high in two floors. The bottom part is five loess cave dwellings covered by gray brick elevation while the upper part is a brick-built bell tower. Figure 10, 11 The total width is about 26 m. The cave in the middle which is 6 m in width was used as the church. The church cave is about 7 m high from inside and 9 m in depth, the altar is on the north. The indoor measurements of the other four smaller caves are 3 m (wide), 3.5 m (high) and 6.4 m (deep) [15].

Fig.10. Sketches of the cave dwelling structure of the Xiaosuangou Catholic Church

Fig. 11. Façade of the Xiaosuangou Catholic Church

5 Conclusion

The article clarified and summed up not only the history, but also the distribution and construction situations of Catholic Churches which were built in northern Shaanxi during 1840 to 1949. A lot of the Catholic churches in northern Shaanxi from that time were designed and constructed under the supervision of the Spanish missionaries from Franciscans.

The construction of Catholic Churches in the northern Shaanxi both had a very strong influence and had brought the unprecedented new architectural formations to the local architecture. Chinese and western architectural style appear in the same building and show the following features.

1) The plan layout of the existing Catholic churches in northern Shaanxi is not as same as the typical basilica plan or Latin cross plan. Some of them have combined the basilica with Gothic towers. However, it is not the same as a typical European basilica as there is no height difference from the central hall to the aisles and the plans are always rectangular without the altar protruding outwards. None of the local churches has the Latin cross plan.
2) Because normally the local Catholic churches are of smaller scale and have relatively simple indoor space arrangement, so the inner space would hardly make people feel small or insignificant as western churches could do. All the churches there only have a wide open indoor space, and we can get a sweeping view the

moment when we enter the entrance. The reason is probably that the traditional Chinese architecture is quite practical; besides, at that time, the economic condition and the construction technology in China both could not achieve the higher standards.

3) The chapiter appears for the first time in the architecture of Northern Shaanxi, but the chapiter in those churches are not normative, there is no completed base or the strict proportions. Only some chapiter forms were used as decoration.
4) Some of the local Catholic churches have the masonry arch structure which was a brand new constructional structure at that time. It shows as the result of the first introduction of western architectural feature. Also, the churches used the local characteristic cave dwelling structure a lot.
5) The elevations of the local Catholic churches are relatively faithfully followed the basic elevation composition of western churches. However, a new elevation formation had been conceived by the local constructors. They combined the foreign composition elements such as the statues of portrait and sculptures and the local architectural form together. Chinese horizontal inscribed board replaced western sculptures, Many engravings of Chinese animals and plants such as peony and plum blossom could be found in many places.

The research of this article provide detailed basis for further conservation research on the existing Catholic Church buildings in northern Shaanxi. During that long period of turbulent history, many of the churches were dismantled. Only a few of them were fortunately conserved due to being used for other purposes or occupied by the government. The existing Catholic Church buildings as important architectural heritages of the Chinese semi-colonial society were lack of attention or protection. Now, when the Chinese government is gradually paying attention to conservation of these architectural heritages, they are short of the related researching materials. This article would be the serious and relatively integrated historical and architectural basis for the restoration on the existing Catholic Church buildings in northern Shaanxi.

Acknowledgments. The authors are grateful for the support and funding from the General Course Construction Project of Literature and Aesthetic Education in Northwestern Polytechnical University (No. 21GH03110202).

References

1. Xu, Z.: Introduction to the History of the Chinese Catholic Missionary. Shanghai Book Store, Shanghai (1990)
2. Zhuo, X., Xu, Z.: The Research on Christianity, vol. II. Beijing Social Sciences Academic Press, Beijing (2000)
3. Xiao, R.: The Investigation on the Spreading of Catholicism in China. Shanghai Book Store, Shanghai (1989)
4. Anon: Continuation of Shaanxi Annals draft. Volume 198 (custom 4). Republic China
5. Charbonnier, J.(Fr.), Geng, S., Zheng, D.: Translated: Histoire des Chréttien de Chine. China Social Sciences Press, Beijing, (1998)

6. Wang, J.: Ethnic—Religion—Christianity. In: Physiographic and Humanities of Northwest China. 24th year of Public China, The 10th. Commercial Press, Beijing (1935)
7. Chen, Y.: 1st Chapter—Shaanxi Geography, Section 4—Religions of Shaanxi; 2nd Chapter—Shaanxi Politics, Section 9—The Secret Country of Northern Shaanxi (Region of Jingbian and Dingbian). In: Shaanxi and Gansu Investigation Report. Series of Northern Magazine, 25th year of Public China (1936)
8. Tang, K., Ma, Z.: The spread of Catholicism in the late Qing Dynasty in Jingbian and Dingbian of Shaanxi Province. Academic Journal of Northwestern Normal University (2004)
9. Zhang, Y.: Commentary on the religions conflict events in Shaanxi from boxer rebellion to 1911. In: Qiao, M. (ed.) Research on the Religions Conflict Events in China between 1840 and 1949. Sichuan Provincial Academy of Social Sciences Publishing House, Sichuan (1987)
10. Anon: Survey of Christian churches and Muslin Mosques in Northern Shaanxi. In: Shaanxi Provincial Archives. Files No. 9 of Republic China, the 4th catalog
11. Ywarood, D.: Architecture of Europe—the Ancient Classical and Byzantine World, 3000BC- AD1453). B.T.Batsford Ltd, London (1992)
12. Chen, K.: Architecture Culturology. Tongji University Press, Shanghai (1996)
13. Liang, S.: A Pictorial History of Chinese Architecture. Sanlia Press, Beijing (2011)
14. Cruickshank, D., Fletcher, S.B., Saint, A., Frampton, K., Jones, B.P.: Sir Banister Fletcher's a History of Architecture. Architectural Press, Oxford
15. Dingbian, J., Jia, Y., Fu, H.: Committee of chorography of Shannxi. In: Suide County and Yan'an Annals. Xi'an press, Xi'an (2000)

Politics of Heritage in the Conservation of Demak Mosque, Indonesia

Yanita Mila Ardiani(✉), Kemas Ridwan, and Yulia Lukito

Department of Architecture, University of Indonesia, Depok, Indonesia
yanita.mila@ui.ac.id

Abstract. Masjid Demak (Demak Mosque), one of the oldest mosques in Java, built-in 1577 AD, had undergone several renovations and restorations as well as political scrutiny in different periods of history. The first renovations happened in the pre-colonial era when the Mosque was adapted to fit with the needs of people's praying space. During the colonial period, especially after the Java war, there were political conflicts concerning the conservation of Demak Mosque between the Dutch and the native aristocrat (Sultan Pakubuwono VII and the Regent of Sumenep). In the Post-colonial era, especially in the New Order period, the Regime used the articulation of traditional Architectures, which were also heritage buildings, as a reference to form a monumental legacy as part of its political identity. This is not the act of conservation; instead of using the simulations or replicating heritage buildings as a strategy to control people's identity. The conservation that belongs to native society becomes an important medium to form the political identity of a particular regime or power. This paper focuses on the issue of politics of the heritage that happens in the Demak Mosque from different periods of time, such as pre-colonial and colonial, until post-colonial time. The historical narratives of the Demak Mosque rely on various sources such as archives, literature, and other media to illustrate the political relationship between the Architectural heritage and power that happened in Demak Mosque.

Keywords: Politics · Heritage · Conservation · Demak · Mosque · Indonesia

1 Introduction

The history of conservation in Indonesia started in the nineteenth century during the colonial era in the Dutch is Indies (the former name of Indonesia). It was during the British interregnum period (1811–1813) that the finding of the Borobudur temple (the biggest Buddhist temple in Indonesia) happened. The British carried out documentation and research on the collection of stories and relics in Borobudur, as well as soil excavation or landfilling. However, this did not last long because, in 1812, the Dutch were again able to seize the Dutch are Indies from the British. According to Bloembergen (2020), the restoration of the Borobudur temple during the British interregnum was also used as a means of western propaganda to attract the hearts of the natives.

After the British, the Dutch also saw the work of the Borobudur restoration as a means of propaganda to support its dominations in order to win the native people's

P. Mendonça and N. D. Cortiços (Eds.): ICAMC 2021, LNCE 226, pp. 464–471, 2022.
https://doi.org/10.1007/978-3-030-94514-5_46

hearts. However, after the Java War in 1825–1930, when the Dutch defeated Pangeran Diponegoro and made people angry, the Dutch shifted its conservation focus from Borobudur temple to Demak Mosque.

This strategy was taken because Demak Mosque was the symbol of natives who were Muslim in the majority, and the Dutch would like to reduce the resistance of native Muslims against the Dutch. But, this conservation plan failed because The King of Java (Sultan Pakubuwono VII) and the Regent of Sumenep knew that the intention behind this conservation was part of Dutch propaganda to win natives' hearts.

1.1 History of Demak Mosque

Demak Mosque is listed as the oldest mosque in Java. Demak Mosque is listed as the oldest mosque in Java. The Roof form describe the form of Indonesia Vernacular Architecture [1]. This mosque was built by Raden Patah and Sunan Kalijaga in the fifteenth century. Raden Patah, who had just conquered the Majapahit Kingdom, built a mosque as a symbol of modern religion at that time. Raden Patah was the first Demak sultan that awarded recognition from the people; actually, Raden Patah was not a native from Demak, but he was the son of King Brawijaya's (Prabu Kertajaya of East Java) with the concubine from Campa, China. Raden Patah was brought to Palembang and awarded nobility from Palembang. After some years, Raden Patah went back to Demak and studied Islam. Raden Patah received recognition from the Demak Bintoro people to become the first Sultan of Demak.

Raden Patah and Sunan Kalijaga built the origin of Demak Mosque in 1479 AD (1401 Saka). After its completion, Demak Mosque gained recognition as the Great Mosque of Demak (The Sultanate Mosque). This different function made Demak had Renovation in 1498 AD and finished in 1506 AD. The Renovation of Demak Mosque is written in the inscription in the entrance of the Mosque: "Hadegipun masjid yasanipun para wali punika dinten Kamis Kliwon malem Jemuwah Legi tanggal 1 Dulkangidah tahun 1428 Saka". It is written that the Renovation of the Great Mosque of demak was completed on Thursday, 1 Dzulqo'dah, in 1506 AD. The completion of the Great Mosque of Demak was celebrated by the Javanese people and Sunan.The architecture of Demak Mosque consists of:

1. Square plan for the main praying room with four pillars (saka guru) that supported the square roof. The architecture of Demak Mosque reflected Javanese vernacular architecture with four main column support the main roof from 4 Sunans in Java
2. Mihrab (The Place to pray for Prayer leader: imam)
3. A terrace that has the columns from the Majapahit building that they won.
4. Pawestren (prayer place for women congregation) located south of the main hall of the mosque.

At the time of determining the direction of this Qibla, there was a debate between Raden Patah and Sunan Kalijaga; finally, Sunan Kalijaga decided to face this Qibla towards the west. This is because, at that time, Islam was still inferior to the strong Javanese influence. There was also the influence of the three roofs of the Hindu temple, which was installed on the 3 layers roof of the Demak Mosque, which meant Brahma, Wisnu, Shiva, were then replaced by 3 Islamic principles, namely: Islam, Iman, Ihsan.

Demak Mosque had many renovations and adding after that in different years. This is held to adapted the Islamic people's activities in the building (Fig. 1). At the right picture describe the origin form of demak Mosque [7]. The Renovation and Restoration happened from Pre Colonialism, Colonialism time until Post Colonialism in the New Regime of Presiden Soeharto. Each Period time has different political motivations during the Demak Mosque restoration. The politics use heritage building as a media to form a political identity as the dominations from the political Regime. Demak Mosque as the vernacular heritage and become media to form political identity during Pre Colonialism, Colonialism time until Post Colonialism in New regime.

Fig. 1. Masjid Demak (Demak Mosque) Source: Picture left Architectural model of Demak Mosque picture retrieved from https://www.delpher.nl. Picture right from: The official picture in Demak Mosque organization

2 Method

The method used in this paper is qualitative. The method used is the historical narrative by conducting a survey of the historic Demak Mosque. Surveys have been carried out. Apart from data from field surveys and interviews, historical data and related theories were also carried out through books and sources about the Demak Mosque. After that, all changes were made to history (historiography). Historiography prudent in historical exploration especially focused on the Politic motivation in Restoration and Conservation of Demak Mosque.

3 Content

As our identity, architecture has a history that must be preserved and requires heritage building preservation work. This has been thought of in ancient times since wars and natural disasters destroyed the Greek Temples, it was repaired and rebuilt by the masons who believed at that time. During the Renaissance, this was expressed by several architects. Alois Reigl, with his aesthetic approach, redraws all historical buildings to preserve them. Violet Le Duc, with the request of Napoleon Bonaparte in his mission to revive Greek and Roman. He restored old Roman and Greek churches and buildings with several approaches adding elements to the aesthetics and strengthening of the structure. This earned Violet Le Duc a World Heritage award from UNESCO for the Carcason castle restoration project. John Ruskin opposed the method of renewing Violet le Duc by keeping the old building without changing it. And Augustus Pugin tried to revive the triumph of Gothic by designing buildings in the UK in Neo-Gothic style, for example, the British Parliament Building.

These restoration practices for old buildings or historic buildings were brought in by colonial countries in their colonies, including Indonesia. By 20 century, many conventions were held to restore historic building essential until the terms Conservation born. Conservation means all processes of looking after a place to retain its cultural significance. It includes maintenance and may, according to circumstance, include preservation, restoration, reconstruction, and adaptation and will be commonly a combination of more than one of these. (The Burra Charter, 1981, p. 1).This conservation activity became a common practice in repairing damaged cities and monuments after the Second World War. Although Conservation Activities are internationally recognized, Zografos (2019) stated that Conservation Architecture Practices and Activities are Western phenomena because they reflect Western values [5]. UNESCO conservation programs to provide awareness about the importance of Conservation to countries in the Third World, one of which is Indonesia. According to S. Della Torre, at this present, the conservation has competences and professionalism differently recognized in the different countries,both from the qualification and the legal point of view [3].

3.1 Timeline Sequences of the Demak Mosque Development and Conservation

The timeline of the Demak Mosque Development can seen in (Fig. 2), and describe as, at the first time in 1420 Saka or 1498 AD, the total restoration of the Great Mosque of Demak was carried out. This history is written in the Cirebon Tradition manuscript. This restoration involved 300 workers, technicians, and undagi plus 1700 volunteers. Its construction was completed in 1506 AD. This Great Mosque was established on Thursday Kliwon Friday evening, 1st Dzulqo'dah - in 1428 Saka or 1506 AD. Then, In 1711, Java king: Sultan Pakubuwana 1 mandated KT.Wiracana to fix Demak Mosque roof and to reduce the heavy weight that come from the roof of the Mosque to the ground through the Coloumn (Saka Guru) and fixed it's broke. In 1767, There built an additional space: Pawestren ($12 \times 3{,}5$M) or a place for prayer specifically for women performed by KRM Tumenggung Aryo Purbaningra.It finished in 1804 which also there was an additional regol or gate and front fence, which was approximately 20 m from the mosque's porch [6]. In 1835, The Restoration by Dutch had been canceled [2]. The Demak Mosque was successfully maintained. Masjid Demak became the symbol of native people who were Muslim in a majority at that time. The Dutch would like to eliminate the resistance of native Muslim people against the Dutch by restoring it. It is the political Colonialism from Dutch to win native people hearts that Muslim in Majority.

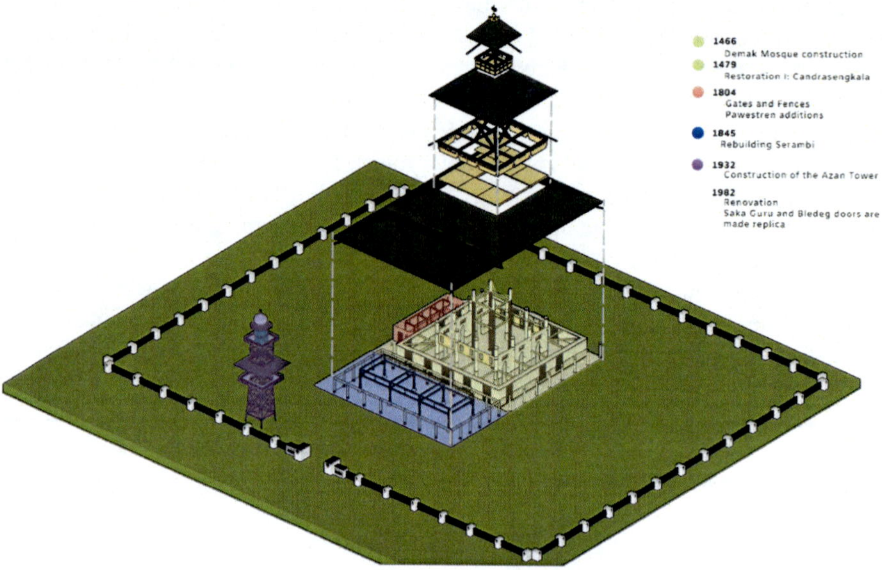

Fig. 2. Ilustration Timeline Development and Conservation of Demak Mossque. Source: Digitalization by Yanita Mila Ardiani (Author)

But, the restoration was canceled because of the defendant from the kings of Java and the Regent of Sumenep that already knew the intentions behind this conservation was part of Dutch propaganda (Fig. 3). This history explains that politics use vernacular heritage as a medium to form a political identity as the dominations from the political regime. Demak Mosque as the heritage building and become media to create political identity during the colonial regime (Bloembergen, M., & Eickhoff, M, 2020).

Fig. 3. Ilustration of timeline in Masjid Demak Restoration Project by Dutch. Source: Digitalization by Yanita Mila Ardiani (Author)

The actors who played in the restoration effort of the Demak Mosque at that time were: The Dutch, who the community would assist in becoming masons. However, the Dutch's intention to propose a regional proposal was recognized by the Sultan of Java and the Regent of Sumenep and then canceled. The other actor that performs The mosque management did not agree with the Dutch plan (Fig. 4).

Fig. 4. Ilustration of the Actor in Masjid Demak Restoration Project Source: Digitalization by Yanita Mila Ardiani (Author)

In 1845, there was the rebuilding of Surambi with columns obtained from the defeated Majapahit crafts. To strengthen Islam's power without forgetting the Hindus from the Majapahit kingdom, many pillars were bought from the Majapahit kingdom to present Islam can combine other cultures inside the mosque and describe as the excellent political between the two Islam and Hindu.In 1855, there was an additional tratag or door loop, a building that connects regol to the mosque's porch by KRM Tumenggung Panji Purbaningrat.

In 1924, there were additional wells and water towers and the renovation of the ablution place located next to the mosque by KR Tumenggung Hariyo Sosro Hadiwijoyo. And then, in 1926, There was a project to add the Azan towers and finished in In 1932. The leader of the Mosque: K.H Abdoerachman as the manager project. It is The tower to call Moslem's prayer time. The tower was made of steel construction with a height of approximately 22 M. According to some sources, and the structure technique was built by the dutch. This year, the dutch took some of the projects inside the Demak mosque. After the Azan tower. Dutch also built the fence, especially in the entrance mosque facing the city square (Alun-alun).

In 1982, the original pillars of the four pillars were renovated. The original door was made into a replica and replaced with a more robust column to support the mosque. The replica of the fourth donation column was put in a museum so that its authenticity was maintained. And in 1987, President Soeharto made the Demak Mosque the

Pancasila Mosque, where the Demak Mosque is a symbol of Indonesian Nationalism. By making Demak Mosque a Pancasila Mosque, Soeharto made the foundation: Yayasan Amal Bakti Muslim Pancasila (YAMP), built as much Mosque in Indonesia (999 unit Mosque) with the same style with Demak Mosque, and called that mosque "Demakan."

3.2 Politics of Heritage in Demak Mosque in Post-colonial Era

Pancasila Mosque for Masjid Demak become the Postcolonial politic that uses the vernacular heritage of Demak mosque as a media to form political identity power. At this time in the Postcolonial era in the new order of Soeharto that try to build an Indonesian nationalism. When Demak Mosque becomes the standard to be copied by many Mosque in Indonesia by YAMP, that is one method in the New order regime to preserve the National heritage building and made it the symbol of Nationalism the Soeharto (New Order regime). This thought of Nationalism in the new order regime is the political identity through National heritage conservation. Even the conservation has a different term with the origin conservation definitions one.

From all the history, It can be said that all that happened in Demak Mosque, from the way it was built, renovated, and conserved, has a political background. Demak Mosque has located in the coastal of North java that Majapahit kingdom conquers in old times until Walisongo spread the new religion Islam in Demak, and Raden patah, after he finished learning Islam Ampel and brought in Demak. The politics of Raden Patah at that time is to combine Islam and Hindu culture. Raden Patah believed that to spread Islam in Java must take a way of culture acculturation and respect. The form of Hindu temple in the Islam religious building to win native's heart at mostly Hindu people in that area and that era. At that time, the political thought is how to win native People's hearts that still Hindu in the majority. The inclusive approach becomes the solution to combine people in the renovations project. The inclusion of development can be seen from the efforts of Raden Patah to break the direction of the Qibla, where at that time Islam was part of Java, not Java part of Islam. This condition because Java was very superior at that time, in its development, it embraced the people in taking the people to embrace Islam finally.

In the gradual renovation (The construction and the addition of function) by embracing the residents to teach the true Islamic principles. The renovations were made gradually from the addition of Pawestren to the Azan tower. From this historical story, it can be concluded that inclusiveness occurred during the mosque construction, where at that time, Islam was a new religion in Java, which was predominantly Hindu. In contrast, it still embraced citizen participation by incorporating many Hindu elements on the three roofs in its structure. In the Dutch era, Demak Mosque, try to be restored in a top-down manner (Exclusivity) and with the Netherlands' will, without involving the Javanese community at that time, was opposed. This happened because The Demak Mosque was usually built inclusively. The People in Central Java did not even agree with the top-down of foreigners who want to interfere in religious matters. However, when President Soeharto, with his exclusivity, finally made the Demak Mosque the standard for mosques throughout Indonesia, the Demak Mosque stood firm at the Javanese's request. Because the Demak Mosque has now changed from Islam is part of Java, then became: Java as part of Islam.

4 Conclusion

The built heritage sector involves much more than interventions to conserve, preserve, rehabilitate or restore a heritage structure [4]. Restoration and Conservation project of Indonesia Heritage was used as an arena for propaganda political identity to win over the natives. One of the things that have been done is the restoration of the Heritage Building: Demak Mosque. The Demak Mosque becomes a medium to form political identity during Pre Colonialism, Colonialism time until Post Colonialism in New regime. In Pre Colonialism, the combinations of Hindu and Islam religion held to win native's heart, that Hindu in Majority. In the Colonial era, the restoration was held by the dutch to win native that Islam in the majority at that time. And in the post-colonial era, in the New Order, President Soeharto saw this as a symbol of Nationalism. Then make this Demak Mosque into the Pancasila Mosque and make it the standard for making mosques in all regions in Indonesia. In this case, Politics and influencing power with conservation and restoration media, one of the witnesses is Demak Mosque as the oldest mosque and a symbol of New Order Nationalism.

References

1. Miksic, J., Tjahjono, G.: Architecture Indonesian Heritage. Architecture. Archipelago press, NY (2003)
2. Bloembergen, M., Eickhoff, M.: The politics of heritage in Indonesia. In: The Politics of Heritage in Indonesia: A Cultural History, 1st edition. Cambridge University Press, Cambridge (2020)
3. Torre, S.D., Moioli, R.: Profession(s) and Professional(s) in the Conservation Process, Professionalism in the Built Heritage Sector, 1st edition. Taylor and Francis Group, London (2019)
4. Van balen, K., Vandesande, A.: Professionalism in the Built Heritage Sector. Taylor and Francis Group. London (2019)
5. Zografos, S.: Architecture and Fire, A Psychoanalytic Approach to Conservation, pp. 59–87. UCL Press, London (2019)
6. Roesmanto, T.: Pelestarian Kenusantaraan Alun Alun Jadi Jawa. Unika Soegijapranata, Semarang (unpublished), Temu ilmiah IPLBI (2018)
7. Architectural model of Demak Mosque picture retrieved from De Maasbode (1939). https://www.delpher.nl/nl/kranten/view?query=Demak+moskee&coll=ddd&identifier=MMKB04:000193770:mpeg21:a0057&resultsidentifier=MMKB04:000193770:mpeg21:a0057&rowid=8. Accessed 2 Oct 2021

The Early Morphology of Banten Cosmopolitan Port City in the 16th Century

R. A. Kusumawardhani[1(✉)], Kemas R. Kurniawan[1], and Susanto Zuhdi[2]

[1] Department of Architecture, Faculty of Engineering, Universitas Indonesia, Kota Depok, Indonesia

[2] Department of History, Faculty of Humanities, Universitas Indonesia, Kota Depok, Indonesia

Abstract. This writing is a result of initial research to study the morphology of Banten Port City located in the west part of Java, Indonesia, during the early stage of the development in the 16th century. The aim is to build the basic knowledge for further research on seeing the historic cosmopolis transformation and urban culture in Indonesian antique cities such as Banten. This research is an architectural and urban history study that relied on 16th-century historical data included local and colonial histories available in various maps, drawings, records, and journals. In conclusion, The Islamic Sultanate of Banten developed this port city and its urban culture as a continuation from previous customs. Built as a fortress city, the division of living area between inside or outside the fortress wall is determined by the citizenship status. Characterized as a city-state that applied urbanism focal concept, Banten followed a centralized pattern where the center became a sacred place. While the dynamic cosmopolitan atmosphere can be felt outside the fortress where the foreigners from different ethnicity and area lived.

Keywords: Banten Port City · Morphology · Urban history

1 Introduction

In the age of 15th–17th century commerce, Banten Port City arose and became the largest cosmopolitan port city in Southeast Asia, that located in a strategic and busy trading routes between the Java Sea and the Sunda Strait. Under the rule of The Sultanate of Banten, the city was able to pose itself as a Cosmopolitan of the East [15] and the main destination for traders from all over the world.

In his long research on cities in South East Asia antique cities, Wheatley (1985) emphasized that in South East Asia urban culture, cities always coincided with a state institution (City-State). City morphology was particularly arranged to show the centralization of power, prosperity, and prestige. He also noticed a highly developed city had a compact pattern to accommodate the citizen population. It is reflected in the spatial hierarchy which followed the hierarchy of power, where the highest power will taking place at the center and encircled with his subordinate according to their rank. He

P. Mendonça and N. D. Cortiços (Eds.): ICAMC 2021, LNCE 226, pp. 472–478, 2022.
https://doi.org/10.1007/978-3-030-94514-5_47

also mentioned about a possibility of a continuous process of previous customs in the city development [20].

In Indonesia, Kuswartojo (2019) said city typology can be differentiated based on the time when the city being built. Before the European influence, the city in Indonesia is having an urbanism focal concept that centralized to a palace, and has no clear boundaries. A city is always attached to a kingdom as its capital, located inland and become the religious center. While after the rise of many Islamic sultanates, City-State not always built inland but many were built in a coastal area, and also functioned as trading center. The expansive network of trading influenced the characteristic of the city to be more heterogenic and multi-ethnic [10].

Some Banten key researchers (Kuswartojo 2019; Santoso 2010; Reid 1988) [10, 16, 19] considered the morphology of Banten Port City as a continuation of urban patterns from the previous custom of Javanese cities. They often make a comparison with Majapahit Kingdom's capital. On the other hand, Michrob (1987) compared the morphology of Sultanate Banten as an Islamic state, with Islamic cities in the Middle East [11]. Moreover, Guillot (2008) believed in the area there was a pre-Islamic urban culture based on archaeological evidence in Banten Girang, which was rarely used as a reference on Banten Port City morphology [7]. Therefore based on the above researches, this writing will explore the morphology of Banten Port City to see its relationship with previous urban cultures, especially Banten Girang, that shaped the early stage of this city development in the 16th century.

This research is an architectural and urban history study that relied on 16th-century historical data included local and colonial histories available in various maps, drawings, records, and journals. The aim is to build the basic knowledge for further research on seeing the historic cosmopolis transformation and urban culture in Indonesian antique cities such as Banten. The morphology of Banten Port City became a precedent for other cities (that were built after) such as Aceh and Yogyakarta (Reid 1998).

2 City on the Archipelago's Trade Routes

Banten became widely known from the voyages of European traders in the early 16th century. Tome Pires, Francisco de Sa, Barros, and Couto were some of the voyagers that brought notes about this port city in the western part of Java. At that time, Banten was still part of the secondary route where traders followed the currents of the Indonesian Ocean. From Aceh on the tip of Sumatra Island, the ship sailed along the Sumatera island's coast stopping at cities such as Barus, Minangkabau, Bengkulu, before turning in the Sunda Strait and entered Banten coastline. The main shipping route at that time was from the Arabian Peninsula to China via the Malay Peninsula with Malacca City as the main port of call.

Trade competition and efforts to obtain spices, especially at low prices, giving Banten the benefit to become an alternative destination at that time. The port of Bantam, which is under the Sunda Pajajaran Kingdom, has already frequently conducted trade to Malacca. Using cargo ships with a capacity of approximately one hundred and fifty tons, sailors from the Kingdom of Sunda Pajajaran traded to Malacca. The main commodities carried were slaves, rice, and pepper. Pepper from the Sunda

Pajajaran Kingdom is said to be of better quantity and quality than the pepper from Kochi, a region in Kerala, India. Banten's position increasingly important after the Sultanate of Banten became the rulers and developed the coastal area in the bay of Banten into a city [18].

3 Babad Banten: Gawe *Kuta Baluwarti Bata Kalawan Kawis*

Local history about the development of Banten Port City is written in Babad Banten or Sajarah Banten, a manuscript about The Sultanate Banten's history since the very beginning [6]. Sultanate of Banten was a combined force from the Sultanate of Demak and Cirebon who controlled the territory of the former Sunda Pajajaran Kingdom since 1527, initially, they continued to use the Banten Girang area as the center of their government. Banten Girang is a city belonging to the Sunda Pajajaran Kingdom which is located on the Cibanten river, approximately ten kilometers from the bay of Banten. When Banten had established itself as an independent sultanate following the collapse of The Sultanate of Demak, the construction of a new city on the coast began.

In Sajarah Banten, the construction of the Banten Port City began with the reign of the first Sultan [6, 14]. Sunan Gunung Jati, who appointed his son Maulana Hassanudin as the first sultan, ordered to build a city on the coast of Banten Bay, which is located near the estuary of the Cibanten river. Naturally, the city is flanked by two large rivers with navigable tributaries. It is also right in the middle of Banten bay, with a calm water and ideal sea level which giving Banten coastal area the ideal to navigate [2, 13].

The construction was started by placing the watu gilang in the center of the city square, then the palace, mosque, and market built around the square. The complex of palace, mosque and city square that was firstly built became the center of Banten port city. The city development took place during the reign of Maulana Hassanudin (1540–1570 AD) which was continued by his successor Maulana Yusuf (1570–1580 AD). During the reign of Maulana Yusuf, further, development was carried out into a city with a strong fortress. In Babad Banten, it's written as *Gawe Kuta Baluwarti Kalawan Kawis*, built a city with walls made of brick and coral.

The development of the port city of Banten in stages is made clear by the records of traders who stopped in Banten (Fig. 1). Diogo do Couto's travel records in 1540 describe the city of Banten as situated in the middle of the coast of a three-mile long bay. The city has a width of 800 fathoms and a length of 400 fathoms. In the middle of the city, there is a clear river which can be navigated by jung and gale, a type of small boat. He also wrote about a fort on one side of the city made of bricks with a thickness of seven palms. The city had two-story high, well-armed, wooden defensive structures. In 1596 when Cornelis de Houtman stopped by in Banten, the entire city was surrounded by a 3-story high brick and fort that was more than one fathoms thick. Some tributaries run through the city and can be navigated by small boats. The spatial pattern of the port city of Banten at that time is quite clearly depicted in De Bry's painting made in 1599 [4]. The city is surrounded by fortress and there is a clear division of spaces according to hierarchy, duty, and nationality. De Bry noted that there are clusters of foreign settlements in the city, namely for Dutch, Chinese, Gujarati, and Bengali.

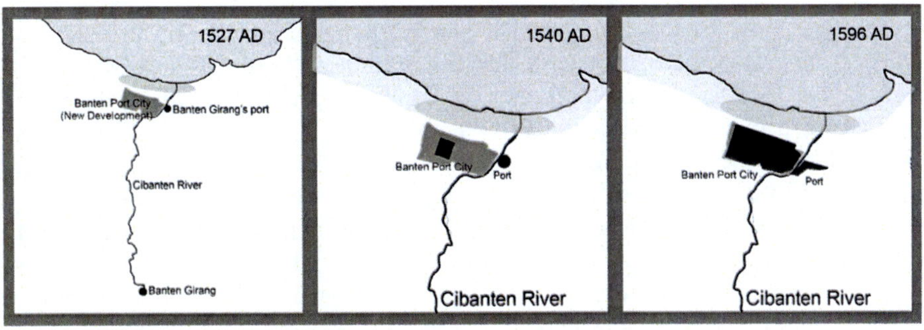

Fig. 1. The city development

4 Coastal City Morphology: Banten Port City

The development of Banten port city in the early stages was very much compared to the capital of Majapahit, which was more famous city in Java and built before Banten. The morphology of the city has a very strong focal urbanism character, that arranged the palace, paseban or alun alun, and shrine as the sacred place in the center [12, 16]. This kind of morphology clearly seen in Banten, although with a simpler pattern, that also found in Demak, an Islamic Kingdom which replaces the power of Majapahit [5]. As an Islamic state, the shrine being replaced with a mosque while taking the same cardinal direction in the west part of paseban. The mosque has similar architecture with Masjid Demak, with the royal tomb on northern side where the first king of Banten, Maulana Hassanudin was buried [1]. The palace built on the south part of paseban while the daily market located on the north side, and on the east side is occupied by royal officers. Since the city used waterways, there were no long straight axis in Banten as found in Majapahit (Fig. 2).

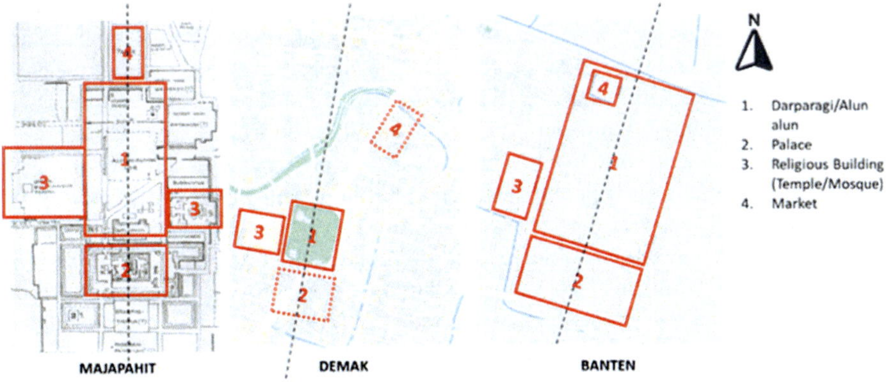

Fig. 2. Mandala comparison between the center of Majapahit, Demak and Banten

The concentric pattern and spatial hierarchy strengthen the theory that this city followed the previous urban customs. Inside the fortress, there was at least four different layers. The first one is the center, as the most sacred place where the palace complex, darparagi, and mosques was built. The second layer was residence of the crown prince as the successor of the kingdom, the closest relatives, and royal officers associated with the daily activities of the king such as the head of the court and the commander of the royal troops. The next layer was the settlement of the royal nobility which was built in the eastern and northern parts of the palace. The settlement pattern of the nobles is in the form of a compound, each nobleman heads a large family who lives in groups in the same neighborhood. Compound is depicted following the pattern found in the palace, which is having one main building and is surrounded by several buildings that circle it. Each compound also has an open field in front of the main building as well as the paseban in front of the king's palace. The fourth layer was located outside the fortress were mostly for foreigners and compound of royal officers associated with the daily activities related to trading and city defences.

Huge resemblances to Majapahit's morphology and the limitless evidence from Banten Girang, made some researchers ignored the existence of this city and the influenced to the development of Banten Port City. The existence of this city is written on Tome Pires journal, he discribed a city and active port in Banten under the rules of Pajajaran Kingdom before the formation of Islamic Sultanate. The archaeological research found Banten Girang as an inland city covering an area of five hectares on a hilly land in a bend of Cibanten River [8]. The city has a fortress made from rock and using the river as a trench as part of the city defence. Some artefacts that still remain are such as the sacred religious object *punden berundak* or stepping pyramid, surveillance cave, and steps (part of a dock) by the river. Although no original buildings being found, probably because the material they use, a small mosque built by The Sultanate of Banten is still standing strong. The location of the mosque we believe is taking place on one of the most sacred place of Banten Girang.

In our comparison, we found some similar morphology between Banten Port City and Banten Girang, one of the most prominent is the selection and/or management of landscapes as part of city protection. The center part of Banten Port City that surrounding by a river as a transportation access and also a barrier to other part of the city, was very much alike with what found in Banten Girang. A fortification system using rock material is another significant aspect of example taken from Banten Girang.

The construction of Banten Port City, was to fill political purpose since most of social and economic activities still taking part in the existing port at the estuary of Cibanten River, at the west part outside the fortress [17]. In here, the atmosphere of cosmopolitan city was more visible. On several etching and map made by De Bry [3, 9], he described a bustling atmosphere with interactions of people from various nationalities. The port and this part of the cities is most likely an expansion of the market and port that previously existed since the time of Banten Girang while under the Sunda Pajajaran Kingdom.

Not only as a trading center, in this part of the city there are some living space and warehouse of the foreigners that was built under the permission of the Sultan. At the end of the 16th century, there were settlements and warehouses belonging to the Dutch, Chinese, Gujarati and Bengali. De Bry's painting shows an exception by mentioning the existence of a Chinese compound in the fort on the north side of the city, most likely this compound belonged to the Chinese who worked under the king. While on the eastern side of the city, an office or compound was built for the harbor master and fleet in a strategic position to observe ships entering and leaving the city. During this period, the position of harbormaster was held by the Chinese while the sea admiral held by the Tamils, showing the Sultanate absorb the cosmopolitan aspect into their government structure.

5 Conclusion

As the results of this study, the morphology Banten Port City mostly followed the previous urban culture, starting from the urban arrangement to the utilization of its landscape. Taking shape as a fortress city, this City-State has a concentric pattern with the center as the most sacred place. Inside the city wall functioned as a political center and a special settlement for royal citizens, followed the morphology of Demak Capital, which is a simplification of Majapahit Capital City. Meanwhile, Banten Girang's influenced appears in the utilization of its landscape and the fortification construction system as a city defence system using the rivers as barriers and brick walls fortresses.

Namely as cosmopolitan city of the east, the cosmopolitan atmosphere can be felt outside the fortress where the foreigners from different ethnicity and area lived. This part of the city was an expansion from existing port since before the construction of Banten Port City. Under The Sultanate, this part of the city rapidly growing and occupied by traders from various ethnicity from the archipelagos and nationality such as Dutch, Chinese, Gujarati and Bengali. Dynamic activities between nations and accommodation provided by the kingdom, especially for commercial activities outside the fort, reinforced the characteristics of Banten Cosmopolitan Port City.

References

1. Abimanyu, S.: Babad Tanah Jawi, pp. 324–326. Penerbit Laksana, Yogyakarta (2014)
2. Baai van Bantam (1596). https://www.rijksmuseum.nl/nl/collectie/RP-P-OB-80.241
3. Bazaar van Bantam (1596). https://www.rijksmuseum.nl/en/collection/RP-P-OB-80.246
4. Bird's-Eye View of the City of Bantam de Bry (1599). https://id.wikipedia.org/wiki/Berkas:Bird%E2%80%99s-eye_View_of_the_City_of_Bantam_de_Bry.jpg
5. De Graaf, H.J., Pigeaud, T.H.G.: Kerajaan-Kerajaan Islam Pertama di Jawa: Kajian Sejarah Politik Abad ke-15 dan ke-16, pp. 79–81. PT Pustaka Utama Grafiti, Jakarta (1985)
6. Djajadiningrat, H.: Tinjauan Kritis Tentang Sajarah Banten, pp. 83–85, 127, 144, 160, 161. Penerbit Djambatan, Jakarta (1983)
7. Guillot, C.: Banten: Sejarah dan Peradaban Abad X – XVII, pp. 46, 59, 110–113. Kepustakaan Populer Gramedia Jakarta (2008)
8. Guillot, C., Nurhakim, L., Wibisono, S.: Banten Sebelum Zaman Islam: Kajian Arkeologi Di Banten Girang 932? – 1526, p. 97. Pusat Penelitian Arkeologi Departemen Pendidikan dan Kebudayaan, Jakarta (1997)
9. Krijgsraad te Bantam (1596). https://www.rijksmuseum.nl/en/collection/RP-P-OB-80.250

10. Kuswartojo, T.: Kaca Benggala: Perkembangan Habitat Manusia di, pp. 3, 17–19. Ukara Lawang Buwana, Bandung (2019)
11. Michrob, H.: A Hypothetical Reconstruction of the Islamic City of Banten, Indonesia, pp. 14, 105. University of Pennsylvania Graduate Program in Historic Preservation, Pennsylvania (1987)
12. Pasaribu, Y.A.: Penataan Ruang Dalam Rangka Pelestarian Kawasan Cagar Budaya: Kajian Kota Kuno Banten Lama. Kalpataru Majalah Arkeologi **28**(2), 17–32 (2019)
13. Plattegrond van Bantam (1596). https://www.rijksmuseum.nl/en/collection/RP-P-OB-80.245
14. Pudjiastuti, T.: Perang, Dagang, Persahabatan: Surat Surat Sultan Banten, pp. 252, 266. Yayasan Obor Indonesia, Jakarta (2007)
15. Reid, A.: Asia Tenggara Dalam Kurun Niaga 1450 – 1680 Jilid II: Jaringan Perdagangan Global Asia Tenggara, pp. 91, 101, 108. Yayasan Pustaka Obor Indonesia, Jakarta (2011)
16. Santoso, J.: Arsitektur-kota Jawa: Kosmos, Kultur & Kuasa, pp. 87–94. Centropolis – Magister Teknik Perencanaan Universitas Tarumanegara, Jakarta (2008)
17. Supriatna, E.: Kemaritiman di Kesultanan Banten Sebuah Perspektif Historis. Jurnal Kemaritiman: Indonesian J. Marit. **1**(1) (2020)
18. The Hakluyt Society: The Suma Oriental of Tome Pires and The Book of Francisco Rodrigues, pp. 166–174. McGill University Library, London (1944)
19. The Spice Map of South East Asia (1594). https://www.raremaps.com/gallery/detail/70838/the-spice-map-insulae-moluccae-celeberrimae-sunt-ob-maxim-plancius-wolfe
20. Wheatley, P.: Nagara and Commandery: Origins of the Southeast Asian Urban Traditions, pp. 6–18. Geography Department University of Chicago, Chicago (1983)

Research on Restoration of Ancient Buildings Based on BIM Technology

Jiayi Chen(✉)

Shanghai Southwest Weiyu Middle School, Shanghai 200233, China

Abstract. As an important part of the three-dimensional digital technology, BIM technology and three-dimensional laser scanning techniques can be used to make information models in combination with the actual situation of the project, and complete the design, construction, management of the project. BIM technology has the advantages of visibility, coordination, simulation, and improves the quality and level of construction projects. This paper studies the use of BIM technology and three-dimensional laser scanning techniques in ancient building protection process. By building a three-dimensional stereoscopic model for the entire ancient building, this paper develops a complete process to improve the overall quality of ancient building repair. In addition, this paper studies the application of BIM technology and three-dimensional laser scanning techniques in ancient building repair protection, and proposes corresponding solution strategies to increase the protection of ancient buildings.

Keywords: Restoration of ancient buildings · BIM · 3D scanning technology

1 Research Background

The Chinese ancient buildings have a long history and a variety of forms. However, the building materials of the ancient buildings are prone to damage such as timber, stone. These destroys have a serious influence on the safety of ancient buildings. In particular, most ancient buildings use the form of wooden architecture, which is easy to corrode. Therefore, in order to give full play to the historical value of the ancient buildings and improve the overall stability and safety of ancient building structures, it is necessary to repair the ancient buildings using advanced technology [1].

With the development of computer technology, the traditional restoration of ancient building has a greatly improvement. Among them, three-dimensional laser technology, point cloud technology and BIM technology are the most applicable. Three-dimensional laser scanning technology is one of high-precision automatic scanning techniques. It records all the information directly into the three-dimensional space to form a three-dimensional model. Point cloud is a large number of dense points after scanning. It is used to extract the internal and external data of the building, and makes the model to be drawn finer. BIM is known as Building Information Model, which integrates digital information of related buildings, and carries out corresponding design and planning management as the basis for the stored information. When processing the

P. Mendonça and N. D. Cortiços (Eds.): ICAMC 2021, LNCE 226, pp. 479–486, 2022.
https://doi.org/10.1007/978-3-030-94514-5_48

building with BIM, the model can be generated according to the basic parameter information and the three-dimensional model obtained. After obtaining the basic information of the building, BIM draws these locations into a three-dimensional building model. The establishment of the three-dimensional model is used to simulate the process of the restoration of ancient building, which prevents damage to ancient buildings during repair [2, 3].

2 Ancient Building Repair Technology

During the ancient building repair process, the collected point cloud data is processed by drone tilt photography, three-dimensional scanning techniques. In addition, a three-dimensional BIM model is established to record the geometric information and form the database, which has a great influence on the protection and repair of ancient buildings. Compared to traditional repair methods, three-dimensional scanning technology, BIM technology and tilt photography technology have many advantages in the restoration of ancient buildings, which can improve repair efficiency, making the restoration of ancient buildings more fine, enabling the protection and repair system of ancient buildings [4–7]. This section will conduct detailed research on the above techniques.

2.1 Information Gathering

UAV Tilt Photography Technology. The use of drones for data collection can quickly complete the collection of data points on the surface of ancient buildings, obtain three-dimensional coordinate point cloud data and store them in a computer. The operation process and planning of the UAV are as follows:

(1) Before designing the route, circle the main buildings in the survey area and the ground sampling distance on the topographic map.
(2) Calculate and determine the aerial photography technical parameters that meet the requirements.
(3) Plan the flight plan, comprehensively design the flight mode of the UAV to meet the requirements of internal processing.
(4) Refer to the on-site investigation and observation, improve the flight plan, and fly according to the plan [8-9]. Figure 1 shows the point cloud data collected by drones.

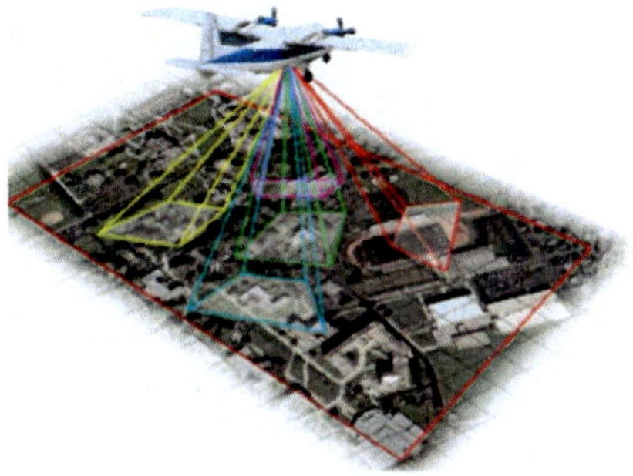

Fig. 1. Data collection of ancient buildings by drone tilt photography technology.

3D Scanning Technology. The quality of data collection directly affects the quality of the result. In order to obtain accurate and low-redundant point cloud data, it is necessary to reduce the human error in the scanning process. After selecting the location of the measuring station, it is necessary to set up the instrument and start scanning. Before scanning, make sure that the scanner battery is fully charged and the tripod is placed in the design position. Place the scanner on the tripod, connect the scanner, power supply and laptop, raise the tripod to an appropriate height according to the actual scanning operation, turn on the power, and the scanner will start to warm up and self-check. The scanning process is generally divided into four stages: rough scanning, fine scanning, fine scanning target and photo acquisition. This scanning job sets two measuring stations. In the scanning process, a handheld digital camera collects photos, and finally obtains three-dimensional coordinate point cloud data through the RISCAN PRO control software.

2.2 Data Processing

RISCAN PRO control software is used for point cloud data processing, and the error data of the spliced point cloud data is manually selected to clear. The main work includes deleting points outside the survey area and obvious noise points. This process is the filtering of point cloud data [10, 11, 12]. After the preprocessing of the point cloud data, the registration and streamlining of the point cloud data need to be performed. The registration of the point cloud data is carried out in the RISCAN PRO control software. Firstly, it is necessary to use the relative method of registration: use four pairs of targets with the same name between the two measuring stations for registration, and then use the Multi Station Adjustment function based on the ICP algorithm in the RISCAN PRO control software to perform the two measuring stations. The standard deviation after registration is 0.0044 m, which meets the requirements for surveying and mapping of ancient buildings. Three-dimensional laser scanning technology and drone tilt photography technology can acquire massive point cloud data of the measurement target in a short time. Massive point cloud data can describe the characteristics of the target in detail. In addition, it brings difficulties to data processing. After the point cloud data of different sites are registered, there will often be a large number of repeated point cloud data in the overlapping area. This leads to the low efficiency of operation, storage and transmission leads to slow data processing speed and low calculation efficiency. At the same time, these redundant data will lead to the quality and efficiency of later surface reconstruction, and the accuracy of data analysis and processing affects the accuracy of model reconstruction. Therefore, in order to achieve efficient data processing, it is necessary to simplify the data so that the simplified data not only retains the useful information on the surface of the object, but also greatly reduces the amount of data. Data reduction mainly includes duplicate data elimination and data simplification. Redundant point cloud data is integrated using the "Cleandata" function of RISCAN PRO software to eliminate redundant data in the point cloud data and greatly reduce the amount of point cloud data. Figures 2 and 3 show the effect comparison chart of the point cloud data after registration and reduction.

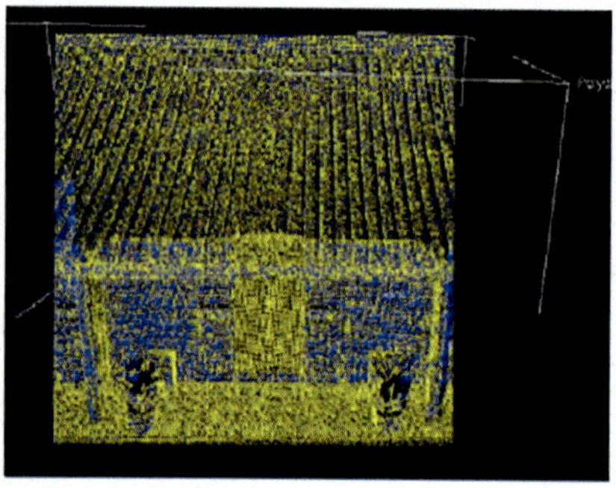

Fig. 2. Two-site cloud data before eliminating redundant data.

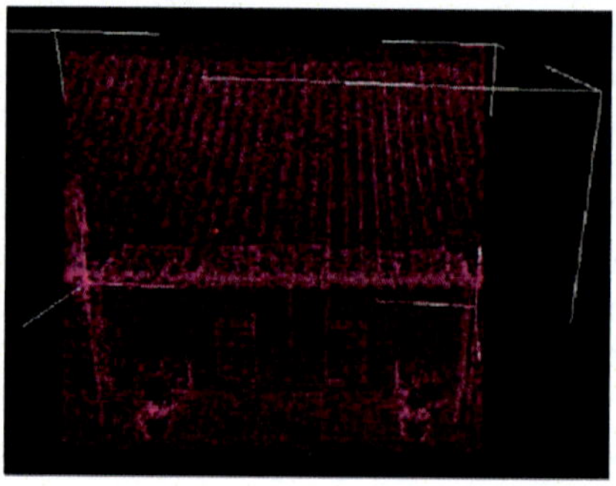

Fig. 3. Point cloud after eliminating redundant data.

2.3 3D Modeling of Ancient Buildings by BIM Technology

The three-dimensional model is constructed in the BIM software Revit. The optimized point cloud data can be imported into the Revit software to perform parametric modeling of ancient buildings. The BIM model contains the geometric information and material information of the component. In Revit software, components appear in the form of "family". The creation of parametric models is accurate, and the modification of ancient building models can be facilitated by adjusting the parameters. Therefore, the restoration model based on real data can be quickly built [13, 14, 15, 16]. The BIM model can output two-dimensional drawings, which provides a better solution for the protection and repair of ancient buildings.

BIM provides microscopic visual spatial data at all stages of the building life cycle, which can facilitate the observation of the ancient building restoration and prevent accidental operations. Spatial visualization can also facilitate the input of numerical simulation and analysis data, which plays a key role in the design process. BIM spatial data management has a great influence on data retrieval. If the ancient building restoration process is complicated, it can retrieve whether part of the data is reasonable and complete. BIM spatial data management will complete geometric description and spatial query in the modeling process of ancient building restoration. The BIM data expression in spatial data management mainly comes from GIS and CG. The former uses simple and effective algorithms to generate and process geometric attributes. The latter is still under development, but has its unique edges. It does not require storage to maintain polygon mesh connectivity and topological consistency with better Performance and lower cost. The lack of IFC format for 3D data identification and modeling in BIM database results in incomplete entity set when exporting in IFC format. Therefore, spatial data management is of great significance. It can ensure the integrity of spatial data and improve the efficiency of query and search. BIM spatial data management can improve the interoperability of data between software, thereby reducing the cost and difficulty of modeling during the restoration of ancient buildings. Although the IFC file format is now widely used in data sharing, different BIM software or different modeling methods may export IFC in an inconsistent way, which is likely to cause data missing in the file and make it difficult for data to flow freely. In addition to considering interoperability from a data format perspective, it is also important to consider interoperability between different stakeholders. This involves complex partial model extraction, and still requires a lot of work for practical applications. The interoperability between different modeling methods and different modeling software can help ancient building repair modeling, facilitate the establishment of data models of damaged parts, and help extract part of the data from the whole.

3 Case Analysis

3.1 Project Overview

In December 2017, a fire broke out in Jiulong Temple in Mianzhu City, Sichuan Province. First, the main hall caught fire, and the Sakyamuni Tower next to it was also ignited. The tower adopts the traditional Chinese mortise and tenon structure. It was built in the second year of Liaoning, Qingning (AD 1056), and is the oldest existing pure wooden pavilion-style building at home and abroad. The total height is 41.31 m, the bottom diameter is 30.27 m, and the total weight is about 7,400 tons. The tower is nine stories high, with one floor smaller than the first floor with a graceful outline.

Fig. 4. BIM model of unburnt tower.

3.2 The Application Process of BIM Technology

Information Collection and Modeling. The Shakya Tower restoration project adopted two scanning methods: unmanned aerial vehicle and station-based scanning. The two scanning methods have their own characteristics, and the combined use is more suitable for the complex information collection of ancient buildings. Point cloud data collection mainly uses drone scanners and standing scanners. Through the cooperation of the two three-dimensional laser scanning methods of the drone and the standing type, the belief data information of the ancient building is collected step by step, and the BIM model is generated using Revit software. The unburned model is shown in Fig. 4. Considering that the restoration of ancient buildings is local, a 3D laser scan was performed on the unburnt tower body this time, and the point cloud model was built using processing software [17].

Effectiveness of the Restoration. After information collection, the 3D laser scanning point cloud model is imported into Revit to build the BIM 3D information component model. The dimensions of the unburnt tower measured by the BIM component model are 32354 mm wide and 25696 mm high. According to the specific requirements of the ancient building restoration management department, after many designs, comparisons and demonstrations, the burned part is finally repaired in the BIM model to generate a restoration rendering of the Sakyamuni Tower.

By establishing the BIM model and comparing the original Sakyamuni Tower with the BIM restoration renderings, the use of 3D laser scanning can accurately collect information and establish a BIM information database, which can effectively save all aspects of ancient building information, in order to guide the next step of restoration of ancient buildings.

Drawing. The project uses the frame family to represent the drawing information, and then puts the saved plane, elevation, section and other information into the frame with accurate position of each drawing. After all the drawings are organized, they can be exported as CAD files for printing. At the same time, Lumion software can be used to generate three-dimensional model animations through layout and rendering.

3.3 Technical Service Innovation

The application of 3D laser scanning technology and BIM technology to the protection and restoration of ancient buildings can give full play to the advantages of these two technologies. Through the realization of "physical restoration", 3D laser scanning fully demonstrates its powerful advantages in the digital restoration of ancient buildings, and also provides a reference for the 3D modeling and restoration of ancient buildings. Using Revit or other software tools to build a 3D model accurately and efficiently by a professional team can make the design intent more intuitive. On the basis of the BIM model, the conflicts between errors, omissions and blanks in drawings and professions were discovered through a three-dimensional method. Establish models and collision detection software to detect various faults and gaps between buildings, structures and equipment professional pipelines, obtain collision detection reports, and propose optimization design suggestions. On the one hand, it can improve the design efficiency and quality of the design unit; on the other hand, it can avoid the loss of construction period delay and investment waste caused by various rework after the construction is completed. Furthermore, the formulation of a complete, detailed, and highly maneuverable technological process to improve the overall quality of the restoration of ancient buildings is of great significance to the protection of ancient buildings.

4 Conclusion

The combination of drone tilt photogrammetry technology, 3D scanning technology, and BIM technology is a new method of digitizing ancient building information. Unmanned tilt photogrammetry technology and three-dimensional scanning technology can quickly and accurately collect geometric information of ancient buildings, and quickly obtain point cloud data of ancient buildings, which reduces the workload of traditional measurement, and the data obtained is highly accurate and practical, etc. advantage. BIM information technology can establish a three-dimensional model of a building. Compared with other three-dimensional drawing software, the model established is closer to the actual situation, has strong visualization ability and high flexibility of use, which is beneficial to improve the quality of ancient building restoration. At the same time, the digital model established by BIM software has certain advantages in storage and resource sharing. The obtained point cloud data establishes a BIM model, which can record and manage building information, and save the most primitive data for future restoration of ancient buildings. The archival protection system of ancient buildings requires BIM model data, point cloud data from three-dimensional scanning, drone tilt photography, and damage and defect information of ancient buildings to provide effective protection and repair for ancient buildings.

References

Sun, W., Liu, Y., Yang, Q.: Research on restoration and protection of ancient buildings based on BIM technology and 3D laser scanning technology. China Bldg. Mat. Sci. Tech. **3**, 144–145 (2019)

Wang, X., Guan, X.: Building information model of ancient buildings based on BIM technology. Shanxi Arch. **45**(11), 190–191 (2019)

Jiang, M.: Research on the application of virtual reality technology in the design of virtual scenic spots—taking Cao Cao transportation road in Bozhou as an example. J. Foshan Univ. Sci. Technol. (Nat. Sci. Ed.) **36**(03), 31–34 (2018)

Chen, Y.: Wooden Structure of Ancient Buildings and Protection of Wooden Cultural Relics. China Constr. Indus. Press (2007)

Lan, Z.: Discuss the application of new materials in the restoration of wooden structures of ancient buildings. Ju She **04**, 35 (2019)

Li, K.: The application of BIM technology in the research and protection of Tai'an ancient buildings. Ju She **25**, 31 (2019)

Sha, Y., Feng, X., Liu, X.: Discussion on the digital technology in the protection of ancient buildings. Doors Windows **11**, 223–224 (2019)

Cai, Z.: Research on digital protection technology of characteristic ancient buildings based on 3D scanning. Sichuan Cement **10**, 140 (2019)

Xiaoyu, W., Chen, W.: Analysis of the application of 3D laser scanning technology in the restoration of ancient buildings. Ju She **26**, 70–147 (2018)

Mark, D.M., Cebrian, J.A.: Octtrees: a useful data structure for the processing of topographic and sub-surface data. ACSM/ASPRS Tech. Papers **1**, 104–113 (1986)

Sárközy, F.: The GIS concept and 3-dimensional modeling. Comput. Environ. Urban Syst. **18**(2), 111–121 (1994)

Wansleeben, M., Verhart, L.: Geographical information systems. Archaeol. Dial. **4**(01), 53–64 (1997)

Chen, H.: GPS data and image acquisition system based on small UAV. University of Electronic Science and Technology of China, Chengdu (2010)

Pan, J., Xiang, S.: Research on the file organization structure of GIS spatial data and attribute data. J. Chongqing Univ. Sci. Technol. Nat. Sci. Ed. **14**(1), 128–133 (2012)

Wang, R., Han, T.: Research on the standardization management of ancient building component information classification and coding based on BIM. Constr. Tech. **44**(24), 105–109 (2015)

Pan, X., Xiao, Y.: Preliminary exploration of ancient building protection ideas based on BIM multi-dimensional technology. Arch. Eng. Tech. Des. (27), 94 (2015)

Liu, J., Hu, W., Li, Z., Cui, C.: Research on ancient building restoration technology based on BIM—taking Sakyamuni tower as an example. Jiangxi Build. Mater. (01), 131–133 (2021)

Research on the Protection of Macau Water-Faring Community (Dan Jia) from the Perspective of Urban Heritage

Yile Chen([✉])

Faculty of Humanities and Arts, Macau University of Science and Technology, Avenida Wai Long, Taipa 999078, Macau, China
2009853GAT30001@student.must.edu.mo

Abstract. In areas with well-developed water systems in China, there are Water-Faring Community (Dan Jia) that can be traced back to seven thousand years ago. For the convenience of labor and living, they live on and coexist with water. Macau was once a city with one-third of the population on the water. Although it has become a world-famous Oriental casino from a small fishing village in the past, in the face of rapid urbanization, it has affected the inheritance and inheritance of the Water-Faring Community (Dan Jia) continue. Based on the perspective of urban heritage, this paper conducts research on the protection of Macau's Water-Faring Community (Dan Jia), summarizes its plight, and puts forward feasible reference suggestions.

Keywords: Water-faring community (Dan Jia) · Urban heritage protection · Macau

The term "Urban heritage" originated in 1931 and was put forward by the Italian architect and restorationist Gustavo Giovannoni in his work on the protection of ancient cities. Nowadays, it mostly refers to historical blocks, historical environment and historical buildings with traditional and regional characteristics that can reflect the historical, scientific, and artistic value of the city. Generally, it is the historical and cultural heritage that has been built in the city.

The historical background of the birth of the concept of urban heritage is very complicated. It originated from the famous historical event "Haussman Paris Urban Planning" in the middle of the 19th century, when he was the mayor of Paris, France during the second period of the Napoleonic Empire, George Eugène Haussmann (1809–1891). Improving the sanitary environment in Paris is a reason to transform Paris on a large scale. However, this has led to a large-scale destruction of the urban fabric of the old city of Paris, and even through this transformation, important historical monuments are completely separated from the surrounding environment. So far, people have mixed opinions on this urban transformation event.

Subsequently, the incident also caused strong repercussions and concerns. Planners and architects from various countries also began to think about salvaging certain important historical monuments. The famous French architectural history and theory scholar Francoise Choay mentioned in his book L'allégorie du Patrimoine that Victor Hugo and his colleague Montalembert once severely rebuked the mayor of Ottoman for

P. Mendonça and N. D. Cortiços (Eds.): ICAMC 2021, LNCE 226, pp. 487–497, 2022.
https://doi.org/10.1007/978-3-030-94514-5_49

his blindly widening roads and destroying the urban texture. They believed that while opening up new roads, it should be Adjust the line type or direction of the road appropriately according to the location of the historical monument to be rescued. At the same time, the British art critic John Ruskin believes that the existence of cities stems from the texture of streets and lanes and is a physical heritage that cities should unconditionally protect.

After World War II, many cities in France were also damaged to varying degrees. Residents everywhere were busy rebuilding their homes. They did not think too much about the protection or restoration of historic buildings in the city. The common practice is to "demolish the old and build the new". Therefore, many historical districts of ancient cities are not immune. But people soon realized that the core value of a city lies in its experience of the characteristics of different eras and the historical blocks formed by the legacy of historical buildings.

Subsequently, André Malraux, then Minister of State and Minister of Culture of the French Presidency, created a law on the protection of historical areas in 1962, the "Loi Malraux", the first time the city, The historical area is included in the scope of heritage protection. After the law was promulgated, the farce of blindly demolition and construction of the city's historic districts has ceased. At the end of the 20th century, this law has undergone many revisions and continuous improvements, further expanding the concept of protected areas and adding natural landscapes and village landscapes. Today, this law has become an important part of the French urban heritage protection system.

Since its development, this concept has supplemented the early definition of "urban heritage" from "memorials", "protected areas", "historical monuments", and "village built heritages". The works on the protection of ancient cities can also be known that the scope of urban heritage application is constantly improving and expanding. It originated from people's understanding of the value of historical buildings, to emphasize the protection of historical buildings, historical blocks, and village landscapes. Therefore, "urban heritage" is not only a single building with historical significance, but also a historic block or even a city.

In China, the concept of "urban heritage" is no longer a narrow description of "historical monuments" in the early foreign countries. It has been extended to the collective terms of "tangible urban heritage" and "intangible urban heritage". "Cultural heritage" and "intangible cultural heritage". Due to historical reasons and different regional legal systems, the classification system of "material cultural heritage" for urban heritage in the Mainland and Hong Kong, Macau and Taiwan is different. Macau is divided into "assessed real property" and "assessed movable property"; Hong Kong was divided into "legal monuments and tentative monuments", "rated historical buildings and locations", etc.; Taiwan calls them "cultural assets" and divides them into "historical sites, historical buildings, settlements", "remains", and "cultural assets". There are six categories of "landscape", "folk custom and related cultural relics", "antiques", and "natural landscapes". In terms of the definition of "intangible cultural heritage", the classification of Hong Kong, Macau and Mainland China is almost the same, while that of Taiwan is summarized as "traditional art".

1 Origin of Water-Faring Community (Dan Jia)

1.1 Records and Textual Research of Water-Faring Community (Dan Jia)'s Historical Data

According to historical records, the ethnic groups living in southern China in ancient times were the Miao and Li nationalities before the Shang Dynasty, the Yue and Baiyue nationalities after the Spring and Autumn Period, and the Nanyue nationality in Lingnan. Zhou Qufei in the Song Dynasty discussed in "Lingwai Daijian · Yanman", "Floating in the rivers and seas, you will be stubborn. There are three kinds of sacs in the Qing Dynasty: one is the fish stubble, which is good for the net; the second is the oyster stubble, which is good at getting oysters without water; Take materials for Mudan(it means that people living on the water who make a living from mountain forest production)and good at getting wood from the forest".

In the late Ming Dynasty, Panyu scholar Qu Dajun compiled the seventh volume of "Cantonese Xinyu", "Human Language", which referred to Water-Faring Community (Dan Jia) as "Water-Faring Community (Dan Jia) thief" and referred to them as one of the various nestless thieves in Guangzhong. The article records that the Water-Faring Community (Dan Jia) in Guangxi, Guangdong, joined the navy in the early Ming Dynasty, and lived a life of adversity in the end of the Ming Dynasty [1].

During the Qianlong period of the Qing Dynasty, the four volumes of "Cantonese Style" compiled by Luojiang Li Tiaoyuan included "Cantonese Songs", "Yue Songs", "Yue Songs" and "Wolf Songs", representing the literature of the four ancient Lingnan tribes. Record. Among them, Li Tiaoyuan has a paragraph after the title of "The Song of Duan": "There are three types of stew: oyster stew, wooden stew, and fish stew. Yu Xunjiang is a fish stew. The beginning is not specified: or the sun, the species of snake, so worship the snake. It is also in the palace. The song is similar to the people: the first person is a floating home and a house, and the gift is not far from Jiangshanger. Guangdong and Guangxi have them".

In October 1978, the archaeological experts of the Guangdong Provincial Museum conducted field research and unearthed a large number of ancient relics within 112 square meters. Among them, there are about 100 wooden tools (pieces); the parietal bones, frontal bones, brow bones, right mandibles and 3 molars of human remains; stone tools including axes, adzes, chisels, arrowheads, rings, balls, sharp stones, etc. Pieces; more than 10 pieces of bone utensils; more than 20 pieces of pottery kettles, pots, basins, spinning wheels, utensil seats, round cakes, etc.; a batch of objects such as bamboo weavers, fragments, straw mat fragments, and rope fragments. As well as the bones and teeth of pigs, cows, deer, sheep, elephants, hedgehogs, herrings, crocodiles and other fish and beasts. At the same time, a number of fishing and hunting tools and a 45.6 m long, 2 to 3 m wide, and more than 2 m thick stacked layered wall of shells, mussels and oyster shells were unearthed. Experts believe that the Maogang site is the largest pre-Qin period water wooden structure building site in Jinjiang River, Guangdong. The wooden structure building site reveals the lifestyle of the original villagers of the Nanyue ethnic group. Therefore, it can be seen that the original inhabitants of the Maogang site should be Yudan(it means that people on the water who live by fishing) or Oyuedan(it means that people living on the water who live by cultivating oysters), that is, today's water inhabitants.

Later, Professor Huang Weizong, the president of the Pearl River Culture Research Association of Guangdong Province, pointed out after field inspections and demonstrations that as early as three thousand years ago, the ancestor of "water dwellers" appeared in Gaoyao Jinli, Zhaoqing City, Guangdong. After thousands of years of evolution and development, the water The residents became Dan family members. The community of residents with "boats as their home and fishing as their industry" was called the "Danjia" because the fishing boats they lived in looked like eggshells, and were later renamed as "Danjia". In Maogang Village of the town, there is still the largest pre-Qin water wooden structure building site near Jianghe in Guangdong. The site has a total area of nearly 50,000 square meters and is rectangular. It is adjacent to the mountain and the water. One end of the mountain is slightly higher for residential use; the section near the water is slightly lower for fishing. The site building is a scaffold structure, and the four walls and upper cover of the scaffold are covered with ancient bark and thatch.

On the whole, the Danjia culture in the Pearl River Delta originated from Gaoyao in Zhaoqing, Guangdong, and many new cultures have been developed and multiplied. At present, there are still more than 20,000 people living on the water in Nan'an, Jindu, Jinli, Xinqiao, Lubu, Xiaoxiang, Dawan and other towns in Gaoyao. Macau is located on the west coast of the Pearl River Delta. In terms of group or community classification, the three most important ethnic groups in Macau are the Portuguese and the Macanese who account for 2.5% of the total population, and the Chinese who account for more than 96%. Among them, the Chinese It can be divided into the original residents of Danjia who account for the total population, modern Chinese immigrants (Chinese living in Australia for more than two generations), returned overseas Chinese from all over the world, and new immigrants from mainland China after 1979. New immigrants are mainly Cantonese Chinese, and some Proportion of Shanghainese and Fujianese.

1.2 Zhuhai Registered Migrant Fishermen and Water-Faring Community (Dan Jia) in Macau

Zhuhai, which is closely connected with Macau, has historically set up an area dominated by fishermen. On December 31, 1952, the Central People's Government Administration Council approved the Central Southern Military and Political Committee Document No. 108, approving the establishment of the "Fishermen County" in Guangdong Province, and the location of the Fishermen County in Tangjia. After experiencing different periods of development and being affected by wars, during the Eastern Jin Dynasty, the Zhuhai area began to have residents from outside the region. Central), the tens of thousands of troops led by him fled and dispersed. Some of the homeless people came to live on nearby deserted islands or were forced to live in the Lingnan Sea, relying on water for their livelihoods and fishing as their business, and became stubborn people. Later generations called them "Lu Tinghu".

After the founding of the People's Republic of China, thanks to the care of the party and the state, we have successively guided and arranged for the people to go ashore, and the contemporary titles of "danjia" and "danmin" have been renamed as "dwellers on the water". Called "fishermen". After the Wanshan Sea Battle ended in August 1950, the fishermen and islands fishing in and around the Pearl River Estuary were fully liberated. However, due to the implementation of border control policies at that time, mobile fishing boats from Hong Kong and Macau were restricted from entering coastal fishing ports in the mainland. In 1953, Zhuhai County was established. In Guangdong Province, adjacent to Hong Kong and Macau, some fishermen are called "Hong Kong and Macau migrant fishermen", referring to the fishermen who work and live between the Pearl River Estuary, Hong Kong and Macau. At that time, the Guangdong Provincial Party Committee and the Provincial Government decided to attract migrant fishermen from Hong Kong and Macau to fishing ports in Zhuhai County and other areas to allow them to live and live in designated ports. This made them unique special fishermen with dual household registrations in the country.

Before the 1960s, migrant fishermen from Hong Kong and Macau mainly lived in boats. Because of the lack of educational opportunities, most fishermen were not well-educated. In the 1970s, fishermen began to live on land. During this period, fishing boat equipment became more and more modern, but fishermen were not able to operate the equipment because of insufficient cultural knowledge. Therefore, the issue of fishermen's education received social attention and attention. For example, Guangdong Province and Hong Kong and Macau have opened literacy classes and night schools for fishermen's children. These schools have adapted to the characteristics of the busy fishing season and are especially popular with fishermen. The government has also begun to organize Hong Kong and Macau migrant fishermen to learn techniques such as trawling and purse seine in the mainland, open various fish skills training courses, and carry out large-scale social welfare work for Hong Kong and Macau migrant fishermen, and have successively established relevant management agencies and formulated a series of policies. Policies, laws and regulations to support the development of the economy of the migrant fishermen [2].

Up to now, Zhuhai has more than 4,200 floating fishermen registered in Hong Kong and Macau, and there are 888 registered floating fishing boats from Hong Kong and Macau. While migrant fishermen from Hong Kong and Macau have a permanent residence in Hong Kong or Macau, they also have a permanent residence in one of the six cities in Guangdong Province. They are dual-registered fishermen. In the whole country, only Guangdong Province implements the special policy of dual household registration of Hong Kong and Macau migrant fishermen. At present, in addition to Zhuhai, there are five cities in Shenzhen, Huizhou, Shanwei, Yangjiang, and Taishan that implement this policy. The Water-Faring Community (Dan Jia) of Macau has also transitioned to the Macau fishermen after the return. Fishing boats mainly dock at anchorages designated by Macau's inner port. When they need to go to sea, they will leave the port in accordance with the guidelines of the Macau Maritime Safety Administration. However, there are still many fishermen's boats that can accommodate one person at the South Sampan Wharf.

1.3 Water-Faring Community (Dan Jia) Cultural Heritage in the Guangdong-Hong Kong-Macau Greater Bay Area

The Water-Faring Community (Dan Jia) in the Guangdong-Hong Kong-Macau Greater Bay Area has a long history. The living utensils, folk arts, residential sites, and existing village landscapes created by it can also be further divided into "material cultural heritage" and "intangible cultural heritage".

On the whole, there are few material cultural heritages of buildings and relics. At present, the only one approved as a cultural relics protection unit in Guangdong Province is the "Maogang Pre-Qin Water Wooden Structure Building Site" in Jinli Town, Gaoyao District, Zhaoqing City. In addition, the "Guantian Hangbei Sand Dune Site" in Yongkou Village, Liuxiang District, Doumen District, Zhuhai City and the "Salt pan Site of the Tai O Stone Age" in the western New Territories of Lantau Island in the Outlying Islands District of the Hong Kong are also relics related to fishermen's agricultural production in ancient times. Because of the age, the scope of the division is not clear enough, and the fishermen's production in the site is only part of it, and it cannot fully represent the people on the water or the people on the water. In terms of intangible cultural heritage, it is mostly the folklore related to Water-Faring Community (Dan Jia) or the production skills of a certain object. Although the folklore appellation of Water-Faring Community (Dan Jia) is not much different in each city, the situation of riverside and seaside in each city is not the same, so the characteristics of development in various places are not completely consistent. The cities with more "urban heritage" of Water-Faring Community (Dan Jia) are Guangzhou, Foshan, Zhaoqing, Zhongshan, Zhuhai and Huizhou. The Water-Faring Community (Dan Jia) is self-contained in terms of personality, language, clothing, residence, marriage customs and religious beliefs, forming a Water-Faring Community (Dan Jia) culture with a rich local customs. Due to the different development contexts of cities in the Bay Area, they have their own characteristics in the development of Water-Faring Community (Dan Jia) culture [3].

2 Distribution of Macau Water-Faring Community (Dan Jia)

2.1 Population of Macau Water-Faring Community (Dan Jia)

In the past, Macau's marine population was dominated by fishermen. It used to be a very large "water community". Fishing boats, sampans, and house boats were moored in groups on the water outside the river from Rua do Almirante Sérgio to Patane. Every Spring Festival is more lively.

On July 20, 1920, the Macau Português government conducted a census in Macau and outlying islands based on occupation, marital status, age and nationality. According to statistics, Macau has a total population of 83,093 people. Among them, the Macau Peninsula currently has 59,689 land residents and 17,283 water residents, an increase compared to 1910; while the outlying islands have a total of 281 land residents and 5,840 water residents, a decrease compared with 1910. In the 1981 census, 11,064 residents stated that they lived on ships. In 1996, the number dropped to 2,995. By 2011, there were only 735 people on the water. This shows the rapid decline of

Macau's fisheries. Nowadays, most of the fishermen who have lived and worked on sailing boats are over 70 years old. According to the 2016 Interim Population Statistics of the Bureau of Statistics, there are only 5 people over 70 years old on the water (Fig. 1).

Fig. 1. Historical pictures of fishing boats in the Inner Harbor of Macau (Image source: Lunwan Media [1]).

The rapid decline of the marine population is not only related to the decline of Macau's fishing industry, but also related to policies in history. As early as the fourteenth year of Guangxu in the Qing Dynasty (February 12, 1888-January 30, 1889), on April 20, because of the Chinese settlement in Honghuali and Pátio das Perpétuas in Bairro de San Kio, Macau, The family boats that migrated live on the shore, and most of them are the homes of thieves and bandits. It is difficult to patrol and inspect the city. Therefore, the Macau Português government limited the Honghuali and Pátio das Perpétuas of Bairro de San Kio in Macau within 30 days. All the home boats in the two surrounding areas will be demolished. On May 30 of that year, it was stipulated that any place within and outside of the Macau Peninsula (including Taipa and Guolu Bay), no matter where the Water-Faring Community (Dan Jia) boat was moved to land resident, within 15 days, relocation as soon as possible is not allowed to take the boat cover thatch. However, with the gradual decrease of the population on the water and fishermen coming ashore to live, some traditional beliefs, traditional living patterns, and life wisdom on the water are also facing the status quo of cultural disappearance.

2.2 Residence Mode of Macau Water-Faring Community (Dan Jia)

People of Water-Faring Community (Dan Jia) shipbuilding is often used for sea operations, but there is also a unique living mode called "house boat". In the mid-19th century, George Chinnery's paintings depict a lot of Water-Faring Community fishermen's lives. In the scene, this kind of house boat is usually used for drying dried fish and shrimp and living in it. Until the 1980s, home boats still existed in Macau.

In the past, the water depth along the old wharf of Coloane Island was relatively shallow, and some of the larger fishing boats of the Water-Faring Community (Dan Jia) could not berth at the shore. They had to rely on barges to return between the fishing boats and the land. The barge looks like a dragon boat, but its length is half that of a dragon boat, and the cost of each ride is 1 pataca. Because of the high cost, unless there are important things, such as going ashore for medical treatment, childbirth, and exchange of fish, the fishermen will not go ashore, and they will always live on boats. At that time, among the Water-Faring Community (Dan Jia) members who were parked on the shore of Coloane, most of the women who were close to giving birth would be sent to the clinic near Coloane Huiji, and returned to the boat after giving birth.

2.3 Water-Faring Community (Dan Jia) and Shops Selling the Fish Catches

Nowadays, there are still many shops selling the fish catches in the inner harbor area of Macau. In the past, the owner of the shops sold the fish from the fishing boat to the shops, and the shops could draw commissions from the shops. In order to have more business, the shops would often lend money to the people to build ships and go fishing in the sea. "Let's release letter" or "for the bow", the shops not only needs to charge interest when lending money to the fisherman, but also ask him to sell the fish to the shops. At the same time, there has also been a "pilot", with the intention that the fish trades let people who are proficient in water use small boats to patrol the sea, and urge the fishermen who fish on the sea to sell the fish to the fish pens. After shops lent capital to the people, it allowed many "poor people" who had no capital and had a hard life "to start their home with a pair of trousers".

According to the historical archives of the Overseas Chinese News in the 1960s, when there was a warm current in the spring, the wholesale price of square cuttlefish in the fish pen was 1 cent per catty and 7 cents per catty. In addition to cuttlefish, there are also many pond fish, wholesale. The price is the same as the cuttlefish price, which is very low. The wholesale price of white hairtail is 4 cents and 3.5 cents per catty, the wholesale price of Daao-fish and Cao Bai-fish 6 cents and 6 cents per catty, and the wholesale price of Cao Bai-fish will be 4 cents and 3 cents per catty 2 and so on. But not every time you catch fish, they can have fish in the shops. The folklore of Water-Faring Community (Dan Jia) fishermen says: "The spring water meets the north wind, and the fisherman is dead". That is to say, if the north wind blows in the spring, the fishermen have difficulty fishing. It is rarely found with fish, so it is necessary to "see the sky and eat", it means live at the mercy of the elements.

3 Traditional Beliefs and Customs of the Water-Faring Community (Dan Jia) in Macau

Macau is connected to Gongbei of Zhuhai to the north and is surrounded by the sea on three sides. In the past, Taipa and Coloane were also islands. Because fishermen moor more along the coast, they also have many temples along the waterfront, which enshrine the god of water in the hearts of fishermen. For example, the Pioneer Temple on the Macau Peninsula enshrines the gods of water "Xiang Ye Yang" and the "Dragon King of the East China Sea"; the Kang Zhenjun Temple on the fifth street of the fifth day of October enshrines the "Hong Sheng King"; the White Pigeon Nest enshrines Zhu Daxian's Bao Gong Temple; Taipa The sea god enshrined in the San Po Temple is the third sister of the Queen of Heaven; in Jiu'ao Village in the northeast of Coloane Island, there is a Sansheng Temple dedicated to the two water gods "Hongsheng" and "Tangong" and Guandi; Heisha Village to the south also has the Dawang Temple, Also served "Hongsheng King". In addition, they also believe in a fire god named Huaguang. In the past, many people on the water in Macau worshiped Huaguang, all for praying to eliminate the fire. Limited by the city's fire-fighting facilities and living environment, it was very important for the people on the water in Macau to prevent fires. They held a glorious ceremony at the Huaguang Festival and carried out the "fire ship", which is to prevent inflammable fires. The hot coals, paper scraps, etc. were bundled and placed at the door. After being collected by Taoist priests, they were concentrated in paper boats and burned in the river [4, 5].

4 Development Dilemma

Before the 1970s, a fisherman's life, old age, illness, and death were all experienced on a fishing boat. In addition to fishing operations, the fishing boat was the fisherman's home for a lifetime at sea. Nowadays, people on the water are gradually coming ashore, and those on the shore have long been alienated from the ocean. According to the prices in Macau at that time, in the 1960s, the shipyard was asked to help order wooden ships with engines, about 30,000, which was enough to land ashore to collect rents. By the 1980s, the cost of a fishing boat increased, soaring to 800,000 patacas., Enough to go ashore to buy two houses. But they firmly believe that the "living boats and dead houses", the houses on the shore are fixed in a certain city, as if they are trapped and cannot move away, but the fishing boats can move around in the sea. For example, it is difficult for Macau to catch high-quality fish. They sailed to Taishan, Jiangmen, and went to Hong Kong for operations. Therefore, for fishermen, a fishing boat is far more meaningful than a house on the shore. But now, the fishing industry in Macau is declining. One is because the seas around Macau are polluted, shipping and fishing are too frequent; the other is because of unstable international oil prices and the loss of employees in the seafood processing industry in Macau. The cost of going out to sea once is far greater than the cost of wholesale from a coastal city near Guangdong and then transferring cross-border trucks into Macau. Fishermen have also encountered bottlenecks in their local livelihood operations.

In terms of management regulations, it has been regulated much more than before, but it has also led to a decrease in the number of fishermen. In the old days, anyone who lived on a boat, went to sea to fish, and used boats as their homes could call themselves "fishermen". Today, those who go to sea must have a "fisherman's certificate" (that is, a person who goes out to sea) to be a fisherman. The fisherman's card is issued by the municipal public security bureau under the jurisdiction of the district and is used to prove that the holder is a boater going to sea. Take the Hong Kong and Macau Migrant Fishermen's Office in Zhuhai City as an example. The applicant for the fisherman's card needs to hold a Hong Kong or Macau resident ID card, a Hong Kong and Macau vessel license, and a fishing license for fishing vessels. To purchase another person's mobile fishing vessel, the owner (shipowner) transfer procedure must be completed in Hong Kong and Macau, accompanied by the original owner (shipowner) "Ship License" (newly-built fishing vessels also need to leave the factory certificate). The Taishan City Hong Kong and Macau Migrant Fishermen Work Office also requires fishermen who are not registered in Taishan, Guangdong to submit a "Certificate of No Criminal Record". When Macau fishermen enter or leave Macau via the South or North sampan wharf of the Inner Harbor, they need to show their ID cards and fishermen. Data such as certificates can be effectively passed. After the global pandemic of the new crown pneumonia virus in 2020, the management of fishermen is also stricter. When fishing boats enter the port, electronic track records must be provided for verification. During the 14-day medical observation, the fishing vessel must be anchored in an independent quarantine area north of the anchorage. If more fishing boats enter the port at the same time, customs officers need time to check the tracks, and fishermen need to wait patiently. After entering the port, fishing boats must catch or replenish their catches at the designated Inner Harbor Piers 5B2, 28, 31A, and 33. After completion, they must anchor at the Inner Harbor anchorage and are not allowed to berth at the pier at will. Non-local fishermen on board are forbidden to land ashore, and all persons who land ashore must present to the customs a valid negative nucleic acid test certificate under the current regulations, and can only enter the South Sampan Wharf.

Judging from the beliefs and customs of the Water-Faring Community (Dan Jia) fishermen, large-scale religious activities such as the Huaguang Festival are no longer carried out in Macau today. Today, the wooden houses in Macau are basically extinct, and the residents on the water may have come ashore. Due to reclamation, there is no place in Bairro de San Kio. In the bay where the boats are anchored, the residents who believe in Huaguang are old, and the younger generation are no longer interested. At the same time, the development of the urban fire protection system makes today's fire protection no longer only asks for the blessing of the gods, but the birthday prayers bring more a kind of spiritual comfort. Therefore, traditional Huaguang birthday beliefs need to be inherited and protected.

5 Summarize

The Water-Faring Community (Dan Jia) is self-contained in terms of personality, language, clothing, residence, marriage customs and beliefs, forming a Water-Faring Community (Dan Jia) culture that is very rich in local customs. From the perspective of urban heritage, we can look at festivals such as the Huaguang Festival, such as the festivals of Macau's Water-Faring Community. It is a dynamic intangible cultural heritage. We can use current technology, surveying and mapping digitization, and architectural restoration to protect their worship temples. But we must also reflect on the protection of a traditional festival of Water-Faring Community fishermen. What really needs to be protected? Who decides what to protect and what needs to be protected?

Although there is currently no way to return to the "brilliant age" when the inner port of Macau is full of fishing boats and the waterfront of Taipa and Coloane are full of Water-Faring Community floating huts, we should not imitate some "fake antique" buildings to satisfy tourist gimmicks, which is easy to mislead. public. If conditions permit, we can try to find historical materials of Water-Faring Community fishermen's previous living space and related shipbuilding skills for textual research, follow-up in-depth research and collation, and present them to the public in the form of exhibitions or publications for scientific and educational propaganda.

As an ancient business card and an ancient ethnic group in the Guangdong-Hong Kong-Macau Greater Bay Area, the unique marine culture of the Water-Faring Community (Dan Jia) settlement and the living mode of the water community have given birth to many outstanding historical, cultural relics, characters, and handicraft skills. Although the current list of cultural relics protection units or intangible cultural heritage is very limited, from the perspective of urban heritage, the existing Water-Faring Community (Dan Jia) should trigger us to spontaneously explore future protection roads and think about future cities. During the development, how to balance the relationship between the water community and the land community, and preserve the excellent Water-Faring Community (Dan Jia) culture.

References

1. The original poster of Xinwen. Haojiang Pioneer King (3). Overseas Chinese News, 10 January 1953. (6)
2. Huang, Y., Wu, W.: Three thousand years of "dan family culture" originated from Gaoyao, Guangdong. Overseas Chinese News, 08 August 2006. (7)
3. Macau Government Gazette, no. 17, 26 April 1888
4. Chen, Y.: Research on the urban renewal planning of the water community (Danjia) from the perspective of urban heritage: Taking the Guangdong-Hong Kong-Macau Greater Bay Area as an example. City University of Macau (2020)
5. Chen, Y.: An analysis of the culture of Danjia in the Guangdong-Hong Kong-Macau Greater Bay Area from the perspective of urban heritage. Innovation based on the Sino-Portuguese platform Development Research. In: Proceedings of the 2020 International Conference on the Development of China and Portuguese-Speaking Countries, p. 14 (2020)

The Theoretical Background of Bastion Castles in Renaissance Architectural Treatises

Olha Tikhonova(✉)

Faculdade de Arquitetura, CIAUD - Research Centre for Architecture, Urbanism and Design, Universidade de Lisboa, Rua Sá Nogueira, Polo Universitário, Alto da Ajuda, 1349-055 Lisbon, Portugal

Abstract. The bastion castle phenomenon has not gotten attention in architectural theory. It lacks knowledge about bastion castle typological classification and has limited sources on its models' representation. When the author searched for theoretical background among scholars' studies to which this phenomenon could be linked or with which the author could partly explain it, she found insufficient information on this topic. Before this study, just one treatise was known and cited as a source among researchers who studied bastion castles in Europe. It is a treatise of Vincenzo Scamozzi *"L'Idea della Architettura Universale"* that describes Vincenzo's design for Zbarazh bastion castle in Ukraine. Another one, the Polish treatise with a section about one bastion castle model, did not get much attention as it was dedicated to military architecture and was mainly a summary of the other well-known European treatises translated into Polish. However, the author found the original manuscripts of four other treatises that talked about bastion castles in archives of several countries and gave their overview here. This overview can be used as a solid background for further studies on bastion castles built in Europe.

Keywords: Bastion castle · Architectural military treatises · Renaissance · Fortified palace · Defensive residence

1 Introduction

The shape of bastion castle represents a particular type of palace, which has directly connected or separated defensive enceinte with bastions on the corners. It was an adaptation of noble residence with defensive enceinte to modern, powerful gunpowder artillery that brought significant changes to fortification design in the middle of the fifteenth century. As this particular type combines two functions: civil and military, it had a very complex development. Its complexity was pushed forward by the diverse development of two main components separately. The theoretical background of those two parts, civil architecture (*architectura civilis*) and military architecture (*architectura militaries*), presented in treatises, have started to follow separated paths since the sixteenth century [1, p. 46]. That is why it was hard to find literature representing two parts of bastion castle combined (civil and military). Thus, the question of bastion castle has not been raised before. Here the author presents an overview of all treatises that she was able to find on this topic in archives of different countries. In the

P. Mendonça and N. D. Cortiços (Eds.): ICAMC 2021, LNCE 226, pp. 498–507, 2022.
https://doi.org/10.1007/978-3-030-94514-5_50

introduction, they all are shown in order from the most known to unknown to researchers. However, a detailed description is given just for the four unknown treatises in the sub-sections. Namely, Sebastiano Serlio (1475 – c. 1554) on building types "Tutte l'opere d'architettura, Libro VI"; Jacques Androuet du Cerceau (1510–1585) "Les plus excellents bastiments de France" (1st Vol. in 1576, 2nd Vol. published in France in 1579); Jacques Perret (active after 1568) "Des fortifications et artifices. Architecture et perspective" (1601); Buonaiuto Lorini, also Bonaiuto, (1545–1611) "Delle fortificationi di Buonaiuto Lorini, nobile fiorentino, libri cinque…" (1596). In sub-sections, they are put in chronological order to underline the development of the ideas about the bastion castle model. The treatises of Vincenzo Scamozzi and Naronowicz-Naroński are discussed briefly in Sect. 2 of this article as they are known among scholars and were written much later than the other four. Moreover, the drawings in the treatise of Naronowicz-Naroński on defensive architecture are almost a copy of Perret's treatise (for example, Perret, 1601, fol. K plate 8, 10 and Naronowicz-Naroński 1659, p. 5, p. 39). Besides, the treatise of Naronowicz-Naroński was dedicated to military architecture and was mainly a summary of the other well-known European treatises translated into Polish.

2 Six Different Treatises that Talk About Bastion Castle

There is one well-known and often used reference concerning bastion castle type among scholars. It is a treatise of an Italian architect and architectural theorist, **Vincenzo Scamozzi** (1548–1616), *"L'Idea dell'Architettura Universale"* (literal translation into English "The Idea of a Universal Architecture") *(1615)*. It shows the project for Polish nobleman [2, p. 99] in Zbarazh city (currently Ukrainian territory). Although the Zbarazh project was conducted as an actual project and not just a model to follow, the bastion castle was constructed using another design.

Another, less-known treatise with information on bastion castle is the work of Polish architect **Józef Naronowicz-Naroński** (c.1616–1678), who created the manual for this type in a treatise written in Polish *"Budownictwo wojenne"* (literal translation into English "Military Building") *(1659)*. The treatise contains a section called *"O Budynkach do Fortece należących"* (literal translation into English "About buildings that belong to the fortress") that has a detailed description of how to build a bastion castle with a small palace inside. Regardless of this fact, the work was cited just among the studies about bastion fortifications. It was not used to reference the comprehensive design that consists of two different functions and represents one unit.

The first treatise, which the author found besides those is an unpublished manuscript with drawings and descriptions made by ***Sebastiano Serlio*** (1475 – c. 1554) on civil types *"Tutte l'opere d'architettura, Libro VI"* (literal translation into English "All the Works of Architecture, Book VI"), completed circa 1550 [3, p. 116]. Serlio was an Italian architect who worked in France. His relocation to another country influenced his vision of architecture. He adapted Italian designs to local French traditions. Serlio presented two types of palaces with defensive features, among other types of dwellings, in the mentioned treatise: villas in the countryside featured with defensive elements on a rectangular plan with courtyard and corner towers, referred in treatise as *"casa a* *modo di fortezza*" ("the house in a fortress way"); and a separate house surrounded by enceinte or curtain wall with bastion fortifications, referred in treatise as *"casa circondata da fortezza"* ("the house surrounded by a fortress").

The concept of Serlio's work was reflected in **Jacques Androuet du Cerceau** (1510–1585) treatise *"Les plus excellents bastiments de France"* (literal translation into English "The most excellent buildings of France") *(1st vol. in 1576, 2nd vol. published in France in 1579)*, wherein the 1st Volume Jacques Androuet du Cerceau presents two castles enhanced by bastion fortifications. One of them is Château de Verneuil-sur-Oise, and another one Château de Maulnes.

Another treatise comes from **Jacques Perret** (active after 1568) *"Des fortifications et artifices. Architecture et perspective"* (literal translation into English "Fortifications and artifices. Architecture and perspective") *(1601)* engraved by Thomas de Leu (active from 1576 to 1614) was less informative as previously described books. The treatise shows very imaginary work that looks more like a picture with perspective views than technical drawing.

Italian architect **Buonaiuto Lorini**, also Bonaiuto, (1545–1611) in his treatise on fortress architecture *"Delle fortificationi di Bvonaivto Lorini, nobile fiorentino, libri cinque..."* (literal translation into English "About the fortifications of Bounaiuto Lorini, Florentine nobleman, five books…") (1596) has shown two plans of houses (palaces) made in a fortress (original name: "*case, overo palazzi fatti in fortezza*") in the third book.

2.1 Sebastiano Serlio "Tutte L'opere D'architettura, Libro VI" (C. 1550)

The theoretical work, which has the most comprehensive approach to bastion castle typology in Europe, was Book VI on civil types written by ***Sebastiano Serlio*** circa 1550 [3, p. 116]. Although Serlio himself avoided giving a defined terminology for the types in the book, we can find several different models of bastion castles between them.

Despite the importance of this work for bastion castle typology, this treatise has not gotten attention among scholars. The brief mention of the existence of Book VI in recent years was made just by Roberta Spallone and Marco Vitali in their analysis on Book VII [4]. It happened because Book VI was Serlio's unpublished work lost for many years (it was first published in Milan in 1966 [5, p. 238]). The complete full copies of the manuscript remained just in two libraries; one is stored in Avery Architectural Library of Columbia University (draft version) and the other in Munich Staatsbibliothek (completed version). Both of them were finished between 1540 and 1553 [6, p. 155]. The first person who identified the manuscript and traced the complete history of Avery's copy of book VI was librarian William Bell Dinsmoor [3]. As the information on this treatise is minimal and originals are in handwritten old Italian language, it was decided to present here some citations from the original manuscript with the author's free translation to explain the purpose of the bastion castle and its elements from Serlio's point of view.

Moreover, the importance of Serlio's treatises is underlined by their excellent performance. They were not only descriptions; they were done as illustrated handbooks for architects. It seems that Serlio was the inventor of an illustrated book, where the texts are a short description of the drawings and not the other way around [5, p. 197]. In his Book VI, Serlio did not provide just an analysis of his contemporary architecture;

he gave a multitudinous number of his drawings and drawings supplied by Baldassare Peruzzi [7, 8]. The concept of Serlio's treaty was reflected in the treatises of two leading Italian theoreticians, Palladio and Vignola, and the leading French architect Jacques Androuet du Cerceau.

The designs of defensive dwellings presented in the book, regardless of the shape of the courtyard, represent the centralized planning composition, where the courtyard is the center of the plan. This type is represented by a residential corpus surrounded by bastion fortifications. It was a mix of the idea of French chateau where we have two components (the keep and enceinte) with modern bastion fortifications of Italian type. This type is represented in Serlio's book VI by six plans in the Munich manuscript (Pls. XXVI-sheet 26, 27; XXVII – sheet 28, 29; XVIII – sheet 30, 31) and three plans (Pls. 28, 30, 31) in Colombia version. The second sheet of each design of the Munich version (sheet 27, 29, 31) looks like a close-up plan and elevation of the keep of the previous general plan when in Columbia we have only general plans and detailed elevation just for Plate 28. However, using the text of the Munich version, we can understand that just sheet 27 is a close-up ground plan of the same design when sheets 29 and 31 are just similar projects to 28 and 30 accordingly. Serlio says about sheet 29 of plate XXVII (Munich version) the following:

"In the previous plan of the house of the prince tyrant, I have not been able to show precisely the apartments in such a small form. Therefore I wanted to show one similar to that one in much greater size, but with better consideration of the particularities which are seen here forward demonstrated, which has a moat around of its sides..." (Free authors translation from Italian) [9, Fol. 28v].

The main difference between the residential corpus in the larger size of sheet 31 from a general plan of the previous sheet (plate XVIII, Munich) is small bastions in each corner of the pentagonal plan. Serlio explains the difference in the following way: first, it was not enough space on the previous plate to show the details, and second, he decided to show this plan with more robust defense using small bastions on the corners as high as the ground floor, when in previous the walls were flat [9, p. 30v].

If we would like to compare the level of defense in these projects in both manuscripts, we need to underline that **project 22** (Pls. XXVI of Munich and Plate 28 in Colombia) has a minor resistance to two others (Et 23, Et 24) because the bastions of that project are not fully developed when in others they are fully developed. However, the Munich version has other defensive advantages. Namely, in the Munich version, the residential corpus is farther inside the defensive enceinte than in Colombia, making it harder to reach the house by a cannonball. Moreover, the Munich version has a moat around the main corpus with four bridges, when in the Colombia version, the keep is situated directly on the ground with the entrance connected to the ground by stairs.

The next project, **Et 23**, has a residential corpus similar to project R17 of Serlio's Book VI, based on the rectangular outline with a central courtyard and four towers of bastion shapes at the corners. The most significant difference of this project from R17 is a curtain wall with fully developed bastions, which surrounds the main corpus. This feature represents the modern defense system in the Renaissance period and provides a good level of defense from cannons. The corpus was placed farther away from the curtain wall in the Munich version than the corpus in the Columbia version. It indicates the exact difference between versions as in the previous project. Respectfully, this difference indicates a more defensive level. However, this difference is not as huge as in project 22.

The following text, written by Serlio as a description of this project just in the Munich version, is suitable to place here in a big piece, as it explains well the purpose for such a protected structure:

"The noble prince of liberal spirit wise: who loves his peoples, who fear God, does not need any fortress because the heart of man and souls of his vassals are impregnable defenses... However, for the Tyrant, cruel, avaricious, robber of the goods of others, rapist of virgins, wives, and widows, an occupier of the goods produced by the sweat of his subjects' brows: all the strongholds of the world could not guarantee him safety. However, oh God immortal, show me the person who has power and does not dominate above his subjects. Nonetheless, I must deal with the dwellings of all ranks of men outside of the city. It is time that I am talking about the tyrant prince, which not only must be surrounded by a large wall guarded by its soldiers, but its own house must also be strong and should be precisely guarded…" (Free authors translation from Italian) [9, Fol. 27v].

On the construction of bastion castle, Serlio says the following:

"I say that after the selection of the place was made, the most appropriate in this regard, that you have to master a square that side by side is not less than 250 crosses. This is steps of the normal man, making the walls no less than 14 feet thick and the same for the earth platform with its strong contra, making its bastions on the corners. It should be emphasized that there are different opinions concerning the forms of bastions. Finally, I should pass on the rest of the measures. The exterior of the fortress will be done in a way demonstrated here in front with vast and deep moats" (Free authors translation from Italian) [9, Fol. 27v].

The third project, which represents a bastion castle in Serlio Book VI, is **Et 24**. As a previous one, it has well-developed bastions of the Italian type. However, the enceinte and the residential corpus are based on the pentagon in this project. The versions from the two manuscripts are very similar and have the same defensive level. Just a planning system in the east part of the main building looks mirrored between Munich and Columbia versions. Serlio, in the description of this project [9, Fol. 29v] says that the best shape of the stronghold after the square is triangular according to defensive purpose. However, ample space is needed to place a palace inside a triangular enceinte, so the space in the acute angles will be wasted. Hence, Serlio thinks that a pentagonal figure is the best after the square form [9, Fol. 29v]. Serlio also underlines that a broad and deep moat should surround the fortified complex with just one entrance through the drawbridge. He proposes that the walls have 15 feet of thickness, and the bastions should be 18 feet long. In addition, for more substantial defensive value, Serlio proposes making a moat around the pentagonal palatial part, at least 15 feet wide, with an entrance to the house by the bridge. The palatial part of this project is particularly interesting as it is based on a drawing plan of villa Caprarola by Peruzzi. Moreover, the project of the Chateau at Maulne (1566) designed by Du Cerceau was based on this plan [10, p. 69].

Summing up, the analysis of the fortified houses presented in both versions of Serlio Book VI shows that houses in the Munich manuscript version are more robust, have additional defensive features, and are presented in more detail. Hence, it is possible to conclude that this type of buildings with defensive features was more needed and should be with higher defense level during the reign of Henry II of France than during King François I. This statement follows from the fact that two manuscripts were performed in different reins as in the Columbia version, in Plate 40 [11, Fol. Plate 40] there is a reference to "*il gran Re Francesco*", when in the Munich version in project XXXIIII is substituted by "*il cristianissimo Re Henrigo*" [9, Fol. 38v], which was firstly underlined in [3, p. 129].

2.2 Jacques Androuet du Cerceau "Les Plus Excellents Bastiments de France" (1st Vol. In 1576)

Jacques Androuet du Cerceau's work falls on hard times in France, conflicts between Roman Catholics and Huguenots (Reformed Protestants) [12, p. 20]. This conflict caused civil and religious wars and military operations. Hence, without doubt, the architecture of noble houses required taking into consideration the critical period. This period affected the planning of splendid buildings in Jacques I du Cerceau's book. The architect used plans, which were based on the country chateau. The models consist of an enclosed bailey with a screen and principal block (*corps de logis*) flanked by pavilions. His architecture, like the designs of his predecessors, was a French interpretation of Italian models. Near 1540 the architect went to Rome, and the Italian architecture made a deep impression upon him [12, p. 21]. The formal organization of his projects is similar to Serlio's "Grand Ferrare", but the use of separate pavilions makes it more similar to the chateau. Androuet du Cerceau became known for his published book "Les Plus Excellents Bastiments de France" [13] and the traces of commissioned works. The book represents drawings of several earlier chateaux, but its importance for this study lies in the two projects of bastion castle, namely château of **Verneuil** (begun in 1568) and **Maulnes** (built between 1566 and 1573). Château de Maulnes resembles Serlio's project of Fortress for a Tyrant Prince from Book VI. However, the designs were constructed without bastion fortifications. It was probably why this treatise was left undiscovered in relation to bastion castle models by other scholars.

The first design of **Verneuil** was very challenging, expensive, and an unusual project prepared by Cerceau for Philippe de Baulainvilliers [12, p. 29]. (For a better understanding of the following text, please see the appropriate pages with design drawings from the referenced book available online in open access at the site https://www.e-rara.ch/zut/content/zoom/15271143). The ground floor design had a big courtyard and two towers on each corner [13, p. 106]. The main block has doubled angle-pavilions on the corners and was surrounded by bastion fortifications with orillons[1] (old Italian school of fortification). The entrance pavilion had a semi-oval complex shape. According to researcher Rosalys Coope [12, p. 26], the first floor of

[1] Orillon – a projecting shoulder of a bastion designed to cover the flank. Sometimes referred to as an orillion.

this pavilion was designed to have a chapel. From the analysis of the perspective view of the main front façade [13, p. 109], it is possible to notice that bastions with orillions were later replaced by rusticated ramparts, which replicate the ground floor's outer line of the main corpus. The main entrance on the other side of the building had three vestibules. Two of them had staircases, which could be approached directly by the outside doors to each staircase separately. However, the project was found to be costly.

Moreover, Philippe de Baulainvilliers must sell an entire estate to the Duc de Nemours in 1575 due to financial problems [12, p. 29]. After that, Cerceau prepared a modified design of Verneuil for Nemours. The second design was more extensive in scale [13, p. 106]; however, it was more compact and uniform. This unity was underlined by replacing doubled-angle pavilions with single four-corner pavilions that look like towers on each corner. Each of them had three stories in the new design when in the first design, the corner pavilions had two main levels and a mansard roof. According to researcher Coope, the domes that covered the pavilions had a typical shape for the architecture of the early seventeenth century [12, p. 30]. In the new design, the enlarged entrance pavilion based on an octagonal plan replaced the oval one. The elevations of the new design were less decorative than in the first one. Rustic pilasters and blind windows replaced the entrance decor, consisting of twin columns and niches. There are no visible details in the second design presented in the book *"Les Plus Excellents Bastiments"*, which indicates how the main block was resolved. However, in another drawing of Verneuil, which comes from the end of the sixteenth century [12, p. 31], and is stored in the Cooper Union Museum in New York, the heart-shaped bastions re-appeared. The construction of Verneuil stopped in 1585 [12, p. 32]. It was completed later but demolished in the eighteenth century. Just foundations remained until nowadays.

The second project of Jacques Androuet du Cerceau that belongs to bastion castle type is the **Château de Maulnes**. The main structure was built in a forest on top of the old, fortified system called "motte of Malunes" between 1566 and 1570. The brief period of peace in France from internal religious wars allowed the construction of the original design. The complex consists of three main structures: a pentagonal residential building connected by the fixed bridge to a covered porch structure, which leads to the semi-circular gatehouse. The whole complex planned to be surrounded by a wall on the rectangular outline with bastions at each corner and semi-bastions ("*Piatta Forma* Bastion") in the middle of each longer curtain (see drawings in [14, pp. 166–167]). The covered porch that leads to a fixed bridge is one floor in height and forms a gallery with five arches. It connects to a short round building similar to a theatre on the opposite side of the bridge. The pentagonal residential building is a massive solid structure with an initiation well (inverted tower) lined with stairs in its center [13]. The house's roof is the pyramid hip type and has a lantern above it on a horizontal platform. The lantern is covered with a dome circumscribed by pyramids and furrows of conduits for the chimneys. The back façade of the house is connected to the nymphaeum (the water fountain), surrounded by stairs forming a small theatre. This element connects a house and a garden, taking almost half of the territory (refer to drawings in [14, pp. 166–167]). The house's footprint is much smaller than the fortifications' footprint, which is beneficial for defensive purposes. However, according to archaeological studies, the fortifications have never been constructed [15].

2.3 Buonaiuto Lorini "Delle Fortificationi Di Bvonaivto Lorini, Nobile Fiorentino, Libri Cinque…" (1596)

In the description of his two plans of houses (palaces) made in a fortress (original: *"case, overo palazzi fatti in fortezza"*), ***Buonaiuto Lorini*** underlines that he presented them to show the possibility of how the modern fortification system with bastions can be easily applied to the palaces or villas to protect the inhabitants that live outside of the city fortress. Primarily these models can serve to defend the houses situated far away from the city and near the borders, in the dangerous places exposed to incursions of robbers. Thus, Lorini presented two versions of buildings for such situations: one bigger and one smaller [16, p. 150].

The first plan consists of four parts, which are formed by four bastions with the moat around. The building connects to the ground by a bridge, as wide as a moat, in its southern part. The bridge leads from the outside to the inner hall surrounded by 16 rooms, divided into four apartments. Such configuration can also be applied to smaller residences, where the bridge leads to the hall room with eight spaces around it [16, pp. 151–152].

2.4 Jacques Perret "Des Fortifications Et Artifices. Architecture Et Perspective" (1601)

Jacques Perret*'s* treatise is divided into three main topics, although they are not strictly ordered in the published work. Namely, he talks about three different groups of buildings in his work: ideal fortified cities, some with citadels; sacral buildings; and private facilities. Particularly interesting for this study are three examples of the singular bastion castles in the treatise.

Perret's comprehensive approach to private buildings presented by buildings for different social statuses of occupants resembles a lot of Serlio's approach to civil architecture. However, Perret's creative approach to the designs with pictorial quality of so-called "military perspectives" and combination of different topics, in time when military architects detached them into a separate field, along with the military sound of the title, put Perret's work to high valued artwork collection (drawings, engravings) rather than became an example of high-quality architectural design acceptable as a model for realization. Although the engravings from *"Des fortifications et artifices. Architecture et perspective"* resembles the approach of Jacques Androuet de Cerceau in his *"Les plus excellents bastiments de France"*, it did not get the same attention and success. Nevertheless, several bastion castles that follow his model were built in Europe. For example, the bastion castle in Pidhirtsi (Ukraine) [17].

One of the two similar bastion castles (Jacques Perret called them palaces) is represented by a palace of two-level solid rectangular form connected by its front façade to the platform with bastions on the corners. The walls of bastions and side curtains connected to the building create a closed courtyard inside with an entrance gate opposite the building entrance, higher than the platform, and can be reached by stairs. The second bastion castle is like the first one, but the palace consists of three parts: two three-level towers in the corners and the lower inner part between the towers.

Another example of a free-standing bastion castle is different from the previous ones. Unfortunately, it is presented just by the plan without a perspective view. The most significant difference is the vast garden inside fortifications and the separation of a residential building from fortifications. Jacques Perret called this plate "Fortified building surrounded by gardens".

3 Conclusions

Each of the six treaties presented here adds valuable knowledge about the bastion castle phenomena in Europe. However, the approach to the representation of this type of building is different in each of them. For example, Vincenzo Scamozzi, in his *"L'Idea dell'Architettura Universale"*, just presented one project that was made for a Polish nobleman, a castle in Zbarazh that finally was not built according to his plan. Jacques Androuet du Cerceau, in his *"Les plus excellents bastiments de France"*, presented two of his designs that belonged to bastion castle type, which were supposed to be built in France, but finally were made without bastion fortifications. In contrast, Sebastiano Serlio took another path. In his work, Serlio showed different models of bastion castle designs, explained why he made them that way, and their main characteristics. Moreover, he was the first who introduced a layout with plans and elevations and a short description of the project, which became very popular among other later treatises. Buonaiuto Lorini also discussed the possible necessity in the defended residences and provided two examples with smaller and bigger plans that can be modified for use when the residence is built far from the city in a dangerous location.

In the line of previously stated treatises, the value of the work of Perret on the first look seems to be just in his high-quality perspective views of several bastion castles. However, after a deep analysis of his work and other treatises, especially the first work in the Polish language about perspectives and military building of Naronowicz-Naroński and built examples of bastion castles in Europe, we can claim that it was not a less critical treatise.

References

1. Dybaś, B.: Fortece Rzeczypospolitej: studium z dziejów budowy fortyfikacji stałych w państwie polsko-litewskim w XVII wieku [Fortresses of the Republic of Poland: a study of the history of the construction of permanent fortifications in the Polish-Lithuanian state in the 17th century]. Wydawnictwo Naukowe Uniwersytetu Mikołaja Kopernika, Toruń (2018)
2. Scamozzi, V., Ottenheym, K., Vroom, W., Derwig, J.: Vincenzo Scamozzi, Venetian Architect: The Idea of a Universal Architecture. III, Villas and country estates. Architectura & Natura Press, Amsterdam (2003)
3. Dinsmoor, W.B.: The literary remains of Sebastiano Serlio. Art Bull. **24**(1), 55–91 (1942)
4. Spallone, R., Vitali, M.: Rectangular ratios in the design of villas from Serlio's manuscript for book VII of architecture. Nexus Netw. J. **21**(2), 293–328 (2019)
5. Murray, P.: The Architecture of the Italian Renaissance, 2nd edn. Thames & Hudson, Norwich (1979)

6. Rosenfeld, M.N.: Sebastiano Serlio's late style in the avery library version of the sixth book on domestic architecture. J. Soc. Archit. Hist. **28**(3), 155–172 (1969)
7. Blunt, A., Beresford, R.: Art and Architecture in France, 1500–1700, 5th edn. Yale University Press, New Haven (1999)
8. Burckhardt, J.: The Architecture of the Italian Renaissance, 2nd edn. Penguin Books, London (1987)
9. Serlio, S.: Sesto libro d'architettura. Delle habitationi fuori e dentrodelle citta [The Sixth book of architecture. Of the habitations outside and inside the cities]. Ex Bibliotheca Serenifs. Viriusq Bauaria Ducum, Munich, Lyon (c.1550)
10. Rosenfeld, M.N.: Sebastiano Serlio on domestic architecture: different dwellings from the meanest hovel to the most ornate palace: the sixteenth-century manuscript of book VI in the Avery Library of Columbia University. Architectural History Foundation, New York (1978)
11. Serlio, S.: Tutte l'opere d'architettura, Libro VI [All architectural works, Book VI. (c.1550)]. (c.1550)
12. Coope, R.: Salomon de Brosse and the development of the classical style in French architecture from 1565 to 1630. Pennsylvania State University, University Park (1972)
13. Cerceau, J.A.: Les plus excellents bastiments de France [The most excellent buildings in France]. Libraire-Editeur, Lévy (1576)
14. Cerceau, J.A., Thomson, D.: Les plus excellents bastiments de France [The most excellent buildings in France.]. Sand & Conti, Paris (1988)
15. Chatenet, M., Henrion, F.: Maulnes: archéologie d'un château de la Renaissance [Maulnes: archeology of a Renaissance castle]. Picard, Paris (2004)
16. Lorini, B.: Delle fortificationi di Buonaiuto Lorini libri cinque: ne'quali si mostra con le piu facili regole la scienza con la pratica, di fortificare le città, [et] altri luoghi sopra diuersi siti ... [Five books of the fortifications of Bonaiuto Lorini: in which science with practice is shown with the easiest rules, of fortifying cities, [and] other places above different sites ...] Appresso Gio, Antonio Rampazetto, Venetia (1596)
17. Tikhonova, O.: Understanding of bastion castles in the XVII century in Galicia Region (Ukraine, Poland). Unpublished Ph.D. thesis, University of Lisbon, Lisbon (2020)

Strategies for University Teaching of Civil Engineering Applying the PBL for Better Academic Performance

Esther Vargas(✉), Arturo Velásquez, Enriqueta Pereyra, Oscar Donayre, and Mario Chauca

Ricardo Palma University, Santiago de Surco, Peru
esther.vargas@urp.edu.pe

Abstract. The purpose of this report is to determine the effectiveness of Project Based Learning (PBL) as a learning strategy in the subjects of the Structures area, in Civil Engineering Programs. A qualitative approach and descriptive design are proposed. Within the teaching strategies, there is a direct relationship between the concepts of mechanics with a practical application of everyday reality, based on observation for its demonstration. The continuous updating of learning in civil engineering opens a challenge to teachers, making it necessary to point out to students that their learning must contain a critical spirit. In the investigations carried out, it is found that the PBL is configured as an effective didactic method, more efficient than the method that combines the expository and the realization of problems by the teacher, significantly raising the academic performance. Finding that in the comparative study between the PBL and the Traditional Expository Method (MET), positive differences were obtained in favor of the PBL in the highest average grade and the highest number of passed. Having found that of the students who pass the control test in the MET of 58%, it rises to 73% in the students who followed the PBL method in the case of students of the Construction course, a similar situation results in the comparison of the students of the irrigation engineering course.

Keywords: Academic performance · Learning strategies · Project Based learning · Traditional expository method

1 Introduction

Project-Based Learning (PBL) is a recognized teaching methodology that ensures educational quality, developing in students the required competencies, applying communication, teamwork and entrepreneurship, qualities that lead them to take on challenges in training and integration as active subjects in society, which raises real problems achieving a solution. Taking into account this social reality, various researchers have pointed to a new learning model which seeks that the student is not a passive agent but an intervening agent, where they learn on their own and interact with others in the search for an answer to situations of the reality that live in their educational environment, in their family and social context, this model is called Project-based Learning, which refers to the learning through which students plan, implement

P. Mendonça and N. D. Cortiços (Eds.): ICAMC 2021, LNCE 226, pp. 508–515, 2022.
https://doi.org/10.1007/978-3-030-94514-5_51

and evaluate projects that have an application in the world real beyond the classroom, and that allows the student to develop their reflective critical thinking, their creativity, their skills and abilities and of course acquire discipline to be competent in the development of activities inherent to their educational context, Blank [2].

When it comes to the LBP, in some cases it refers to Problems and in other cases to Projects, the difference being that Problems refers to the understanding of reality, while Project must be a proposal for action or intervention of the reality. Thus, in the first six (06) subjects of the Structures area of the Civil Engineering Career - Ricardo Palma University, shown in Table 1, the methodology based on PBL, we consider that the application of PBL should be oriented at the level of Problems, because they are subjects conceived as basic and therefore are not associated with projects; While in the seven (07) most advanced subjects, we consider that the application of the PBL should be oriented at the Project level, because they are subjects related to buildings and all buildings are projects.

Table 1. Level in PBL subjects of the structures area - EPIC URP.

Subjects of the structures area	Nivel LBP (problems)	Nivel LBP (proyects)	Learning intention
Static	X		Understanding of reality
Dynamic	X		
Strength materials I and II	X		
Analysis structural I and II	X		
Reinforced concrete I and II		X	Proposal to act, intervene or modify reality
Steel design		X	
Earthquake resistant engineering and natural disasters		X	
Pre-stressed concrete		X	
Bridges and works of Art		X	
Design of composite structures		X	

Source: ourselves

Therefore, we consider that Problem-Based Learning (PBL) is more related to basic levels of human understanding, while Project-Based Learning (PBL) is conceived as a more complex level of knowledge, therefore teachers should consider this difference in their selection of specific application cases.

In addition, the appearance of the computer has revolutionized the methods of calculating structures, replacing manual methods with the matrix method and the finite element method. This has been logically reflected in the teaching of subjects in the Structures area, in which manual methods gradually lost weight, Vázquez-Boza [3]. However, the application of computational methods does not imply an improvement of the student in the conceptual design of structures or the recognition of structural patterns, competences highly valued professionally in civil engineering [4]. Delgado et al. [5] conclude that learning to calculate structures should focus on the main tasks that students will perform: designing, modeling and calculating the structure, verifying and interpreting the results, as well as improving the design based on them.

On the other hand, it is important to highlight that the contents of the various subjects in the Structures area are duly connected so that they achieve the competencies sought in a Structures Project: "Ability to analyze and understand how the characteristics of the structures influence their behavior. Ability to apply knowledge about the resistant operation of structures to size them following the corresponding regulations and using analytical and numerical calculation methods". Faced with this problem, Mills [6] proposes the use of PBL in subjects related to the project of structures as a methodology for the integration of knowledge. This methodology has been very satisfactory for students in Civil Engineering subjects, with a very good perception regarding the acquisition of knowledge, collaborative work and motivation [7].

2 Discussion

2.1 Applications of PBL Methodology

Rodríguez [1], demonstrated the existence of positive differences for the PBL that would justify its implementation compared to the Traditional Expository Methodology (MET), in the application to the subjects involved in their research, reaching the conclusion that the students participating in a program Teaching based on an PBL obtained better overall academic performance results compared to other students who participated in a learning program following the MET.

Next, Fig. 1 is presented, with the flow diagram, proposed by the University of Alicante, which shows the stages to follow for a PBL project, which can be adapted to specific subjects. It is emphasized in the learning stages, the importance of having control points where the understanding of the project is verified, sufficient availability of information and verification of the quality of the solution.

In previous experiences, it has been found that the application of the LBP methodology in the teaching of subjects in the Structures area has allowed a notable increase in motivation and in the involvement of students in these subjects. This improvement translates into an increase in class attendance and in the receptivity of students, who perceive the methodology as useful and relevant for their training.

Thus, in the research carried out by Enrique de Justo Moscardo [9], who applied the LBP for Structures students, who found in the survey to measure the level of satisfaction with the application of the LBP, that 93.9% of the students, stated that they were between moderately satisfied (value 3) and very satisfied (value 5), as detailed in the following Table 2:

Although the LBP methodology requires greater dedication from teachers, compared to MET, it has the advantage of allowing a greater number of students to achieve or achieve significant learning, consequently, it is possible to improve the academic performance rate.

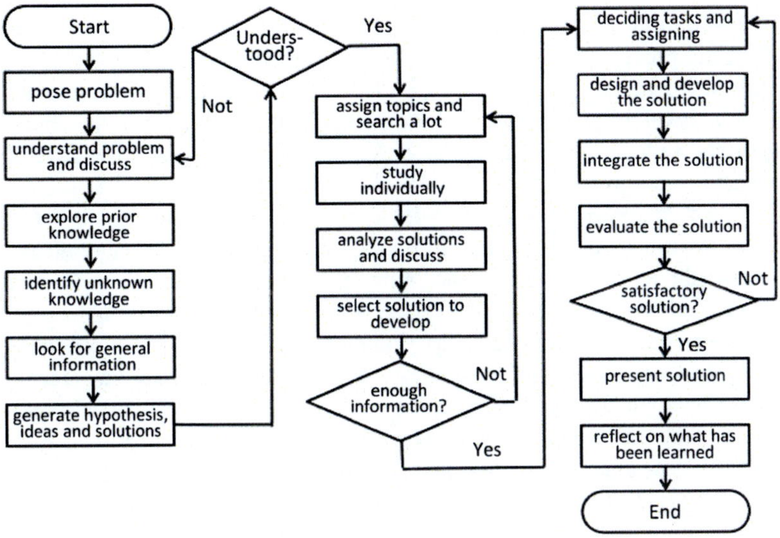

Fig. 1. Base flow diagram to suit subjects with PBL Source: University of Alicante [8], 2016.

Table 2. Frequencies of satisfaction responses of the surveyed students.

Global valuation				
		Frecuency	Percent	Accumulated percent
Values	1	4	1.5	1.5
	2	12	4.5	6.1
	3	46	17.4	23.5
	4	158	59.8	83.3
	5	44	16.7	100.0
Total		276	100.0	

Source: Enrique de Justo Moscardo (2013)

2.2 Characteristics of the PBL Methodology

The implementation of the LBP can be summarized in the expression "hands to work". Since the student goes from a passive-receptive attitude, to a required participation in each session, requiring the effort of the student, for which he must be duly motivated. Indicating that the practical applications will be presented in the future professional life. He must also be reminded that it is precisely the work in groups, within the PBL sessions, they are collaborative and that they constitute a model of said future professional life. Motivation will always be a key factor in PBL.

Project Based Learning (PBL) promotes the acquisition of critical thinking and self-learning skills, collaborative work, building their knowledge based on collaboration and interaction of attitudes towards a common goal.

Addine, González and Recarey [10] consider that the PBL meet or meet the demands of the process, where the educator "has to link his educational message with life, otherwise it will come to him empty, abstract, devoid of significance. and therefore he will not be involved in the learning task, he must take advantage of the experiential learning of his students, rely on it for future learning; prevent by all means the divorce between theory and practice".

Another of the principles that guide a developmental teaching is established between the relationship of the cognitive and the affective. Understanding this relationship contributes to the very conception of a problem-based teaching-learning process as a method. We consider the relationship of developer learning as a process, and developer teaching procedures as contextualized elements of problem-based learning. In this connection with the context, the assessment of what the student learns during the investigative process should not be lacking, as has been stated previously, not only the content is important, with which it will be possible to instruct; It must be ensured that he is able to visualize and give value to what he learns and thereby contribute to his education. Figure 2 shows the didactic processes and procedures that can be related to the PBL.

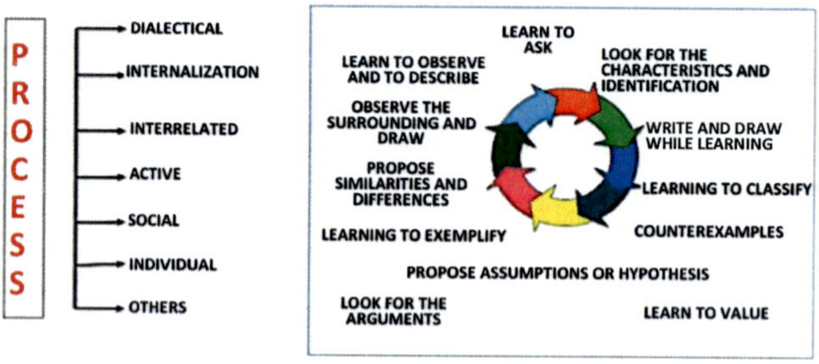

Fig. 2. Learning and developer didactic procedures that can be related to the PBL. Source: Edgar Marcelo and Jacinto Bolívar Méndez Urresta, LBP (2021).

3 Putting the PBL into Practice

- Among the aspects to consider for the implementation and monitoring of the PBL, as investigated, it will be convenient to consider:
- The LBP teaching classrooms must have enough space for the work groups to be made up of 5 subgroups of 4 students, with a 70 to 80 m^2 room it would be ideal since they can work without transferring information between subgroups. The desirable thing would be that the tables and chairs can be moved.
- The schedule of the PBL application sessions will take into account holidays. The teacher, based on her experience and according to the characteristics of the corresponding academic year and calendar, will program the case studies that will be given in the semester. It will be convenient to previously have a collection of cases, including plans, project information, technical reports, studies, certificates and works books, photographic material, and others.

- Creation of teams or subgroups, where students belong to the same Group assigned to the teacher. The case-subgroup relationship is fundamental, as well as the resolution-case, which must be continuously monitored.
- It is convenient to avoid anticipating the solutions of the exercise in question, and in turn, to promote the ability to resolve it. Each resolution takes a given time, and before culminating, no solution should be given. If there are any subgroups that are particularly lagging behind, no excess time should be allowed in the resolution.
- Establish rules for preliminary and final reviews: for example, it is common that, after a problem of some difficulty, individual students appear wanting to know the solution beforehand. The teacher must be firm about the calendar, and not give the global solution for the case under study ahead of time. However, in complex problems, the deadline for resolutions can be advanced and some partial solutions can be verified. This will always be done in class for all students.
- Consider the level of difficulty of the proposed problems according to the abilities of the students. It is not advisable to start the application of the PBL with the most difficult topics, nor to use the basic topics, more typical of a MET. It is better to program topics of intermediate difficulty, which include technical applications, to start implementing a PBL.

4 PBL Results for a Civil Engineering Subject

The results of the research in the thesis of Dr. César Rodríguez Gonzales, where the statistical analysis indicates that it has been oriented to the favourable or unfavourable evaluation of the PBL, leaving aside other cross-cutting issues. Research results include quantitative and qualitative information. Due to the accent marked by the main objective and the dependent variable, the quantitative information will be preponderant to verify, or falsify, the hypothesis "the differential treatment with PBL does not generate significant differences in the academic performance of the students, with respect to the MET". The alternative hypothesis "the differential treatment with LBP generates significant (positive) differences in the academic performance of the students, with respect to the MET" is the one that would formally affirm the initial question posed at the beginning of the research: "Are there significant differences in the learning in Irrigation and Construction Engineering between students?

The research showed significant differences in the learning of both subjects, if the LBP (Experimental Group GE) or MET (Control Group GC) is applied, finding that the qualifications improved by 23% in Irrigation Engineering, 34% in Constructions, as shown in Table 3.

Likewise, in the research it was found that there is a notable increase in the passing percentages that can be raised in the order of 58% to 73% in the respective subjects, applying MET and then PBL as shown in Fig. 3.

Successful results are also shown, in Fig. 4(a) results by didactic method according to subject and (b) total results by didactic method.

Table 3. Result of the analysis compared between PBL and MET

Global average mark			
Constructions		Irrigation engineering	
GC	GE	GC	GE
4,7	6,3	5,2	6,4

Source. Thesis Dr. César Rodríguez Gonzales
Note: the minimum passing grade is 5.0, CG: Control Group (MET) and EG: Experimental Group (LBP)

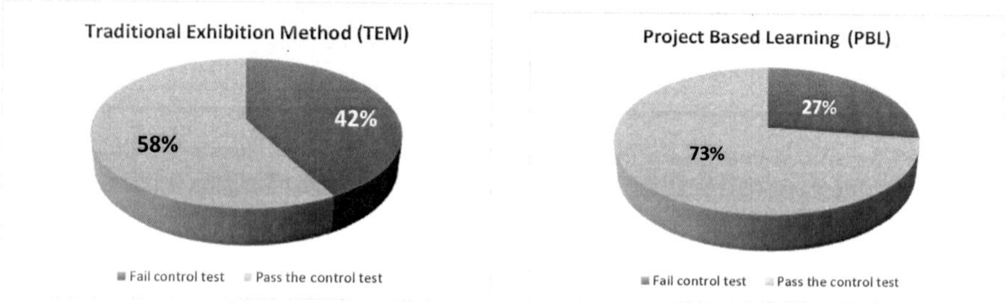

(a) (b)

Fig. 3. (a) Percentages of results applying MET (b) Percentages of results applying LBP

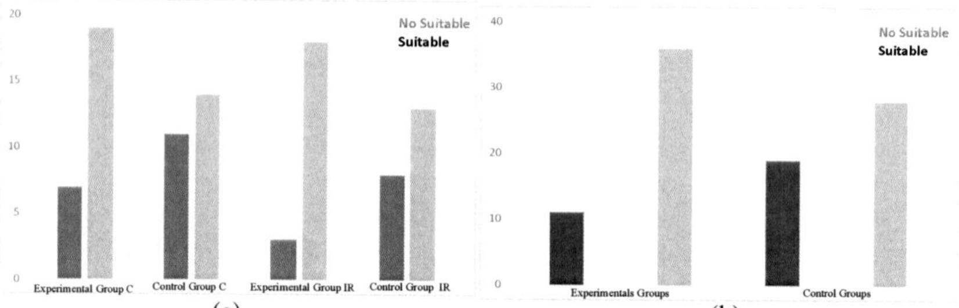

(a) (b)

Fig. 4. (a) Result by didactic method according to subject (b) Total results by didactic method

5 Conclusions

- The application of the PBL should be oriented at the Project level, in subjects related to buildings and civil infrastructures, which are essentially projects, while in the basic subjects the application of the PBL should be oriented to problems.
- It is recommended that, in all LBP, three (03) control points are established where the compression of the project, the sufficient availability of information and the verification of the quality of the solution are verified.

- The design to apply the PBL requires careful programming, in which an initial period is considered where the MET is applied to cover the fundamental or basic topics.
- It has been found that the application of the LBP methodology allows achieving a notable increase in the percentages of passed in the subjects of the career, with respect to the application of the MET, which according to studies can rise in the order of 58% to 73%, with the consequent improvement in the academic performance of student.

References

1. Rodríguez, C.: Application of problem-based learning in irrigation and construction engineering university students (Ph.D. thesis). Sevilla University, Sevilla, Spain (2017)
2. Blank, W.: Authentic instruction. In: Blank, W.E., Harwell, S. (eds.) Promising practices for connecting high school to the real world, pp. 15–21. University of South Florida, Tampa (1997). (ERIC Document Reproduction Service No. ED407586) Blumenfeld, P., Marx, R., Krajcik, J.S., Soloway, E.
3. Vázquez-Boza, M., Justo Moscardo, E.D., Delgado Trujillo, A.: The evolution of structural engineering education in the era of computer. In: 3rd International Conference on Mechanical Models in Structural Engineering, pp. 712–729. CMMoST 2015, Seville (2015). Víctor Compán Cardiel [etc.]
4. Johnson, D., May, I.M.: The teaching of structural analysis. Struct. Eng. **86**(22), 32–39 (2008)
5. Delgado Trujillo, A., Justo Moscardo, E.D., Molina Huelva, M., Rodríguez-Mayorga, E.: Relevant features in steel structures teaching in building construction by project-based learning (2015)
6. Mills, J.E.: A case study of project-based learning in structural engineering. In: Proceedings of the 2002 American Society for Engineering Education Annual Conference & Exposition, pp. 1–18 (2002)
7. El-Maaddawy, T., El-Hassan, H., Al Jassmi, H.: Student perceptions of the use of project-based learning in civil engineering courses. In: 2018 IEEE Global Engineering Education Conference (EDUCON), pp. 243–250. IEEE (2018)
8. University of Alicante: example of flow diagram (2016). http://www.dccia.ua.es/pe18/LBP_espanol/un_ejemplo_de_diagrama_de_flujo.html
9. Justo Moscardo, E.D.: Design and evaluation of a program for learning building Structures through LBP. (Unpublished Doctoral thesis). University of Seville, Seville (2013)
10. Urresta, E.M.M., Urresta, J.B.M.: LBP-Technical University of the North- Ecuador (2021)

Palmela Mansion in Rua dos Anjos in Lisbon

V. P. Matos(✉)

Lisbon School of Architecture, University of Lisbon, Rua Sá Nogueira, Pólo Universitário, Alto da Ajuda, 1349-063 Lisbon, Portugal

Abstract. Palmela's mansion, one of the current edifices of LisbonBreaks agency, is located in an area that underwent radical urbanistic transformations in the late 19th and the early 20th centuries. The edifice has an eclectic main façade, which hides the high technology that the mansion presents. The descriptive memory shows a profound knowledge of the technologies of the epoch, disseminated by architectural magazines, namely A Construção Moderna, regarding the type of laminated iron, hydraulic screed, the layer of folded brick, and the concrete bed anchored to the walls by the beams that support the storage room gallery. These building materials allow to conceive more open and fluid spaces, to control more the width and master the natural daylight, fundamental for spaces, as ateliers, foreseen in the 1914's project of the edifice.

Keywords: Mansion · Lisbon · Architectural magazines · Building materials · Real estate market

1 Introduction

At the turn of the 20th century different revivalisms coexisted: the Neo-Romanic, the Neo-Gothic, the Neo-Manuelino, the Neo-Moorish styles which had in themselves the principles of modernity from which emerged the new architecture of the 20th century.

This epoch was a period where it was necessary to reinforce the Portuguese soul and identity to overcome the hardship originated by the British Ultimatum and the *Rose-Coloured Map* in 1890, which forced to a political and geographical repositioning of Portugal in Africa, Europe and the world. It was in this context that in, 1860–1867, was made a statue of Luís de Camões, author of the epic poem *Os Lusíadas*, considered at the end of the 19th century an iconic piece of writing of the Portuguese literature, containing the modern ideas and the spirit of arms suitable to a modern Portugal.

The late 19th century and the early 20th century were troubled times in Portugal. On 1 February 1908 occurred the regicide: the king D. Carlos I and the Prince Luís Filipe, heir to the crown, were murdered. King D. Manuel II would reign until the advent of the Republic, on 5 October 1910. After this event, king Manuel II would be exiled in the United Kingdom, where he would marry Princess Augusta Vitória de Hohenzollern-Sigmaringen. This epoch was also marked by the separation of the State from the Church. The consensus between Republicans and Catholics was only achieved after the First World War [1]. The estates of the Church were nationalized. In 1910, the royal estate buildings were integrated into the State. Without the governance of the Portuguese crown in Portugal, many noble buildings, palaces and mansions,

P. Mendonça and N. D. Cortiços (Eds.): ICAMC 2021, LNCE 226, pp. 516–528, 2022.
https://doi.org/10.1007/978-3-030-94514-5_52

were also incorporated into the State or acquired new functions, for instance, as City Halls and Civil Parishes.

Currently, most mansions in Portugal are a heritage under tremendous real estate pressure, since they are a prestigious asset to companies to have in their portfolio. Many of them have already been refurbished into hotel, hostels, apartments, public buildings and state edifices, or buildings chosen for government public acts occur.

Palmela's mansion (1914), in *Rua dos Anjos* (Anjos street) is an edifice, currently flanked by two buildings, implanted in an area which would see urbanistic transformations and a residential growth with the plan of *Avenida dos Anjos* (Anjos Avenue), and adjacent streets. The edifice's descriptive memory and the technical drawings evidence a profound knowledge of the technologies of the epoch, disseminated by architectural magazines such as, *A Construção Moderna* (The Modern Construction), regarding the type of materials and the constructive systems used: laminated iron, hydraulic lime floor screed, folded brick layer and concrete bed. The degree of detail reinforces the role that the Portuguese architectural magazines played in the dissemination of the most advanced technologies at the time, in technically describing the building materials and the constructive systems, and in presenting architectural and engineering projects.

2 Mansion in Lisbon Region

The term mansion, disseminated in the 19th century, named a more graceful and smaller edifice than the palace, adapted to the functions of a single family house, dwelt by industrials, bankers and business men. The mansion is mainly an urban edifice.

The mansion may have corner pilasters, and the façades may show classical elements, and may be coated partial or totally, with glazed tiles. Their façades layout differs across the country, partially as a result of their building materials.

In the late 19th century and the early 20th century, this heritage obeyed to the general legislation (Matos 2020). Clever ways of solving the problems posed by this legislation (e.g. 3/6/1859 Act and 20//10/1865 Act) are seen in Lisbon's mansions. Now most of these buildings are private owned or public edifices [2].

The analysis of Sistema de Informação para o Património Arquitetónico's (Information System for the Architectonic Heritage) inventory, permitted us to conclude that regarding their planimetry the mansions in Lisbon Region fit into the following main types: complex plan, irregular plan, L shape plan, T shape plan, composed plan, longitudinal plan, rectangular plan, U shape plan, square plan, L shape plan with a rectangular central patio, rectangular plan with a longitudinal axis; longitudinal plan, rectangular plan with a central patio; polygonal plan, Inverted L shape plan, and pentagonal plan.

3 The Importance of the Architectural Magazines of the Early 20th Century

The periodical *A Construcção Moderna* (*The Modern Construction*) (1900–1909), published in Lisbon, was the first Portuguese magazine on the technical and aesthetical aspects of the construction and on the modernization of the architecture targeting a technical public. In the late 19th century and the early 20th century similar architectural magazines had appeared in Europe and in United States. After the advent of *A Construcção Moderna*, other magazines emerged in Portugal, namely *A Architectura Portugueza* (1908–1929), *Arquitectura* (1927–1939), *Arquitectura portuguesa, cerâmica e edificação – Reunida* (1935–1951), and *Revista oficial do Sindicato Nacional dos Arquitetos* (1936–1942). These magazines were spaces of creation and circulation of ideas and culture, disseminating the architecture of the epoch, promoting food for thought in the field of architecture. The knowledge was disseminated and constructive details were published becoming very useful to the specialized readers.

The Portuguese architecture in the early 20th century is characterized by a change in what regards the aesthetics and the building materials to which an importation of the knowledge and the technologic developments were added. New urbanistic concerns and socioeconomic shifts influenced the private and public orders of the construction of buildings. The Portuguese architectural magazines played an important role in technically describing the building materials and the constructive systems, presenting architectural and engineering projects and different branches of the arts, disseminating congresses, legislation, and professional architectural contents. These magazines played an important role in the debate on the urban planning, on heritage and on the restoration of buildings. They presented to a wider public low-income dwellings, single-family houses, collective dwellings, and significant buildings and infrastructures.

A Construcção Moderna had a significant role in decisively contributing to the construction of an architectonic culture in the early 20th century. It fostered the multidisciplinary debate, disseminated information on the new building materials and constructive systems, promoted the debate between technicians, informed the readers of upcoming or realized exhibitions and conferences. Additionally, it answered to questions asked by the readers on the building materials, their use, processes of construction and calculation methods.

The magazine focused mainly on the interpretation of the building materials and the presentation of the ideas conveyed which supported the architectural production of the epoch. In a cultural context marked by the 19th romanticism, the modernity was introduced in Portugal by two cultural and artistically vectors. One vector was represented by the Academy, with French influence, with references to the rationalism and the beaux-arts and the cosmopolitism of the *haussmannian* city. The other vector was represented by the 19th century English scenario: the Arts and Crafts, the Gothic Revival, and the cultural Anglo-Saxon traditionalism [3].

In the beginning of the 20th century the iron, the glass and the reinforced concrete were new building materials used and experimented in Europe. In the half of the 19th century appeared in Europe the first edifices using the iron as a structural element (present in factories, bridges, harbor warehouses). *A Construcção Moderna* published articles on iron and its uses; it presented various projects where the potentiality of iron was shown, described constructive systems and where they were used, and even include formulas for conceiving a bridge. The magazine demonstrated the use of this building material in architecture the in the artistic metalworking.

The glass was a building material that had less articles than the iron and the reinforced concrete. The articles published in the magazine focused on the thermic properties of the glass, its production process, the existing types of glass and its uses in architecture, furniture, decorative arts and folding screens.

The reference to the reinforced concrete appeared on the first number of the magazine, in 1900, mentioning the *constructive system of Henebique*. Several article detailed the potentialities, advantages, disadvantages and the use of this building material. In 1905 the magazine published an article on the Portuguese patent of the *Henebique System* [4].

4 The Place

Palmela's mansion is implanted in an area that underwent radical urbanistic transformations in the late 19th century and early 20th century. At that epoch several plans were approved for different areas of Lisbon to improve and modernize the city. The old *Rua dos Anjos* was one of those chosen as a strategic area (Fig. 1, Fig. 2). The final version of the project of *Avenida dos Anjos e ruas adjacentes* (Anjos Avenue and adjacents streets) was presented on 17 February, 1892. The matrix of the preliminary project is located between 1870 and 1888, a period of stability in of Portuguese liberal régime, marked by economic growth. However, 1892 corresponds to an epoch of crises of the liberal period and of bankruptcy of the country.

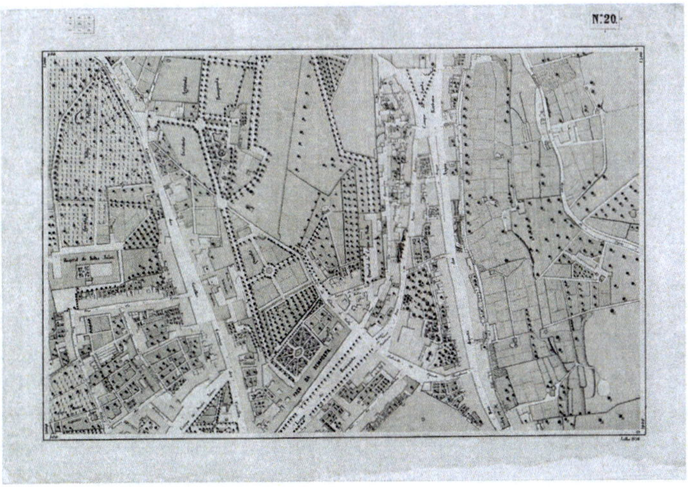

Fig. 1. Atlas of the topographic charter of Lisbon n 20, Lisbon Municipal Archive, PT/AMLSB/CMLSB/UROB-PU/05/01/22)

Fig. 2. Atlas of the topographic charter of Lisbon n 13; Lisbon Municipal Archive, PT/AMLSB/CMLSB/UROB-PU/05/01/15)

In the previous plan of *Avenida dos Anjos e ruas adjacentes*, approved on 9 March 1887, the construction of three longitudinal streets was foreseen, linking *Rua dos Anjos*, near Intendente, to *Estrada da Circunvalação* (Circunvallation Road), currently Rua Morais Soares.

The opening of the new avenue also foresaw a set of small streets linking the old urban mesh with the new one, and the growth of the plan of Anjos Avenue, currently *Avenida Almirante Reis* (Almirante Reis Avenue). This one was presented in 1892 and enlarged in 1895. In contrast with the pre-project, this plan exhibited an additional connection at the right of the new avenue with *Rua Direita da Palma*, the construction of a new square which replaced *Largo do Intendente* (Intendente square), never built, and the abandon of a eastern street, parallel to the new avenue, with extension foreseen of 665 m, which presented a ramp from Areeiro to the road Sacavém, ending, at South, in the current *Praça João do Rio*.

A residential urban area was developed around the axis *Rua Direita dos Anjos/Arroios/Estrada de Sacavém*, a main path to access to *Baixa* (Downtown), in the last decades of the 19th century. In the hillside of *Anjos* valley, urban plots were made in *Cruz do Tabuado*, near the present *Hospital da Estefania (Estefania Hospital)*. The focus of this project was the creation of new streets in the urban mesh.

The goal of *Avenida dos Anjos* project was to create a new axis, to get closer places that were distant from *Baixa* (Downtown), such as *Largo de Arroios* (Fig. 3).

The avenue was built in the late 19th century, and in the boarder of the *Avenida dos Anjos* the old large properties were acquired and capitalized. The expropriation process did not occurred as in *Avenidas Novas* (1900–1910), where the City Hall took charge of the expropriation and urbanization processes. In Anjos Avenue the operations were capitalist, the urban plots were smaller and the streets narrower (12m or less). The goal was to sale the terrains and divide them in urban plots no matter the size and shape.

Fig. 3. Plan of Anjos Avenue and adjacents streets, 1900, Lisbon Municipal Archive, PT/AMLSB/CMLSBAH/PURB/002/01245

Hence, these cheaper and smaller plots targeted mainly the bourgeoisie and the middle class.

The expropriation foreseen in this plan mainly concerned the terrains occupied by *Avenida dos Anjos* and adjacents streets, as well as slopes of the landfills and trenches. Because there was not a precise legislation on the expropriations, sometimes they bitterly dragged on in time.

Additionally, the peak of the financial Portuguese crisis and the declaration of bankruptcy, in 1892, would contribute to formally and structurally limit the plan of *Anjos Avenue* topography, namely the connection of the neighboring *Vale da Avenida/ Rua Marques da Silva* and *Rua Damasceno Monteiro* with the paths of *Penha de França* and *Nossa Senhora do Monte*, among others.

To accomplish this important urban axis seven transversal streets were foreseen, which would connect with the existing ones. In some cases, in the bordering streets, the connections between new streets and the old ones was assured by ramps, as in *Regueirão dos Anjos* e in *Largo* do *Forno do Tijolo* and *Rua Forno do Tijolo* [5–7].

The construction of *Rua dos Anjos*, between 1898 and 1908, implied the transformation of the area.

To accomplish the urbanism of the new streets some modifications were made. The buildings saw their façades shortened. A fountain, the *Bica dos Anjos* (Fountain of

Anjos) was demolished (1900) so that the tramway could pass. A new fountain, in largo do Intendente, replaced the old one. Furthermore, the *Igreja dos Anjos* (Anjos Church) was demolished in 1908 to open Avenida Rainha Dona Amélia (former Avenida dos Anjos, renamed as Avenida Almirante Reis), and a new one was built and inaugurated in 1911 (Fig. 4).

Fig. 4. The old Anjos Church and the old place of Regueirão dos Anjos (before 1908); Lisbon Municipal Archive, PT/AMLSB/CMLSBAH/PCSP/004/BAR/000809

In 20 July 1903 was inaugurated the street, baptized as Avenida D. Amélia, which connected Rua da Palma to the old *Estrada da Circunvalação*. However the street remained uncompleted, due to the difficulty of reaching an agreement with Desterro Hospital on the concession of an urban plot which included the nurseries area. Therefore an extension of 120 m of the avenue was unbuilt. Additionally the crossroad with *Rua direita dos Anjos* was not solved efficiently and the extension of the avenue with *Estrada da Circunvalação* had not begun.

5 Palmela Mansion

Palmela Mansion in Rua dos Anjos 82/Rua Regueirão dos Anjos 51–53, in Lisbon is an edifice flanked by two buildings, implanted in an urban street, exhibiting a T shaped plan, with two recessed bodies in the main façade and a flat rear façade (Fig. 5).

The plan of this mansion, owned by D. Caetano de Bragança, with statement of responsibility of the work of 1914, was conceived in accordance with those of the beginning of the 20th century, in which the car was a relevant element in the solution of the urban plot, decisive for the existence of a garage or an annex for the car, as it also happens in Rua Andrade Corvo, now headquarters of Lisbon Metropolitan, and in Viscountess de Valmor Mansion.

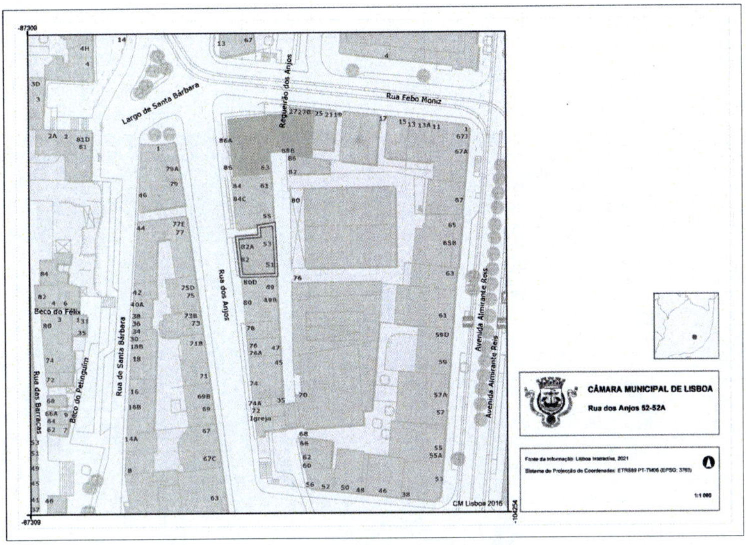

Fig. 5. Plan of *Rua dos Anjos* where the mansion is located

In 1914, Palmela's Mansion plans exhibited the following rooms: the 1rst pavement included the garage, the basement and the storage galleries. The 2nd pavement comprised, near the front façade, three rooms, a cabinet, an atelier, and, near the rear façade, a help room and a starching room (Fig. 6). In the 3rd pavement, i. e., the ground floor, there were the main rooms of the edifice including the socializing spaces, near the main façade, the entrance, the stairs, the cabinet, a billiard room, a living room, a small room, a pantry, near the rear façade, and a corridor in between, a dining room, a kitchen, a sowing room, and a drawing room. The 4th pavement, 1rst floor, was devoted to the bedrooms and related with hygiene and dressing rooms. The 1rst floor presented four bedrooms, a W.C., two rooms for toilette, another for clothing, and an oratory room connected to one of the bedrooms. The 5th pavement was the terrace. These specializations of the dwelling are similar to those found in Monserrate Palace, in Quinta de Monserrate, which include in its set of rooms several drawing rooms, bedrooms, library, and a terrace area. However, the Palmela Mansion in *Rua dos Anjos* 82 is a profoundly urban building marked by the advent of the car. In the borders of the main façade, preceded by the gateways there were the main entrance of the house and the entrance for the garage, that justifies the semi-basement floor. In the rear façade there was another entrance for the garage and the storage room, which had interior stairs communicating with the horizontal circulation of the ground floor, in the proximity of the service areas and the ironing room. The main façade, strongly scenographic, presents the coat of arms of the Palmela family, which contrasts with the rear façade, less ornamented, without elements that ennobles it, making of the street a secondary one. In the rear façade, in the 1rst floor, in the alignment of the central portal, a balcony was constructed, presently adulterated. In the plan of 1914 the building presented a terrace roof, which has been kept. The body of the stairs that access the terrace appears as a small tower above the terrace floor.

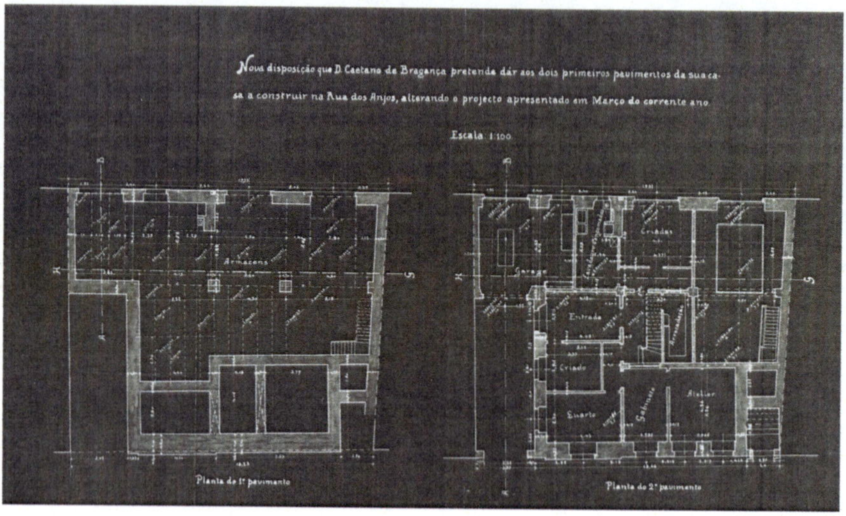

Fig. 6. Plans of the first two floors of Palmela's Mansion

The main façade of the edifice exhibits three bodies, two recessed regarding the main central body. In the front façade, the façade's panel of the main body presents a socle with a bush-hammered appliance potting to a fine peak until the main floor pavement. The lateral façades of this main body exhibit a polished stone socle. The façade panel in the main façade presents in the ground floor a coating in stonework with square recessed joint (*silharia fendida*) and the majority of the windows on this flor are ended by a cornice. The noble floor exhibits a balcony window and windows ended by pediments, which are absent from the remaining openings in *Rua dos Anjos*. The balcony window is ennobled by a stone balcony crowned by a pediment in which the coat of arms is placed. The platband is constituted by acroteria and balustrades. The façade's panels of noble floor are plastered and painted (Fig. 7).

The rear façade has a medium bush-hammered stone socle to medium peak, interrupted by the limestone straight door openings. Some of the openings in the ground floor have an arched lintel. The rear façade is plastered and painted (Fig. 8).

The upper floors present windows. In the noble floor, there is a very adulterated balcony, constituted by the blind panels and two windows. The platband of the rear façade is a full platband that is, a blind wall; which is plastered and painted.

Fig. 7. Palmela's Mansion main façade

Fig. 8. Palmela's Mansion rear façade

6 Palmela Mansion's Constructive Systems

The edifice presents an eclectic main façade, a fairly modest rear façade which hide the high technology that the mansion exhibits.

The descriptive memory, 17 March 1914, of Palmela mansion, in *Rua dos dos Anjos* 82-82A/ *Regueirão dos Anjos* 51–53, shows how an important house was detailed. The foundations would lay on the rock and would have as the rising walls the thickness detailed on the technical drawings. The pavements would be made of brick vault on the terrace and on the floors over the storage room and the garage, with iron beam of 160 × 80 × 8 mm. The part of the basement which was not over the garage and the storage room would be paved by hydraulic screed. The remaining floor would be made of wood. The terrace would be paved with mosaic tile over a layer of folded brick. The frontals would be of Marroco brick with ground-sills and plumb beams made of laminated iron of 140 × 80 × 8 mm. The frontal over the storage room would lay on a reinforced beam 0,53 m high, supported at the center by a plumb beam made of laminated irons, laying over a concrete bed and anchored to the walls by the beams which support the storage room gallery. The partitions would be plastered. Above the ceilings of the first floor, there was a space destined to isolate the rooms from the heat and humidity from the terrace, only accessible for cleaning and repairing. All the materials used would have to be of prime quality and according the best methods of the building art. The mortars composition would be variable according their use. However the proportion of lime should not be inferior in 2:5 of sand (Fig. 9, Fig. 10).

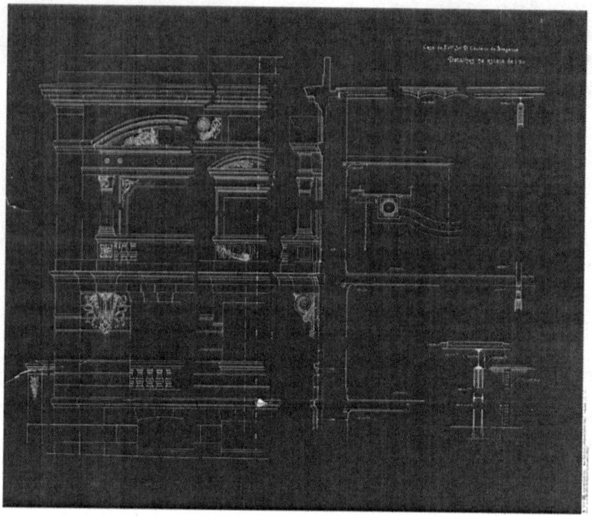

Fig. 9. Constructive details of Palmela's Mansion

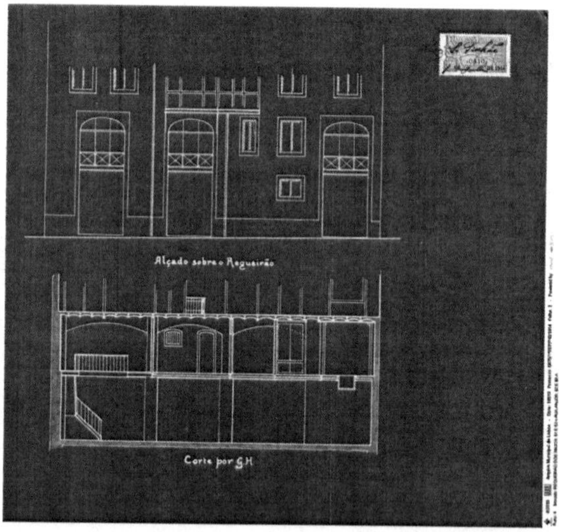

Fig. 10. Section and of rear façade of the mansion

7 Lisbon Breaks

Palmela's mansion was converted into a local housing edifice, by LisbonBreaks, an apartment rental agency with active presence in today's Portuguese real estate market.

LisbonBreaks was created to give to the clients short and long term renting periods, in a modern and cozy environment combining tradition and youth. The company manages different types of apartments fully furnished and with free internet ranging from T2 typology (two bedrooms per dwelling) to apartments with capacity for 30 persons. The apartments are spread by the *Lisbon Metropolitan Area*, in the areas of *Anjos, Intendente, Castelo, Bairro Alto, Santa Apolónia, Paço de Arcos, Carcavelos, and Parede.*

Palmela Mansion is one of the edifices of LisbonBreaks. The building was renovated and the old characteristics were preserved, namely the abundant glazed tiles in the interior architecture. The Mansion has five bedrooms and two bathrooms available for renting for a minimum of 3 nights and maximum of 364 nights.

This is one of the various steps that the real estate in Lisbon has been taken to face the degradation and abandon of Lisbon's abundant heritage, well succeeded for the moment.

The Portuguese luxury market is mainly targeted by the international segment, and because of that it suffered with the mobility restrictions during the covid 19 pandemic. In the first trimester of 2020 the prices of the luxury market supported the levels achieved in 2019 above 7000€ m2 but during the year 2021 there was a decrease. In the second trimester of 2020, in the historical centre of Lisbon, a decrease of 14,4% was registered in relation to the first half of the previous year. For the time being, it seems to be hard to maintain the pre-covid 19 levels of activity. However, the luxury market is known to suffer less with the crises.

Lisbon is an appealing city for the tourists and to their short and long term dwellers. The apartments most commonly looked for are those overlooking Lisbon sightseeing's. In terms of typologies the most asked for are T2 and T3 with a view and/ or a garden. The clients of the luxury real estate market; Portuguese and foreigners, are looking for the less repeatable, more exclusive and high quality pattern dwellings.

The people who look for these historical buildings for renting in the real estate market look for edifices with recognized location within the city area, with top quality building materials, ample dwelling areas, private parking lots, sometimes with facilities such as swimming pool and gymnasium. They are also looking for a multifunctional and renovated dwelling equipped with, high technology permitting them to be in telewoking regime. The history, oldness and charisma of the building give charm to the dwelling, reinforcing its attractiveness in the real estate market. These are standard references which can be found in Palmela's Mansion and makes it such an appealing edifice to stay for a short or a longer term stay.

8 Conclusions

At the turn of the 20th century different revivalisms coexisted which had in themselves the principles of modernity from which emerged the new architecture of the 20th century. The architecture incorporated the latest building materials, adequately described in the technical drawings and descriptive memories of the building projects. The innovative interior architecture contrasted with a more exterior traditional appearance of the edifices. In Palmela mansion (1914), the open interior space, conceived with brick vaults on the floors over the old storage room and the garage, and iron beam of $160 \times 80 \times 8$, contrasts with the more traditional look of the main façade.

The descriptive memory evidences a the profound knowledge of the technologies of the epoch disseminated by architectural magazines such as, *A Construção Moderna*, regarding the type of laminated iron, the hydraulic screed, the layer of folded brick, and the concrete bed anchored to the walls by the beams which support the storage room gallery. These building materials permitted to conceive more open and fluid spaces in the residential architecture, allowing to control the width of the openings in the façades, and master the natural daylight in the interior architecture, fundamental in spaces as the ateliers, foreseen in the plan of this building.

Palmela's mansion (1914), in *Rua dos Anjos*, in Lisbon is an edifice currently flanked by two buildings, with a T shaped plan, which belonged to D. Caetano de Bragança. The mansion contemplated already a garage for the car in its 1914 plans. An important division of the building is the oratory room, signaling that the religion marked routine of days in the mansion. The plans of the pavements placed the socializing spaces near the main façade, and put near the rear façade the ones related the deeper levels of the private life, or related to the personnel which helped to daily functioning of the house. Exception was made to the pavement solely dedicated to the bedrooms.

The mansion was refurbished and currently belongs to the LisbonBreaks agency, offering short or long stays for travelers who which to stay, for leisure or for work, in a charismatic and exclusive mansion, in the core of Lisbon.

References

1. Matos, L.S.: A separação do Estado e da Igreja, Concórdia e conflito, entre a Primeira República e o Catolicismo, 1st edn., pp. 32–33. D. Quixote, Alfragide (2010)
2. Matos, V.P.: Ribeiro da Cunha Mansion a Principe Real project. In: Proceedings of the 7th International Conference on Heritage and Sustainable Development, Coimbra, vol. 1, pp. 529–537. Green Lines Institute for Sustainable Development, Barcelos (2020)
3. Nunes, P.S.: A cultura arquitectónica em Portugal na dobra do século XIX para o século XX: sinais de influência estrangeiras nas páginas d' A Construção Moderna. Revistas de Arquitectura: Arquivo(s) da Modernidade coord M D Mesquita, part II, chap. 1 pp. 229–253. Caleidoscópio, Casal de Cambra (2011)
4. Andrade, L., Aleixo, S., Faustino, P.: Materiais e técnicas de construção do início do século XX em Portugal na revista A Construcção Moderna. In: 3rd International Congress on Luso-Brazilian Construction History, pp. 1294–1307. Núcleo de Tecnologia da Preservação e da Restauração, Salvador (2019)
5. Lourenço, T.B.: Entre o Desterro e Arroios. O desenvolvimento urbanístico dos bairros do primeiro troço da Avenida Almirante Reis (Avenida dos Anjos). Cad. Arq. Munic. **12**, 89–116 (2019)
6. Ribeiro, L.: O projecto da Avenida dos Anjos – algumas considerações gerais. Cad. Arq. Munic. **5**, 64–107 (2001)
7. Ribeiro, M.L.: As vésperas da modernidade: do Intendente à Rua dos Anjos (1898–1908). Cad. Arq. Munic. **4**, 105–134 (2000). 1.ª série, Lisboa

Author Index

P. Mendonça and N. D. Cortiços (Eds.): ICAMC 2021, LNCE 226, pp. 529–531, 2022.
https://doi.org/10.1007/978-3-030-94514-5